WIT
D0687276
UTSA LIBRARIES

Multi-scale Modelling for Structures and Composites

by

G. PANASENKO

A C.I.P. Catalogue record for this book is available from the Library of Congress.

ISBN 1-4020-2981-0 (HB)
ISBN 1-4020-2982-9 (e-book)

Published by Springer,
P.O. Box 17, 3300 AA Dordrecht, The Netherlands.

Sold and distributed in North, Central and South America
by Springer,
101 Philip Drive, Norwell, MA 02061, U.S.A.

In all other countries, sold and distributed
by Springer,
P.O. Box 322, 3300 AH Dordrecht, The Netherlands.

Printed on acid-free paper

Printed in the Netherlands.

Contents

PREFACE.

Rod structures are widely used in modern engineering. These are bars, beams, frames and trusses of structures, gridwork, network, framework and other constructions. A variety of theories based on the Kirchhoff-Love, Kirchhoff-Clebsch and other hypotheses are applied for their analysis. Structural mechanics software based on material strength theory methods also exists. At the same time the questions concerning the limits of applicability of these hypotheses and theories and the possibilities of their refinement are very important. In this connection we develop the multi-scale asymptotic analysis of equations of mathematical physics, and in particular the elasticity equations of set in the rod structures (without these hypotheses and simplifying assumptions being imposed) . Problems with one small parameter (the ratio of bar diameter to its length) as well as problems with two and more small parameters (periodic framework systems, where the second parameter represents the ratio of a period to the characteristic space dimension of the problem, weakly compressible bars, etc.) are studied.

The homogenization technique for partial differential equations described in the book by N.Bakhvalov and G.Panasenko [16] and the boundary layer techniques are used as a main tool in these investigations. The physical processes are simulated by partial differential equations set in "thin" domains containing a small parameter. The asymptotic analysis is applied for investigation of these partial differential equations.

The multi-scale models are developed according to two main schemes:

- the up-scaling procedure of asymptotic derivation of macroscopic models from the microscopic ones (ie., the homogenization approach) and

- the hybrid multi-scale models, combining two scales inside one model, making a microscopic zoom inside the macroscopic model (ie., the asymptotic partial domain decomposition, partial homogenization).

The present monograph consists of six chapters. The first chapter is introductory. It presents the main notions and methods of the book, advantages and disadvantages of these methods. In the second chapter we consider the three-dimensional conductivity problem as well as theory for elasticity problems and other equations of mathematical physics in a thin cylindrical domain (bar) with non-homogeneous structure. Full asymptotic expansions of solutions are constructed for the small parameter equal to the ratio of the bar diameter to the length; boundary layers are investigated and one-dimensional equations for bars are derived. Then we consider the problem of junction of two heterogeneous rods. The connection conditions for rods result from the analysis of boundary layers arising in the neighborhood of the bounds of the rods. Time dependent models are considered.

The third chapter is devoted to the similar analysis of a heterogeneous layer (plate). Full asymptotic expansions of solutions are constructed for the small parameter equal to the ratio of the plate thickness to the length of the plate; boundary layers are studied and two-dimensional equations for plates are derived.

Similar questions for systems with finite number of bars are studied in the Chapter 4. The asymptotic analysis of a structure with finite number of bars is developed first for nonlinear equations in some weak norms (L-convergence method), and then the detailed asymptotic analysis is done, the asymptotic expansions of solutions are constructed. The Korn inequality or, more specifically, the investigation of how the constant depends on the small parameters in this inequality is essential for the justification. Finally the results obtained by means of L-convergence method are applied to shape design of finite rod structures. The code OPTIFOR implements this algorithm. We discuss some model examples. The similar analysis is developed for some problems of flow in a system of tubes described by Stokes and Navier-Stokes equations.

In Chapter 5, periodic framework structures (lattice-like domains) are considered and their homogenization is carried out, i.e., systems with great number of bars are investigated. The question of the existence of homogenized models and the convergence of the exact solutions to them is studied. In some cases the first approximation of the asymptotic theory may be also obtained by means of classical structural mechanics. At the same time, the asymptotic theory gives the opportunity to obtain corrections to the structural mechanics models. It is possible to obtain analytical formulas for these corrections in some examples.

Chapter 6 is devoted to a new multi-scale method of solution of different problems with small parameter. It is the method of asymptotic domain decomposition. The direct numerical solution of partial derivative equations in finite rod structures is very expensive because the complicated geometry demands a large number of nodes in the grid. The complete asymptotic expansions are often cumbersome. So we propose a hybrid numerical-asymptotic method which uses a combined 3D-1D models: it is three-dimensional in the boundary layer domain and it is one-dimensional outside of the boundary layer domain. We cut the rods at some distance from the ends of the rods, we keep the dimension three in the neighborhood of the ends and we reduce dimension on the truncated (main) part of rods. So the principal idea of the method is to extract the subdomain of singular behavior of the solution and to reduce dimension of the problem in the subdomain of regular behavior of the solution. Of course the most important question is: what are the interface conditions between 3D and 1D parts? We formulate two approaches of construction of such hybrid models and justify the closeness of the partially decomposed model and initial model. We analyze such hybrid models for conductivity and elasticity equations stated in rod structures as well as Stokes and Navier-Stokes equations stated in a system of thin tubes.

We consider below two versions of the method of asymptotic partial decomposition of domain. The first version is "differential", i.e. we work with the differential formulation of the initial problem , we obtain the 1D differential equation in the reduced part of the rod structure and we add the differential interface conditions on the boundary between 3D and 1D parts . Of course we can pass to a variational formulation of the partially decomposed problem but it is generated by a differential one.

The second version is a direct variational approach, when the 3D integral identity for the original problem is restated for a special subspace of functions having a form of the ansatz of the asymptotic solution in the regular thin part of the rod structure. We give some examples of application of this method.

As a rule the results are presented in the form of theorems. The obtained asymptotic approximations are justified: estimates of their closeness to exact solutions are proved.

The book is destined to graduate and postgraduate students, specialists in applied mathematics, specialists in mechanics, engineers, university professors. It is partially based on courses of lectures delivered by the author at the Department of Mechanics and Mathematics at Moscow State University Lomonossov and at the Mathematical Department of the University Jean Monnet (Saint Etienne, France). Knowledge acquired after the first three years of advanced mathematical training at a college or university is sufficient to read the most of the book. Presentation of the material is based on the principle "from simple to complicated," every chapter beginning with an elementary example to illustrate the main idea of the method to be described.

Notations

$\varepsilon, \mu, \omega^{-1}$ -small parameters

G - a domain in $\mathbb{R}^s$, i.e., an open connected set (usually bounded) in the s-dimensional space where s is equal to 2 or 3

∂G -boundary of the domain G

$\bar{G}$ -closure of the domain G, i.e., $\bar{G} = G \cup \partial G$

B_μ -finite rod structure i.e., a connected union of a finite number of thin cylinders (the diameter of the base is of order of μ and the height is of order of 1); e is a segment inside of a cylinder constituting B_μ, concluded between the bases

$B_{\varepsilon,\mu}$ -lattice structure i.e., a connected ε-periodic union of an infinite number of thin cylinders (the diameter of the base is of order of product $\varepsilon\mu$ and the height is of order of ε)

$x = (x_1, ..., x_s)$ -point in $\mathbb{R}^s$, slow (macroscopic) variable

$\xi = (\xi_1, ..., \xi_s)$ -fast (microscopic) variable, $\xi = x/\varepsilon$ or $\xi = x/\mu$

$z = (z_1, ..., z_s)$ -fast (microscopic) variable in a boundary layer, $z = x/\mu$ or $z = x/(\varepsilon\mu)$

$L^2(G)$ (the same that $L_2(G)$) space of functions with bounded norm $\|u\|_{L^2(G)} = \sqrt{\int_G u(x)dx}$

$C(\bar{G})$ -space of functions continuous in $\bar{G}$, $\|u\|_{C(\bar{G})} = \sup_{x\in\bar{G}} |u(x)|$

$C^n(\bar{G})$ -space of functions n times differentiable in $\bar{G}$

$H^1(G)$, $H^K(G)$, $H^1_0(G)$ -are the classical Sobolev spaces

$u|_{\partial G}$ -trace on ∂G

$[u]|_\Sigma$ -means the difference of the limit values of function u on the two sides of surface Σ

$\langle f(x_1, ..., x_s, \xi_1, ..., \xi_s)\rangle$ is a mean over the period, i.e.,

$$\int_0^1 ... \int_0^1 f(x_1, ..., x_s, \xi_1, ..., \xi_s) d\xi_1 ... d\xi_s$$

(a, b) -inner product of the vectors a and b

div, div_x, div_ξ -divergence with respect to the variables x or ξ

$grad$, $grad_x$, $grad_\xi$ -gradient with respect to the variables x or ξ

∇, ∇_x, ∇_ξ -the same

rot, rot_x, rot_ξ -curl with respect to the variables x or ξ

Δ, Δ_x, Δ_ξ -Laplacian with respect to the variables x or ξ

$\tilde{\mathcal{I}}(\xi)$ - matrix of rigid displacements (translations and rotations)

$i = (i_1, ..., i_l)$ is a multi-index, $|i| = l$ is its length, $i_j \in \{1, ..., s\}$

$D^i = \frac{\partial^l}{\partial x_{i_1} ... \partial x_{i_l}}$ (partial derivative)

$f(x, \varepsilon) \sim \sum_{j=0}^{\infty} \varepsilon^j g_j(x, \varepsilon)$ is an asymptotic expansion (a.e.) of $f(x, \varepsilon)$, i.e., for any real N there exists M such that, for all $m \geq M$ the relation holds:

$$f(x,\varepsilon) - \sum_{j=0}^{m} \varepsilon^j g_j(x,\varepsilon) = O(\varepsilon^N)$$

$u,\ u_\mu,\ u_\varepsilon,\ u_{\varepsilon,\mu}$ -exact solution of the problem (which depends on some parameters)

$u^{(\infty)},\ u_\mu^{(\infty)},\ u_\varepsilon^{(\infty)},\ u_{\varepsilon,\mu}^{(\infty)}$ -formal asymptotic solution (f.a.s.) which is normally proved to be an asymptotic expansion of the exact solution of the problem

$u^{(K)},\ u_\mu^{(K)},\ u_\varepsilon^{(K)},\ u_{\varepsilon,\mu}^{(K)}$ - the asymptotic approximation of order K

v or ω the formal asymptotic solution of the homogenized problem

$\hat{.}$ - effective coefficient symbol

δ_{ij} - the Kronecker symbol

Chapter 1

Introduction: Basic Notions and Methods

1.1 What is an inhomogeneous rod?

We define an inhomogeneous rod as a long cylinder constituted of a unidirectional chain of recurrent elements (cells). Each cell consists of some subdomains occupied by different materials: the inclusions of one compound and the matrix of another compound which fills the space between the inclusions . The dimensions of the cell are of the same order as the diameters of the subdomains.

Let L be the length of the rod and μ the length of the recurrent cell. We assume that the diameter of the cross section of the cylinder is of the same order as the length of the recurrent cell and that $\mu << L$. If we take $L = 1$, then $\mu << 1$, and μ can be considered as a small dimensionless parameter.

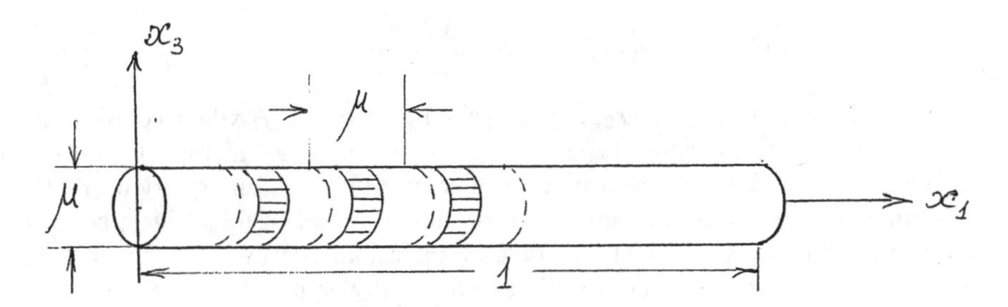

Figure 1.1.1. An inhomogeneous rod

Fields and processes in inhomogeneous rods with periodic structure are described by partial differential equations with periodic coefficients or by systems of thereof. Thus , the temperature in the rod is described by the Poisson equation

$$div(K_\mu(x_1,x_2,x_3)\ grad\ u) = f(x_1,x_2,x_3) \tag{1.1.1}$$

everywhere in the rod (except of the matrix-inclusions contact surfaces). Here $K_\mu(x_1,x_2,x_3)$ is the conductivity coefficient at a point (x_1,x_2,x_3) : if this point falls on an inclusion, K_μ takes the value K_I, the inclusion conductivity coefficient; otherwise K_μ is the conductivity coefficient of the matrix K_M; the function $f(x_1,x_2,x_3)$ is the density of heat sources in the composite rod; and $u(x_1,x_2,x_3)$ is the temperature at the point (x_1,x_2,x_3).

Continuity conditions are satisfied on the matrix-inclusion contact surfaces for temperature

$$[u] = 0 \tag{1.1.2}$$

and for the heat-flux density:

$$[K_\mu(x_1,x_2,x_3)\frac{\partial u}{\partial n}] = 0, \tag{1.1.3}$$

the square brackets [ ] denote the jump of the function in transition through the interface; and $\frac{\partial u}{\partial n}$ is the normal derivative to the interface.

On the lateral boundary of the rod the insulation condition is imposed:

$$K_\mu(x_1,x_2,x_3)\frac{\partial u}{\partial n} = 0. \tag{1.1.4}$$

The conductivity $K_\mu(x_1,x_2,x_3)$ changes by $|K_I - K_M|$ when coordinates (x_1,x_2,x_3) change by a value of the order $\mu << 1$. Thus the function $K_\mu(x_1,x_2,x_3)$ oscillates rapidly. This makes a numerical solution of such an equation practically impossible, because too small a mesh has to be taken for at least several nodes of the difference scheme to fall on each inclusion. However, since $\mu << 1$, problem (1.1.1)-(1.1.4) can be solved asymptotically as $\mu \to 0$.

Without loss of generality it can be assumed that f depends on x_1 only. We assume that it is a smooth function.

In the same way, a steady state elasticity problem can be formulated for the inhomogeneous rod. It consists of the elasticity equations

$$\sum_{i,j=1}^{3}\frac{\partial}{\partial x_i}(A_{ij}^\mu(x_1,x_2,x_3)\frac{\partial u}{\partial x_j}) = f(x_1,x_2,x_3) \tag{1.1.5}$$

everywhere in the rod (except of the matrix-inclusions contact surfaces). Here the 3×3 matrix-valued functions

$$A_{ij}^\mu(x_1,x_2,x_3) = \|a_{ij}^{\mu kl}(x_1,x_2,x_3)\|$$

determine the rigidity tensor at (x_1, x_2, x_3) : if the point belongs to the inclusion, then the tensor $a_{ij}^{\mu kl}(x_1, x_2, x_3)$ coincides with the rigidity tensor a_{Iij}^{kl} of the inclusion material; otherwise $a_{ij}^{\mu kl}(x_1, x_2, x_3)$ equals the matrix rigidity tensor a_{Mij}^{kl}; the three-dimensional vector-valued unknown function $u(x_1, x_2, x_3)$ defines the displacement of the point which had the coordinates (x_1, x_2, x_3) in equilibrium; and $f(x_1, x_2, x_3)$ is a mass-force vector.

The continuity conditions are satisfied on the inclusion surfaces for the displacement vector and for the normal component of the stress tensor:

$$[u] = 0 \tag{1.1.6}$$

$$[\sum_{i,j=1}^{3} n_i A_{ij}^{\mu}(x_1, x_2, x_3) \frac{\partial u}{\partial x_j}] = 0, \tag{1.1.7}$$

where n_i are the direction cosines of the outside normal to the inclusion surface.

On the lateral boundary we impose the free boundary condition:

$$\sum_{i,j=1}^{3} n_i A_{ij}^{\mu}(x_1, x_2, x_3) \frac{\partial u}{\partial x_j} = 0, \tag{1.1.8}$$

where n_i are the direction cosines of the outside normal to the lateral boundary.

1.2 What are effective coefficients?

The purpose of our reasoning is to obtain equations whose coefficients are not rapidly oscillating while their solutions are close to those of the original equations. Usually these new equations are of reduced dimension (for example in case of the inhomogeneous rod they have dimension 1 with respect to space variables). These new equations are called homogenized equations, and their coefficients are the effective coefficients.

Consider problem (1.1.1)-(1.1.4) in a rod occupying a domain $G_\mu = (0,1) \times (-\frac{\mu}{2}, \frac{\mu}{2})^2$ with the right hand side f depending on x_1 only. Let $K_\mu(x_1, x_2, x_3)$ be a μ-periodic in x_1 function:

$$K_\mu(x_1, x_2, x_3) = K(\frac{x_1}{\mu}, \frac{x_2}{\mu}, \frac{x_3}{\mu}),$$

where $K_\mu(\xi_1, \xi_2, \xi_3)$ is a 1-periodic in ξ_1 positive piecewise-smooth function (in case of composite materials K is a piecewise-constant function).

Complete the formulation of the problem by boundary conditions

$$u = 0 \tag{1.2.1}$$

on the ends $x_1 = 0$ and $x_1 = 1$. Then a solution $u(x_1, x_2, x_3)$ of this problem is close to the solution $v_0(x_1)$of the homogenized 1D problem

$$\hat{K}\frac{d^2 v_0}{dx_1^2} = f(x_1), \quad x_1 \in (0,1), \tag{1.2.2}$$

$$v_0(0) = 0, \quad v_0(1) = 0. \tag{1.2.3}$$

It will be shown in Chapter 2 that the error is of order $\sqrt{\mu}$, i.e.

$$\frac{\|u - v_0\|_{L^2(G_\mu)}}{\sqrt{mes\ G_\mu}} = O(\sqrt{\mu}).$$

Here $\hat{K}$ is the effective conductivity of the rod (normalized with respect to the cross-section).

The elasticity equation (1.1.5) has a similar property, but the structure of the 1D homogenized problem is more complex. Consider the problem (1.1.5)-(1.1.8),(1.2.1) in the rod $G_\mu = (0,1) \times (-\frac{\mu}{2}, \frac{\mu}{2})^2$ with the right hand side having the form

$$(\psi_1(x_1)\ ,\ \mu^2\psi_2(x_1) - \sqrt{6}\mu^{-1}x_3\psi_4(x_1)\ ,\ \mu^2\psi_3(x_1) + \sqrt{6}\mu^{-1}x_2\psi_4(x_1))^*,$$

where $\psi_1, \psi_2, \psi_3, \psi_4 \in C^\infty([0,1])$.

This right-hand side (mass forces) has a tensile-compressive part (the first component of the vector ψ, bending parts (the second and the third components of vector $\mu^2\psi$) and a torsional part (the fourth component of vector ψ multiplied by a torsional rigid displacement $(0, -\sqrt{6}\mu^{-1}x_3\ ,\ \sqrt{6}\mu^{-1}x_2)^*$.

Assuming in addition some symmetry of the coefficients with respect to the planes $\{x_2 = 0\}$ and $\{x_3 = 0\}$ and assuming an isotropy of the compounds we obtain the homogenized 1D system of equations

$$\bar{E}^{(3)}\frac{d^2\omega_1}{dx_1^2} = \psi_1(x_1), \tag{1.2.4}$$

$$-\bar{J}_2^{(3)}\frac{d^4\omega_2}{dx_1^4} = \psi_2(x_1), \tag{1.2.5}$$

$$-\bar{J}_3^{(3)}\frac{d^4\omega_3}{dx_1^4} = \psi_3(x_1), \tag{1.2.6}$$

$$\bar{M}^{(3)}\frac{d^2\omega_4}{dx_1^2} = \psi_4(x_1), \quad x_1 \in (0,1), \tag{1.2.7}$$

with the boundary conditions

$$\omega_i(0) = 0, \quad \omega_i(1) = 0, \quad i = 1,2,3,4,$$

$$\frac{d\omega_j}{dx_1}(0) = 0, \quad \frac{d\omega_j}{dx_1}(0) = 0, \quad j = 2,3.$$

It will be shown in Chapter 2 that the difference between the exact solution u and the approximation

$$u_a = (\omega_1(x_1)\ ,\ \omega_2(x_1) - \sqrt{6}\mu^{-1}x_3\omega_4(x_1)\ ,\ \omega_3(x_1) + \sqrt{6}\mu^{-1}x_2\omega_4(x_1))^*$$

is of order $\sqrt{\mu}$, i.e.,

$$\frac{\|u - u_a\|_{L^2(G_\mu)}}{\sqrt{mes\ G_\mu}} = O(\sqrt{\mu}).$$

The coefficients $\bar{E}^{(3)}, \bar{J}_2^{(3)}, \bar{J}_3^{(3)}, \bar{M}^{(3)}$ are called effective rigidities of the rod (the tensile-compressive, two bending and the torsional rigidities respectively).

1.3 A scheme for calculating effective coefficients

The theoretical constructions given below yield the following algorithm for calculating the effective (homogenized) coefficients in case of the examples of the previous section.

1. Effective conductivity. Consider the following *cell problem:* find a 1-periodic in ξ_1 solution of Poisson equation

$$div(K(\xi_1,\xi_2,\xi_3)grad(N_1+\xi_1)) = 0, \quad (\xi_1,\xi_2,\xi_3) \in I\!R \times (-\frac{1}{2},\frac{1}{2})^2 \qquad (1.3.1)$$

with the boundary condition

$$K(\xi_1,\xi_2,\xi_3)\frac{\partial N_1}{\partial n} = 0 \qquad (1.3.2)$$

on the lateral boundary.

On the coefficient discontinuity surfaces N_1 satisfies the interface conditions:

$$[N_1] = 0 \qquad (1.3.3)$$

and

$$[K(\xi_1,\xi_2,\xi_3)\frac{\partial(N_1+\xi_1)}{\partial n}] = 0. \qquad (1.3.4)$$

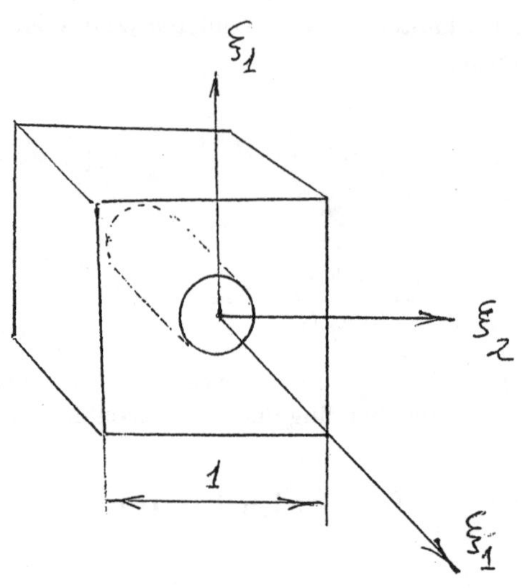

Figure 1.3.1. A periodic cell

This problem can be solved numerically, by the finite element method, or by finite difference methods (see for example the code EFMODUL, Chapter 2).

The effective conductivity coefficient is

$$\hat{K} = \int_{(-\frac{1}{2},\frac{1}{2})^3} K(\xi_1,\xi_2,\xi_3) \frac{\partial(N_1+\xi_1)}{\partial\xi_1} \, d\xi_1 \, d\xi_2 \, d\xi_3. \tag{1.3.5}$$

Note that with the volume fractions of the inclusions and the matrix being equal, (1.3.4) produces the result which differs from the arithmetic mean coefficient $(K_I + K_M)/2$, and from the harmonic mean, $((K_I^{-1} + K_M^{-1})/2)^{-1}$, of K_I and K_M. The greater is the difference between K_I and K_M, the more $\hat{K}$ and these means differ.

2. Effective elasticity.

Consider first the four problems:

1) find a 1-periodic in ξ_1 vector-valued three dimensional solution $N_1^{(1)}(\xi_1,\xi_2,\xi_3)$ of elasticity equation:

$$\sum_{i,j=1}^{3} \frac{\partial}{\partial\xi_i}(A_{ij}(\xi_1,\xi_2,\xi_3)\frac{\partial(N_1^{(1)}+\xi_1\mathbf{e_1})}{\partial\xi_j}) = 0, \quad (\xi_1,\xi_2,\xi_3) \in I\!R \times (-\frac{1}{2},\frac{1}{2})^2 \tag{1.3.6}$$

with the boundary condition

$$\sum_{i,j=1}^{3} n_i(\xi_1,\xi_2,\xi_3)A_{ij}(\xi_1,\xi_2,\xi_3)\frac{\partial(N_1^{(1)}+\xi_1\mathbf{e_1})}{\partial\xi_j} = 0, \tag{1.3.7}$$

on the lateral boundary where the outside normal has the components $n_i(\xi_1,\xi_2,\xi_3)$. Here $\mathbf{e_1} = (1,0,0)^*$.

On the coefficient discontinuity surfaces $N_1^{(1)}$ satisfies the interface conditions:

$$[N_1^{(1)}] = 0 \tag{1.3.8}$$

and

$$[\sum_{i,j=1}^{3} n_i(\xi_1,\xi_2,\xi_3)A_{ij}(\xi_1,\xi_2,\xi_3)\frac{\partial(N_1^{(1)}+\xi_1\mathbf{e_1})}{\partial\xi_j}] = 0. \tag{1.3.9}$$

We calculate the effective longitudinal tensile-compressive rigidity by the following formula:

$$\bar{E}^{(3)} = \int_{(-\frac{1}{2},\frac{1}{2})^3}\sum_{j,k=1}^{3} a_{1j}^{1k}(\xi_1,\xi_2,\xi_3)\frac{\partial(n_1^{(k1)}+\xi_1\delta_{1k})}{\partial\xi_j}\,d\xi_1\,d\xi_2\,d\xi_3, \tag{1.3.10}$$

where a_{ij}^{lk} are the components of the matrix A_{ij} and $n_1^{(k1)}$ is the k−th component of the vector $N_1^{(1)}$.

2) The second problem is as follows:

find a 1-periodic in ξ_1 vector-valued three dimensional solution $N_1^{(4)}(\xi_1,\xi_2,\xi_3)$ of elasticity equation:

$$\sum_{i,j=1}^{3}\frac{\partial}{\partial\xi_i}(A_{ij}(\xi_1,\xi_2,\xi_3)\frac{\partial(N_1^{(4)}+\xi_1\eta(\xi_1,\xi_2,\xi_3))}{\partial\xi_j})$$

$$= 0, \quad (\xi_1,\xi_2,\xi_3)\in I\!R\times(-\frac{1}{2},\frac{1}{2})^2 \tag{1.3.11}$$

with the boundary condition

$$\sum_{i,j=1}^{3} n_i(\xi_1,\xi_2,\xi_3)A_{ij}(\xi_1,\xi_2,\xi_3)\frac{\partial(N_1^{(4)}+\xi_1\eta(\xi_1,\xi_2,\xi_3))}{\partial\xi_j} = 0, \tag{1.3.12}$$

on the lateral boundary . Here $\eta(\xi_1,\xi_2,\xi_3) = \sqrt{6}(0,-\xi_3,\xi_2)^*$.

On the coefficient discontinuity surfaces $N_1^{(4)}$ satisfies the interface conditions:

$$[N_1^{(4)}] = 0 \tag{1.3.13}$$

and

$$[\sum_{i,j=1}^{3} n_i(\xi_1,\xi_2,\xi_3)A_{ij}(\xi_1,\xi_2,\xi_3)\frac{\partial(N_1^{(4)}+\xi_1\eta(\xi_1,\xi_2,\xi_3))}{\partial\xi_j}] = 0. \tag{1.3.14}$$

We calculate the effective longitudinal torsional rigidity by the following formula:

$$\bar{M}^{(3)} = \int_{(-\frac{1}{2},\frac{1}{2})^3} \sum_{j,k=1}^{3} a_{1j}^{1k}(\xi_1,\xi_2,\xi_3) \frac{\partial(n_1^{(k4)} + \xi_1\eta_k(\xi_1,\xi_2,\xi_3))}{\partial\xi_j} \, d\xi_1 \, d\xi_2 \, d\xi_3, \tag{1.3.15}$$

where $n_1^{(k4)}$ is the $k-$th component of the vector $N_1^{(4)}$, η_k is the $k-$th component of the vector η.

3/4) The third and the fourth problems are as follows:

for $q = 2,3$ find a 1-periodic in ξ_1 vector-valued three dimensional solution $N_2^{(q)}(\xi_1,\xi_2,\xi_3)$ of elasticity equation:

$$\sum_{i,j=1}^{3} \frac{\partial}{\partial\xi_i}(A_{ij}(\xi_1,\xi_2,\xi_3)\frac{\partial N_2^{(q)}}{\partial\xi_j} - A_{i1}\mathbf{e_1}\xi_q) = 0, \quad (\xi_1,\xi_2,\xi_3) \in I\!R \times (-\frac{1}{2},\frac{1}{2})^2 \tag{1.3.16}$$

with the boundary condition

$$\sum_{i,j=1}^{3} n_i(\xi_1,\xi_2,\xi_3)(A_{ij}(\xi_1,\xi_2,\xi_3)\frac{\partial N_2^{(q)}}{\partial\xi_j} - A_{i1}\mathbf{e_1}\xi_q) = 0, \tag{1.3.17}$$

on the lateral boundary.

On the coefficient discontinuity surfaces $N_2^{(q)}$ satisfies the interface conditions:

$$[N_2^{(q)}] = 0 \tag{1.3.18}$$

and

$$[\sum_{i,j=1}^{3} n_i(\xi_1,\xi_2,\xi_3)(A_{ij}(\xi_1,\xi_2,\xi_3)\frac{\partial N_2^{(q)}}{\partial\xi_j} - A_{i1}\mathbf{e_1}\xi_q)] = 0. \tag{1.3.19}$$

We calculate the effective bending rigidities by the following formula:

$$\bar{J}^{(q)} = \int_{(-\frac{1}{2},\frac{1}{2})^3} (\sum_{j,k=1}^{3} a_{1j}^{1k}(\xi_1,\xi_2,\xi_3)\frac{\partial n_2^{(kq)}}{\partial\xi_j} - a_{11}^{11}\xi_q)\xi_q \, d\xi_1 \, d\xi_2 \, d\xi_3, \tag{1.3.20}$$

where $n_2^{(kq)}$ is the $k-$th component of the vector $N_2^{(q)}$.

Problems (1.3.6)-(1.3.20) can be solved numerically, by the finite element method, or by finite difference methods.

1.4 Microscopic structure of a field

In section 1.2 we noted that an exact solution of the original problem is close to that of the averaged one. Nevertheless, we shall see that the difference between derivatives of the exact solution and those of the homogenized problem are great, while the derivatives of the exact solution are parts of expressions for heat fluxes or the stress tensor. Therefore, it does not suffice just to know effective coefficients to be able to determine correctly fluxes and stresses in a heterogeneous bar. However, if problems (1.3.11)-(1.3.20) are solved, the formula

$$\sigma_{\mathbf{i}} = \sum_{j=1}^{3} A_{ij}^{\mu}(x_1, x_2, x_3)\frac{\partial u}{\partial x_j} \simeq$$

$$(\sum_{j=1}^{3} A_{ij}(\xi_1,\xi_2,\xi_3)(\frac{\partial(N_1^{(1)}+\xi_1\mathbf{e_1})}{\partial \xi_j})\frac{d\omega_1}{dx_1} + \frac{\partial(N_1^{(4)}+\xi_1\eta(\xi_1,\xi_2,\xi_3))}{\partial \xi_j}\frac{d\omega_4}{dx_1} +$$

$$\mu\sum_{q=2}^{3}\sum_{j=1}^{3}(A_{ij}(\xi_1,\xi_2,\xi_3)\frac{\partial N_2^{(q)}}{\partial \xi_j} - A_{i1}(\xi_1,\xi_2,\xi_3)\mathbf{e_1}\xi_q)\frac{d^2\omega_q}{dx_1^2})|_{(\xi_1,\xi_2,\xi_3)=(\frac{x_1}{\mu},\frac{x_2}{\mu},\frac{x_3}{\mu})},$$

$$i = 1, 2, 3, \qquad (1.4.1)$$

enables stresses to be determined with accuracy of order μ in interior points of a rod; the last sum of order μ determines the main part of stresses with accuracy of order μ^2 in absence of the torsional forces ($\psi_4 = 0$) and of the tensile-compressive forces ($\psi_1 = 0$).

Formula (1.4.1) takes into account the macroscopic as well as microscopic structure of the stress field and provides an approach to solving problems of strength calculations for heterogeneous bars. Similar formulas can be obtained for heat flows problems.

In vicinities of the ends of the rod there are boundary layers which can not be described in a macroscopic scale. Some special methods for their description are proposed in chapters 2,4,5 and 6.

1.5 What is the homogenization method?

The main asymptotic tool of the book is the homogenization method. It is a hybrid of the multi-scale method combined with the averaging method by Krylov, Bogoliubov, and Mitropolsky. Solution is sought in the form of series in powers of a small parameter μ with coefficients depending both on the variables x_i (usually referred to as slow or macroscopic variables) and $\xi_i = x_i$ (fast or microscopic variables). The slow variables correspond to the global structure of the fields and the fast variables to their local structure. Such a series is substituted into the original equations. By equating the coefficients of the powers of μ to zero, we get equations for the terms of the expansion. Usually we introduce an unknown function describing the macroscopic properties which does not depend on the rapid variables and we express the terms of the expansion as a sum of products of some functions of rapid variables multiplied by derivatives of this new unknown function. The homogenization method (as well as the boundary layer method) is described in Chapter 2.

1.6 What is a finite rod structure?

A finite number of joint rods form a finite rod structure. Its macroscopic model is described by the equations of type (1.2.2), (1.2.4)-(1.2.7) with some special conditions of junction. For example in case of two joint co-linear rods with effective rigidities $\bar{E}_{+}^{(3)}$, $\bar{J}_{2+}^{(3)}, \bar{J}_{3+}^{(3)}, \bar{M}_{+}^{(3)}$ and $\bar{E}_{-}^{(3)}$, $\bar{J}_{2-}^{(3)}, \bar{J}_{3-}^{(3)}, \bar{M}_{-}^{(3)}$ respectively the junction conditions are:

$$\omega_i(+0) = \omega_i(-0), \quad i = 1, 2, 3, 4, \tag{1.6.1}$$

$$\frac{d\omega_j}{dx_1}(+0) = \frac{d\omega_j}{dx_1}(-0), \quad j = 2, 3, \tag{1.6.2}$$

$$\bar{E}_{+}^{(3)} \frac{d\omega_1}{dx_1}(+0) = \hat{E}_{-}^{(3)} \frac{d\omega_1}{dx_1}(-0), \tag{1.6.3}$$

$$\bar{J}_{2+}^{(3)} \frac{d^3\omega_2}{dx_1^3}(+0) = \bar{J}_{2-}^{(3)} \frac{d^3\omega_2}{dx_1^3}(-0), \tag{1.6.4}$$

$$\bar{J}_{3+}^{(3)} \frac{d^3\omega_3}{dx_1^3}(+0) = \bar{J}_{3-}^{(3)} \frac{d^3\omega_3}{dx_1^3}(-0), \tag{1.6.5}$$

$$\bar{M}_{+}^{(3)} \frac{d\omega_4}{dx_1}(+0) = \hat{M}_{-}^{(3)} \frac{d\omega_4}{dx_1}(-0), \tag{1.6.6}$$

$$\bar{J}_{2+}^{(3)} \frac{d^2\omega_2}{dx_1^2}(+0) = \bar{J}_{2-}^{(3)} \frac{d^2\omega_2}{dx_1^2}(-0), \tag{1.6.7}$$

$$\bar{J}_{3+}^{(3)} \frac{d^2\omega_3}{dx_1^2}(+0) = \bar{J}_{3-}^{(3)} \frac{d^2\omega_3}{dx_1^2}(-0), \tag{1.6.8}$$

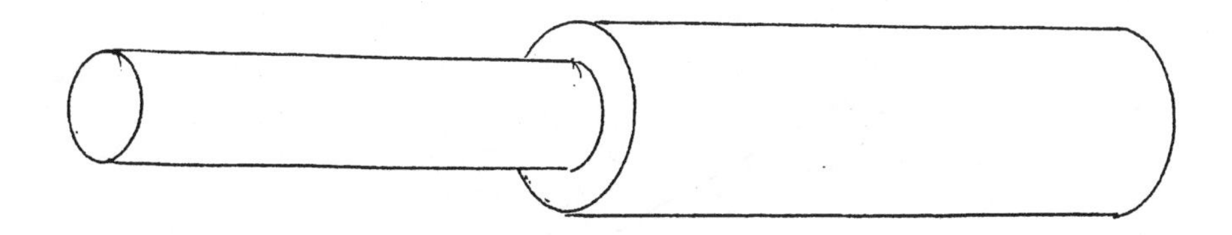

Figure 1.6.1. A junction of two rods

Here the junction surface is $\{x_1 = 0\}$.

This result is generalized in Chapter 4 for more general finite rod structures.

Let $\beta_1, ..., \beta_J$ be bounded domains in $I\!R^{s-1}$, $(s = 2, 3)$ with a piecewise smooth boundary, $B_{j,\mu}$ $(j = 1, ..., J)$ the cylinders, defined by

$$B_{j,\mu} = \{x \in R^s \mid (x_2/\mu, ..., x_s/\mu) \in \beta_j, \ x_1 \in I\!R\},$$

and $B^{\alpha}_{h,j,\mu}$ the cylinder obtained from $B_{j,\mu}$ by orthogonal transformation $\tilde{\Pi}$ with the matrix α^T, $\alpha = (\alpha_{il})$ and a translation $h = (h_1, ..., h_s)^T$. Let e^{α}_h be an s-dimensional vector obtained by means of $\tilde{\Pi}$ and h from the vector $i(h, \alpha)$ of the axis Ox_1 with the beginning at the point O .

Definition 1.6.1. *Let B be a union of all segments e^{α}_h when α belongs to a set $\Delta \subset I\!R^{s^2}$ and h to a set $H_\alpha \subset I\!R^s$, and these sets are independent of μ. Let B be such that any two segments e^{α}_h can have only one common point which is the end point for both segments. The set B is called skeleton, the end points of e^{α}_h are called nodes.*

We associate the cylinder $B^{\alpha}_{h,j,\mu}$ with every e^{α}_h and denote by $\tilde{B}^{\alpha}_{h,j,\mu}$ the part of $B^{\alpha}_{h,j,\mu}$ enclosed between the two planes passing through the ends of segment e^{α}_h and perpendicular to it (we assume that the bases belong to $\tilde{B}^{\alpha}_{h,j,\mu}$). Suppose that Δ and H_α are finite sets and B is connected.

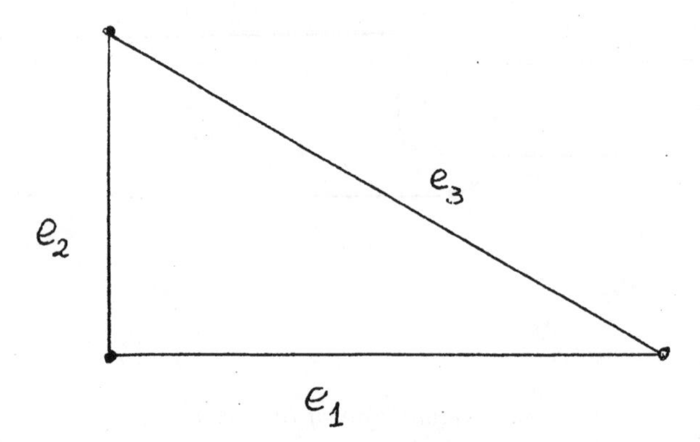

Figure 1.6.2. A skeleton B containing three segments

Definition 1.6.2. *By a finite rod structure we understand the set of interior points of the union*

$$B_\mu = \cup_{\alpha\in\Delta} \cup_{h\in H_\alpha} \tilde{B}^{\alpha}_{h,j,\mu}.$$

We assume that B_μ is connected, satisfies the cone condition and that μ is a small parameter.

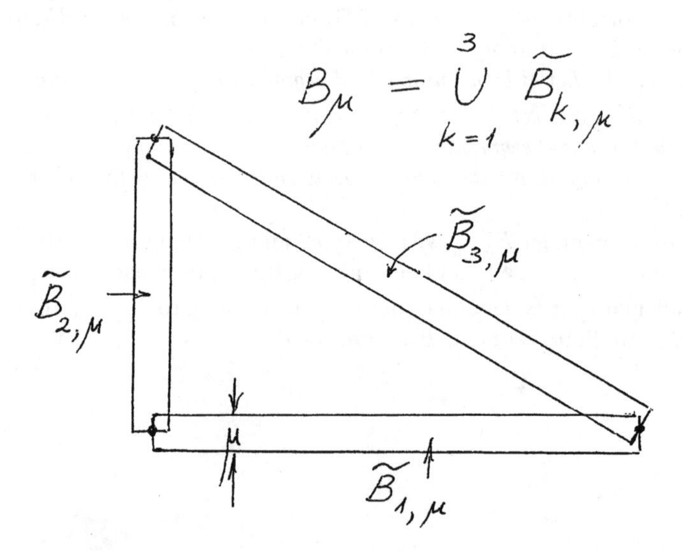

Figure 1.6.3. The corresponding finite rod structure B_μ.

The asymptotic approach reduces the elasticity problem stated in such domain to a set of equations of type (1.2.4)-(1.2.7) associated with the segments

e_h^α; these equations are coupled by some junction conditions similar to conditions (1.6.1)-(1.6.8). An asymptotic expansion of the exact solution is built and justified.

1.7 What is a lattice structure?

A great number of joint rods form a lattice structure.

Lattice-like domains simulate some widely adopted engineering constructions such as frameworks of houses, trusses of bridges, industrial installations, supports of electric power lines, spaceship grids, etc. as well as some capillary or fissured systems. The methods of solving the system of equations of structural mechanics presently used for calculating the frameworks have the drawback that the number of equations increases with number of nodes of lattice. This greatly impedes the calculation of lattices with large number of nodes, especially when investigating non-stationary and nonlinear processes. The homogenization techniques and splitting principle for the homogenized operator proposed in [134,136,16] allowed essentially simplify the process of calculation of such structures.

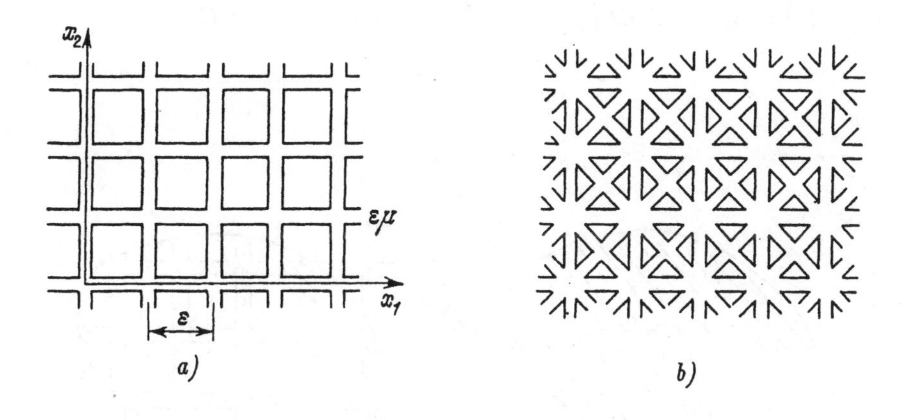

Figure 1.7.1. Two examples of lattice structures (the left one is a rectangular lattice).

Consider the simplest model of lattice structures: two-dimensional rectangular lattice.

Definition 1.7.1. *The union*

$$B_{\varepsilon,\mu} = \cup_{k=-\infty}^{+\infty} \left(\{(x_1,x_2) \in R^2 \mid \mid x_2 - k\varepsilon \mid < \varepsilon\mu/2 \} \right.$$

$$\left. \cup \{(x_1,x_2) \in R^2 \mid \mid x_1 - k\varepsilon \mid < \varepsilon\mu/2 \} \right)$$

is called the two-dimensional rectangular lattice.

Thus the rectangular lattice is a union of thin strips of the width $\varepsilon\mu$ stretched in each coordinate direction and forming the $\varepsilon-$ periodic system in each dimension. We also denote

$$B^j_{\varepsilon,\mu} = \cup_{k=-\infty}^{+\infty} \{(x_1,x_2) \in R^2 \mid | x_{3-j} - k\varepsilon | < \varepsilon\mu/2 \}$$

the unions of horizontal ($j=1$) and vertical ($j=2$) strips, so

$$B_{\varepsilon,\mu} = B^1_{\varepsilon,\mu} \cup B^2_{\varepsilon,\mu}.$$

Let G be a domain with the boundary $\partial G \in C^\infty$ which is independent of ε and μ. In the domain $B_{\varepsilon,\mu} \cap G$ we consider the Poisson equation

$$-div\ (\ A\ grad\ u_{\varepsilon,\mu}) = f(x)\ , \tag{1.7.1}$$

with the boundary conditions

$$(\ A\ grad\ u_{\varepsilon,\mu}\ ,\ n\ ) = 0,\ for\ x \in \partial B_{\varepsilon,\mu} \cap G, \tag{1.7.2}$$

$$u_{\varepsilon,\mu} = 0,\ for\ x \in \bar{B}_{\varepsilon,\mu} \cap \partial G, \tag{1.7.3}$$

here $x = (x_1, x_2)$, $A = (a_{ij})$ is a constant $(2 \times 2)-$matrix independent of ε and μ , $A = A^T > 0$, i.e. $a_{ij} = a_{ji}$ and A is positive, $f \in C^1(G)$.

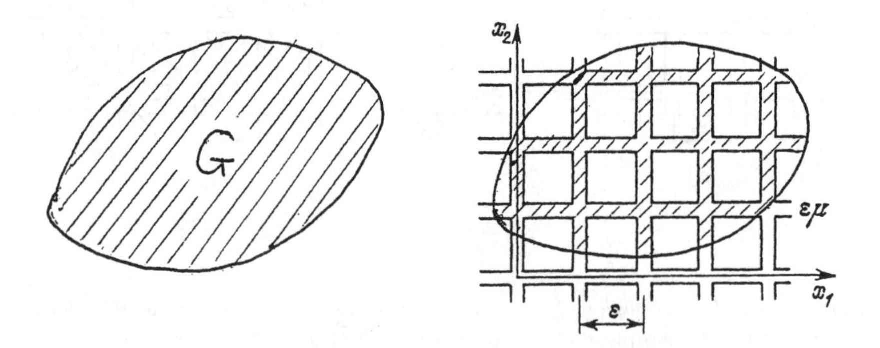

Figure 1.7.2. Domain G independent of small parameters and its intersection with the rectangular lattice.

Problems (1.7.1)-(1.7.3) simulate a stationary heat field in the lattice structure $B_{\varepsilon,\mu} \cap G$ under the conditions of thermal insulation on the boundary of G

with A being the heat conductivity tensor for the material of which the framework is made ("cut out"). For $a_{ij} = \delta_{ij} D$ (1.7.1)-(1.7.3) may be interpreted also as a problem of diffusion of a substance in a fissured rock filled with water, with D being the diffusion coefficient of the substance in the water. In this case, the framework $B_{\varepsilon,\mu}$ models a rectangular system of cracks $\varepsilon\mu$ wide, filled with water. This model is very approximate since the system of cracks in a rock has a less regular structure . The model of a random framework is more suitable for its description.

Numerical solution of problems (1.7.1)-(1.7.3), with $\varepsilon << 1$, $\mu << 1$, is very difficult since the step size of the grid must have an order much less than ε. The implementation of the standard homogenization procedure is also impeded, since the problem on a cell depends on the small parameter μ, and in order to solve it numerically, we must select the step size of the grid to be much less than μ. Hence, an asymptotic investigation of the problem is needed. The result of this investigation obtained by means of homogenization techniques can be presented in a form of the following theorem, proved by the author [134,136,16] and later generalized in [47],[48]. But first we shall formulate a definition of L-convergence.

Definition 1.7.2. *Let $u_{\varepsilon,\mu}(x)$ is a sequence of functions from $L_2(B_{\varepsilon,\mu} \cap G)$, $u_0(x) \in L_2(G)$. One says that $u_{\varepsilon,\mu}$ L - converges to $u_0(x)$ at $B_{\varepsilon,\mu} \cap G$ if and only if*

$$\frac{\|u_{\varepsilon,\mu} - u_0\|_{L_2(B_{\varepsilon,\mu} \cap G)}}{\sqrt{meas(B_{\varepsilon,\mu} \cap G)}} \to 0, \quad (\varepsilon, \mu \to 0).$$

The normalization factor $1/\sqrt{meas(B_{\varepsilon,\mu} \cap G)}$ is necessary because $\|1\|_{L_2(B_{\varepsilon,\mu} \cap G)} = \sqrt{meas(B_{\varepsilon,\mu} \cap G)}$. Notice that L-convergence is not a convergence in common sense because the domain depends on small parameters.

Theorem 1.7.1. *Let $\hat{A} = (\hat{a}_{ij})$,*

$$\hat{a}_{11} = 0.5(a_{11} - a_{12}a_{22}^{-1}a_{21}), \ \hat{a}_{12} = 0,$$

$$\hat{a}_{22} = 0.5(a_{22} - a_{21}a_{11}^{-1}a_{12}), \ \hat{a}_{21} = 0.$$

Let $u_0(x)$ is the solution of the homogenized problem

$$-div \ (\ \hat{A} \ grad \ u_0) \ = \ f(x) \ , \ x \in G, \quad u_0|_{\partial G} \ = \ 0. \tag{1.7.4}$$

Then $u_{\varepsilon,\mu}$ L-converges to u_0, and

$$\frac{\|u_{\varepsilon,\mu} - u_0\|_{L_2(B_{\varepsilon,\mu} \cap G)}}{\sqrt{meas(B_{\varepsilon,\mu} \cap G)}} = O(\sqrt{\varepsilon} + \sqrt{\mu}), \quad (\varepsilon, \mu \to 0).$$

This means that the original complicated problem (1.7.1)-(1.7.3) stated in the lattice like domain depending on small parameters is asymptotically reduced to the homogenized problem (1.7.4) stated in a regular domain G independent of small parameters. Problem (1.7.4) is a standard partial derivative equation problem for numerical analysis tools.

In Chapter 5 we consider more general lattice structures.

1.8 Advantages and disadvantages of the asymptotic approach

1. The asymptotic approach (and especially the homogenization techniques and the boundary layer method) is an effective tool for investigating both macroscopic and microscopic properties of inhomogeneous rods, plates and structures. Even the first approximation yields an error of order ε, where ε is the ratio of the characteristic microscopic size and the characteristic macroscopic size of the problem. An asymptotic expansion of the solution enables accuracy of order ε^n to be reached for any integer n.

2. The asymptotic approach is universal. It is applicable to a lot of processes that might occur in heterogeneous rods, plates and structures, such as elastic vibrations, heat propagation, diffusion, fluid flow (in thin tubes and tube structures), electromagnetic oscillations, and radiation among others. It is possible to consider both linear and non-linear models, differential as well as operator equations. The boundary layer method can also take into account boundary effects, in some cases to any degree of accuracy.

3. The asymptotic results are rigorously justified for many models. Theorems have been proved about the solvability of the problems which are elementary steps of an algorithm for constructing an asymptotic solution. Estimates of the method's accuracy have been obtained to determine the order of error in a particular approximation and consequently, to draw distinct limits for application of this approximation.

4. The main disadvantage of the homogenization method is that the numerical solutions of cell problems of types (1.3.1)-(1.3.4) and (1.3.6)-(1.3.20) is required. Three-dimensional problems on a cell take a lot of computer time and memory capacity.

It should be borne in mind, however, that the many existing "forks", mixing rules, and engineering formulas frequently produce very rough approximations to effective characteristics. Thus, for composite materials with widely differing characteristics of their components, the error of the Voight-Reuss-Hill approximations is comparable to the effective characteristics themselves and may exceed 100 percent (see [16]).

Thus, citing A.A. Iliushin, this difficulty is the price to pay for the exactitude of the homogenization method. Nevertheless, there exist a certain number of solvers for cell problems, for example code EFMODUL described in Chapter 2.

5. The traditional methods do not describe the microscopic details of processes, such as local strains and stresses, and do not take into account boundary layer effects, while an asymptotic approach analyzes these details.

The asymptotic approach permits sometimes to establish the limits of applicability for some approximate engineering methods and to improve considerably their accuracy when necessary.

What does it mean that "ε is small"? In breef, the answer depends on the demanded accuracy, i.e. on the value of the admissible error. Assume that an asymptotic method gives an error of order $O(\varepsilon^n)$ and this value does not exceed the admissible error. Then one can say that ε is small enough. Although an additional study is required to make sure that a constant in $O(\varepsilon^n)$ is not too great. To this end a numerical experiment could be developed for some finite value of ε in order to estimate this constant.

1.9 Appendices

1.A1. Appendix 1: what is the Poincaré-Friedrichs-Korn inequalities?

These inequalities are the important tools in the mathematical elasticity theory, and normally they have to be proved in order to apply the Lax-Milgram lemma and to get the a priori estimates. We will consider here these inequalities in the simpliest model case, when the domain is the square $Q = (0,1)^2$.

Lemma 1.A1.1. (the Poincaré-Friedrichs inequality) *There exists a positive constant C_{PF} such that for any $u \in H^1(Q)$, such that $u|_{x_1=0} = 0$, the inequality holds :*

$$\int_Q u^2(x)dx \leq C \int_Q (\nabla u)^2 dx,$$

ie.

$$\|u\|^2_{L^2(Q)} \leq C\|\nabla u\|^2_{L^2(Q)}$$

.

Proof. All proofs will be given for the $C^2(\overline{Q})-$smooth functions. The declared assertions can be go further by a limit passage in the H^1 - norm. We have from the Newton-Leibnitz formula

$$I = \int_Q u^2(x_1,x_2)dx_1dx_2 = \int_Q \left(\int_0^{x_1} \frac{\partial u}{\partial t}(t,x_2)dt\right)^2 dx_1dx_2.$$

Applying the Cauchy-Schwartz-Buniakowskii inequality, we get

$$\begin{aligned} I &\leq \int_Q x_1 \int_0^{x_1} \left(\frac{\partial u}{\partial t}(t,x_2)\right)^2 dt\, dx_1dx_2 \\ &\leq \int_Q x_1 \int_0^{1} \left(\frac{\partial u}{\partial t}(t,x_2)\right)^2 dt\, dx_1dx_2 \\ &\leq \frac{1}{2}\int_Q \left(\frac{\partial u}{\partial x_1}\right)^2 dx_1dx_2 \leq \frac{1}{2}\|\nabla u\|^2_{L^2(Q)}. \end{aligned}$$

The theorem is proved with $C_{PF} = 1/2$.

Lemma 1.A1.2. (the Poincaré inequality) *There exists a positive constant C_P such that for any $u \in H^1(Q)$, the inequality holds*

$$\|u\|^2_{L^2(Q)} \leq C_P\|\nabla u\|^2_{L^2(Q)} + \left(\int_Q u dx\right)^2 /(mesQ).$$

Proof. Let $x = (x_1, x_2)$ and $y = (y_1, y_2)$ be two points of Q. Applying the Newton-Leubnitz formula, we get

$$u(y_1, y_2) - u(x_1, x_2) = \int_{x_1}^{y_1} \frac{\partial u}{\partial t}(t, x_2)dt + \int_{x_2}^{y_2} \frac{\partial u}{\partial t}(y_1, t)dt,$$

so

$$\int_Q \int_Q (u(y) - u(x))^2 dxdy = \int_Q \int_Q \left(\int_{x_1}^{y_1} \frac{\partial u}{\partial t}(t, x_2)dt + \int_{x_2}^{y_2} \frac{\partial u}{\partial t}(y_1, t)dt \right)^2 dxdy \leq$$

$$\leq 2 \int_Q \int_Q \left(\left(\int_{x_1}^{y_1} \frac{\partial u}{\partial t}(t, x_2)dt \right)^2 + \left(\int_{x_2}^{y_2} \frac{\partial u}{\partial t}(y_1, t)dt \right)^2 \right) dxdy.$$

As in the proof of Lemma 1.A1.1 this integral can be estimated by

$$2 \int_Q \int_Q \left(\int_0^1 \left(\frac{\partial u}{\partial t}(t, x_2) \right)^2 dt + \int_0^1 \left(\frac{\partial u}{\partial t}(y_1, t) \right)^2 dt \right) dxdy$$

and so by

$$2 \|\nabla u\|^2_{L^2(Q)}.$$

So, we have

$$\int_Q \int_Q (u^2(y) - 2u(x)u(y) + u^2(x))dxdy \leq 2\|\nabla u\|^2_{L^2(Q)},$$

i.e.

$$2 \int_Q u^2(x)dx - 2 \left(\int_Q u(x)dx \right)^2 \leq 2\|\nabla u\|^2_{L^2(Q)}$$

This estimate implies the statement of the lemma with $C_P = 1$.

Now consider vector-valued functions $u = (u_1, u_2) \in (H^1(Q))^2$; then ∇u is a 2×2 - matrix, $\|\nabla u\|^2_{L^2(Q)} = \sum_{i,j=1}^{2} \int_Q \left(\frac{\partial u_i}{\partial x_j} \right)^2 dx$; denote

$$e_i^j(u) = \frac{1}{2} \left(\frac{\partial u_i}{\partial x_j} + \frac{\partial u_j}{\partial x_i} \right) ; \quad e(u) = \sum_{i,j=1}^{2} (e_i^j(u))^2 ; \quad E_Q(u) = \int_Q e(u)dx.$$

Evidently, $E_Q(u) \leq \|\nabla u\|^2_{L^2(Q)}$.

Lemma 1.A1.3. (the Korn inequality) *There exists a positive constant C_K such that for any $u \in H_0^1(Q)$, the inequality holds*

$$\|\nabla u\|^2_{L^2(Q)} \leq C_K E_Q(u).$$

Proof. Consider $\int_Q (e_1^2(u))^2 dx = \frac{1}{4} \int_Q \left(\frac{\partial u_1}{\partial x_2} \right)^2 + 2 \frac{\partial u_1}{\partial x_2} . \frac{\partial u_2}{\partial x_1} + \left(\frac{\partial u_2}{\partial x_1} \right)^2 dx;$

Let us transform

$$\int_Q \frac{\partial u_1}{\partial x_2}\frac{\partial u_2}{\partial x_1}dx = -\int_Q \frac{\partial^2 u_1}{\partial x_1 \partial x_2}u_2 = \int_Q \frac{\partial u_1}{\partial x_1}\frac{\partial u_2}{\partial x_2}$$

for $u \in C^2(Q)$, $u = 0$ on ∂Q. So

$$E_Q(u) = \int_Q \left(\frac{\partial u_1}{\partial x_1}\right)^2 + \left(\frac{\partial u_2}{\partial x_2}\right)^2 + \frac{1}{2}\left(\frac{\partial u_1}{\partial x_2}\right)^2 + \frac{1}{2}\left(\frac{\partial u_2}{\partial x_1}\right)^2 + \frac{\partial u_1}{\partial x_1}\frac{\partial u_2}{\partial x_2}dx \geq$$

$$\geq \frac{1}{2}\int_Q (\nabla u)^2 + (div\ u)^2 dx \geq \frac{1}{2}\|\nabla u\|^2_{L^2(Q)}.$$

So the lemma is proved with $C_K = 2$.

The same assertion (and the same proof) takes place for periodic functions of $(H^1(Q))^2$.

These three inequalities hold true for a large class of domains (see books and reviews [16], [25], [54], [55], [58], [65], [75], [76], [85], [105], [119], [129], [170]).

Of course if the domain depends on one or some small parameters, the constants C_{PF}, C_P and C_K can also depend on this or these parameters. For the thin domains this dependency was studied in [144] and later in [28], [37] for the Poincare and the Poincare-Friedrichs inequalities, and in [18], [43], [68], [74], [114], [116]- [118], [119], [144] for the Korn inequality. In the appendices to Chapter 4 and Chapter 5 we will give the proof of these inequalities for the finite rod structures and for the lattice structures.

Chapter 2

Heterogeneous Rod

Here we consider the three-dimensional conductivity problem in a cylindrical domain (a rod) having a heterogeneous structure. We assume that the characteristic size of heterogeneities is of the same order as the diameter. Assume that it is much less than the length of the rod. So the ratio of the diameter to the length of the rod is the small parameter of the same order as the ratio of the characteristic size of heterogeneity to the length of the rod. We assume that the length of the rod has a finite value.

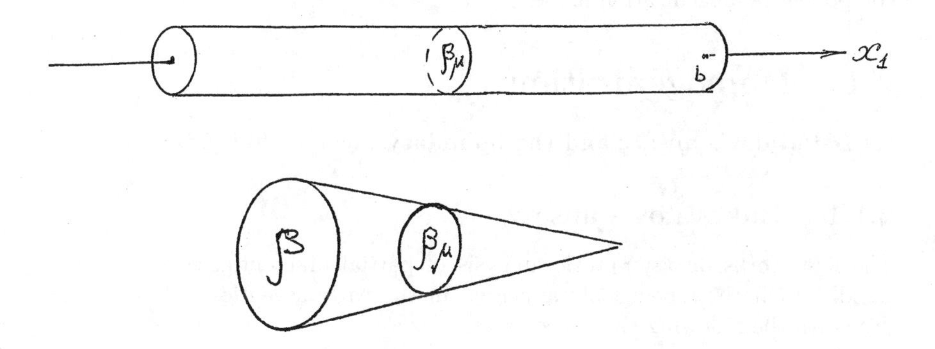

Figure 2.1. A rod.

So the geometrical model of a rod is a cylinder C_μ defined as

$$C_\mu = (0, b) \times \beta_\mu,$$

where $\beta_\mu = \{(x_2, x_3) \in I\!R^2;\ (x_2/\mu, x_3/\mu) \in \beta\}$, β is a bounded domain in $I\!R^2$ with piecewise smooth boundary; C_μ is supposed to satisfy the cone condition [55]. We assume that $(0,0) \in \beta$ and μ is a small parameter. The microscopical heterogeneity of the rod is simulated by a special dependence of the coefficients of the material on space variable $x = (x_1, x_2, x_3)$; these coefficients are functions of x/μ.

An asymptotic analysis of the mathematical model is developed as μ tends to zero. To this end we apply the homogenization technique and the boundary layer technique described in section 2.1. Then in section 2.2 we study the conductivity equation; an expansion of a solution is constructed and justified and the one-dimensional homogenized model of the rod is obtained. In section 2.3 the three-dimensional elasticity problem for a rod is studied. Then the results of the asymptotic analysis of conductivity equation and elasticity equation are generalized for a non-stationary case (section 2.4) and for a non-linear case (section 2.5). Sections 2.6 and 2.7 are devoted to a case when the material contains the second parameter: compressibility or a ratio of Young modulus of the compounds of the rod. When this ratio is sufficiently large (i.e. in case of contrasting compounds) the scale effect takes place: the one-dimensional homogenized model of the rod is more complex. It is analogous to the multi-component homogenization model.

2.1 Homogenization

(N.Bakhvalov's ansatz and the boundary layer technique)

2.1.1 Bakhvalov's ansatz

The first works on asymptotic analysis of partial differential equations with rapidly oscillating coefficients appeared in the early seventies. These papers deal with the following elliptic equation :

$$\sum_{k,m=1}^{s} \frac{\partial}{\partial x_k}\left(A_{km}\left(\frac{x}{\varepsilon}\right)\frac{\partial u}{\partial x_m}\right) = f(x). \tag{2.1.1}$$

Here $\varepsilon > 0$ is a small parameter, $x = (x_1, \ldots, x_s) \in I\!R^s$ $A_{km}(\xi)$, $\xi = (\xi_1, \ldots, \xi_s) \in I\!R^s$ are periodic functions of each of the ξ_i with period 1 (1-periodic functions), and $f(x)$ is the right-hand side. Equation (2.1.1) may be considered for either a scalar valued or a vector valued unknown function $u(x)$. In the latter case $u(x)$ and $f(x)$ are n-dimensional vector valued functions and $A_{km}(\xi)$ are $n \times n$ matrices. Here the elasticity equation is the most interesting case, when the matrices A_{km} satisfy some special additional conditions (see below). Direct numerical solution of problems of type (2.1.1) is complicated. Indeed, the rapid oscillations in the coefficients require a closely spaced mesh with number of nodes much greater than ε^{-s} (this mesh should be "sensitive" to characteristic variations of the coefficients). We note the pioneer works on the asymptotic analysis of equations with rapidly oscillating coefficients by E. Sanchez-Palencia [177], E. De Giorgi and S. Spagnollo [49], N. Bakhvalov [12]-[14], J. L. Lions, A. Bensoussan, and G. Papanicolaou [22], O. Oleinik [127], V. Berdichevsky [23], I. Babuska [8], and V. Marchenko and E. Khruslov [103]. In particular, in [13] the series permitting the construction of the complete asymptotic expansion appeared for the first time. It has the form

$$u^{(\infty)} = \Big(\sum_{l=0}^{\infty} \varepsilon^l \sum_{|i|=l} N_i(\xi) D^i v(x)\Big)\Big|_{\xi=x/\varepsilon}, \tag{2.1.2}$$

where $i = (i_1, \ldots, i_l)$, $i_j \in \{1, \ldots, s\}$, is a multi-index. The number $l = |i|$ of components of the multi-index will be called its *length*. The derivatives $D^i v = \frac{\partial^l v}{\partial x_{i_1} \ldots \partial x_{i_l}}$, $N_i(\xi)$ are 1-periodic functions of $\xi_1, \ldots, \xi_s$; the function $v(x)$ does not depend of x/ε. In the second sum the summation over is all multi-indices i of length l. The derivations varying in the order of differentiation are assumed to be different. For example, $D^{(1,1)} = \frac{\partial^2}{\partial x_1^2}$, $D^{(1,2)} = \frac{\partial^2}{\partial x_1 \partial x_2}$, $D^{(1,2,1)} = \frac{\partial^3}{\partial x_1 \partial x_2 \partial x_1}$.

In the vector case in (2.1.1) $v(x)$ is a column n-vector and $N_i(\xi)$ are $n \times n-$matrices. The form of the series (2.1.2) is close to the asymptotic series of the averaging technique introduced by N. Krylov, N. Bogoljubov, and Yu. Mitropol'sky [29].

Remark 2.1.1. Let us remind the notion of a formal asymptotic solution (ansatz); for more details we refer to [16], Chapter 2. It is a *formal* series. We don't care about its convergence, but for any given natural N there exists such a partial sum $\sum_{l=0}^{J}$ of this series which satisfies the equation (the problem) with an error of order $O(\varepsilon^N)$. All operations, for example, the substitution into the equation , are meant in the sense of existence of such partial sums which satisfy all passages with an error of order $O(\varepsilon^N)$. Such passages will be further denoted as well by symbol $\sim$.

2.1.2 An example of formal asymptotic solution

Consider first the one-dimensional case, $s = 1$. Equation (2.1.1) takes form

$$\frac{d}{dx}\Big(A\big(\frac{x}{\varepsilon}\big)\frac{d}{dx}u\Big) = f(x). \tag{2.1.1$'$}$$

Then the expansion (2.1.2) transforms into form

$$u^{(\infty)} = \Big(\sum_{l=0}^{\infty} \varepsilon^l N_l(\xi) D^l v(x)\Big)\Big|_{\xi=x/\varepsilon}, \tag{2.1.2$'$}$$

where $D^l = \frac{d^l}{dx^l}$.

Substituting (2.1.2') into equation (2.1.1') and collecting similar terms, we obtain the equation

$$\sum_{l=0}^{\infty} \varepsilon^{l-2} (H_l(\xi) D^l v(x))\big|_{\xi=x/\varepsilon} = f(x), \tag{2.1.3}$$

where

$$H_l(\xi) = L_{\xi\xi} N_l + \frac{d}{d\xi}(A(\xi) N_{l-1}) + A(\xi) \frac{dN_{l-1}}{d\xi} + A(\xi) N_{l-2}, \tag{2.1.4}$$

$$H_1(\xi) = L_{\xi\xi} N_1 + \frac{d}{d\xi}(A(\xi) N_0) + A(\xi) \frac{dN_0}{d\xi}, \quad l = 1, \tag{2.1.5}$$

$$H_0(\xi) = L_{\xi\xi} N_0, \quad l = 0. \tag{2.1.6}$$

Here

$$L_{\xi\xi} = \frac{d}{d\xi}(A(\xi) \frac{d}{d\xi}).$$

Note that (2.1.4) can be extended to the case $l = 1, = 0$ if $N_l = 0$ is assumed for $l < 0$, $N_0 = 1$ (in vectorial case $N_0 = I$ is the identity matrix).

We assume that the series (2.1.2) begins with v, $u^{(\infty)} = v + O(\varepsilon)$. Then

$$H_1(\xi) = \frac{d}{d\xi}\Big(A(\xi) \frac{d}{d\xi} N_1 + A(\xi)\Big), \quad H_0(\xi) = 0.$$

Using the notation

$$T_l(\xi) = \frac{d}{d\xi}(A(\xi) N_{l-1}) + A(\xi) \frac{dN_{l-1}}{d\xi} + A(\xi) N_{l-2}, \tag{2.1.7}$$

one can rewrite (2.1.4)–(2.1.6) in the form

$$H_l(\xi) = L_{\xi\xi} N_l + T_l(\xi).$$

Let us require that

$$H_l(\xi) = h_l, \tag{2.1.8}$$

where $h_l = \text{const}$ are constants such that

$$h_0 = h_1 = 0. \tag{2.1.9}$$

If N_l and h_l satisfying (2.1.8) and (2.1.9) are obtained, then problem (2.1.3) is reduced to the following problem with constant coefficients for v:

$$\sum_{l=2}^{\infty} \varepsilon^{l-2} h_l D^l v = f(x). \tag{2.1.10}$$

The formal asymptotic solution to problem (2.1.10) can be obtained in the form of a regular series

$$v = \sum_{q=0}^{\infty} \varepsilon^q v_q(x). \tag{2.1.11}$$

Let us construct N_l and h_l by induction. Let all N_l of priority $l < k$ be found. Let us define $T_k(\xi)$ by (2.1.7) (N_l with $l < 0$ are equal to zero).

Let the equation

$$L_{\xi\xi} N = F(\xi) \tag{2.1.12}$$

with 1-periodic $n \times n$ matrix right-hand side $F(\xi) \in L_2^{\mathrm{loc}}$ be solvable in the class of 1-periodic matrix functions $N(\xi)$ if and only if $\langle F \rangle = 0$ (here and below in this section $\langle F(\xi) \rangle$ denotes the average $\int_{(0,1)^s} F(\xi)\, d\xi$). If the Fredholm theorems are valid for equation (2.1.12), this assumption means that the homogeneous equation adjoint to (2.1.12) has only constant solutions.

Then we must set

$$h_k = \langle T_k(\xi) \rangle. \tag{2.1.13}$$

In this case there exists a 1-periodic solution to the problem

$$L_{\xi\xi} N_k + T_k(\xi) = h_k. \tag{2.1.14}$$

Solving this problem, we obtain N_k . The induction step is now complete.

The induction basis is $k = 0$, $N_0 = 1$. Note that

$$h_0 = 0, \quad h_1 = \left\langle \frac{d}{d\xi} A(\xi) \right\rangle = 0. \tag{2.1.15}$$

Taking into account (2.1.8) and (2.1.9), we rewrite (2.1.3) in form (2.1.10). This equation with respect to v is called the *infinite-order homogenized equation*. If the principal part $\widehat{L}$ of the operator on the left-hand side is elliptic, then, according to the above, the formal asymptotic solution can be obtained in the form (2.1.11); here v_q are the solutions to the recurrent sequence of problems:

$$\widehat{L} v_q = f_q(x), \tag{2.1.16}$$

where

$$\widehat{L} = h_2 D^2, \quad f_q = -\sum_{l=3}^{q+2} h_l D^l v_{q+2-l} \quad (q > 0), \quad f_0 = f.$$

Thus the construction of the formal asymptotic solution by means of the homogenization method (N. Bakhvalov's version) consists in substituting series (2.1.2) into equation (2.1.1), collecting similar terms, and putting the rapidly oscillating coefficients $H_l(\xi)$ of the derivations $D^l v$ equal to constants h_l. These constants are determined from the problems that give the solvability conditions for N_l.

2.1.3 The boundary conditions corrector

The method presented above is designed for equations in the entire space $\mathbb{R}$. Consider equation (2.1.1) in the layer $x_1 \in (0, b)$, where b/ε is a positive integer and b is a number of the order of 1. Let the boundary conditions

$$u\big|_{x=0} = g_0, \quad u\big|_{x=b} = g_1 \tag{2.1.17}$$

be imposed.

Substituting (2.1.2') into (2.1.1') and (2.1.17) and taking into account the fact that $u^{(\infty)}$ is as constructed above, we obtain

$$\sum_{l=0}^{\infty} \varepsilon^{l-2} H_l\Big(\frac{x}{\varepsilon}\Big) D^l v = f(x), \tag{2.1.18}$$

$$\sum_{l=0}^{\infty} \varepsilon^l N_l\Big(0\Big) D^l v\big|_{x=0} = g_0, \tag{2.1.19}$$

$$\sum_{l=0}^{\infty} \varepsilon^l N_l\Big(\frac{b}{\varepsilon}\Big) D^l v\big|_{x=b} = g_1, \tag{2.1.20}$$

where $N_l\Big(0\Big) = N_l\Big(\frac{b}{\varepsilon}\Big)$ due to the periodicity of N_l (we remind that $\frac{b}{\varepsilon}$ is an integer). Denote $h_l^0 = N_l(0)$.

Then problem (2.1.18)-(2.1.20) is reduced to the following problem with constant coefficients for v:

$$\sum_{l=2}^{\infty} \varepsilon^{l-2} h_l D^l v = f(x), \quad x \in (0, b),$$

$$\sum_{l=0}^{\infty} \varepsilon^l h_l^0 D^l v\big|_{x=0} = g_0,$$

$$\sum_{l=0}^{\infty} \varepsilon^l h_l^1 D^l v\big|_{x=b} = g_1.$$

This problem can be solved by the standard technique of expanding v into a regular series (2.1.11). This procedure again gives (2.1.16) for v_q with the boundary conditions

$$h_0^0 v_q\big|_{x=0} = g_{0q}, \quad h_0^0 v_q\big|_{x=b} = g_{1q},$$

where $g_{r0} = g_r$, $g_{rq} = -\sum_{l=1}^{q} h_l^0 D^l v_{q-l}\big|_{x=0,b}$ $(q > 0)$, $r = 0, 1$.

Thus the formal asymptotic solution to problem (2.1.1'),(2.1.17) is completely constructed. The justification of the presented formalism is carried out in [16].

2.1.4 Introduction to the boundary layer technique

The boundary layer method consists of two stages. The first stage is construction of an expansion of a solution in entire space without boundary conditions. The second stage is construction of some correctors to the expansion decaying at any finite distance from the boundary.

Consider the following model problem:

$$\varepsilon^2 u'' - u = f(x), \quad x \in (0,1), \tag{2.1.21}$$

$$u(0) = g_1, u(1) = g_2, \tag{2.1.22}$$

where $f \in C^\infty([0,1])$.

At the first stage we consider an ansatz

$$\bar{u} = \sum_{l=0}^{\infty} \varepsilon^{2l} \bar{u}_l(x). \tag{2.1.23}$$

Substituting it into equation (2.1.21), we obtain

$$\sum_{l=0}^{\infty} \varepsilon^{2l}(-\bar{u}_l(x) + \bar{u}''_{l-1}(x)) = f(x), \tag{2.1.24}$$

where $\bar{u}_{-1} = 0$.

Relation (2.1.24) generates a chain of equations

$$-\bar{u}_l(x) + \bar{u}''_{l-1}(x) = f(x)\delta_{l0},$$

i.e.,

$$\bar{u}_l(x) = \bar{u}''_{l-1}(x) - f(x)\delta_{l0}, \; l = 0, 1, ... \tag{2.1.25}$$

This chain defines all terms $\bar{u}_l$ of the ansatz (2.1.23), i.e.,

$$\bar{u}_0(x) = -f(x),$$
$$\bar{u}_1(x) = \bar{u}''_0(x) = -f''(x),$$
$$\dots$$
$$\bar{u}_l(x) = -f^{(2l)}(x). \tag{2.1.26}$$

At the second stage we consider the boundary layer correctors

$$u^0 = \sum_{l=0}^{\infty} \varepsilon^{2l} u_l^0(\frac{x}{\varepsilon})$$

and

$$u^1 = \sum_{l=0}^{\infty} \varepsilon^{2l} u_l^1(\frac{x-1}{\varepsilon}),$$

where the terms u_l^0 and u_l^1 are defined on $\mathbb{R}_+$ and $\mathbb{R}_-$ respectively and

$$\lim_{\xi\to+\infty} u_l^0(\xi) = 0, \quad \lim_{\xi\to-\infty} u_l^1(\xi) = 0.$$

We substitute the sum

$$u^{(\infty)} = \bar{u} + u^0 + u^1 \tag{2.1.27}$$

into equation (2.1.21) and into boundary conditions (2.1.22). We have

$$\sum_{l=0}^{\infty} \varepsilon^{2l}(-\bar{u}_l(x) + \bar{u}_{l-1}''(x)) +$$

$$\sum_{l=0}^{\infty} \varepsilon^{2l}(-u_l^0(\xi) + \frac{d^2}{d\xi^2}u_l^0(\xi)) +$$

$$\sum_{l=0}^{\infty} \varepsilon^{2l}(-u_l^1(\eta) + \frac{d^2}{d\eta^2}u_l^1(\eta)) = f(x), \tag{2.1.28}$$

where $\xi = \frac{x}{\varepsilon}$, $\eta = \frac{x-1}{\varepsilon}$; we have also

$$\sum_{l=0}^{\infty} \varepsilon^{2l}(\bar{u}_l(0) + u_l^0(0) + u_l^1(-\frac{1}{\varepsilon})) = g_0, \tag{2.1.29}$$

and

$$\sum_{l=0}^{\infty} \varepsilon^{2l}(\bar{u}_l(1) + u_l^0(\frac{1}{\varepsilon}) + u_l^1(0)) = g_1. \tag{2.1.30}$$

We shall see later that the decaying rate for the functions u_l^0 and u_l^1 is exponential. It means that there exist such positive constants C_l and c_l that the estimates hold:

$$|u_l^0(\xi)| \leq C_l e^{-c_l \xi},$$

$$|u_l^1(\eta)| \leq C_l e^{-c_l |\eta|}.$$

So for any K, $u_l^0(\frac{1}{\varepsilon})$ and $u_l^1(-\frac{1}{\varepsilon})$ are values of order $O(\varepsilon^K)$; therefore these terms can be neglected in equations (2.1.29) and (2.1.30) with an accuracy $O(\varepsilon^K)$. Then we obtain two independent chains of *boundary layer problems:*

$$-u_l^0(\xi) + \frac{d^2}{d\xi^2}u_l^0(\xi) = 0, \quad \xi > 0, \quad u_l^0(0) = g_0\delta_{l0} - \bar{u}_l(0), \; l = 0, 1, \dots, \tag{2.1.31}$$

and

$$-u_l^1(\eta) + \frac{d^2}{d\eta^2}u_l^1(\eta) = 0, \ \ \eta < 0, \ \ u_l^1(0) = g_1\delta_{l0} - \bar{u}_l(1), \ l = 0, 1, \dots . \quad (2.1.32)$$

Indeed, these problems (2.1.31) and (2.1.32) have exponentially decaying solutions:

$$u_l^0(\xi) = (g_0\delta_{l0} - \bar{u}_l(0))e^{-\xi}, u_l^1(\eta) = (g_1\delta_{l0} - \bar{u}_l(1))e^{\eta}.$$

So all the terms of ansatz (2.1.27) are determined, i.e., the asymptotic expansion $u^{(\infty)}$ of solution to problem (2.1.21),(2.1.22) is presented in a form of sum (2.1.27) of the so called regular expansion

$$\bar{u}(x) = -\sum_{l=0}^{\infty} \varepsilon^{2l} f^{(l)}(x)$$

and two boundary layers

$$u^0(x) = \sum_{l=0}^{\infty} \varepsilon^{2l}(g_0\delta_{l0} - f^{(l)}(0))e^{-\frac{x}{\varepsilon}},$$

and

$$u^1(x) = \sum_{l=0}^{\infty} \varepsilon^{2l}(g_1\delta_{l0} - f^{(l)}(1))e^{-\frac{x-1}{\varepsilon}}.$$

These boundary layers are "concentrated" near the boundary of the segment $[0, 1]$ only and they are exponentially small at any finite distance from the boundary.

Thus the boundary layer technique means construction of some correctors to an expansion of a solution in entire space; these correctors are located just near the boundary and rapidly decay when a point is at a finite distance from the boundary.

2.1.5 Homogenization in $\mathbb{R}^s$

Here below we describe the homogenization technique in the case of s dimensions. We consider equation (2.1.1) in whole space $\mathbb{R}^s$. Consider ansatz (2.1.2).

Substituting (2.1.2) into equation (2.1.1) and collecting similar terms, we obtain the equation

$$\sum_{l=0}^{\infty} \varepsilon^{l-2} \sum_{|i|=l} (H_i(\xi)D^i v(x))\big|_{\xi=x/\varepsilon} = f(x), \quad (2.1.33)$$

where

$$H_i(\xi) = L_{\xi\xi} N_{i_1\dots i_l} + \frac{\partial}{\partial \xi_k}(A_{ki_1}(\xi) N_{i_2\dots i_l}) + A_{i_1 j}(\xi)\frac{\partial N_{i_2\dots i_l}}{\partial \xi_j} + A_{i_1 i_2}(\xi) N_{i_3\dots i_l}, \tag{2.1.34}$$

$$H_{i_1}(\xi) = L_{\xi\xi} N_{i_1} + \frac{\partial}{\partial \xi_k}(A_{ki_1}(\xi) N_\emptyset) + A_{i_1 j}(\xi)\frac{\partial N_\emptyset}{\partial \xi_j}, \quad |i| = 1, \tag{2.1.35}$$

$$H_\emptyset(\xi) = L_{\xi\xi} N_\emptyset, \quad |i| = 0. \tag{2.1.36}$$

Throughout the book summation over repeated indices from 1 to s is implied unless otherwise specified; here

$$L_{\xi\xi} = \frac{\partial}{\partial \xi_k}\Big(A_{km}(\xi)\frac{\partial}{\partial \xi_m}\Big).$$

Note that (2.1.34) can be extended to the case $|i| = 1, |i| = 0$ if $N_i = 0$ is formally assumed for $|i| < 0$, and $N_\emptyset = 1$ or (in vectorial case)$N_\emptyset = I$, i.e. the identity matrix. Thus, we assume that the series (2.1.2) begins with v, $u^{(\infty)} = v + O(\varepsilon)$. Then

$$H_{i_1}(\xi) = \frac{\partial}{\partial \xi_k}\Big(A_{kj}(\xi)\frac{\partial N_{i_2}}{\partial \xi_j} + A_{ki_1}(\xi)\Big), \quad H_\emptyset(\xi) = 0.$$

Using the notation

$$T_i(\xi) = \frac{\partial}{\partial \xi_k}(A_{ki_1}(\xi) N_{i_2\dots i_l}) + A_{i_1 j}(\xi)\frac{\partial N_{i_2\dots i_l}}{\partial \xi_j} + A_{i_1 i_2}(\xi) N_{i_3\dots i_l} \tag{2.1.37}$$

one can rewrite (2.1.34)–(2.1.36) in the form

$$H_i(\xi) = L_{\xi\xi} N_i + T_i(\xi).$$

Let us arrange N_i in a sequence to satisfy the relationship $|j| < |i|$ (the priority of N_i is assumed to be higher than the priority of N_j); the functions N_i with the same index length $|i|$ can be ordered in an arbitrary way. Note that the functions T_i are defined by the functions N_j with priority $|j| < |i|$. This condition may be used for defining N_i inductively in terms of the functions N_j of lower priority if some requirements on $H_i(\xi)$ are imposed.

Let us require that

$$H_i(\xi) = h_i, \tag{2.1.38}$$

where $h_i = \text{const}$;

$$h_\emptyset = h_{i_1} = 0. \tag{2.1.39}$$

If N_i and h_i satisfying (2.1.38) and (2.1.39) are obtained, then problem (2.1.33) is reduced to the following problem with constant coefficients for v:

$$\sum_{l=2}^{\infty} \varepsilon^{l-2} \sum_{|i|=l} h_i D^i v = f(x). \tag{2.1.40}$$

The formal asymptotic solution to problem (2.1.40) can be obtained in the form of a regular series

$$v = \sum_{q=0}^{\infty} \varepsilon^q v_q(x). \tag{2.1.41}$$

Let us construct N_i and h_i by induction. Let all N_i of priority $|i| < k$ be found. Let us define $T_i(\xi)$ for all i with $|i| = k$ by equations (2.1.37) (N_i with $|i| < 0$ are equal to zero).

Let the equation

$$L_{\xi\xi} N = F(\xi) \tag{2.1.42}$$

with 1-periodic $n \times n$ matrix right-hand side $F(\xi) \in L_2^{\text{loc}}$ be solvable in the class of 1-periodic matrix functions $N(\xi)$ if and only if $\langle F \rangle = 0$ (here and below in this section $\langle F(\xi) \rangle$ denotes the average $\int_{(0,1)^s} F(\xi)\, d\xi$). If the Fredholm theorems are valid for equation (2.1.42), this assumption means that the homogeneous equation adjoint to (2.1.42) has only constant solutions. In particular, this is the case for a single elliptic equation and for the system of equations of elasticity theory.

Then we must set

$$h_i = \langle T_i(\xi) \rangle. \tag{2.1.43}$$

In this case there exists a 1-periodic solution to the problem

$$L_{\xi\xi} N_i + T_i(\xi) = h_i. \tag{2.1.44}$$

Solving this problem, we obtain N_i for each multi-index i of length $|i| = k$. The induction step is complete.

The induction basis is $k = 0$, $N_\emptyset = 1$ or I. Note that

$$h_\emptyset = 0, \quad h_{i_1} = \left\langle \frac{\partial}{\partial \xi_k} A_{ki_1}(\xi) \right\rangle = 0. \tag{2.1.45}$$

Taking into account (2.1.38) and (2.1.39), we rewrite (2.1.33) in form (2.1.40). This equation with respect to v is called the *infinite-order homogenized equation*. If the principal part $\widehat{L}$ of the operator on the left-hand side is elliptic, then, according to the above, the formal asymptotic solution can be obtained in the form (2.1.41); here v_q are the solutions to the recurrent sequence of problems:

$$\widehat{L} v_q = f_q(x), \tag{2.1.46}$$

where

$$\widehat{L} = \sum_{k,p=1}^{s} h_{kp} \frac{\partial^2}{\partial x_k \partial x_p}, \quad f_q = -\sum_{l=3}^{q+2} \sum_{|i|=l} h_i D^i v_{q+2-l} \quad (q > 0), \quad f_0 = f.$$

Thus the construction of the formal asymptotic solution by means of the homogenization method (N. Bakhvalov's version) consists in substituting series (2.1.2) into equation (2.1.1), collecting similar terms, and putting the rapidly oscillating coefficients $H_i(\xi)$ of the derivations $D^i v$ equal to constants h_i. These constants are determined from the problems that give the solvability conditions for N_i.

In the monograph by N. Bakhvalov and G. Panasenko [16], the justification of the formalism presented above is carried out and estimates for the difference between the partial sums of the series (2.1.2) and the exact solution are obtained for the case in which (2.1.1) is an elliptic equation or the system of equations of elasticity theory.

2.1.6 Boundary layer correctors to homogenization in $\mathbb{R}^s$

The formal asymptotic solution for the simplest boundary value problem was first constructed by the author [132],[133] (see also [16],[100]). Consider equation (2.1.1) in the layer $x_1 \in (0,b)$, where b/ε is a positive integer and b is a number of the order of 1. Let the boundary conditions

$$u\big|_{x_1=0} = g_0(x'), \quad u\big|_{x_1=b} = g_1(x') \tag{2.1.47}$$

be set. Here and below for any s-dimensional vector a we set $a' = (a_2, \dots, a_s)^*$, i.e. a' is a column vector comprising all components of a except for the first component. Suppose that the right-hand sides f, g_0, and g_1 are T-periodic with respect to their arguments (T is a number of the order of 1 such that T/ε is a positive integer). The solution is also sought in the class of T-periodic functions of $x_2, \dots, x_s$.

The formal asymptotic solution to problem (2.1.1), (2.1.47) is sought in the form of Bakhvalov's series u_B with the corrector $u_p^0 + u_p^1$ of boundary-layer type,

$$u^{(\infty)} = u_B + u_p^0 + u_p^1, \tag{2.1.48}$$

where u_B is defined by equation (2.1.2) and

$$u_p^0 = \Big(\sum_{l=0}^{\infty} \varepsilon^l \sum_{|i|=l} N_i^0(\xi) D^i v(x)\Big)\Big|_{\xi_j = x_j/\varepsilon}, \tag{2.1.49}$$

$$u_p^1 = \Big(\sum_{l=0}^{\infty} \varepsilon^l \sum_{|i|=l} N_i^1(\xi) D^i v(x)\Big)\Big|_{\xi_1=(x_1-b)/\varepsilon, \xi_j = x_j/\varepsilon, \quad j=2,\dots,s.} \tag{2.1.50}$$

Substituting (2.1.48) into (2.1.1) and (2.1.47) and taking into account the fact that u_B is as constructed above, we obtain

$$\sum_{l=0}^{\infty} \varepsilon^{l-2} \sum_{|i|=l} \bar{H}_i\Big(\frac{x}{\varepsilon}\Big) D^i v = f(x), \tag{2.1.51}$$

$$\sum_{l=0}^{\infty} \varepsilon^{l} \sum_{|i|=l} \bar{N}_i\Big(0, \frac{x'}{\varepsilon}\Big) D^i v\big|_{x_1=0} = g_0(x'), \tag{2.1.52}$$

$$\sum_{l=0}^{\infty} \varepsilon^l \sum_{|i|=l} \bar{N}_i\Big(\frac{b}{\varepsilon}, \frac{x'}{\varepsilon}\Big) D^i v\big|_{x_1=b} = g_1(x'), \tag{2.1.53}$$

where

$$\bar{H}_i\Big(\frac{x}{\varepsilon}\Big) = H_i\Big(\frac{x}{\varepsilon}\Big) + H_i^0\Big(\frac{x}{\varepsilon}\Big) + H_i^1\Big(\frac{x_1 - b}{\varepsilon}, \frac{x'}{\varepsilon}\Big),$$

$$\bar{N}_i\Big(\frac{x}{\varepsilon}\Big) = N_i\Big(\frac{x}{\varepsilon}\Big) + N_i^0\Big(\frac{x}{\varepsilon}\Big) + N_i^1\Big(\frac{x_1 - b}{\varepsilon}, \frac{x'}{\varepsilon}\Big)$$

$N_i(\xi)$ and $H_i(\xi)$ are the 1-periodic functions constructed above, and $N_i^r(\xi)$ and $H_i^r(\xi)$ are 1-periodic functions of $\xi_2, \dots, \xi_s$ exponentially stabilizing to zero as $\xi_1 \to \pm\infty$ (+ for $r = 0$, − for $r = 1$). Here

$$H_i^r(\xi) = L_{\xi\xi} N_{i_1 \dots i_l}^r + \frac{\partial}{\partial \xi_k}\big(A_{ki_1}(\xi) N_{i_2 \dots i_l}^r\big) +$$

$$+ A_{i_1 j}(\xi) \frac{\partial N_{i_2 \dots i_l}^r}{\partial \xi_j} + A_{i_1 i_2}(\xi) N_{i_3 \dots i_l}^r, \tag{2.1.54}$$

where $i = (i_1, \dots, i_l)$ with negative length $|i|$ is assumed to be zero.

Denoting

$$T_i^r = \frac{\partial}{\partial \xi_k}\big(A_{ki_1}(\xi) N_{i_2 \dots i_l}^r\big) + A_{i_1 j}(\xi) \frac{\partial N_{i_2 \dots i_l}^r}{\partial \xi_j} + A_{i_1 i_2}(\xi) N_{i_3 \dots i_l}^r,$$

one can reduce (24) to the form $H_i^r(\xi) = L_{\xi\xi} N_i^r + T_i^r(\xi)$.

Let us arrange N_i^r in a sequence as was done above for N_i. Namely, let $N_i^r \succ N_j^r$ whenever $|i| > |j|$. Note that T_i^r is defined by the functions N_j^r with priority $|j| < |i|$. This condition may be used for the recurrent definition of N_i^r in terms of N_j^r with lower priority if certain requirements on $H_i^r(\xi)$, $\bar{N}_i(0, \xi')$, and $\bar{N}_i(b/\varepsilon, \xi')$ are imposed.

Let $H_i^r(\xi) = 0$, $r = 0, 1$, $\bar{N}_i(0, \xi') = \text{const}$, and $\bar{N}_i(b/\varepsilon, \xi') = \text{const}$. Note that, according to the construction of the previous subsection, $H_i(\xi) = h_i = \text{const}$. Denote the constants $\bar{N}_i(0, \xi')$ by h_i^0 and $\bar{N}_i(b/\varepsilon, \xi')$ by h_i^1. If N_i^r and h_i^r satisfying these conditions are found, then problem (2.1.51)–(2.1.53) is reduced to the following problem with constant coefficients for v:

$$\sum_{l=2}^{\infty} \varepsilon^{l-2} \sum_{|i|=l} h_i D^i v = f(x), \quad x_1 \in (0, b), \tag{2.1.55}$$

$$\sum_{l=0}^{\infty} \varepsilon^l \sum_{|i|=l} h_i^0 D^i v\big|_{x_1=0} = g_0(x'), \tag{2.1.56}$$

$$\sum_{l=0}^{\infty} \varepsilon^l \sum_{|i|=l} h_i^1 D^i v\big|_{x_1=b} = g_1(x'). \tag{2.1.57}$$

This problem can be solved by the standard technique of expanding v into a regular series (2.1.41). This again gives (2.1.46) for v_q with the boundary conditions

$$h^0_\emptyset v_q\big|_{x_1=0} = g_{0q}(x'), \quad h^1_\emptyset v_q\big|_{x_1=b} = g_{1q}(x'), \tag{2.1.58}$$

where $g_{r0} = g_r$, $g_{rq} = -\sum_{l=1}^{q}\sum_{|i|=l} h_i^r D^i v_{q-l}\big|_{x_1=0,b}$ $(q > 0)$.

Hence, in order to construct the formal asymptotic solution to the original problem, we only need to construct functions N_i^r and constants h_i^r satisfying the equations

$$H_i^r(\xi) = 0, \quad r = 1, 2,$$
$$\bar{N}_i(0, \xi') = h_i^0 + O(e^{-c/\varepsilon}),$$
$$\bar{N}_i\Big(\frac{b}{\varepsilon}, \xi'\Big) = h_i^1 + O(e^{-c/\varepsilon}),$$

where c is a positive constant.

The functions N_i^r and the constants h_i^r are constructed by induction. Namely, we set $N_\emptyset^r = 0$ and $h_\emptyset^r = 1$ or I in the vectorial case. Let all N_i^r of priority $|i| \le k$ be found. Let us define $T_i^r(\xi)$ for all i with $|i| = k+1$. Choose a constant h_i^0 ($|i| = k+1$) such that the 1-periodic in $\xi_2, \dots, \xi_s$ solution to the problem

$$L_{\xi\xi} N_i^0 + T_i^0(\xi) = 0, \quad \xi_1 > 0, \tag{2.1.59}$$

$$N_i^0(0, \xi') = -N_i(0, \xi') + h_i^0, \tag{2.1.60}$$

exponentially tending to zero as $\xi_1 \to +\infty$, exists, and choose a constant h_i^1 such that the similar solution to the problem

$$L_{\xi\xi} N_i^1 + T_i^1(\xi) = 0, \quad \xi_1 < 0, \tag{2.1.61}$$

$$N_i^1(0, \xi') = -N_i(0, \xi') + h_i^1, \tag{2.1.62}$$

exponentially tending to zero as $\xi_1 \to -\infty$, exists (the possibility of such a choice is assumed). The possibility of such a choice in the case of the elliptic equation is proved [89], [90], and for the system of equations of elasticity theory this is proved in [129], [130].

Note that $N_i(0, \xi') = N_i(b/\varepsilon, \xi')$, since b/ε is an integer and N_i is 1-periodic in ξ_1. In this connection, condition (2.1.62) is equivalent to the condition

$$N_i^1(0, \xi') = -N_i\Big(\frac{b}{\varepsilon}, \xi'\Big) + h_i^1.$$

To determine the constants h_i^0, we first solve the problem

$$L_{\xi\xi} \widetilde{N}_i^0 + T_i^0(\xi) = 0, \quad \xi_1 > 0, \quad \widetilde{N}_i^0(0, \xi') = -N_i(0, \xi')$$

in the class of bounded functions 1-periodic in ξ'. Let the solution to this problem stabilize to a constant C_r^0 (this is true for elliptic equations). Next, let $h_i^0 = -\lim_{\xi_1 \to +\infty} \widetilde{N}_i^0$, and then

$$N_i^0 = \widetilde{N}_i^0 - \lim_{\xi_1 \to +\infty} \widetilde{N}_i^0, \quad |i| = k+1$$

is the solution to problem (2.1.59), (2.1.60). Problem (2.1.61), (2.1.62) can be solved in a similar manner. The induction step is complete.

Remark 2.1.2. By construction, we obtain $u^{(\infty)}$ asymptotically satisfying equation (2.1.1). At the same time, the boundary conditions are satisfied asymptotically not for $u^{(\infty)}$, but for $u_B + u_p^0$ at the left end and for $u_B + u_p^1$ at the right end. But in view of the exponential decay of $N_i^r(\xi)$ for any N we have

$$(u^{(\infty)} - u_B - u_p^0)\big|_{x_1=0} = u_p^1\big|_{x_1=0} = O(\varepsilon^N)$$

and similarly

$$(u^{(\infty)} - u_B - u_p^1)\big|_{x_1=b} = u_p^0\big|_{x_1=b} = O(\varepsilon^N).$$

Thus, $u^{(\infty)}$ is a formal asymptotic solution to problem (2.1.1), (2.1.47). The justification of the presented formalism is carried out in [16].

Remark 2.1.3. (On the justification procedure). In this section 2.1 we have constructed several asymptotic solutions to some problems depending on small parameter ε. It means that having a problem

$$L_\varepsilon u_\varepsilon = f_\varepsilon \tag{2.1.63}$$

stated for an unknown function u_ε with an ε−dependent linear operator L_ε and right hand side f_ε (which stands for the data of this problem), we have constructed (instead of this unknown u_ε) an asymptotic approximation $u_\varepsilon^{(K)}$ such that it satisfies (2.1.63) with some remainder $r_\varepsilon^{(K)}$ of order $O(\varepsilon^K)$, i.e. $\|r_\varepsilon^{(K)}\|_{H_d} = O(\varepsilon^K)$. Here $L_\varepsilon : H_s \to H_d$, H_s, H_d are some normed spaces. It means that

$$L_\varepsilon u_\varepsilon = f_\varepsilon + r_\varepsilon^{(K)}. \tag{2.1.64}$$

Assume that for any $f_\varepsilon \in H_d$ problem (2.1.63) has a unique solution and that an a priori estimate holds:

$$\|u_\varepsilon\|_{H_s} \leq C\|f_\varepsilon\|_{H_d} \tag{2.1.65}$$

with a constant C independent of f_ε and ε.

Then subtracting (2.1.63) from (2.1.64) and applying estimate (2.1.65) to the difference $u_\varepsilon^{(K)} - u_\varepsilon$ (that is a solution to equation $L_\varepsilon(u_\varepsilon^{(K)} - u_\varepsilon) = r_\varepsilon^{(K)}$) we get

$$\|u_\varepsilon^{(K)} - u_\varepsilon\|_{H_s} \leq C\|r_\varepsilon^{(K)}\|_{H_d},$$

i.e.,

$$\|u_\varepsilon^{(K)} - u_\varepsilon\|_{H_s} = O(\varepsilon^K). \tag{2.1.66}$$

This is the main idea of justifications of asymptotic procedures in this book. Sometimes the constant C in (2.1.65) depends on ε in such a way that there exists $\alpha > 0$ such that $C \leq C_0 varepsilon^{-\alpha}$ where C_0 does not depend on ε. Then we get

$$\|u_\varepsilon^{(K)} - u_\varepsilon\|_{H_s} = O(\varepsilon^{K-\alpha}). \tag{2.1.66'}$$

instead of (2.1.66). If estimate (2.1.66) holds true for any K and α is a positive integer number and if

$$\|u_\varepsilon^{(K+\alpha)} - u_\varepsilon^{(K)}\|_{H_s} = O(\varepsilon^K), \tag{2.1.67}$$

then estimate (2.1.66') can be improved. Indeed from (2.1.66) for $K+\alpha$ we get

$$\|u_\varepsilon^{(K+\alpha)} - u_\varepsilon\|_{H_s} = O(\varepsilon^K). \tag{2.1.68}$$

Comparing (2.1.67) to (2.1.68)and applying the triangular inequality we get estimate (2.1.66).

If L_ε is not linear then we must try to prove that for any $f_{1\varepsilon}, f_{2\varepsilon} \in H_d$ problems

$$L_\varepsilon u_{i\varepsilon} = f_{i\varepsilon}, \quad i = 1, 2,$$

have a unique solution and that there exists a constant C independent of ε and $f_{i\varepsilon}$, such that,

$$\|u_{1\varepsilon} - u_{2\varepsilon}\|_{H_s} \leq C\|f_{1\varepsilon} - f_{2\varepsilon}\|_{H_d}.$$

The spaces H_s and H_d may also depend on ε.

These ideas will be applied in particular in the next section, as well as further in the book.

2.2 Steady-state conductivity of a rod

2.2.1 Statement of the problem

Definition 2.2.1. *Let β be a bounded domain in $\mathbb{R}^2$ with piecewise smooth boundary, and let U_μ be the cylinder $\{x_1 \in \mathbb{R},\ x'/\mu \in \beta\}$, satisfying the strong cone condition [55]. The cylinder $C_\mu = U_\mu \cap \{x_1 \in (0, b)\}$ is called a three-dimensional* ($s = 3$) *rod.*

Here $\mu > 0$ is a small parameter. We recall that $x' = (x_2, x_3)$.

Definition 2.2.1.' *In the two-dimensional case* ($s = 2$) $U_\mu = \mathbb{R}\times(-\mu/2, \mu/2)$, *and the rectangle $C_\mu = U_\mu \cap \{x_1 \in (0, b)\} = (0, b) \times (-\mu/2, \mu/2)$ is called a two-dimensional rod.*

Next, for $s = 2$, we assume that $\beta = (-1/2, 1/2)$.

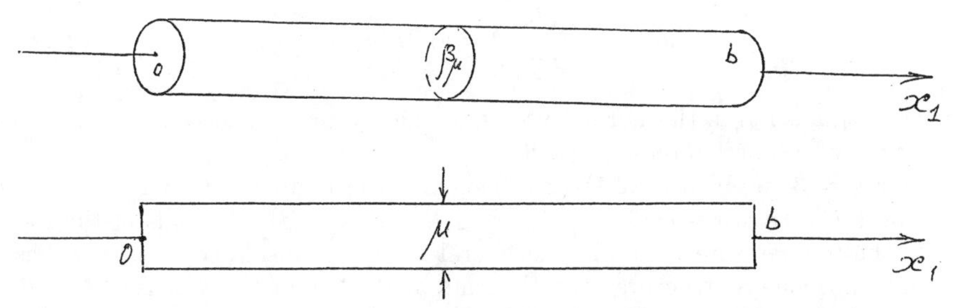

Figure 2.2.1. A rod (three-dimensional and two-dimensional)

We consider the stationary conductivity equation

$$Pu \equiv \sum_{i,j=1}^{s} \frac{\partial}{\partial x_i}\Big(A_{ij}\Big(\frac{x}{\mu}\Big)\frac{\partial u}{\partial x_j}\Big) = F_\mu(x), \quad x \in C_\mu \tag{2.2.1}$$

with the boundary conditions

$$\frac{\partial u}{\partial \nu} \equiv \sum_{i,j=1}^{s} A_{ij}\Big(\frac{x}{\mu}\Big)\frac{\partial u}{\partial x_j} n_i = 0, \quad x \in \partial U_\mu, \tag{2.2.2}$$

$$u = 0, \quad x_1 = 0 \quad \text{or} \quad x_1 = b. \tag{2.2.3}$$

Here $(n_1, \ldots, n_s)$ is a vector normal to the boundary U_μ and $A_{ij}(\xi)$ are functions satisfying the following conditions:

1)for any $\xi \in I\!R^s$,

$$A_{ij}(\xi) = A_{ji}(\xi),$$

and

2) there exists $\kappa > 0$ such that, for any $\eta_i \in I\!R$, $i = 1, ..., s$, the following inequality holds:

$$\sum_{i,j,k,l=1}^{s} A_{ij}(\xi)\eta_i\eta_j \geq \kappa \sum_{i=1}^{s} (\eta_i)^2, \quad \kappa > 0.$$

It is assumed that the functions $A_{ij}(\xi)$ are 1-periodic in ξ_1, infinitely differentiable everywhere outside a set Σ (consisting of smooth nonintersecting surfaces Σ_l) up to this set Σ; we assume that there is only a finite number of surfaces Σ_l that intersect with the cylinder $[0, b] \times \beta$; $\Sigma \cap ([0, 1] \times \partial\beta) = \emptyset$, and on Σ the coefficients have jump discontinuities. If $A_{ij}(\xi)$ depend on ξ_1, we require that b/μ be an integer. The right-hand side F_μ has the following structure:

$$F_\mu = F\Big(\frac{x}{\mu}\Big)\Psi(x_1);$$

here $F(\xi)$ is a 1-periodic matrix function of ξ_1, which has the same smoothness as the coefficients (see above) and that $\psi \in C^\infty([0, b])$.

On the interfaces of discontinuity of the coefficients, the natural conjugation conditions are assumed:

$$[u] = 0; \ \Big[\sum_{i,j=1}^{s} A_{ij} \partial u / \partial x_j n_i \Big] = 0, \tag{2.2.4}$$

where $(n_1, \ldots, n_s)$ is the normal vector to the interface of discontinuity, $[\cdot]$ is the jump of a function on this interface.

For $s = 3$, problem (2.2.1)–(2.2.3) models the temperature field in a rod of length b with cross-section $\beta_\mu = \{x' \in R^2, \ x'/\mu \in \beta\}$; the ends of the rod have the temperature zero and the lateral surface is insulated. The rod has an inhomogeneous structure: the elements of the conductivity tensor A_{ij} are μ-periodic functions of the longitudinal coordinate x_1, they also depend on the transverse coordinates x'. The right-hand side (sources) is the product of the fast-oscillating function $F(x/\mu)$ by the slowly varying factor $\psi(x_1)$.

For $s = 2$, the problem models a two-dimensional temperature field in a plate of length b and thickness μ.

We construct formal asymptotic solutions (f.a.s.) to problem (2.2.1)–(2.2.3) in two stages, as in section 2.1. In the first stage we construct f.a.s. to equation (2.2.1) with condition (2.2.2), using an analog of N. Bakhvalov's series; on the second stage we construct the boundary layer corrector. For the reader's convenience, we first establish a formalism for each stage and then justify the asymptotic behavior, i.e., prove different assertions stated in constructing f.a.s. Furthermore, we prove that f.a.s. is an asymptotic expansion of an exact solution to the problem and establish an error estimate for the partial sum of the series of f.a.s.

2.2.2 Inner expansion

Let us represent the right-hand side F as the sum

$$F = \bar{F} + \widetilde{F}, \tag{2.2.5}$$

where $\bar{F} = \langle F \rangle$, $\langle \widetilde{F} \rangle = 0$. From now on in this section $\langle F(\xi, x) \rangle$ denotes the average

$$\int_{(0,1)\times\beta} \frac{F(\xi, x) d\xi}{\operatorname{mes} \beta}.$$

We seek f.a.s. in the form of an analog of Bakhvalov's ansatz,

$$u^{(\infty)} = \sum_{l=0}^{\infty} \mu^l N_l \Big(\frac{x}{\mu}\Big) \frac{d^l \omega(x_1)}{dx_1^l} + \sum_{l=0}^{\infty} \mu^{l+2} M_l \Big(\frac{x}{\mu}\Big) \frac{d^l \psi(x_1)}{dx_1^l}, \tag{2.2.6}$$

where N_l, M_l are 1-periodic in ξ_1 functions, $\omega(x_1)$ is a C^∞ - function. Substituting series (2.2.6) into (2.2.1), (2.2.2), and (2.2.4) and collecting terms with like powers of μ yields

$$Pu^{(\infty)} - F\psi = \sum_{l=0}^{\infty} \mu^{l-2} H_l^N(\xi) \frac{d^l \omega}{dx_1^l} + \sum_{l=0}^{\infty} \mu^l H_l^M(\xi) \frac{d^l \psi}{dx_1^l} \quad - F\psi - \widetilde{F}\psi, \tag{2.2.7}$$

where

$$H_l^N(\xi) = L_{\xi\xi} N_l + T_l^N(\xi),$$

$$L_{\xi\xi} = \sum_{j,m=1}^{s} \frac{\partial}{\partial \xi_j}\Big(A_{jm}\frac{\partial}{\partial \xi_m}\Big),$$

$$T_l^N(\xi) = \sum_{j=1}^{s} \frac{\partial}{\partial \xi_j}(A_{j1}N_{l-1}) + \sum_{j=1}^{s} A_{1j}\frac{\partial N_{l-1}}{\partial \xi_j} + A_{11}N_{l-2}; \tag{2.2.8}$$

$$\frac{\partial u^{(\infty)}}{\partial \nu} = \sum_{l=0}^{\infty} \mu^{l-1} G_l^N(\xi)\frac{d^l\omega}{dx_1^l} + \sum_{l=0}^{\infty} \mu^{l+1} G_l^M(\xi)\frac{d^l\psi}{dx_1^l}, \tag{2.2.9}$$

$$G_l^N = \sum_{m=1}^{s}\Big(\sum_{j=1}^{s} A_{mj}\frac{\partial N_l}{\partial \xi_j} + A_{m1}N_{l-1}\Big)n_m; \tag{2.2.10}$$

H_l^M, T_l^M, G_l^M are obtained by replacing N with M. Here, as in section 2.1, $N_l = 0$ whenever $l < 0$; $N_0 = 1$.

We require that, as in the procedure from section 2.1,

$$H_l^N(\xi) = h_l^N, \quad l \geq 0, \quad H_l^M(\xi) = h_l^M, \quad l > 0, \quad H_0^M(\xi) = \widetilde{F}(\xi),$$

where h_l^N and h_l^M are constants. Moreover, we require that $G_l^N(\xi) = 0$, $G_l^M(\xi) = 0$ on the surface $(0,1)\times\partial\beta$ and that $[N] = 0$, $[G_l^N] = 0$; $[M] = 0$, $[G_l^M] = 0$ on the discontinuity interfaces Σ of the coefficients $A_{ij}(\xi)$. We obtain the following recurrent chain of problems for N_l, M_l:

$$L_{\xi\xi}N_l = -T_l^N(\xi) + h_l^N, \quad \xi \in (0,1)\times\beta, \tag{2.2.11}$$

$$\partial/\partial\nu_\xi N_l = -\sum_{m=2}^{s} A_{m1}N_{l-1}n_m, \quad \xi \in (0,1)\times\partial\beta, \tag{2.2.12}$$

$$[N_l]\big|_\Sigma = 0, \quad [\partial N_l/\partial\nu_\xi]\big|_\Sigma = -[\sum_{m=1}^{s} A_{m1}N_{l-1}n_m]\big|_\Sigma. \tag{2.2.13}$$

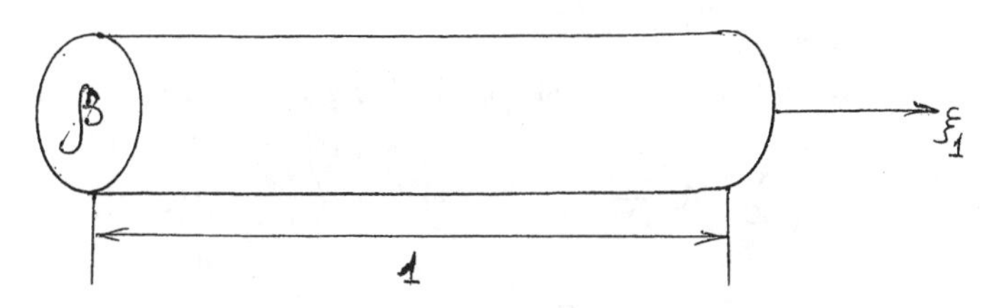

Figure 2.2.2. A periodic cell

Here, as in section 2.1, h_l^N are chosen from the solvability conditions for (2.2.11), (2.2.12), (2.2.13) (see Lemma 2.2.1):

$$\langle(-T_l^N + \Phi h_l^N)\rangle = -\langle\sum_{m=2}^{s} A_{m1}N_{l-1}n_m\rangle_{(0,1)\times\partial\beta} - \langle\sum_{m=1}^{s}[A_{m1}N_{l-1}n_m]\rangle_\Sigma,$$

where for the $(s-1)$-dimensional hyper-surface Γ we have

$$\langle\cdot\rangle_\Gamma = \int_\Gamma \frac{d\xi}{\operatorname{mes}\beta},$$

i.e.,

$$h_l^N = \Big\langle \sum_{j=1}^{s} A_{1j} \frac{\partial N_{l-1}}{\partial \xi_j} + A_{11} N_{l-2} \Big\rangle. \tag{2.2.14}$$

M_l are the solutions of the same problems with N_l replaced by M_l. However, M_0 is the solution to the problem

$$L_{\xi\xi} M_0 = \widetilde{F}, \qquad \frac{\partial M_0}{\partial \nu_\xi}\Big|_{\xi' \in \partial\beta} = 0, \qquad [M_0]\big|_\Sigma = 0, \qquad \Big[\frac{\partial M_0}{\partial \nu_\xi}\Big]\Big|_\Sigma = 0. \tag{2.2.15}$$

Thus, the algorithm for constructing the matrices N_l and M_l is inductive. Suppose that $N_l = 0$, $M_l = 0$ for $l < 0$, $N_0 = 1$, and M_0 is the solution to problem (2.2.15). If $l > 0$, then N_l and M_l are the solutions to problems (2.2.11)–(2.2.14). The right-hand sides of these problems contain N_m and M_m with indices $m < l$, which permits one to define them successively. We have

$$h_0^M = 0, \quad h_0^N = 0, \quad h_1^N = 0, \qquad h_2^N = \Big\langle \sum_{j=1}^{s} A_{1j} \frac{\partial N_1}{\partial \xi_j} + A_{11} \Big\rangle. \tag{2.2.16}$$

Here a 1-periodic solution N_l to problem (2.2.11)-(2.2.13) is understood in the following sense. Denote $Q = (0,1) \times \beta$. Let $H^1_{per\ \xi_1}(Q)$ be the completion in the norm of $H^1(Q)$ of the space of 1-periodic in ξ_1 infinitely differentiable functions $f(\xi)$ defined at $I\!R \times \beta$.

A solution$U \in H^1_{per\ \xi_1}(Q)$ to the problem

$$L_{\xi\xi} U = F_0(\xi) + \sum_{j=1}^{s} \frac{\partial F_j}{\partial \xi_j}, \quad \xi \in I\!R \times \beta,$$

$$\frac{\partial U}{\partial \nu_\xi} = \sum_{j=2}^{s} F_j n_j, \quad \xi \in I\!R \times \partial\beta,$$

is understood as a function U satisfying the integral identity

$$-\Big\langle \sum_{i,j=1}^{s} A_{ij} \frac{\partial U}{\partial \xi_j} \frac{\partial \Psi}{\partial \xi_i} \Big\rangle = \langle F_0 \Psi \rangle - \Big\langle \sum_{j=1}^{s} F_j \frac{\partial \Psi}{\partial \xi_j} \Big\rangle$$

for any function $\Psi \in H^1_{per\ \xi_1}(Q)$. Here the right-hand sides $F_0, F_1, \ldots, F_s$ belong to $L_2((-\mathcal{A}, \mathcal{A}) \times \beta)$ for each $\mathcal{A}$ and are 1-periodic in ξ_1.

Lemma 2.2.1. *This problem has a solution if and only if*

$$\langle F_0 \rangle = 0.$$

Proof.

The proof is based on the Riesz representation theorem for a bounded linear functional on a Hilbert space. Namely, the right-hand side of the integral identity represents the linear functional $G(\Psi)$ that is bounded in the norm of $H^1(Q)$ by the inequality

$$|G(\Psi)| \leq \left(\sum_{j=0}^{s} \|F_j\|_{L_2(Q)}\right) \|\Psi\|_{H^1(Q)}.$$

On the other hand, for functions $\Psi \in H^1(Q)$ that satisfy $\langle\Psi\rangle = 0$, the Poincaré inequality $\langle\sum_{i=1}^{s}\left(\frac{\partial\Psi}{\partial\xi_i}\right)^2\rangle \geq c_1\|\Psi\|^2_{H^1(Q)}$, $c_1 > 0$, holds (cf. [85])

Since

$$\left\langle \sum_{i,j=1}^{s} A_{ij}\frac{\partial\Psi}{\partial\xi_j}\frac{\partial\Psi}{\partial\xi_i}\right\rangle \geq \kappa\langle\sum_{i=1}^{s}\left(\frac{\partial\Psi}{\partial\xi_i}\right)^2\rangle,$$

the norm

$$\|\Psi\|_1 = \sqrt{\left\langle \sum_{i,j=1}^{s} A_{ij}\frac{\partial\Psi}{\partial\xi_j}\frac{\partial\Psi}{\partial\xi_i}\right\rangle}$$

is equivalent to the $H^1(Q)$-norm on the subspace of functions of $H^1_{per\ \xi_1}(Q)$ with vanishing average and this norm corresponds to the inner product

$$[\Psi,\Theta]_1 = \left\langle \sum_{i,j=1}^{s} A_{ij}\frac{\partial\Psi}{\partial\xi_j}\frac{\partial\Theta}{\partial\xi_i}\right\rangle.$$

Then we see that the functional $G(\Psi)$ is bounded in the norm $\|\Psi\|_1$ on this subspace as well, and therefore it can be represented in the form of an inner product $[U,\Psi]_1 = G(\Psi)$, where U is an element of the above subspace. Hence, on the subspace of vector functions $\Psi \in H^1_{per\ \xi_1}(Q)$ such that $\langle\Psi\rangle = 0$, the integral identity holds for some vector function U from this subspace. Let $\Psi(\xi) \in H^1_{per\ \xi_1}(Q)$. Represent $\Psi(\xi)$ in the form $\Psi(\xi) = \Psi_0(\xi) + h$, where h is a constant $h = \langle\Psi\rangle$ and

$$\langle\Psi_0\rangle = 0.$$

Then

$$[U,\Psi]_1 = [U,\Psi_0]_1 + [U,h]_1 = G(\Psi_0) + \left\langle \sum_{i,j=1}^{s} A_{ij}\frac{\partial h}{\partial\xi_j}\frac{\partial U}{\partial\xi_i}\right\rangle = G(\Psi_0) = G(\Psi).$$

So the integral identity holds for all functions $\Psi(\xi) \in H^1_{per\ \xi_1}(Q)$.

This proves the lemma.

Remark 2.2.1. A solution U having vanishing average is unique. This follows from the Riesz theorem for $G(\Psi)$.

Theorem 2.2.1. $h_2 > 0$.

Proof. Let us obtain a new representation for the matrix

$$h_2^N = \Big\langle \sum_{j=1}^{s} A_{1j} \frac{\partial N_1}{\partial \xi_j} + A_{11} \Big\rangle =$$

$$= \Big\langle \sum_{j=1}^{s} A_{1j} \Big(\frac{\partial N_1}{\partial \xi_j} + \frac{\partial \xi_1}{\partial \xi_j} \Big) \Big\rangle.$$

Since

$$\sum_{m=1}^{s} \frac{\partial \xi_1}{\partial \xi_m} A_{mj} = A_{1j},$$

we obtain

$$h_2^N = \sum_{m,j=1}^{s} \Big\langle \frac{\partial \xi_1}{\partial \xi_m} A_{mj} \frac{\partial}{\partial \xi_j} (N_1 + \xi_1) \Big\rangle. \tag{2.2.17}$$

On the other hand, it follows from the integral identity for problem (2.2.11)–(2.2.14) with the test function N_1 that

$$0 = \Big\langle \sum_{m,j=1}^{s} \frac{\partial}{\partial \xi_m} N_1 A_{mj} \frac{\partial}{\partial \xi_j} (N_1 + \xi_1) \Big\rangle. \tag{2.2.18}$$

Summing (2.2.17) and (2.2.18), we obtain the following representation for h_2^N:

$$h_2^N = \Big\langle \sum_{m,j=1}^{s} \frac{\partial}{\partial \xi_m} (N_1 + \xi_1) A_{mj} \frac{\partial}{\partial \xi_j} (N_1 + \xi_1) \Big\rangle \geq$$

$$\kappa \sum_{j=1}^{s} \Big\langle \Big(\frac{\partial}{\partial \xi_j} (N_1 + \xi_1) \Big)^2 \Big\rangle \geq \kappa \sum_{j=1}^{s} \Big\langle \frac{\partial}{\partial \xi_j} (N_1 + \xi_1) \Big\rangle^2 / \;= \kappa / \;> 0.$$

Theorem 2.2.1 is proved.

Equation (2.2.7) then takes the form

$$Pu^{(\infty)} - F\psi =$$

$$= h_2^N \frac{d^2 \omega}{dx_1^2} - \bar{F}\psi + \sum_{k=3}^{\infty} \mu^{k-2} h_k^N \frac{d^k \omega}{dx_1^k} + \sum_{l=1}^{\infty} \mu^l h_l^M \frac{d^l \psi}{dx_1^l} = 0. \tag{2.2.19}$$

Problem (2.2.19) can be regarded as a homogenized equation of infinite order for ω. The f.a.s. to this problem is sought as the series

$$\omega = \sum_{j=0}^{\infty} \mu^j \omega_j(x_1), \tag{2.2.20}$$

where $\omega_j(x_1)$ are independent of μ. Substitution of series (2.2.20) into (2.2.19) gives the following recurrent chain of equations for ω_j:

$$h_2^N \frac{d^2 \omega_j}{dx_1^2} = f_j(x_1), \tag{2.2.21}$$

where f_j depend on ω_{j_1}, $j_1 < j$, and their derivatives. Indeed,

$$h_2^N \frac{d^2\omega}{dx_1^2} - \bar{F}\psi + \sum_{k=3}^{\infty}\sum_{j=0}^{\infty} \mu^{k+j-2} h_k^N \frac{d^k\omega}{dx_1^k} + \sum_{l=1}^{\infty} \mu^l h_l^M \frac{d^l\psi}{dx_1^l} =$$

$$= \sum_{l=0}^{\infty} \mu^l \left(h_2^N \frac{d^2\omega_l}{dx_1^2} + \sum_{r=0}^{l-1} h_{l-r+2}^N \frac{d^{l-r+2}\omega_r}{dx_1^{l-r+2}} + h_l^M \frac{d^l\psi}{dx_1^l} \right) = 0,$$

where $h_0^M = -\bar{F}$.

So,

$$f_j(x_1) = -\sum_{r=0}^{j-1} h_{j-r+2}^N \frac{d^{j-r+2}\omega_r}{dx_1^{j-r+2}} - h_j^M \frac{d^j\psi}{dx_1^j}.$$

This completes the construction of the f.a.s. for problem (2.2.1), (2.2.2).

2.2.3 Boundary layer corrector.

We construct the f.a.s. to problem (2.2.1)–(2.2.3) in the form

$$u^{(\infty)} = u_B + u_P^0 + u_P^1, \tag{2.2.22}$$

where u_B is defined by equation (2.2.6) and

$$u_P^0 = \left(\sum_{l=0}^{\infty} \mu^l N_l^0(\xi) \frac{d^l\omega}{dx_1^l} + \sum_{l=0}^{\infty} \mu^{l+2} M_l^0(\xi) \frac{d^l\psi}{dx_1^l} \right)\Big|_{\xi=x/\mu},$$

$$u_P^1 = \left(\sum_{l=0}^{\infty} \mu^l N_l^1(\xi) \frac{d^l\omega}{dx_1^l} + \sum_{l=0}^{\infty} \mu^{l+2} M_l^1(\xi) \frac{d^l\psi}{dx_1^l} \right)\Big|_{\xi_1=(x_1-b)/\mu\xi'=x'/\mu}. \tag{2.2.23}$$

Substituting (2.2.22) into (2.2.1)–(2.2.3) and taking into account the fact that u_B is as constructed above, we obtain the asymptotic equations

$$\sum_{l=0}^{\infty} \mu^{l-2} \bar{H}_l^N\left(\frac{x}{\mu}\right) \frac{d^l\omega}{dx_1^l} + \sum_{l=0}^{\infty} \mu^l \bar{H}_l^M\left(\frac{x}{\mu}\right) \frac{d^l\psi}{dx_1^l} - \bar{F}\psi - \widetilde{F}\psi = 0, \quad x \in C_\mu, \tag{2.2.24}$$

$$\sum_{l=0}^{\infty} \mu^{l-1} \bar{G}_l^N\left(\frac{x}{\mu}\right) \frac{d^l\omega}{dx_1^l} + \sum_{l=0}^{\infty} \mu^{l+1} \bar{G}_l^M\left(\frac{x}{\mu}\right) \frac{d^l\psi}{dx_1^l} = 0, \quad x \in \partial C_\mu \cap \partial U_\mu, \tag{2.2.25}$$

$$\sum_{l=0}^{\infty} \mu^l [\bar{N}_l] \frac{d^l\omega}{dx_1^l} + \sum_{l=0}^{\infty} \mu^{l+2} [\bar{M}_l] \frac{d^l\psi}{dx_1^l} = 0, \quad \frac{x}{\mu} \in \Sigma,$$

$$\sum_{l=0}^{\infty} \mu^{l-1} \left[\bar{G}_l^N\left(\frac{x}{\mu}\right)\right] \frac{d^l\omega}{dx_1^l} + \sum_{l=0}^{\infty} \mu^{l+1} \left[\bar{G}_l^M\left(\frac{x}{\mu}\right)\right] \frac{d^l\psi}{dx_1^l} = 0, \quad \frac{x}{\mu} \in \Sigma, \tag{2.2.26}$$

$$\sum_{l=0}^{\infty} \mu^l \bar{N}_l(0,\xi') \frac{d^l\omega}{dx_1^l}\Big|_{x_1=0} + \sum_{l=0}^{\infty} \mu^{l+2} \bar{M}_l(0,\xi') \frac{d^l\psi}{dx_1^l}\Big|_{x_1=0} = 0,$$

$$\sum_{l=0}^{\infty}\mu^l\bar{N}_l\Big(\frac{b}{\mu},\xi'\Big)\frac{d^l\omega}{dx_1^l}\Big|_{x_1=b}+\sum_{l=0}^{\infty}\mu^{l+2}\bar{M}_l\Big(\frac{b}{\mu},\xi'\Big)\frac{d^l\psi}{dx_1^l}\Big|_{x_1=b}=0, \qquad (2.2.27)$$

where

$$\bar{H}_l^N(x/\mu)=H_l^N(x/\mu)+H_l^{N0}(x/\mu)+H_l^{N1}(x_1-b/\mu,x'/\mu),$$

$$\bar{G}_l^N(x/\mu)=G_l^N(x/\mu)+G_l^{N0}(x/\mu)+G_l^{N1}(x_1-b/\mu,x'/\mu),$$

$$\bar{N}_l(x/\mu)=N_l(x/\mu)+N_l^0(x/\mu)+N_l^1(x_1-b/\mu,x'/\mu),\ H_l^{Nr}(\xi)=L_{\xi\xi}N_l^r+T_l^{Nr}(\xi),$$

$$T_l^{Nr}(\xi)=\sum_{j=1}^{s}\partial/\partial\xi_j(A_{j1}N_{l-1}^r)+\sum_{j=1}^{s}A_{1j}\partial N_{l-1}^r/\partial\xi_j+A_{11}N_{l-2}^r;$$

$$G_l^{Nr}=\sum_{m=1}^{s}\left(A_{mj}\partial N_l^r/\partial\xi_j+A_{m1}N_{l-1}^r\right)n_m \qquad (2.2.28)$$

and $\bar{H}_l^M$, $\bar{G}_l^M$, $\bar{M}_l$, T_l^{Mr}, and G_l^{Mr} are obtained by replacing N with M. Here $N_l(\xi)$, $M_l(\xi)$, $H_l^{(N,M)}(\xi)$, and $G_l^{(N,M)}(\xi)$ are the 1-periodic functions in ξ_1 constructed above and $N_l^r(\xi)$, $H_l^{(N,M)r}(\xi)$, and $G_l^{(N,M)r}(\xi)$ stabilize exponentially to zero as $\xi_1\to\pm\infty$ ($+$ for $r=0$ and $-$ for $r=1$). The superscript (N,M) means that the assertion is true for both N and M. The exponential stabilization to zero implies that

$$\int_{(\sigma,\sigma+1)\times\beta}|N_l^r(\xi)|d\xi,\quad \int_{(\sigma,\sigma+1)\times\beta}|M_l^r(\xi)|d\xi\le c_1e^{-c_2|\sigma|}\quad \text{as}\quad \sigma\to\pm\infty$$

(as above, $+$ for $r=0$ and $-$ for $r=1$). Here c_1 and c_2 are positive constants independent of σ.

We arrange N_l^r and M_l^r in ascending order with respect to the index l and note that $T_l^{(N,M)r}$ is defined by the functions $N_{l_1}^r$ (or $M_{l_1}^r$) for $l_1<l$. We require that $H_l^{(N,M)r}$, $\bar{N}_l(0,\xi')$, $\bar{M}_l(0,\xi')$, $\bar{N}_l(b/\varepsilon,\xi')$, and $\bar{M}_l(b/\varepsilon,\xi')$ satisfy

$$H_l^{(N,M)r}=0,\quad G_l^{(N,M)r}=0,\qquad \bar{N}_l(0,\xi')=h_l^{N0}+O(e^{-c/\mu}),$$

$$\bar{M}_l(0,\xi')=h_l^{M0}+O(e^{-c/\mu}),\qquad \bar{N}_l\Big(\frac{b}{\mu},\xi'\Big)=h_l^{N1}+O(e^{-c/\mu}),$$

$$\bar{M}_l\Big(\frac{b}{\mu},\xi'\Big)=h_l^{M1}+O(e^{-c/\mu}),\quad c>0. \qquad (2.2.29)$$

The constants $h_l^{(N,M)r}$ are to be defined from the conditions of the existence of functions N_l^r and M_l^r exponentially stabilizing to zero.

We construct the functions N_l^r, M_l^r satisfying equations (2.2.29) by induction on l. Assume that N_l^r, $M_l^r=0$ for $l<0$. Let all $N_{l_1}^r$, $M_{l_1}^r$ be constructed for $l_1<l$. We choose a constant h_l^{N0} such that there exists a solution, exponentially stabilizing to zero, to the problem

$$L_{\xi\xi}N_l^0+T_l^{N0}=0,\quad \xi_1>0,\quad \xi'\in\beta,$$

$$\frac{\partial N_l^0}{\partial \nu_\xi} = -\sum_{m=2}^{s} A_{m1} N_{l-1} n_m, \quad \xi' \in \partial\beta,$$
$$N_l^0(0,\xi') = -N_l(0,\xi') + h_l^{N0}, \tag{2.2.30}$$

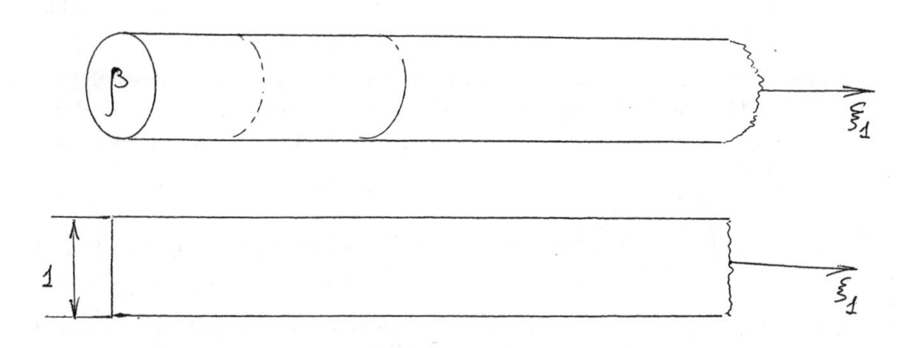

Figure 2.2.3. A boundary layer problem.

and choose a constant h_l^{N1} such that there exists a solution exponentially stabilizing to zero to the problem

$$L_{\xi\xi} N_l^1 + T_l^{N1} = 0, \quad \xi_1 < 0, \quad \xi' \in \beta,$$
$$\frac{\partial N_l^1}{\partial \nu_\xi} = -\sum_{m=2}^{s} A_{m1} N_{l-1}^1 n_m, \quad \xi' \in \partial\beta,$$
$$N_l^1(0,\xi') = -N_l(0,\xi') + h_l^{N1}. \tag{2.2.31}$$

The functions M_l^r and constants h_l^{Mr} are defined similarly (by replacing N with M in equations (2.2.30) and (2.2.31)).

The following theorem provides the possibility of such a choice of the constants $h_l^{(N,M)r}$.

Theorem 2.2.2. *Let $A_{ij}(\xi)$ satisfy the above conditions, let $F_j(\xi) \in L^2([0,+\infty)\times\beta)$, $j = 0,1,\ldots,s$, satisfy*

$$\int_{(\sigma,\sigma+1)\times\beta} |F_j(\xi)| d\xi \le c_1 e^{-c_2\sigma}, \quad \sigma > 0, \tag{2.2.32}$$

and let $u_0(\xi') \in H^{1/2}(\{0\}\times\beta)$ be given. Then there exists a solution $u(\xi)$ to the problem

$$L_{\xi\xi} u = F_0(\xi) + \sum_{j=1}^{s} \frac{\partial F_j}{\partial \xi_j}, \quad \xi_1 > 0, \quad \xi' \in \beta, \tag{2.2.33}$$

$$\frac{\partial u}{\partial \nu_\xi} = \sum_{j=2}^{s} F_j n_j, \quad \xi' \in \partial\beta, \tag{2.2.34}$$

$$u\big|_{\xi_1=0} = u_0(\xi') \tag{2.2.35}$$

such that for some constant h, the following inequality holds

$$\int_{(\sigma,\sigma+1)\times\beta} |u-h| d\xi \le c_1 e^{-c_2\sigma}, \quad c_1, c_2 > 0, \quad \sigma > 0. \tag{2.2.36}$$

Remark 2.2.2. Let us explain some terms from Theorem 2.2.2 (see [89],[90], [129],[130]). The space $H^{1/2}(\{0\}\times\beta)$ is thought of as the space of traces $V(\xi')$ on the set $\{\xi_1 = 0,\ \xi' \in \beta\}$ of functions from $H^1((0,1)\times\beta)$ equipped with the norm

$$\|V\|^2_{1/2} = \inf_v \Big\{ \int_{(0,1)\times\beta} \sum_{j=1}^{s} (\partial v/\partial x_j)^2 + |v|^2\, d\xi,\ v \in H^1((0,1)\times\beta),\ v\big|_{\xi_1=0} = V(\xi')\Big\}.$$

A solution to problem (2.2.33)–(2.2.35) is sought in the form of a function $u(\xi) \in H^{1\,\mathrm{loc}}$ whose trace is equal to $u_0(\xi')$ for $\xi_1 = 0$. For any function $\Psi(\xi) \in C^\infty$ with support in the cylinder $(\delta, D)\times\beta$, $\delta > 0$, the function $u(\xi)$ must satisfy the integral identity

$$-\sum_{i,j=1}^{s} \int_{(0,+\infty)\times\beta} A_{ij} \frac{\partial u}{\partial \xi_j}\frac{\partial \Psi}{\partial \xi_i} d\xi = \int_{(0,+\infty)\times\beta} F_0 \Psi d\xi - \sum_{j=1}^{s} \int_{(0,+\infty)\times\beta} F_j \frac{\partial \Psi}{\partial \xi_j} d\xi.$$

Theorem 2.2.2 follows directly from Theorems 4 and 5 in [130]. Their proofs are "corrected" for the right-hand side having specific representation (2.2.33) and (2.2.34) and the momenta in Theorem 5 are chosen so that they are equal to zero at infinity .

To determine the h_l^{N0} from (2.2.30), it suffices to first solve the problem

$$L_{\xi\xi}\widetilde{N}_l^0 + T_l^{N0} = 0, \quad \xi_1 > 0, \quad \xi' \in \beta,$$

$$\frac{\partial \widetilde{N}_l^0}{\partial \nu_\xi} = -\sum_{m=2}^{s} n_m A_{m1} N_{l-1}^0, \quad \xi' \in \partial\beta,$$

$$\widetilde{N}_l^0(0,\xi') = -N_l(0,\xi'). \tag{2.2.37}$$

According to Theorem 2.2.2, $\widetilde{N}_l^0$ stabilizes to a constant h. If one subtracts h from $\widetilde{N}_l^0(\xi)$, then the difference $N_l^0 = \widetilde{N}_l^0(\xi) - h$ will stabilize to zero and satisfy (2.2.30) for $h_l^{N0} = -h$. Problem (2.2.31) and the problems for M_l^r and h_l^{Mr} can be solved in the same manner.

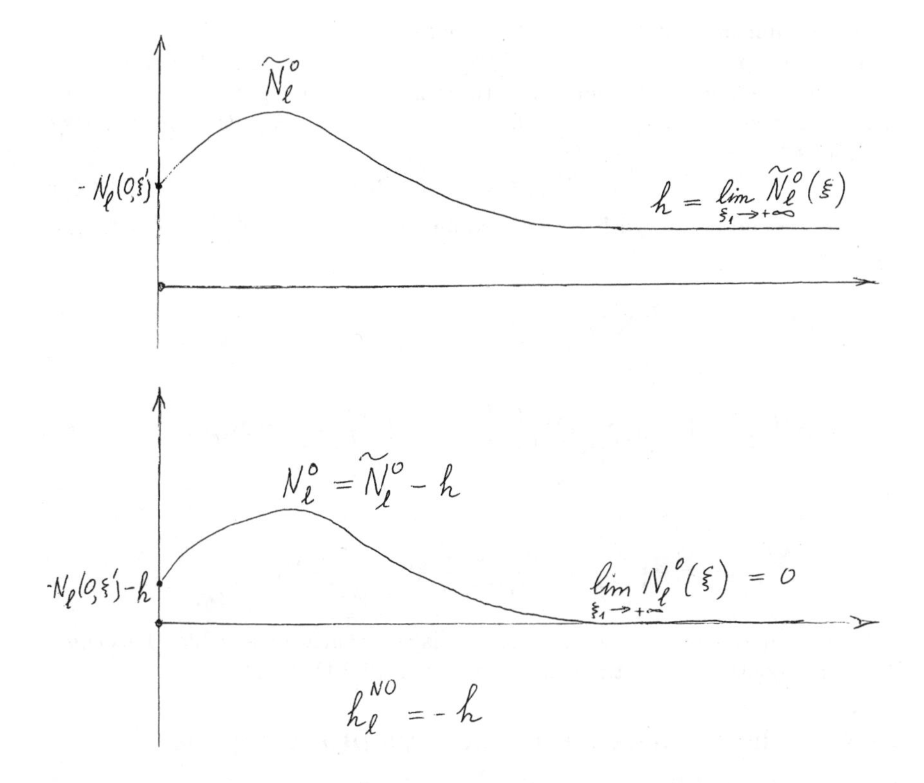

Figure 2.2.4. How to determine h_l^{N0}.

It follows from Theorem 2.2.2 that the energy e of the columns N_l^r and M_l^r also exponentially stabilizes to zero.

After the successive determination of all N_l^r and M_l^r by induction (note that $N_0^r = 0$) we obtain a homogenized problem of infinite order for the function $\omega(x_1)$. This problem is given by equation (2.2.17) with the boundary conditions

$$\left(\omega + \sum_{l=1}^{\infty} \mu^l h_l^{Nr} \frac{d^l \omega}{dx_1^l} + \sum_{l=0}^{\infty} \mu^{l+2} h_l^{Mr} \frac{d^l \psi}{dx_1^l}\right)\Big|_{x_1=rb} = 0, \quad r = 0, 1, \tag{2.2.38}$$

It is significant that h_0^{Nr} and h_1^{Nr} are equal to 1.

The f.a.s. to problem (2.2.19), (2.2.38) and (2.2.39) is represented in the form of the series (2.2.20) and, being substituted into (2.2.19), (2.2.38) and (2.2.39), leads to the recurrent chain of equations (2.2.21) for ω_j with the boundary conditions

$$\omega_j\Big|_{x_1=rb} = g_{jr}, \quad r=0,1, \tag{2.2.39}$$

Here g_{jr} depend on ω_{j_1} with $j_1 < j$, and on the derivatives of these functions.

Indeed,

$$\Big(\sum_{j=0}^{\infty}\mu^j\omega_j + \sum_{l=1}^{\infty}\sum_{j=0}^{\infty}\mu^{l+j}h_l^{Nr}\frac{d^l\omega_j}{dx_1^l} + \sum_{l=0}^{\infty}\mu^{l+2}h_l^{Mr}\frac{d^l\psi}{dx_1^l}\Big)|_{x_1=rb} =$$

$$= \Big(\sum_{j=0}^{\infty}\mu^j\Big(\omega_j + \sum_{p=0}^{j-1}h_{j-p}^{Nr}\frac{d^{j-p}\omega_p}{dx_1^{j-p}} + h_{j-2}^{Mr}\frac{d^{j-2}\psi}{dx_1^{j-2}}\Big)\Big)|_{x_1=rb} = 0,$$

so,

$$g_{jr} = -\Big(\sum_{p=0}^{j-1}h_{j-p}^{Nr}\frac{d^{j-p}\omega_p}{dx_1^{j-p}} - h_{j-2}^{Mr}\frac{d^{j-2}\psi}{dx_1^{j-2}}\Big)|_{x_1=rb}.$$

The chain of problems (2.2.21),(2.2.39) is consecutively solvable. This completes the construction of the f.a.s. for problem (2.2.1)-(2.2.3).

2.2.4 The justification of the asymptotic expansion

Here we prove the following main theorem.

Theorem 2.2.3. *Let $K \in \{0,1,2,...\}$. Denote χ a function from $C^{(K+2)}([0,b])$ such that $\chi(t)=1$ when $t\in[0,b/3]$, $\chi(t)=0$ when $t\in[2b/3,b]$. Consider a function*

$$u^{(K)}(x) = u_B^{(K)}(x) + u_P^{0(K)}(x)\chi(x_1) + u_P^{1(K)}(x)\chi(b-x_1) + \rho(x_1),$$

where

$$u_B^{(K)} = \sum_{l=0}^{K+1}\mu^l N_l\Big(\frac{x}{\mu}\Big)\frac{d^l\omega^{(K)}(x_1)}{dx_1^l} + \sum_{l=0}^{K-1}\mu^{l+2}M_l\Big(\frac{x}{\mu}\Big)\frac{d^l\psi(x_1)}{dx_1^l},$$

$$u_P^{0(K)} = \Big(\sum_{l=0}^{K+1}\mu^l N_l^0(\xi)\frac{d^l\omega^{(K)}}{dx_1^l} + \sum_{l=0}^{K-1}\mu^{l+2}M_l^0(\xi)\frac{d^l\psi}{dx_1^l}\Big)\Big|_{\xi=x/\mu},$$

$$u_P^{1(K)} = \Big(\sum_{l=0}^{K+1}\mu^l N_l^1(\xi)\frac{d^l\omega^{(K)}}{dx_1^l} + \sum_{l=0}^{K-1}\mu^{l+2}M_l^1(\xi)\frac{d^l\psi}{dx_1^l}\Big)|_{\xi_1=(x_1-b)/\mu \xi'=x'/\mu},$$

$$\omega^{(K)} = \sum_{j=0}^{K}\mu^j\omega_j(x_1);$$

$$\rho(x_1) = (1 - x_1/b)q_0 + (x_1/b)q_1,$$

$$q_r = \Big(\sum_{l=K+1}^{2K+1} \mu^l \sum_{j+p=l, 0\le j\le K, 0\le p\le K+1} h_p^{Nr} \frac{d^p\omega_j}{dx_1^p} + \mu^{K+1} h_{K-1}^{Mr} \frac{d^{K-1}\psi}{dx_1^{K-1}} \Big)|_{x_1=rb}, \; r = 0, 1,$$

$N_l, N_l^0, N_l^1, M_l, M_l^0, M_l^1, \omega_j$ are defined in subsection 2.3. Then the estimate holds

$$\|u^{(K)} - u\|_{H^1(C_\mu)} \;=\; O(\mu^K)\sqrt{mes\ C_\mu}.$$

Proof.
Let us estimate the discrepancy functional

$$I(\phi) \;=\; \int_{C_\mu} \sum_{m,j=1}^{s} A_{mj}\Big(\frac{x}{\mu}\Big) \frac{\partial(u - u^{(K)})}{\partial x_j} \frac{\partial \phi}{\partial x_m}\, dx,$$

where $\phi \in H^1(C_\mu),\ \phi = 0$ for $x_1 = 0$ or b.

1. At the first stage represent it in the form

$$I(\phi) \;=\; \int_{C_\mu} F\psi\phi\, dx \;-\; J_B(\phi) \;-\; J_0(\phi) \;-\; J_1(\phi) \;-\; J_2(\phi),$$

where

$$J_B(\phi) \;=\; \int_{C_\mu} \sum_{m,j=1}^{s} A_{mj}\Big(\frac{x}{\mu}\Big) \frac{\partial u_B^{(K)}}{\partial x_j} \frac{\partial \phi}{\partial x_m}\, dx,$$

$$J_0(\phi) \;=\; \int_{C_\mu} \sum_{m,j=1}^{s} A_{mj}\Big(\frac{x}{\mu}\Big) \frac{\partial(u_P^{0(K)}(x)\chi(x_1))}{\partial x_j} \frac{\partial \phi}{\partial x_m}\, dx,$$

$$J_1(\phi) \;=\; \int_{C_\mu} \sum_{m,j=1}^{s} A_{mj}\Big(\frac{x}{\mu}\Big) \frac{\partial(u_P^{1(K)}(x)\chi(b - x_1))}{\partial x_j} \frac{\partial \phi}{\partial x_m}\, dx,$$

$$J_2(\phi) \;=\; \int_{C_\mu} \sum_{m,j=1}^{s} A_{mj}\Big(\frac{x}{\mu}\Big) \frac{\partial \rho(x_1)}{\partial x_j} \frac{\partial \phi}{\partial x_m}\, dx \;=$$

$$=\; \frac{q_1 - q_0}{b} \int_{C_\mu} \sum_{m=1}^{s} A_{m1}\Big(\frac{x}{\mu}\Big) \frac{\partial \phi}{\partial x_m}\, dx;$$

so,

$$|J_2(\phi)| \;\le\; c_0\mu^{K+1}\|\phi\|_{H^1(C_\mu)}\sqrt{mes\ C_\mu}.$$

The boundary layer functions N_l^r, M_l^r decay exponentially as $|\xi_1| \to +\infty$, so (2.2.3) implies the following estimates

$$|\int_{C_\mu \cap \{x_1 \geq b/3\}} \sum_{m,j=1}^{s} A_{mj}\Big(\frac{x}{\mu}\Big) \frac{\partial (u_P^{0(K)}(x)\chi(x_1))}{\partial x_j} \frac{\partial \phi}{\partial x_m}\, dx| \leq c_1 e^{-c_2/\mu} \|\phi\|_{H^1(C_\mu)},$$

$$|\int_{C_\mu \cap \{x_1 \leq 2b/3\}} \sum_{m,j=1}^{s} A_{mj}\Big(\frac{x}{\mu}\Big) \frac{\partial (u_P^{1(K)}(x)\chi(b-x_1))}{\partial x_j} \frac{\partial \phi}{\partial x_m}\, dx| \leq c_1 e^{-c_2/\mu} \|\phi\|_{H^1(C_\mu)}$$

with some constants $c_1, c_2 > 0$ independent of μ.

So

$$I(\phi) = \int_{C_\mu} F\psi\phi\, dx - J_B(\phi) - \hat{J}_0(\phi) - \hat{J}_1(\phi) +$$

$$+ (\hat{J}_0(\phi) - J_0(\phi)) + (\hat{J}_1(\phi) - J_1(\phi)) - J_2(\phi),$$

where

$$\hat{J}_r(\phi) = \int_{C_\mu} \sum_{m,j=1}^{s} A_{mj}\Big(\frac{x}{\mu}\Big) \frac{\partial u_P^{r(K)}(x)}{\partial x_j} \frac{\partial \phi}{\partial x_m}\, dx, \ r = 1, 2,$$

and

$$|\hat{J}_r(\phi) - J_r(\phi)| \leq c_1 e^{-c_2/\mu} \|\phi\|_{H^1(C_\mu)}.$$

Consider now

$$J_B(\phi) =$$

$$= \sum_{l=0}^{K+1} \sum_{m=1}^{s} \int_{C_\mu} \mu^{l-1} \Big(\sum_{j=1}^{s} A_{mj}\Big(\frac{x}{\mu}\Big) \frac{\partial N_l(\xi)}{\partial \xi_j}|_{\xi = x/\mu} +$$

$$+ A_{m1}\Big(\frac{x}{\mu}\Big) N_{l-1}\Big(\frac{x}{\mu}\Big)\Big) \frac{d^l \omega^{(K)}(x_1)}{dx_1^l} \frac{\partial \phi}{\partial x_m}\, dx +$$

$$+ \sum_{l=0}^{K-1} \sum_{m=1}^{s} \int_{C_\mu} \mu^{l+1} \Big(\sum_{j=1}^{s} A_{mj}\Big(\frac{x}{\mu}\Big) \frac{\partial M_l(\xi)}{\partial \xi_j}|_{\xi = x/\mu} +$$

$$+ A_{m1}\Big(\frac{x}{\mu}\Big) M_{l-1}\Big(\frac{x}{\mu}\Big)\Big) \frac{d^l \psi(x_1)}{dx_1^l} \frac{\partial \phi}{\partial x_m}\, dx +$$

$$+ \sum_{m=1}^{s} \int_{C_\mu} \mu^{K+1} A_{m1}\Big(\frac{x}{\mu}\Big) N_{K+1} \frac{d^{K+2} \omega^{(K)}(x_1)}{dx_1^{K+2}} \frac{\partial \phi}{\partial x_m}\, dx +$$

$$+ \sum_{m=1}^{s} \int_{C_\mu} \mu^{K+1} A_{m1}\Big(\frac{x}{\mu}\Big) M_{K-1} \frac{d^{K+2} \psi(x_1)}{dx_1^{K+2}} \frac{\partial \phi}{\partial x_m}\, dx.$$

Denote

$$A_{ml}^{BN}(\xi) = \sum_{j=1}^{s} A_{mj}(\xi)\frac{\partial N_l(\xi)}{\partial \xi_j} + A_{m1}(\xi)N_{l-1}(\xi),$$

$$A_{ml}^{BM}(\xi) = \sum_{j=1}^{s} A_{mj}(\xi)\frac{\partial M_l(\xi)}{\partial \xi_j} + A_{m1}(\xi)M_{l-1}(\xi),$$

$$\Delta_1^N(\phi) = \sum_{m=1}^{s}\int_{C_\mu} \mu^{K+1}A_{m1}(x/\mu)N_{K+1}\frac{d^{K+2}\omega^{(K)}(x_1)}{dx_1^{K+2}}\frac{\partial\phi}{\partial x_m}\,dx,$$

$$\Delta_1^M(\phi) = \sum_{m=1}^{s}\int_{C_\mu} \mu^{K+1}A_{m1}(x/\mu)M_{K-1}\frac{d^{K+2}\psi(x_1)}{dx_1^{K+2}}\frac{\partial\phi}{\partial x_m}\,dx.$$

We have,

$$|\Delta_1^N(\phi)|\,,|\Delta_1^M(\phi)| \le c_3\mu^{K+1}\|\phi\|_{H^1(C_\mu)}\sqrt{mes\ C_\mu},$$

with a constant $c_3 > 0$ independent of μ.

We get

$$J_B(\phi) =$$

$$= \sum_{l=0}^{K+1}\sum_{m=1}^{s}\int_{C_\mu} \mu^{l-1}A_{ml}^{BN}(x/\mu)\frac{d^l\omega^{(K)}(x_1)}{dx_1^l}\frac{\partial\phi}{\partial x_m}\,dx +$$

$$+ \sum_{l=0}^{K-1}\sum_{m=1}^{s}\int_{C_\mu} \mu^{l+1}A_{ml}^{BM}(x/\mu)\frac{d^l\psi(x_1)}{dx_1^l}\frac{\partial\phi}{\partial x_m}\,dx +$$

$$+ \Delta_1^N(\phi) + \Delta_1^M(\phi).$$

2. At the second stage consider the sum

$$\tilde{J}_B^N(\phi) =$$

$$= \sum_{l=0}^{K+1}\sum_{m=1}^{s}\int_{C_\mu} \mu^{l-1}A_{ml}^{BN}(x/\mu)\frac{d^l\omega^{(K)}(x_1)}{dx_1^l}\frac{\partial\phi}{\partial x_m}\,dx =$$

$$= \sum_{l=0}^{K+1}\sum_{m=1}^{s}\int_{C_\mu} \mu^{l-1}A_{ml}^{BN}(x/\mu)\frac{\partial}{\partial x_m}\Big(\phi\frac{d^l\omega^{(K)}(x_1)}{dx_1^l}\Big)\,dx -$$

$$- \sum_{l=0}^{K+1}\int_{C_\mu} \mu^{l-1}A_{1l}^{BN}(x/\mu)\phi\frac{d^{l+1}\omega^{(K)}(x_1)}{dx_1^{l+1}}\,dx =$$

$$= \sum_{l=0}^{K+1} \sum_{m=1}^{s} \int_{\mu^{-1}C_\mu} \mu^{s+l-2} A_{ml}^{BN}(\xi) \frac{\partial}{\partial \xi_m} \Big(\phi \frac{d^l \omega^{(K)}(x_1)}{dx_1^l} \Big)|_{x=\mu\xi} \, d\xi \; -$$

$$- \sum_{l=0}^{K+1} \int_{\mu^{-1}C_\mu} \mu^{s+l-1} A_{1l}^{BN}(\xi) \Big(\phi \frac{d^{l+1} \omega^{(K)}(x_1)}{dx_1^{l+1}} \Big)|_{x=\mu\xi} \, d\xi,$$

where

$$\mu^{-1}C_\mu \;=\; \{\xi \in I\!R^s \mid \mu\xi \in C_\mu\} \;=\; (0, b/\mu) \times \beta.$$

Finally

$$\tilde{J}_B^N(\phi) \;=\; \sum_{l=0}^{K+1} \int_{\mu^{-1}C_\mu} \mu^{s+l-2} \sum_{m=1}^{s} A_{ml}^{BN}(\xi) \frac{\partial}{\partial \xi_m} \Big((\phi \frac{d^l \omega^{(K)}(x_1)}{dx_1^l} \Big)|_{x=\mu\xi} \; -$$

$$- \; A_{1,l-1}^{BN}(\xi) \Big(\phi \frac{d^l \omega^{(K)}(x_1)}{dx_1^l} \Big)|_{x=\mu\xi} \, d\xi \; -$$

$$- \; \mu^{s+K} \int_{\mu^{-1}C_\mu} A_{1,K+1}^{BN}(\xi) \Big(\phi \frac{d^{K+2} \omega^{(K)}(x_1)}{dx_1^{K+2}} \Big)|_{x=\mu\xi} \, d\xi,$$

where $A_{ml}^{BN} = 0$ when $l < 0$ (by convention).

Denote

$$\Delta_2^N(\phi) \;=\; \mu^{s+K} \int_{\mu^{-1}C_\mu} A_{1,K+1}^{BN}(\xi) \Big(\phi \frac{d^{K+2} \omega^{(K)}(x_1)}{dx_1^{K+2}} \Big)|_{x=\mu\xi} \, d\xi.$$

3. The variational formulation for problem (2.2.11)-(2.2.13) gives:

$$\forall \tilde{\phi}(\xi) \in H^1_{per \; \xi_1}(Q),$$

$$\Big\langle \sum_{m=1}^{s} A_{ml}^{BN}(\xi) \frac{\partial \tilde{\phi}}{\partial \xi_m} - A_{1,l-1}^{BN}(\xi) \tilde{\phi}(\xi) \Big\rangle = \langle h_l^N \tilde{\phi}(\xi) \rangle.$$

Then this identity holds, in particular, for any function $\tilde{\phi} \in H^1_{per \; \xi_1}(Q)$, vanishing when $\xi_1 = 0$ (and therefore when $\xi_1 = 1$). So, this identity holds for any function $\tilde{\phi} \in H^1(Q)$, vanishing when $\xi_1 = 0$ and when $\xi_1 = 1$, as well as for any function $\tilde{\phi} \in H^1((a, a+1) \times \beta)$, vanishing when $\xi_1 = a$ and when $\xi_1 = a+1$; in the last case the average $\langle \; \rangle$ is defined as $(mes \; \beta)^{-1} \int_{(a,a+1)\times\beta} \; d\xi$.

Let $\tilde{\phi}$ be a function from $H^1(\mu^{-1}C_\mu)$ vanishing when $\xi_1 = 0$ and when $\xi_1 = b/\mu$. We represent it in the form

$$\tilde{\phi}(\xi) \;=\; \tilde{\phi}(\xi) \; sin^2(\pi\xi_1) \; + \; \tilde{\phi}(\xi) \; cos^2(\pi\xi_1),$$

where $\tilde{\phi}(\xi)\ sin^2(\pi\xi_1)$ vanishes when $\xi_1 = i,\ i \in \{0, 1, ..., b/\mu\}$ and $\tilde{\phi}(\xi)\ cos^2(\pi\xi_1)$ vanishes when $\xi_1 = i - 1/2,\ i \in \{1, ..., b/\mu\}$ as well as when $\xi_1 = 0$ or $\xi_1 = b/\mu$.

Therefore, by simple addition the identity can be generalized as

$$\int_{\mu^{-1}C_\mu} \sum_{m=1}^{s} A_{ml}^{BN}(\xi)\frac{\partial\tilde{\phi}}{\partial\xi_m} - A_{1,l-1}^{BN}(\xi)\tilde{\phi}(\xi)\ d\xi = \int_{\mu^{-1}C_\mu} h_l^N\tilde{\phi}(\xi)\ d\xi$$

for any $\tilde{\phi}(\xi) \in H^1(\mu^{-1}C_\mu)$, vanishing when $\xi_1 = 0$ or $\xi_1 = b/\mu$; for instance, it remains valid for $\tilde{\phi}(\xi) = \left(\phi\frac{d^l\omega^{(K)}(x_1)}{dx_1^l}\right)|_{x=\mu\xi}$.

Thus,

$$\tilde{J}_B^N(\phi) \;=\; \sum_{l=0}^{K+1}\int_{\mu^{-1}C_\mu} \mu^{s+l-2}h_l^N\left(\phi\frac{d^l\omega^{(K)}(x_1)}{dx_1^l}\right)|_{x=\mu\xi}\ d\xi \;+\; \Delta_2^N(\phi) \;=$$

$$=\; \sum_{l=0}^{K+1}\int_{C_\mu} \mu^{l-2}h_l^N\phi(x)\frac{d^l\omega^{(K)}}{dx_1^l}\ dx \;+\; \Delta_2^N(\phi),$$

where

$$|\Delta_2^N(\phi)| \;\leq\; c_4^{(N)}\mu^K\|\phi\|_{H^1(C_\mu)}\sqrt{mes\ C_\mu},$$

with a constant $c_4^{(N)} > 0$ independent of μ.

4. The same reasoning for

$$\tilde{J}_B^M(\phi) = \sum_{l=0}^{K-1}\sum_{m=1}^{s}\int_{C_\mu} \mu^{l+1}A_{ml}^{BM}(x/\mu)\frac{d^l\psi(x_1)}{dx_1^l}\frac{\partial\phi}{\partial x_m}\ dx$$

gives

$$\tilde{J}_B^M(\phi) = \int_{C_\mu} \tilde{F}\phi + \sum_{l=1}^{K-1}\mu^l h_l^M\phi(x)\frac{d^l\psi}{dx_1^l}\ dx \;+\; \Delta_2^M(\phi),$$

where

$$|\Delta_2^M(\phi)| \;\leq\; c_4^{(M)}\mu^K\|\phi\|_{H^1(C_\mu)}\sqrt{mes\ C_\mu},$$

with a constant $c_4^{(M)} > 0$ independent of μ.

5. Replacing N_l by N_l^0 in the expressions A_{ml}^{BN} and replacing M_l by M_l^0 in the expressions A_{ml}^{BM} we define A_{ml}^{0N} and A_{ml}^{0M} respectively. Applying the same reasoning as at stages 2 and 3 we obtain the relation

$$\hat{J}_0(\phi) \;=\; \sum_{l=0}^{K+1}\int_{\mu^{-1}C_\mu} \mu^{s+l-2}\sum_{m=1}^{s}A_{ml}^{0N}(\xi)\frac{\partial}{\partial\xi_m}\left((\phi\frac{d^l\omega^{(K)}(x_1)}{dx_1^l}\right)|_{x=\mu\xi} \;-$$

$$-\; A_{1,l-1}^{0N}(\xi)\left(\phi\frac{d^l\omega^{(K)}(x_1)}{dx_1^l}\right)|_{x=\mu\xi}\ d\xi \;-$$

$$- \mu^{s+K} \int_{\mu^{-1}C_\mu} A^{0N}_{1,K+1}(\xi)\Big(\phi \frac{d^{K+2}\omega^{(K)}(x_1)}{dx_1^{K+2}}\Big)|_{x=\mu\xi}\ d\xi\ +$$

$$+ \sum_{l=0}^{K-1} \int_{\mu^{-1}C_\mu} \mu^{s+l} \sum_{m=1}^{s} A^{0M}_{ml}(\xi) \frac{\partial}{\partial \xi_m}\Big((\phi \frac{d^l\omega^{(K)}(x_1)}{dx_1^l}\Big)|_{x=\mu\xi}\ -$$

$$- A^{0M}_{1,l-1}(\xi)\Big(\phi \frac{d^l\omega^{(K)}(x_1)}{dx_1^l}\Big)|_{x=\mu\xi}\ d\xi\ -$$

$$- \mu^{s+K} \int_{\mu^{-1}C_\mu} A^{0M}_{1,K+1}(\xi)\Big(\phi \frac{d^K\psi(x_1)}{dx_1^K}\Big)|_{x=\mu\xi}\ d\xi.$$

Variational formulation for problem (2.2.30) gives

$$\int_{\mu^{-1}C_\mu} \sum_{m=1}^{s} A^{0N}_{ml}(\xi)\frac{\partial\tilde{\phi}}{\partial\xi_m} - A^{0N}_{1,l-1}(\xi)\tilde{\phi}(\xi)\ d\xi = 0$$

for any function $\tilde{\phi} \in H^1(\mu^{-1}C_\mu)$ vanishing when $\xi_1 = 0$ or $\xi_1 = b/\mu$.

Analogously,

$$\int_{\mu^{-1}C_\mu} \sum_{m=1}^{s} A^{0M}_{ml}(\xi)\frac{\partial\tilde{\phi}}{\partial\xi_m} - A^{0M}_{1,l-1}(\xi)\tilde{\phi}(\xi)\ d\xi = 0$$

for any function $\tilde{\phi} \in H^1(\mu^{-1}C_\mu)$ vanishing when $\xi_1 = 0$ or $\xi_1 = b/\mu$.

So,

$$|\hat{J}_0(\phi)| \ \le\ c_5\mu^K \|\phi\|_{H^1(C_\mu)}\sqrt{mes\ C_\mu},$$

with a constant $c_5 > 0$ independent of μ.

Similarly,

$$|\hat{J}_1(\phi)| \ \le\ c_6\mu^K \|\phi\|_{H^1(C_\mu)}\sqrt{mes\ C_\mu},$$

with a constant $c_6 > 0$ independent of μ.

Thus,

$$I(\phi) \ =\ \int_{C_\mu} \Big(F\psi - \sum_{l=0}^{K+1} \mu^{l-2} h^N_l \frac{d^l\omega^{(K)}}{dx_1^l}\ -$$

$$-\tilde{F}\psi - \sum_{l=1}^{K-1} \mu^l h^M_l \frac{d^l\psi}{dx_1^l}\Big)\phi(x)\ dx\ +\ \Delta_3(\phi),$$

where $\Delta_3(\phi)$ is a linear functional of ϕ such that

$$|\Delta_3(\phi)| \ \le\ c_7\mu^K \|\phi\|_{H^1(C_\mu)}\sqrt{mes\ C_\mu},$$

with a constant $c_7 > 0$ independent of μ.

Note that $F\psi - \tilde{F}\psi = \langle F\rangle\psi$.

Consider the expression

$$B(x_1) = \langle F\rangle\psi - \sum_{l=0}^{K+1}\mu^{l-2}h_l^N\frac{d^l\omega^{(K)}}{dx_1^l} -$$

$$-\tilde{F}\psi - \sum_{l=1}^{K-1}\mu^l h_l^M\frac{d^l\psi}{dx_1^l} =$$

$$= -\Big(h_2^N\frac{d^2\omega^{(K)}}{dx_1^2} - \bar{F}\psi + \sum_{l=3}^{K+1}\mu^{l-2}h_l^N\frac{d^l\omega^{(K)}}{dx_1^l} - \sum_{l=1}^{K-1}\mu^l h_l^M\frac{d^l\psi}{dx_1^l}\Big).$$

Substitute $\omega^{(K)}(x_1) = \sum_{j=0}^{K}\mu^j\omega_j(x_1)$ and remind (2.2.21). Then there exist a constant c_8 independent of μ such that

$$|B(x_1)| \le c_8\mu^K.$$

Thus

$$|I(\phi)| \le |\int_{C_\mu}B(x_1)\phi\,dx| + |\Delta_3(\phi)| \le$$

$$\le c_8\mu^K\|\phi\|_{H^1(C_\mu)}\sqrt{mes\ C_\mu} + |\Delta_3(\phi)| \le$$

$$\le (c_8+c_7)\mu^K\|\phi\|_{H^1(C_\mu)}\sqrt{mes\ C_\mu} \le$$

$$\le c_9\mu^K\|\phi\|_{H^1(C_\mu)}\sqrt{mes\ C_\mu},$$

with a constant $c_9 > 0$ independent of μ.

On the other hand, $u^{(K)} \in H^1(C_\mu)$ and it vanishes when $x_1 = 0$ or $x_1 = b$. Indeed,

$$u^{(K)}|_{x_1=0} = (u_B^{(K)} + u_P^{0(K)})|_{x_1=0} + q_0 =$$

$$= \sum_{l=0}^{K+1}\mu^l\Big(N_l\Big(\frac{x}{\mu}\Big) + N_l^0\Big(\frac{x}{\mu}\Big)\Big)|_{x_1=0}\frac{d^l\omega^{(K)}(x_1)}{dx_1^l}|_{x_1=0} +$$

$$+\sum_{l=0}^{K-1}\mu^{l+2}\Big(M_l\Big(\frac{x}{\mu}\Big) + M_l^0\Big(\frac{x}{\mu}\Big)\Big)|_{x_1=0}\frac{d^l\psi(x_1)}{dx_1^l}|_{x_1=0} + q_0 =$$

$$= \sum_{l=0}^{K+1}\mu^l h_l^{N0}\frac{d^l\omega^{(K)}}{dx_1^l}|_{x_1=0} + \sum_{l=0}^{K-1}\mu^{l+2}h_l^{M0}\frac{d^l\psi}{dx_1^l}|_{x_1=0} + q_0 = 0$$

due to relations (2.2.39). Similarly, $u^{(K)}|_{x_1=b} = 0$.

Taking $\phi = u - u^{(K)}$ we obtain the inequality

$$\kappa\|u - u^{(K)}\|^2_{H^1(C_\mu)} \le I(u - u^{(K)}) =$$

$$= \int_{C_\mu} \sum_{m,j=1}^{s} A_{mj}\left(\frac{x}{\mu}\right) \frac{\partial(u-u^{(K)})}{\partial x_j} \frac{\partial(u-u^{(K)})}{\partial x_m}\, dx \le$$

$$\le c_{10}\mu^K \|u-u^{(K)}\|_{H^1(C_\mu)} \sqrt{mes\ C_\mu}, \tag{2.2.40}$$

with a constant $c_{10} > 0$ independent of μ.

So,

$$\|u-u^{(K)}\|_{H^1(C_\mu)} \le c_{10}/\kappa\mu^K \sqrt{mes\ C_\mu};$$

this completes the proof of the theorem.

Corollary 2.2.1. *The estimate holds*

$$\|u-u_B^{(K-1)} - u_P^{0(K-1)} - u_P^{1(K-1)}\|_{H^1(C_\mu)} = O(\mu^K)\sqrt{mes\ C_\mu},\ K=1,2,....$$

Indeed, the H^1−norm of the differences $u^{0(K)} - u^{0(K)}\chi(x_1)$ and $u^{1(K)} - u^{1(K)}\chi(b-x_1)$ and the H^1−norm of ρ are of order $O(\mu^K)\sqrt{mes\ C_\mu}$. Moreover, $\|u_B^{(K)} - u_B^{(K-1)}\|_{H^1(C_\mu)}$, $\|u_P^{0(K)} - u_P^{0(K-1)}\|_{H^1(C_\mu)}$, $\|u_P^{1(K)} - u_P^{1(K-1)}\|_{H^1(C_\mu)} = O(\mu^K)\sqrt{mes\ C_\mu}$.

So when one replaces in the estimate of the theorem $u^{(K)}$ by $u_B^{(K-1)} + u_P^{0(K-1)} + u_P^{1(K-1)}$, the order of this estimate does not change. It proves the corollary.

Remark 2.2.3. The function $u_B^{(K-1)} + u_P^{0(K-1)} + u_P^{1(K-1)}$ does not vanish at the ends of the rod; however, it has the order $O(\mu^K)$ there.

Remark 2.2.4. In the above proof we have estimated some linear functionals: $\Delta_j^N(\phi), \Delta_j^M(\phi),\ j=1,2,\ \Delta_3(\phi), J_2(\phi), \hat{J}_0(\phi), \hat{J}_1(\phi)$, as well as the differences $\hat{J}_r(\phi) - J_r(\phi),\ r=1,2$, by the upper born $c\mu^K\|\phi\|_{H^1(C_\mu)}\sqrt{mes\ C_\mu}$ (for sufficiently small μ). Finally, we have obtained the same upper born for the functional $I(\phi)$. One can easily check that all these functionals can be presented in a form of the H^1−inner product

$$\int_{C_\mu} \Psi_{0,\mu}(x)\phi\, dx + \sum_{m=1}^{s} \int_{C_\mu} \Psi_{m,\mu}(x) \frac{\partial\phi}{\partial x_m}\, dx, \tag{2.2.41}$$

where functions $\Psi_{0,\mu}, \Psi_{m,\mu}, m=1,...,s$, are estimated as

$$\|\Psi_{0,\mu}\|_{L^2(C_\mu)}, \|\Psi_{m,\mu}\|_{L^2(C_\mu)} \le c\mu^K\sqrt{mes\ C_\mu}. \tag{2.2.42}$$

with a constant $c > 0$ independent of μ.

2.3 Steady state elasticity equation in a rod

Constructions of section 2.2.3 repeat some ideas of section 2.2.2 but the elasticity system of equations is essentially more complex than the conductivity equation.

2.3.1 Formulation of the problem

We consider the elasticity system of equations (see [55])

$$Pu \equiv \sum_{i,j=1}^{s} \frac{\partial}{\partial x_i}\Big(A_{ij}\Big(\frac{x}{\mu}\Big)\frac{\partial u}{\partial x_j}\Big) = F_\mu(x), \quad x \in C_\mu \tag{2.3.1}$$

with the boundary conditions

$$\frac{\partial u}{\partial \nu} \equiv \sum_{i,j=1}^{s} A_{ij}\Big(\frac{x}{\mu}\Big)\frac{\partial u}{\partial x_j} n_i = 0, \quad x \in \partial U_\mu, \tag{2.3.2}$$

$$u = 0, \quad x_1 = 0 \quad \text{or} \quad x_1 = b. \tag{2.3.3}$$

Here $(n_1, \dots, n_s)$ is an exterior vector normal to the boundary U_μ and $A_{ij}(\xi)$ are $(s \times s)$-matrices whose elements $a_{ij}^{kl}(\xi)$ satisfy the conditions

$$a_{ij}^{kl}(\xi) = a_{il}^{kj}(\xi) = a_{ji}^{lk}(\xi),$$

and there exists a constant $\kappa > 0$ such that for any symmetric matrix η_i^k, for any $\xi \in I\!R^s$ the following inequality holds:

$$\sum_{i,j,k,l=1}^{s} a_{ij}^{kl}(\xi)\eta_i^k\eta_j^l \geq \kappa \sum_{i,k=1}^{s} (\eta_i^k)^2, \quad \kappa > 0.$$

It is assumed that the functions $A_{ij}(\xi)$ are 1-periodic in ξ_1, infinitely differentiable everywhere outside of a set Σ (consisting of smooth nonintersecting surfaces Σ_l) up to Σ; we assume that there is only a finite number of surfaces Σ_l that intersect with the cylinder $[0, b] \times \beta$; $\Sigma \cap ([0,1] \times \partial\beta) = \emptyset$, $a_{ij}^{kl} \in C^\infty$ and on Σ the coefficients have jump discontinuities. If a_{ij}^{kl} depend on ξ_1, we require that b/μ be an integer. The right-hand side F_μ and the unknown u are s-dimensional vector-functions, and F_μ has the following structure:

$$F_\mu = \Phi\Big(\frac{x'}{\mu}\Big)F\Big(\frac{x}{\mu}\Big)\Psi(x_1);$$

here $\Phi(\xi')$ is the matrix of rigid displacements; for $s = 2$, $\Phi(\xi') = I$ is the 2×2 identity matrix, and for $s = 3$ we have

$$\Phi(\xi') = \begin{pmatrix} 1 & 0 & 0 & 0 \\ 0 & 1 & 0 & -a\xi_3 \\ 0 & 0 & 1 & a\xi_2 \end{pmatrix}, \quad a = \Big(\int_\beta \frac{(\xi_2^2 + \xi_3^2)\, d\xi_2 d\xi_3}{\text{mes}\,\beta}\Big)^{-1/2};$$

$F(\xi)$ is a $d \times d$ square matrix with $d = 2$ if $s = 2$ and $d = 4$ if $s = 3$; $\psi(x_1)$ is a d-dimensional vector function.

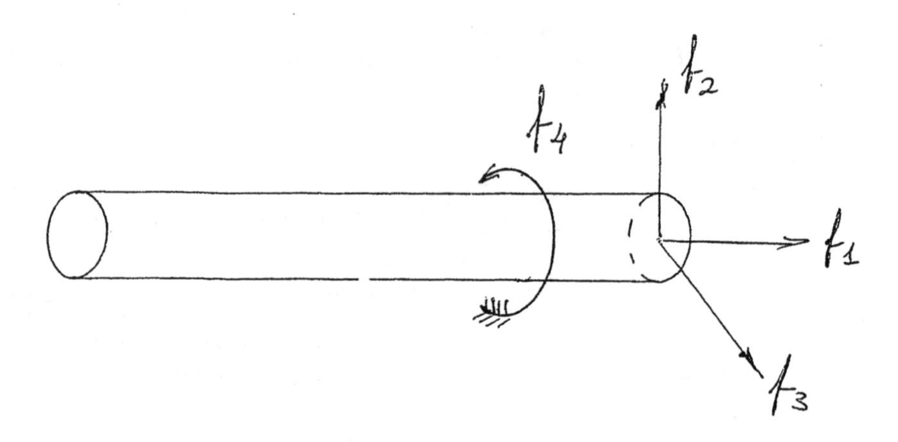

Figure 2.3.1. Right hand side force structure

It is assumed that $F(\xi)$ is a 1-periodic matrix function of ξ_1, which has the same smoothness as the coefficients (see above) and that $\psi \in C^\infty([0,b])$.

On the interfaces of discontinuity of the coefficients, the natural conjugation conditions are assumed:

$$[u] = 0; \quad \Big[\sum_{i,j=1}^{s} A_{ij} \partial u / \partial x_j n_i \Big] = 0, \tag{2.3.4}$$

where $(n_1, \ldots, n_s)$ is the normal vector to the interface of discontinuity, $[\cdot]$ is the jump of a function on this interface.

For $s = 3$, problem (2.3.1)–(2.3.3) models the stress-strain state of a rod of length b with cross-section $\beta_\mu = \{x' \in R^2, \ x'/\mu \in \beta\}$; the ends of the rod are clamped and the lateral surface is free. The rod has an inhomogeneous structure: the elements of the elasticity tensor a_{ij}^{kl} are μ-periodic functions of the longitudinal coordinate x_1, they also depend on the transverse coordinates x'. The right-hand side (mass forces) has a tensile-pressing part (the first component of the vector $F(x/\mu)\psi(x_1)$), bending parts (the second and third components of $F(x/\mu)\psi(x_1)$) and a torsional part (the fourth component of $F(x/\mu)\psi(x_1)$), each part is the product of the fast-oscillating function $F(x/\mu)$ by the slowly varying factor $\psi(x_1)$.

For $s = 2$, the problem models a 2-D stress-strain state of the plate of length b and thickness μ. The right-hand side here has both tensile-pressing and bending components.

We construct formal asymptotic solutions (f.a.s.) to problem (2.3.1)–(2.3.3) in two stages, as in section 1. At the first stage we construct f.a.s. to equation (2.3.1) with condition (2.3.2), using an analog of Bakhvalov's ansatz; at the second stage we construct the boundary layer corrector. For the reader's convenience, we first establish a formalism for each stage and then justify the asymptotic expansion, i.e., prove different assertions stated in constructing f.a.s. Furthermore, we prove that f.a.s. is an asymptotic expansion of an exact solution to the problem and establish an error estimate for the partial sum of the series of f.a.s. This justification is developed for the rods with the symmetry of a periodic cell with respect to two co-ordinate planes (Condition A in section 2.3.2).

2.3.2 Inner expansion

Let us represent the right-hand side F as the sum

$$F = \bar{F} + \widetilde{F}, \tag{2.3.5}$$

where $\bar{F} = \langle \Phi^* \Phi F \rangle$, $\langle \Phi^* \Phi \widetilde{F} \rangle = 0$. From now on in this section $\langle F(\xi, x) \rangle$ denotes the average

$$\int_{(0,1)\times\beta} \frac{F(\xi, x) d\xi}{\mathrm{mes}\, \beta}.$$

We seek f.a.s. in the form of an analog of Bakhvalov's ansatz,

$$u^{(\infty)} = \sum_{l=0}^{\infty} \mu^l N_l\Big(\frac{x}{\mu}\Big) \frac{d^l \omega(x_1)}{dx_1^l} + \sum_{l=0}^{\infty} \mu^{l+2} M_l\Big(\frac{x}{\mu}\Big) \frac{d^l \psi(x_1)}{dx_1^l}, \tag{2.3.6}$$

where N_l, M_l are 1-periodic in ξ_1 matrix functions, $\omega(x_1)$ is a d - vector function. Substituting series (2.3.6) into (2.3.1), (2.3.2), and (2.3.4) and collecting terms with like powers of μ yields

$$Pu^{(\infty)} - \Phi F \psi = \sum_{l=0}^{\infty} \mu^{l-2} H_l^N(\xi) \frac{d^l \omega}{dx_1^l} + \sum_{l=0}^{\infty} \mu^l H_l^M(\xi) \frac{d^l \psi}{dx_1^l} \quad - \Phi F \psi - \Phi \widetilde{F} \psi, \tag{2.3.7}$$

where

$$H_l^N(\xi) = L_{\xi\xi} N_l + T_l^N(\xi),$$

$$L_{\xi\xi} = \sum_{j,m=1}^{s} \frac{\partial}{\partial \xi_j}\Big(A_{jm} \frac{\partial}{\partial \xi_m}\Big),$$

$$T_l^N(\xi) = \sum_{j=1}^{s} \frac{\partial}{\partial \xi_j}(A_{j1} N_{l-1}) + \sum_{j=1}^{s} A_{1j} \frac{\partial N_{l-1}}{\partial \xi_j} + A_{11} N_{l-2}; \tag{2.3.8}$$

$$\frac{\partial u^{(\infty)}}{\partial \nu} = \sum_{l=0}^{\infty} \mu^{l-1} G_l^N(\xi) \frac{d^l \omega}{dx_1^l} + \sum_{l=0}^{\infty} \mu^{l+1} G_l^M(\xi) \frac{d^l \psi}{dx_1^l}, \tag{2.3.9}$$

$$G_l^N = \sum_{m=1}^{s} \Big(\sum_{j=1}^{s} A_{mj} \frac{\partial N_l}{\partial \xi_j} + A_{m1} N_{l-1}\Big) n_m; \tag{2.3.10}$$

H_l^M, T_l^M, G_l^M are obtained by replacing N with M. Here, as in section 2.2, $N_l = 0$ whenever $l < 0$.

We require that, as in the procedure from section 2.2,

$$H_l^N(\xi) = \Phi(\xi')h_l^N, \quad l \geq 0, \quad H_l^M(\xi) = \Phi(\xi')h_l^M, \quad l > 0, \quad H_0^M(\xi) = \Phi(\xi')\widetilde{F}(\xi),$$

where h_l^N and h_l^M are constant $d \times d$ matrices. Moreover, we require that $G_l^N(\xi) = 0$, $G_l^M(\xi) = 0$ on the surface $(0,1) \times \partial\beta$ and that $[N] = 0$, $[G_l^N] = 0$; $[M] = 0$, $[G_l^M] = 0$ on the discontinuity interfaces Σ of the coefficients $a_{ij}^{kl}(\xi)$. We obtain the following recurrent chain of problems for N_l, M_l:

$$L_{\xi\xi}N_l = -T_l^N(\xi) + \Phi h_l^N, \quad \xi \in (0,1) \times \beta, \tag{2.3.11}$$

$$\partial/\partial\nu_\xi N_l = -\sum_{m=2}^{s} A_{m1}N_{l-1}n_m, \quad \xi \in (0,1) \times \partial\beta, \tag{2.3.12}$$

$$[N_l]\big|_\Sigma = 0, \quad [\partial N_l/\partial\nu_\xi]\big|_\Sigma = -\big[\sum_{m=1}^{s} A_{m1}N_{l-1}n_m\big]\big|_\Sigma. \tag{2.3.13}$$

Here, as in section 2.2, h_l^N are chosen from the solvability conditions for (2.3.11), (2.3.12), (2.3.13) (see further Lemma 2.3.1):

$$\langle\Phi^*(-T_l^N + \Phi h_l^N)\rangle = -\langle\sum_{m=2}^{s}\Phi^* A_{m1}N_{l-1}n_m\rangle_{(0,1)\times\partial\beta} - \langle\sum_{m=1}^{s}\Phi^*[A_{m1}N_{l-1}n_m]\rangle_\Sigma,$$

where for the $(s-1)$-dimensional hyper-surface Γ we have

$$\langle\cdot\rangle_\Gamma = \int_\Gamma \frac{d\xi}{\operatorname{mes}\beta},$$

i.e.,

$$h_l^N = \Big\langle\Phi^*\Big(\sum_{j=1}^{s} A_{1j}\frac{\partial N_{l-1}}{\partial\xi_j} + A_{11}N_{l-2}\Big)\Big\rangle. \tag{2.3.14}$$

M_l are the solutions of the same problems with N_l replaced by M_l. However, M_0 is the solution to the problem

$$L_{\xi\xi}M_0 = \Phi\widetilde{F}, \qquad \frac{\partial M_0}{\partial\nu_\xi}\Big|_{\xi'\in\partial\beta} = 0, \qquad [M_0]\big|_\Sigma = 0, \qquad \Big[\frac{\partial M_0}{\partial\nu_\xi}\Big]\Big|_\Sigma = 0. \tag{2.3.15}$$

Thus, the algorithm for constructing the matrices N_l and M_l is inductive. Suppose that $N_l = 0$, $M_l = 0$ for $l < 0$, $N_0 = \Phi$, and M_0 is the solution to problem (2.3.15). If $l > 0$, then N_l and M_l are the solutions to problems (2.3.11)–(2.3.14). The right-hand sides of these problems contain N_m and M_m with indices $m < l$, which permits one to define them successively. We have

$$h_0^M = 0, \quad h_0^N = 0, \quad h_1^N = 0, \qquad h_2^N = \Big\langle\Phi^*\Big(\sum_{j=1}^{s} A_{1j}\frac{\partial N_1}{\partial\xi_j} + A_{11}\Phi\Big)\Big\rangle. \tag{2.3.16}$$

Here a 1-periodic solution N_l to problem (2.3.11)-(2.3.13) is understood in the same sense as in section 2.2. Let $\left(H^1_{per\ \xi_1}(Q)\right)^{s\times s}$ be space of matrix valued functions $s\times s$ with the components from $H^1_{per\ \xi_1}(Q)$.

A solution$\bar{U}\in\left(H^1_{per\ \xi_1}(Q)\right)^{s\times s}$ to the problem

$$L_{\xi\xi}\bar{U} = \bar{F}_0(\xi)+\sum_{j=1}^{s}\frac{\partial\bar{F}_j}{\partial\xi_j},\quad \xi\in I\!R\times\beta,$$

$$\frac{\partial\bar{U}}{\partial\nu_\xi} = \sum_{j=2}^{s}\bar{F}_j n_j,\quad \xi\in I\!R\times\partial\beta,$$

is understood as an $s\times s$ matrix valued function U satisfying the integral identity

$$-\left\langle\sum_{i,j=1}^{s}\frac{\partial\Psi^*}{\partial\xi_i}A_{ij}\frac{\partial\bar{U}}{\partial\xi_j}\right\rangle\rangle=\langle\Psi^*\bar{F}_0\rangle-\left\langle\sum_{j=1}^{s}\frac{\partial\Psi^*}{\partial\xi_j}\bar{F}_j\right\rangle$$

for any $s\times s$ matrix valued function $\Psi\in\left(H^1_{per\ \xi_1}(Q)\right)^{s\times s}$. Here the right-hand sides $\bar{F}_0,\bar{F}_1,\ldots,\bar{F}_s$ are $s\times s$ matrix valued functions such that their components belong to $L^2((-\mathcal{A},\mathcal{A})\times\beta)$ for each $\mathcal{A}$ and are 1-periodic in ξ_1.

This problem can be easily reduced to s independent problems for columns of matrix $\bar{U}$ with right hand sides equal to corresponding columns of the matrices $F_0,F_1,\ldots,F_s$. In this case for each column U of the matrix $\bar{U}$ we have the following variational formulation:

find a column $U\in\left(H^1_{per\ \xi_1}(Q)\right)^s$ such that for any column $\Psi\in\left(H^1_{per\ \xi_1}(Q)\right)^s$, the integral identity holds

$$-\left\langle\left(\sum_{i,j=1}^{s}A_{ij}\frac{\partial U}{\partial\xi_j},\frac{\partial\Psi}{\partial\xi_i}\right)\right\rangle=\langle(F_0,\Psi)\rangle-\left\langle\sum_{j=1}^{s}\left(F_j,\frac{\partial\Psi}{\partial\xi_j}\right)\right\rangle.$$

Here the right-hand sides $F_0,F_1,\ldots,F_s$ are vector valued functions (corresponding columns of the matrices $\bar{F}_0,\bar{F}_1,\ldots,\bar{F}_s$) such that their components belong to $L^2((-\mathcal{A},\mathcal{A})\times\beta)$ for each $\mathcal{A}$ and are 1-periodic in ξ_1.

Lemma 2.3.1. *Assume that the right-hand sides $F_1,\ldots,F_s$ are 1-periodic s-dimensional vector functions of ξ_1 with components satisfying $F^j_i=F^i_j$. The problem has a solution if and only if*

$$\langle\Phi^*F_0\rangle=0.$$

Proof.

The proof similar to the proof of lemma 2.2.1 and it is also based on the Riesz representation theorem for a bounded linear functional on a Hilbert space.

Namely, the right-hand side of the integral identity (2.2.82) represents the linear functional $G(\Psi)$ that is bounded in the norm of $\left(H^1(Q)\right)^s$ by the inequality

$$|G(\Psi)| \leq \left(\sum_{j=0}^{s} \|F_j\|_{(L^2(Q))^s}\right) \|\Psi\|_{(H^1(Q))^s},$$

On the other hand, for vector functions $\Psi \in \left(H^1(Q)\right)^s$ that satisfy $\langle \Phi^* \Psi \rangle = 0$ and the Korn inequality $\langle e(\Psi) \rangle \geq c_1 \|\Psi\|^2_{(H^1(Q))^s}, \quad c_1 > 0$, holds (it can be proved by repeating the proof of [55](see also [129]), Theorem 12.11 and the remark to this theorem literally). Here we write

$$e(\Psi) = \sum_{i,j=1}^{s} \left(\frac{\partial \Psi^i}{\partial \xi_j} + \frac{\partial \Psi^j}{\partial \xi_i}\right)^2.$$

Since

$$\left\langle \sum_{i,j=1}^{s} \left(A_{ij} \frac{\partial \Psi}{\partial \xi_j}, \frac{\partial \Psi}{\partial \xi_i}\right) \right\rangle \geq \kappa \langle e(\Psi) \rangle,$$

the norm

$$\|\Psi\|_1 = \sqrt{\left\langle \sum_{i,j=1}^{s} \left(A_{ij} \frac{\partial \Psi}{\partial \xi_j}, \frac{\partial \Psi}{\partial \xi_i}\right) \right\rangle}$$

is equivalent to the H^1-norm on the subspace of vector functions of $\left(H^1_{per\ \xi_1}(Q)\right)^s$ orthogonal to the columns of the matrix Φ and this norm corresponds to the inner product

$$[\Psi, \Theta]_1 = \left\langle \sum_{i,j=1}^{s} \left(A_{ij} \frac{\partial \Psi}{\partial \xi_j}, \frac{\partial \Theta}{\partial \xi_i}\right) \right\rangle.$$

Then we see that the functional $G(\Psi)$ is bounded in the norm $\|\Psi\|_1$ on this subspace as well, and therefore it can be represented in the form of an inner product $[U, \Psi]_1 = G(\Psi)$, where U is an element of the above subspace. Hence, on the subspace of vector functions $\Psi \in \left(H^1_{per\ \xi_1}(Q)\right)^s$ such that $\langle \Phi^* \Psi \rangle = 0$, the mentioned above integral identity holds for some vector function U from this subspace.

Let $\Psi(\xi)$ be any vector valued function of space $\left(H^1_{per\ \xi_1}(Q)\right)^s$. Represent $\Psi(\xi)$ in the form $\Psi(\xi) = \Psi_0(\xi) + \Phi h$, where h is a constant d-dimensional vector and

$$\langle \Phi^* \Psi_0 \rangle = 0 \quad (h = -\langle \Phi^* \Phi \rangle \langle \Phi^* \Psi \rangle = -\langle \Phi^* \Psi \rangle).$$

Then

$$[U, \Psi]_1 = [U, \Psi_0]_1 + [U, \Phi h]_1 = G(\Psi_0) + \left\langle \sum_{i,j=1}^{s} \left(A_{ij} \frac{\partial}{\partial \xi_j}(\Phi h), \frac{\partial U}{\partial \xi_i}\right) \right\rangle =$$

$$= G(\Psi_0) = G(\Psi).$$

Indeed,

$$\langle (F_0, \Phi h) \rangle - \sum_{j=1}^{s} \left\langle \left(F_j, \frac{\partial(\Phi h)}{\partial \xi_j} \right) \right\rangle = \langle h^* \Phi^* F_0 \rangle - \sum_{j,k=1}^{s} \left\langle F_j^k \frac{\partial(\Phi h)^k}{\partial \xi_j} \right\rangle = 0$$

because

$$F_j^k = F_k^j, \quad \frac{\partial(\Phi h)^k}{\partial \xi_j} = -\frac{\partial(\Phi h)^j}{\partial \xi_k}.$$

This proves the lemma. □

Remark 2.3.1. A solution U that is orthogonal to the rigid-body displacements is unique. This follows from the Riesz theorem for $G(\Psi)$.

Lemma 2.3.1 implies formula (2.3.14) for the calculation of $h_l^{(N,M)}$.

We introduce now the reflection operator S_α $R^s \to R^s$. For $s = 3$ we have

$$S_2\xi = (\xi_1, -\xi_2, \xi_3), S_3\xi = (\xi_1, \xi_2, -\xi_3).$$

For $s = 2$ we have $S_2\xi = (\xi_1, -\xi_2)$. It is assumed that $\forall \alpha = 2, \dots, s$ $S_\alpha((0,1) \times \beta) = (0,1) \times \beta$. Consider the following conditions:

Condition A:

$$\forall \alpha = 2, \dots, s \quad a_{ij}^{kl}(S_\alpha \xi) = (-1)^{\delta_{\alpha i} + \delta_{\alpha k} + \delta_{\alpha j} + \delta_{\alpha l}} a_{ij}^{kl}(\xi).$$

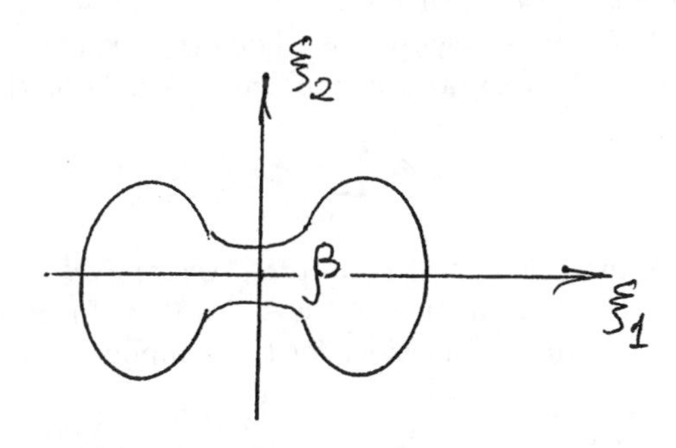

Figure 2.3.2. Symmetry of a cross-section

Condition B:

$$a_{ij}^{kl}(\xi) = (-1)^{\delta_{1i} + \delta_{1k} + \delta_{1j} + \delta_{1l}} a_{ij}^{kl}(\xi),$$

and a_{ij}^{kl} is independent of ξ_1.

If condition A is satisfied, then the following assertion is valid.

Theorem 2.3.1 *The matrix h_l^N is diagonal. If $s = 3$, then*

$$h_2^N = \begin{pmatrix} \bar{E}^{(3)} & 0 & 0 & 0 \\ 0 & 0 & 0 & 0 \\ 0 & 0 & 0 & 0 \\ 0 & 0 & 0 & \bar{M}^{(3)} \end{pmatrix}, \quad h_3^N = \begin{pmatrix} C_1 & 0 & 0 & 0 \\ 0 & 0 & 0 & 0 \\ 0 & 0 & 0 & 0 \\ 0 & 0 & 0 & C_2 \end{pmatrix},$$

$$h_4^N = \begin{pmatrix} C_3 & 0 & 0 & 0 \\ 0 & -\bar{J}_2^{(3)} & 0 & 0 \\ 0 & 0 & -\bar{J}_3^{(3)} & 0 \\ 0 & 0 & 0 & C_4 \end{pmatrix},$$

but if $s = 2$, then

$$h_2^N = \begin{pmatrix} \bar{E}^{(2)} & 0 \\ 0 & 0 \end{pmatrix}, \quad h_3^N = \begin{pmatrix} C_5 & 0 \\ 0 & 0 \end{pmatrix}, \quad h_4^N = \begin{pmatrix} C_6 & 0 \\ 0 & -\bar{J}_2^{(2)} \end{pmatrix},$$

where $\bar{E}^{(s)}$, $\bar{J}_2^{(s)}$, $\bar{J}_3^{(s)}$, $\bar{M}^{(s)} > 0$, C_1, C_2, C_3, C_4, C_5, C_6 are some constants.

(This theorem as well as following theorems 2.3.2-2.3.7 will be proved in subsection 2.3.6.)

Then (2.3.7) takes the form

$$Pu^{(\infty)} - \Phi F \psi =$$

$$= \Phi\Big(h_2^N \frac{d^2\omega}{dx_1^2} + \mu h_3^N \frac{d^3\omega}{dx_1^3} + \mu^2 h_4^N \frac{d^4\omega}{dx_1^4} - \bar{F}\psi + \sum_{k=5}^{\infty} \mu^{k-2} h_k^N \frac{d^k\omega}{dx_1^k} + \sum_{l=1}^{\infty} \mu^l h_l^M \frac{d^l\psi}{dx_1^l}\Big) =$$

$$= 0. \tag{2.3.17}$$

If condition B is satisfied, then the first sum contains only even powers of μ.

Problem (2.3.17) can be regarded as a homogenized equation of infinite order for the d-vector ω. The f.a.s. to this problem is sought as the series

$$\omega = \sum_{j=-2}^{\infty} \mu^j \omega_j(x_1), \tag{2.3.18}$$

where $\omega_j(x_1)$ are independent of μ, ω_j are d-vectors, $\omega_j^1 = 0$ ($\omega_j^4 = 0$ if $s = 3$) for $j = -2, -1$. The substitution of the series (2.3.18) into (2.3.17) gives the following recurrent chain of equations for the components ω_j^k of the vectors ω_j:

$$\bar{E}^{(s)} \frac{d^2\omega_j^1}{dx_1^2} = f_j^1(x_1), \quad -\bar{J}_2^{(s)} \frac{d^4\omega_j^2}{dx_1^4} = f_j^2(x_1)$$

and, in addition, for $s = 3$, $-\bar{J}_3^{(s)} d^4\omega_j^3/dx_1^4 = f_j^3(x_1)$

$$\bar{M}^{(s)} \frac{d^2\omega_j^4}{dx_1^2} = f_j^4(x_1), \tag{2.3.19}$$

where f_j^r depend on $\omega_{j_1}^r$, $j_1 < j$, and their derivatives. The f.a.s. to problem (2.3.1), (2.3.2), (2.3.4) is thus constructed.

2.3.3 Boundary layer corrector.

We construct the f.a.s. to problem (2.3.1)–(2.3.4) in the form,

$$u^{(\infty)} = u_B + u_P^0 + u_P^1, \tag{2.3.20}$$

where u_B is defined by equation (2.3.6) and

$$u_P^0 = \Big(\sum_{l=0}^{\infty} \mu^l N_l^0(\xi) \frac{d^l \omega}{dx_1^l} + \sum_{l=0}^{\infty} \mu^{l+2} M_l^0(\xi) \frac{d^l \psi}{dx_1^l}\Big)\Big|_{\xi=x/\mu},$$

$$u_P^1 = \Big(\sum_{l=0}^{\infty} \mu^l N_l^1(\xi) \frac{d^l \omega}{dx_1^l} + \sum_{l=0}^{\infty} \mu^{l+2} M_l^1(\xi) \frac{d^l \psi}{dx_1^l}\Big)|_{\xi_1=(x_1-b)/\mu \xi'=x'/\mu}. \tag{2.3.21}$$

Substituting (2.3.20) into (2.3.1)–(2.3.4) and taking into account the fact that u_B is as constructed above, we obtain the asymptotic equations

$$\sum_{l=0}^{\infty} \mu^{l-2} \bar{H}_l^N \Big(\frac{x}{\mu}\Big) \frac{d^l \omega}{dx_1^l} + \sum_{l=0}^{\infty} \mu^l \bar{H}_l^M \Big(\frac{x}{\mu}\Big) \frac{d^l \psi}{dx_1^l} - \Phi \bar{F} \psi - \Phi \widetilde{F} \psi = 0, \quad x \in C_\mu, \tag{2.3.22}$$

$$\sum_{l=0}^{\infty} \mu^{l-1} \bar{G}_l^N \Big(\frac{x}{\mu}\Big) \frac{d^l \omega}{dx_1^l} + \sum_{l=0}^{\infty} \mu^{l+1} \bar{G}_l^M \Big(\frac{x}{\mu}\Big) \frac{d^l \psi}{dx_1^l} = 0, \quad x \in \partial C_\mu \cap \partial U_\mu, \tag{2.3.23}$$

$$\sum_{l=0}^{\infty} \mu^l [\bar{N}_l] \frac{d^l \omega}{dx_1^l} + \sum_{l=0}^{\infty} \mu^{l+2} [\bar{M}_l] \frac{d^l \psi}{dx_1^l} = 0, \quad \frac{x}{\mu} \in \Sigma, \tag{2.3.24}$$

$$\sum_{l=0}^{\infty} \mu^{l-1} \Big[\bar{G}_l^N \Big(\frac{x}{\mu}\Big)\Big] \frac{d^l \omega}{dx_1^l} + \sum_{l=0}^{\infty} \mu^{l+1} \Big[\bar{G}_l^M \Big(\frac{x}{\mu}\Big)\Big] \frac{d^l \psi}{dx_1^l} = 0, \quad \frac{x}{\mu} \in \Sigma, \tag{2.3.25}$$

$$\sum_{l=0}^{\infty} \mu^l \bar{N}_l(0, \xi') \frac{d^l \omega}{dx_1^l}\Big|_{x_1=0} + \sum_{l=0}^{\infty} \mu^{l+2} \bar{M}_l(0, \xi') \frac{d^l \psi}{dx_1^l}\Big|_{x_1=0} = 0, \tag{2.3.26}$$

$$\sum_{l=0}^{\infty} \mu^l \bar{N}_l \Big(\frac{b}{\mu}, \xi'\Big) \frac{d^l \omega}{dx_1^l}\Big|_{x_1=b} + \sum_{l=0}^{\infty} \mu^{l+2} \bar{M}_l \Big(\frac{b}{\mu}, \xi'\Big) \frac{d^l \psi}{dx_1^l}\Big|_{x_1=b} = 0, \tag{2.3.27}$$

where

$$\bar{H}_l^N(x/\mu) = H_l^N(x/\mu) + H_l^{N0}(x/\mu) + H_l^{N1}(x_1 - b/\mu, x'/\mu),$$

$$\bar{G}_l^N(x/\mu) = G_l^N(x/\mu) + G_l^{N0}(x/\mu) + G_l^{N1}(x_1 - b/\mu, x'/\mu),$$

$$\bar{N}_l(x/\mu) = N_l(x/\mu) + N_l^0(x/\mu) + N_l^1(x_1 - b/\mu, x'/\mu),$$

$$H_l^{Nr}(\xi) = L_{\xi\xi} N_l^r + T_l^{Nr}(\xi),$$

$$T_l^{Nr}(\xi) = \sum_{j=1}^{s} \partial/\partial \xi_j (A_{j1} N_{l-1}^r) + \sum_{j=1}^{s} A_{1j} \partial N_{l-1}^r / \partial \xi_j + A_{11} N_{l-2}^r;$$

$$G_l^{Nr} = \sum_{m=1}^{s} \left(A_{mj}\partial N_l^r/\partial\xi_j + A_{m1}N_{l-1}^r\right)n_m \tag{2.3.28}$$

and $\bar{H}_l^M$, $\bar{G}_l^M$, $\bar{M}_l$, T_l^{Mr}, and G_l^{Mr} are obtained by replacing N with M. Here $N_l(\xi)$, $M_l(\xi)$, $H_l^{(N,M)}(\xi)$, and $G_l^{(N,M)}(\xi)$ are the 1-periodic functions in ξ_1 constructed above and $N_l^r(\xi)$, $H_l^{(N,M)r}(\xi)$, and $G_l^{(N,M)r}(\xi)$ exponentially stabilize to zero as $\xi_1 \to \pm\infty$ (+ for $r = 0$ and $-$ for $r = 1$). The superscript (N, M) means that the assertion is true for both N and M. The exponential stabilization to zero implies that

$$\int_{(\sigma,\sigma+1)\times\beta} |N_l^r(\xi)|d\xi, \quad \int_{(\sigma,\sigma+1)\times\beta} |M_l^r(\xi)|d\xi \le c_1 e^{-c_2|\sigma|} \quad \text{as} \quad \sigma \to \pm\infty$$

(as above, + for $r = 0$ and $-$ for $r = 1$). Here c_1 and c_2 are positive constants independent of σ.

We arrange N_l^r and M_l^r in ascending order with respect to the index l and note that $T_l^{(N,M)r}$ is defined by the functions $N_{l_1}^r$ (or $M_{l_1}^r$) for $l_1 < l$. We require that $H_l^{(N,M)r}$, $\bar{N}_l(0, \xi')$, $\bar{M}_l(0, \xi')$, $\bar{N}_l(b/\varepsilon, \xi')$, and $\bar{M}_l(b/\varepsilon, \xi')$ satisfy

$$H_l^{(N,M)r} = 0, \quad G_l^{(N,M)r} = 0, \qquad \bar{N}_l(0, \xi') = \widetilde{\Phi}(0, \xi')h_l^{N0} + O(e^{-c/\mu}),$$

$$\bar{M}_l(0, \xi') = \widetilde{\Phi}(0, \xi')h_l^{M0} + O(e^{-c/\mu}), \qquad \bar{N}_l\left(\frac{b}{\mu}, \xi'\right) = \widetilde{\Phi}(0, \xi')h_l^{N1} + O(e^{-c/\mu}),$$

$$\bar{M}_l\left(\frac{b}{\mu}, \xi'\right) = \widetilde{\Phi}(0, \xi')h_l^{M1} + O(e^{-c/\mu}), \quad c > 0, \tag{2.3.29}$$

where $\widetilde{\Phi}(\xi)$ is the extended $s \times \bar{d}$ matrix of rigid displacements,

$$\widetilde{\Phi}(\xi) = \begin{pmatrix} 1 & 0 & -\xi_2 \\ 0 & 1 & \xi_1 \end{pmatrix},$$

for $s = 2$ and $\bar{d} = 3$, and

$$\widetilde{\Phi}(\xi) = \begin{pmatrix} 1 & 0 & 0 & 0 & -\xi_2 & -\xi_3 \\ 0 & 1 & 0 & -a\xi_3 & \xi_1 & 0 \\ 0 & 0 & 1 & a\xi_2 & 0 & \xi_1 \end{pmatrix},$$

for $s = 3$ and $\bar{d} = 6$. The constant $\bar{d} \times d$ matrices $h_l^{(N,M)r}$ are to be defined from the conditions of the existence of matrix functions N_l^r and M_l^r exponentially stabilizing to zero.

We construct the functions N_l^r, M_l^r satisfying equations (2.3.29) by induction on l. Assume that N_l^r, $M_l^r = 0$ for $l < 0$. Let all $N_{l_1}^r$, $M_{l_1}^r$ be constructed for $l_1 < l$. We choose a constant $(\bar{d} \times d)$-matrix h_l^{N0} such that there exists a solution, exponentially stabilizing to zero, to the problem

$$L_{\xi\xi}N_l^0 + T_l^{N0} = 0, \quad \xi_1 > 0, \quad \xi' \in \beta,$$

$$\frac{\partial N_l^0}{\partial \nu_\xi} = -\sum_{m=2}^{s} A_{m1} N_{l-1} n_m, \quad \xi' \in \partial\beta,$$

$$N_l^0(0,\xi') = -N_l(0,\xi') + \widetilde{\Phi}(0,\xi') h_l^{N0}, \tag{2.3.30}$$

and choose a constant matrix h_l^{N1} such that there exists a solution exponentially stabilizing to zero to the problem

$$L_{\xi\xi} N_l^1 + T_l^{N1} = 0, \quad \xi_1 < 0, \quad \xi' \in \beta,$$

$$\frac{\partial N_l^1}{\partial \nu_\xi} = -\sum_{m=2}^{s} A_{m1} N_{l-1}^1 n_m, \quad \xi' \in \partial\beta,$$

$$N_l^1(0,\xi') = -N_l(0,\xi') + \widetilde{\Phi}(0,\xi') h_l^{N1}. \tag{2.3.31}$$

The matrices M_l^r and h_l^{Mr} are defined similarly (by replacing N with M in equations (2.3.30) and (2.3.31)).

The following theorem provides the possibility of such a choice of the constants $h_l^{(N,M)r}$.

Theorem 2.3.2. *Let $A_{ij}(\xi)$ satisfy the above conditions, let s-vectors $F_j(\xi) \in L^2([0,+\infty) \times \beta)$, $j = 0,1,\ldots,s$, satisfy*

$$\int_{(\sigma,\sigma+1)\times\beta} |F_j(\xi)| d\xi \le c_1 e^{-c_2\sigma}, \quad \sigma > 0, \tag{2.3.32}$$

and let an s-vector $u_0(\xi') \in H^{1/2}(\{0\}\times\beta)$ be given. Then there exists a solution $u(\xi)$ to the problem

$$L_{\xi\xi} u = F_0(\xi) + \sum_{j=1}^{s} \frac{\partial F_j}{\partial \xi_j}, \quad \xi_1 > 0, \quad \xi' \in \beta, \tag{2.3.33}$$

$$\frac{\partial u}{\partial \nu_\xi} = \sum_{j=2}^{s} F_j n_j, \quad \xi' \in \partial\beta, \tag{2.3.34}$$

$$u\big|_{\xi_1=0} = u_0(\xi') \tag{2.3.35}$$

such that for some rigid displacement $w(\xi) = \widetilde{\Phi}(\xi) h$, where h is a constant $\bar{d}$-vector, the following inequality holds

$$\int_{(\sigma,\sigma+1)\times\beta} |u - w| d\xi \le c_1 e^{-c_2\sigma}, \quad c_1, c_2 > 0, \quad \sigma > 0. \tag{2.3.36}$$

Remark 2.3.2. Let us explain some terms from Theorem 2.3.2 (see [129],[130]). The space $H^{1/2}(\{0\}\times\beta)$ is thought of as the space of traces $V(\xi')$ on the set $\{\xi_1 = 0, \ \xi' \in \beta\}$ of vector functions from $H^1((0,1)\times\beta)$ equipped with the norm

$$\|V\|_{1/2}^2 = \inf_{\mathbf{v}} \Big\{ \int_{(0,1)\times\beta} e(\mathbf{v}) + |\mathbf{v}|^2 \, d\xi, \quad \mathbf{v} \in W_2^1((0,1)\times\beta), \quad \mathbf{v}\big|_{\xi_1=0} = V(\xi') \Big\},$$

$$e(\mathbf{v}) = \sum_{i,j=1}^{s} \left(\frac{\partial v^i}{\partial x_j} + \frac{\partial v^j}{\partial x_i} \right)^2, \quad \mathbf{v} = (v^1, \ldots, v^s)^*.$$

A solution to problem (2.3.33)–(2.3.35) is viewed as an s-vector function $u(\xi) \in H^{1\,\mathrm{loc}}$ whose trace is equal to $u_0(\xi')$ for $\xi_1 = 0$. For any vector function $\Psi(\xi) \in\rightarrow C^\infty$ with support in the cylinder $(\delta, D) \times \beta$, $\delta > 0$, the function $u(\xi)$ must satisfy the integral identity

$$-\sum_{i,j=1}^{s} \int_{(0,+\infty)\times\beta} \left(A_{ij} \frac{\partial u}{\partial \xi_j}, \frac{\partial \Psi}{\partial \xi_i} \right) d\xi = \int_{(0,+\infty)\times\beta} (F_0, \Psi)\, d\xi -$$

$$-\sum_{j=1}^{s} \int_{(0,+\infty)\times\beta} \left(F_j, \frac{\partial \Psi}{\partial \xi_j} \right) d\xi.$$

Theorem 2.3.2 follows directly from Theorems 4 and 5 in [130]. Their proofs are "corrected" for the right-hand side having specific representation (2.3.33) and (2.3.34) and the momenta in Theorem 5 are chosen so that they are equal to zero at infinity (i.e., in the notation of [130], the momenta are

$$P^r(0,u) = -\int_{(0,+\infty)\times\beta} (F_0, \eta^r) d\xi + \int_{(0,+\infty)\times\beta} \sum_{j=1}^{s} \left(F_j, \frac{\partial \eta^r}{\partial \xi_j} \right) d\xi,$$

where η^r, $r = 1, \ldots, \bar{d}$ are elements of the basis of the rigid displacements space).

To determine the matrix h_l^{N0} from (2.3.30), it suffices to first solve the problem

$$L_{\xi\xi} \widetilde{N}_l^0 + T_l^{N0} = 0, \quad \xi_1 > 0, \quad \xi' \in \beta,$$

$$\frac{\partial \widetilde{N}_l^0}{\partial \nu_\xi} = -\sum_{m=2}^{s} n_m A_{m1} N_{l-1}^0, \quad \xi' \in \partial\beta,$$

$$\widetilde{N}_l^0(0, \xi') = -N_l(0, \xi'). \tag{2.3.37}$$

According to Theorem 2.3.2, each vector-column of the matrix $\widetilde{N}_l^0$ stabilizes to a rigid displacement, so that the entire matrix $\widetilde{N}_l^0(\xi)$ stabilizes to some matrix $\widetilde{\Phi} h$, where h is a constant $\bar{d} \times d$-matrix. If one subtracts $\widetilde{\Phi} h$ from $\widetilde{N}_l^0(\xi)$, then the difference $N_l^0 = \widetilde{N}_l^0(\xi) - \widetilde{\Phi} h$ will stabilize to zero and satisfy (2.3.30) for $h_l^{N0} = -h$. Problem (2.3.31) and the problems for M_l^r and h_l^{Mr} can be solved in the same manner.

It follows from [130] that the energy e of the columns N_l^r and M_l^r also exponentially stabilizes to zero.

After the successive determination of all N_l^r and M_l^r by induction (note that $N_0^r = 0$) we obtain a homogenized problem of infinite order for the d-vector function $\omega(x_1)$. This problem is given by equation (2.3.17) with the boundary conditions

$$\left(\omega + \sum_{l=1}^{\infty} \mu^l h_l^{Nr(\mathrm{cut}\, d)} \frac{d^l \omega}{dx_1^l} + \sum_{l=0}^{\infty} \mu^{l+2} h_l^{Mr(\mathrm{cut}\, d)} \frac{d^l \psi}{dx_1^l} \right) \Bigg|_{x_1 = rb} = 0, \tag{2.3.38}$$

$$\left(\mu\frac{d(\omega^2,\ldots,\omega^s)^*}{dx_1}+\sum_{l=2}^{\infty}\mu^l h_l^{Nr(\mathrm{oth})}\frac{d^l\omega}{dx_1^l}+\sum_{l=0}^{\infty}\mu^{l+2}h_l^{Mr(\mathrm{oth})}\frac{d^l\psi}{dx_1^l}\right)\Big|_{x_1=rb}=$$

$$=0,\quad r=0,1, \tag{2.3.39}$$

where the superscript (cut d) denotes the first d rows of a matrix and the superscript (oth) denotes the remaining $\bar{d}-d$ rows (we recall that $d=2$ and $\bar{d}=3$ if $s=2$; $d=4$ and $\bar{d}=6$ if $s=3$); $\omega^1,\omega^2,\ldots,\omega^d$ are the components of the vector ω.

It is significant that the matrices h_0^{Nr} and h_1^{Nr} have the following structure: for $s=2$:

$$h_0^{Nr}=\begin{pmatrix}1&0\\0&1\\0&0\end{pmatrix},$$

and the third row of h_1^{Nr} is equal to $(0,1)$. For $s=3$ we have

$$h_0^{Nr}=\begin{pmatrix}1&0&0&0\\0&1&0&0\\0&0&1&0\\0&0&0&1\\0&0&0&0\\0&0&0&0\end{pmatrix},$$

and the fifth and sixth rows of h_1^{Nr} are equal to $(0,1,0,0)$ and $(0,0,1,0)$, respectively. The proof of this assertion (Lemma 2.3.6) will be given in the sequel.

The f.a.s. to problem (2.3.17), (2.3.38) and (2.3.39) is represented in the form of the series (2.3.18) and, being substituted into (2.3.17), (2.3.38) and (2.3.39), leads to the recurrent chain of equations (2.3.19) for the components ω_l^k of the vectors ω_j with the boundary conditions

$$\omega_j\big|_{x_1=rb}=g_{jr},\quad r=0,1,$$

$$\frac{d\omega_j^q}{dx_1}\Big|_{x_1=rb}=g_{jr}^{d+q},\quad r=0,1,\quad q=2,\ldots,s. \tag{2.3.40}$$

Here f_j^r, g_{jr}^{d+q}, $q=1,\ldots,\bar{d}$ depend on $\omega_{j_1}^r$ when $j_1<j$, and on the derivatives of these functions. Thus, the f.a.s. to problem (2.3.1)–(2.3.4) is constructed.

2.3.4 The boundary layer corrector when the left end of the bar is free

Consider problem (2.3.1)–(2.3.3) with condition (2.3.3) if $x_1=0$, replaced by the free surface condition

$$\frac{\partial u}{\partial\nu}\equiv-\sum_{j=1}^{s}A_{1j}\left(\frac{x}{\mu}\right)\frac{\partial u}{\partial x_j}=0. \tag{2.3.3$'$}$$

Then, when constructing the boundary layer corrector, condition (2.3.26) must be replaced by

$$\sum_{l=0}^{\infty}\mu^{l-1}\bar{G}_l^N\Big(0,\frac{x'}{\mu}\Big)\frac{d^l\omega}{dx_1^l}\Big|_{x_1=0}+\sum_{l=0}^{\infty}\mu^{l+1}\bar{G}_l^M\Big(0,\frac{x'}{\mu}\Big)\frac{d^l\psi}{dx_1^l}\Big|_{x_1=0}=0 \qquad (2.3.26')$$

and the requirements (2.3.29) on $\bar{N}_l(0,\xi')$ and $\bar{M}_l(0,\xi')$ must be replaced by

$$\bar{G}_l^N(0,\xi') = \widetilde{\Phi}h_l^{N0}+O(e^{-c/\mu}), \bar{G}_l^M(0,\xi') = \widetilde{\Phi}h_l^{M0}+O(e^{-c/\mu}), \quad c>0. \qquad (2.3.29')$$

The construction of the functions N_l^r and M_l^r is carried out, as above, by induction on l; however, problem (2.3.30) is replaced by

$$L_{\xi\xi}N_l^0+T_l^{N0}=0, \quad \xi_1>0, \quad \xi'\in\beta,$$

$$\frac{\partial N_l^0}{\partial\nu_\xi}=-\sum_{m=2}^{s}A_{m1}N_{l-1}^0 n_m, \quad \xi'\in\partial\beta,$$

$$\frac{\partial N_l^0}{\partial\nu_\xi}=-\frac{\partial N_l}{\partial\nu_\xi}+A_{11}(N_{l-1}^0+N_{l-1})+\widetilde{\Phi}(0,\xi')h_l^{N0}, \quad \xi_1=0,$$

where the $\bar{d}\times d$ constant matrix h_l^{N0} must be chosen so as to ensure that there exists a solution N_l^0 exponentially stabilizing to zero. The matrices M_l^0 and h_l^{M0} are defined in the usual way.

The possibility of such a choice of constants $h_l^{(N,M)0}$ is a consequence of the following theorem.

Theorem 2.3.3. *Let $A_{ij}(\xi)$ satisfy the conditions imposed at the beginning of the section, let the s-vectors*

$$F_j(\xi)\in L^2([0,+\infty)\times\beta), \quad j=0,1,\dots,s,$$

satisfy (2.3.32), let an s-vector $u_0(\xi')\in L^2(\beta)$, and let the $\bar{d}$-dimensional constant vector h be given by

$$h=\langle\widetilde{\Phi}^*(0,\xi')\widetilde{\Phi}(0,\xi')\rangle^{-1}\Big(\frac{1}{\operatorname{mes}\beta}\int_{(0,+\infty)\times\beta}\widetilde{\Phi}^*(\xi)F_0(\xi)\,d\xi-$$

$$-\langle\widetilde{\Phi}^*(0,\xi')\times u_0(\xi')\rangle\Big). \qquad (2.3.41)$$

Then there exists a solution $u(\xi)$ to the problem

$$L_{\xi\xi}u=F_0(\xi)+\sum_{j=1}^{s}\frac{\partial F_j}{\partial\xi_j}, \quad \xi_1>0, \quad \xi'\in\beta, \qquad (2.3.42)$$

$$\frac{\partial u}{\partial\nu_\xi}=\sum_{j=2}^{s}F_j n_j, \quad \xi'\in\partial\beta, \qquad (2.3.43)$$

$$\frac{\partial u}{\partial \nu_\xi} = -F_1 + u_0(\xi') + \widetilde{\Phi}(0,\xi')h, \quad \xi_1 = 0 \tag{2.3.44}$$

exponentially stabilizing to zero,

$$\int_{(\sigma,\sigma+1)\times\beta} |u|\, d\xi \le c_1 e^{-c_2\sigma}, \quad c_1, c_2 > 0.$$

The solution to problem (2.3.42)–(2.3.44) is thought of as an s-vector function $u(\xi) \in H^{1\,\mathrm{loc}}$ satisfying the following integral identity for any vector function $\Psi(\xi) \in C^\infty([0,+\infty)\times\bar{\beta})$ equal to zero for sufficiently large ξ_1:

$$-\sum_{i,j=1}^{s} \int_{(0,+\infty)\times\beta} \Big(A_{ij}\frac{\partial u}{\partial \xi_j}, \frac{\partial \Psi}{\partial \xi_i}\Big)\, d\xi =$$

$$= \int_{(0,+\infty)\times\beta} (F_0, \Psi)\, d\xi - \sum_{j=1}^{s} \int_{(0,+\infty)\times\beta} \Big(F_j, \frac{\partial \Psi}{\partial \xi_j}\Big)\, d\xi -$$

$$- \int_\beta (\widetilde{\Phi}(0,\xi')h + u_0, \Psi)\, d\xi'.$$

Theorem 2.3.3 follows from Theorem 4 and from an analog of Theorem 5 in [130]. It follows from Theorem 2.3.3 that

$$h_l^{N0} = -\langle \widetilde{\Phi}^*(0,\xi')\widetilde{\Phi}(0,\xi')\rangle^{-1}\Big(\frac{1}{\operatorname{mes}\beta}\int_{(0,+\infty)\times\beta} \widetilde{\Phi}^*(\xi)\Big(\sum_{j=1}^{s} A_{1j}\frac{\partial N_{l-1}^0}{\partial \xi_j} +$$

$$+ A_{11}N_{l-2}^0\Big)\, d\xi + \Big\langle \widetilde{\Phi}^*(0,\xi') \times \Big(\sum_{j=1}^{s} A_{1j}\frac{\partial N_l}{\partial \xi_j} + A_{11}N_{l-1}\Big)\Big\rangle\Big) \tag{2.3.45}$$

and a similar equation holds for h_l^{M0}.

Thus, (2.3.26) takes the form

$$\widetilde{\Phi}\Big(\sum_{l=0}^{\infty} \mu_{l-1}h_l^{N0}\frac{d^l\omega}{dx_1^l}\Big|_{x_1=0} + \sum_{l=0}^{\infty} \mu_{l+1}h_l^{M0}\frac{d^l\psi}{dx_1^l}\Big|_{x_1=0}\Big) = 0. \tag{2.3.46}$$

We have $N_0^0 = 0$, $T_1^{N0} = 0$.

The second (and the third if $s=3$) row of the matrix N_1^0 is a solution to the homogeneous problem and therefore is equal to zero. Hence, the second (and the third) row of T_2^{N0} is zero as well. In what follows conditions A are assumed to be satisfied.

Theorem 2.3.4. *The matrices* $h_l^{N0(\mathrm{cut}\, d)}$ *are diagonal, and* $h_0^{N0} = 0$. *For* $s = 3$ *we have*

$$h_1^{N0} = -\begin{pmatrix} \bar{E}^{(3)} & 0 & 0 & 0 \\ 0 & 0 & 0 & 0 \\ 0 & 0 & 0 & 0 \\ 0 & 0 & 0 & \bar{M}^{(3)} \\ 0 & 0 & 0 & 0 \\ 0 & 0 & 0 & 0 \end{pmatrix},$$

$$h_2^{N0} = -\begin{pmatrix} C_1 & 0 & 0 & 0 \\ 0 & 0 & 0 & 0 \\ 0 & 0 & 0 & 0 \\ 0 & 0 & 0 & C_2 \\ 0 & \bar{J}_2^{(3)}/\langle\xi_2^2\rangle & 0 & 0 \\ 0 & 0 & \bar{J}_3^{(3)}/\langle\xi_3^2\rangle & 0 \end{pmatrix},$$

$$h_3^{N0(\text{cut } d)} = -\begin{pmatrix} C_3 & 0 & 0 & 0 \\ 0 & -\bar{J}_2^{(3)} & 0 & 0 \\ 0 & 0 & -\bar{J}_3^{(3)} & 0 \\ 0 & 0 & 0 & C_4 \end{pmatrix},$$

where $\bar{E}^{(3)}$, $\bar{M}^{(3)}$, $\bar{J}_2^{(3)}$, $\bar{J}_3^{(3)} > 0$ are the constants from Theorem 1, and C_1, C_2, C_3, C_4 are some constants.

For $s = 2$ we have

$$h_1^{N0} = -\begin{pmatrix} \bar{E}^{(2)} & 0 \\ 0 & 0 \\ 0 & 0 \end{pmatrix},$$

$$h_2^{N0} = -\begin{pmatrix} C_5 & 0 \\ 0 & 0 \\ 0 & \bar{J}_2^{(2)}/\langle\xi_2^2\rangle \end{pmatrix},$$

$$h_3^{N0(\text{cut } d)} = -\begin{pmatrix} C_6 & 0 \\ 0 & -\bar{J}_2^{(2)} \end{pmatrix},$$

where $\bar{E}^{(2)}$, $\bar{J}_2^{(2)} > 0$ are the constants from Theorem 1, and C_5, C_6 are some constants.

It follows from Theorem 2.3.4 that upon substituting the series (2.3.18) into the homogenized problem of infinite order {(2.3.17), (2.3.38) for $r = 1$ (i.e., $x_1 = b$) and (2.3.46)} we obtain a recurrent chain of problems for the functions ω_j. They consist of equation (2.3.19), for $x_1 = b$ of the boundary conditions (2.3.40) ($r = 1$), and for $x_1 = 0$ of the boundary conditions

$$\bar{E}^{(s)}\frac{d\omega_j^1}{dx_1} = g_{j0}^1, \quad -\bar{J}_2^{(s)}\frac{d^3\omega_j^2}{dx_1^3} = g_{j0}^2, \quad \frac{\bar{J}_2^{(s)}}{\langle\xi_2^2\rangle}\frac{d^2\omega_j^2}{dx_1^2} = g_{j0}^3$$

and, moreover, of

$$-\bar{J}_3^{(s)}\frac{d^3\omega_j^3}{dx_1^3} = g_{j0}^4, \quad \bar{M}^{(s)}\frac{d\omega_j^4}{dx_1} = g_{j0}^5, \quad \frac{\bar{J}_3^{(s)}}{\langle\xi_3^2\rangle}\frac{d^2\omega_j^s}{dx_1^2} = g_{j0}^6 \tag{2.3.47}$$

if $s = 3$, where g_{j0}^k depend on ω_{j_1} if $j_1 < j$.

2.3.5 The boundary layer corrector for the two bar contact problem

Consider the following problem, which is, in some sense, a model problem for systems consisting of a finite number of bars. Namely, let the two bars

$$C_\mu^+ = U_\mu \cap \{x_1 \in (0, b_1)\} \quad \text{and} \quad C_\mu^- = U_\mu \cap \{x_1 \in (-b_2, 0)\},$$

be given, where b_1 and b_2 are integer multiples of μ and each bar has specific coefficients $A_{ij}(\xi)$, i.e.,

$$A_{ij}(\xi) = A_{ij}^+(\xi) \quad \text{if} \quad \xi_1 > 0,$$

$$A_{ij}(\xi) = A_{ij}^-(\xi) \quad \text{if} \quad \xi_1 < 0,$$

where $A_{ij}^\pm(\xi)$ are 1-periodic in ξ_1 matrix functions satisfying the conditions stated at the beginning of this section. Consider the elasticity system of equations (2.3.1) on $C_\mu^+ \cup C_\mu^-$ with the boundary conditions (2.3.2) for $x \in \partial U_\mu$ and (2.3.3) for $x_1 = b_1$ or $-b_2$, i.e.,

$$\begin{aligned} Pu &= F_\mu(x), \quad x \in C_\mu^+ \cup C_\mu^-, \\ \frac{\partial u}{\partial \nu} &= 0, \quad x \in \partial U_\mu, \\ u = 0 \quad &\text{when} \quad x_1 = b_1 \quad \text{or} \quad x_1 = -b_2. \end{aligned} \tag{2.3.48}$$

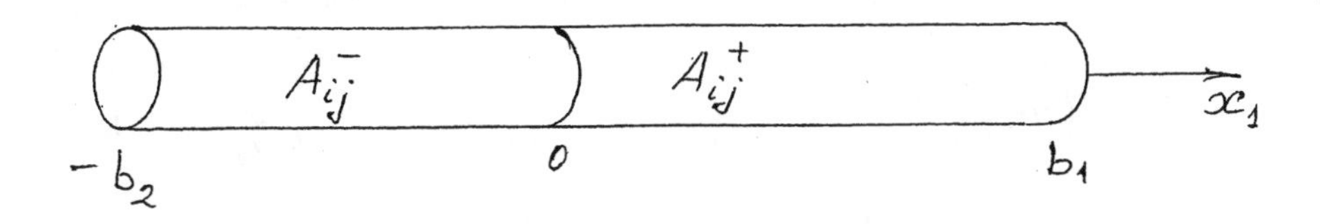

Figure 2.3.3 Two rod system

On the discontinuity interfaces of the coefficients (in particular, for $x_1 = 0$) the conjugation conditions (2.3.4) are imposed. Such a problem models the stress-strain state of a bar consisting of two parts with different microstructure.

The construction of a f.a.s. to problem (2.3.48), (2.3.4) is again carried out in two stages. In the first stage, we construct inner expansions for C_μ^+ and C_μ^-, and in the second stage we construct the boundary layer correctors. As above, they are constructed independently for the cross-sections $\{x_1 = -b_2\}$, $\{x_1 = b_1\}$, and $\{x_1 = 0\}$. The construction of the boundary layers near the ends $\{x_1 = -b_2\}$ and $\{x_1 = b_1\}$ is described above (see item 2.3). In what follows, we dwell on the construction of the boundary layer corrector near the cross-section $\{x_1 = 0\}$.

We seek the f.a.s. in the form

$$u^{(\infty)} = u_B^- + u_P^{-1} + u_P^0 + u_P^{+1} \quad \text{for} \quad x_1 < 0,$$

$$u^{(\infty)} = u_B^+ + u_P^{-1} + u_P^0 + u_P^{+1} \quad \text{for} \quad x_1 > 0, \tag{2.3.49}$$

where u_B^- and u_B^+ are the corresponding inner decompositions of type (2.3.6), the functions u_P^{-1} and u_P^{+1} are the boundary layer correctors near the ends $\{x_1 = -b_2\}$ and $\{x_1 = b_1\}$, respectively (see (2.3.21)), and u_P^0 is the boundary layer corrector near the contact interface

$$u_P^0 = \Big(\sum_{l=0}^{\infty} \mu^l \Big(N_l^{+0}(\xi) \frac{d^l\omega}{dx_1^l}\Big|_{x_1=+0} + N_l^{-0}(\xi) \frac{d^l\omega}{dx_1^l}\Big|_{x_1=-0}\Big)$$

$$+ \sum_{l=0}^{\infty} \mu^{l+2} \Big(M_l^{+0}(\xi) \frac{d^l\psi}{dx_1^l}\Big|_{x_1=+0} + M_l^{-0}(\xi) \frac{d^l\psi}{dx_1^l}\Big|_{x_1=-0}\Big)\Big|_{\xi=x/\mu}. \tag{2.3.50}$$

The substitution of (2.3.49), (2.3.50) into (2.3.48), (2.3.4) gives the asymptotic equations

$$\sum_{l=0}^{\infty} \mu^{l+2} \Big(\Big(\bar{H}_l^N\Big(\frac{x}{\mu}\Big) \frac{d^l\omega}{dx_1^l} + H_l^{N+0}\Big(\frac{x}{\mu}\Big) \frac{d^l\omega}{dx_1^l}\Big|_{x_1=+0} + H_l^{N-0}\Big(\frac{x}{\mu}\Big) \frac{d^l\omega}{dx_1^l}\Big|_{x_1=-0}\Big)$$

$$+ \sum_{l=0}^{\infty} \mu^l \Big(\Big(\bar{H}_l^M\Big(\frac{x}{\mu}\Big) \frac{d^l\psi}{dx_1^l} + H_l^{M+0}\Big(\frac{x}{\mu}\Big) \frac{d^l\psi}{dx_1^l}\Big|_{x_1=+0} + H_l^{M-0}\Big(\frac{x}{\mu}\Big) \frac{d^l\psi}{dx_1^l}\Big|_{x_1=-0}\Big)$$

$$-\Phi \bar{F} \psi - \Phi \widetilde{F} \psi = 0, \quad x \in C_\mu^+ \cup C_\mu^-; \tag{2.3.51}$$

$$G =$$

$$= \sum_{l=0}^{\infty} \mu^{l-1} \Big(\Big(\bar{G}_l^N\Big(\frac{x}{\mu}\Big) \frac{d^l\omega}{dx_1^l} + G_l^{N+0}\Big(\frac{x}{\mu}\Big) \frac{d^l\omega}{dx_1^l}\Big|_{x_1=+0} + G_l^{N-0}\Big(\frac{x}{\mu}\Big) \frac{d^l\omega}{dx_1^l}\Big|_{x_1=-0}\Big)$$

$$+ \sum_{l=0}^{\infty} \mu^{l+1} \Big(\Big(\bar{G}_l^M\Big(\frac{x}{\mu}\Big) \frac{d^l\psi}{dx_1^l} + G_l^{M+0}\Big(\frac{x}{\mu}\Big) \frac{d^l\psi}{dx_1^l}\Big|_{x_1=+0} +$$

$$+ G_l^{M-0}\Big(\frac{x}{\mu}\Big) \frac{d^l\psi}{dx_1^l}\Big|_{x_1=-0}\Big), \tag{2.3.52}$$

$$\sum_{l=0}^{\infty}\mu^l\Big(\Big(\Big[\bar{N}_l\frac{d^l\omega}{dx_1^l}\Big]+[N_l^{+0}]\frac{d^l\omega}{dx_1^l}\Big|_{x_1=+0}+[N_l^{-0}]\frac{d^l\omega}{dx_1^l}\Big|_{x_1=-0}\Big)$$

$$+\sum_{l=0}^{\infty}\mu^{l+2}\Big(\Big(\Big[\bar{M}_l\frac{d^l\psi}{dx_1^l}\Big]+[M_l^{+0}]\frac{d^l\psi}{dx_1^l}\Big|_{x_1=+0}+[M_l^{-0}]\frac{d^l\psi}{dx_1^l}\Big|_{x_1=-0}\Big)=0, \quad (2.3.53)$$

$$[G]=0, \quad \frac{x}{\mu}\in\Sigma \quad \text{or} \quad x_1=0, \quad (2.3.54)$$

$$\sum_{l=0}^{\infty}\mu^l\Big(\Big(\bar{N}_l\frac{d^l\omega}{dx_1^l}+N_l^{+0}\frac{d^l\omega}{dx_1^l}\Big|_{x_1=+0}+N_l^{-0}\frac{d^l\omega}{dx_1^l}\Big|_{x_1=-0}\Big)$$

$$+\sum_{l=0}^{\infty}\mu^{l+2}\Big(\Big(\bar{M}_l\frac{d^l\psi}{dx_1^l}+M_l^{+0}\frac{d^l\psi}{dx_1^l}\Big|_{x_1=+0}+M_l^{-0}\frac{d^l\psi}{dx_1^l}\Big|_{x_1=-0}\Big)=0, \quad (2.3.55)$$

for $\xi_1=-b_2/\mu$, $x_1=-b_2$ and for $\xi_1=b_1/\mu$, $x_1=b_1$, where

$$\bar{N}_l\Big(\frac{x}{\mu}\Big)=N_l\Big(\frac{x}{\mu}\Big)+N_l^-\Big(\frac{x_1+b_2}{\mu},\frac{x'}{\mu}\Big)+N_l^+\Big(\frac{x_1-b_1}{\mu},\frac{x'}{\mu}\Big), \quad N_l(\xi)$$

are the matrix valued functions of the inner expansion (specific for each of the subspaces $\xi_1>0$ and $\xi_1<0$), while $N_l^-(\xi)$, $N_l^+(\xi)$ are the boundary layer correctors near the ends of the bar satisfying estimates of type (2.3.28). The matrices $\bar{H}_l^{(N,M)}$ and $\bar{G}_l^{(N,M)}$ have a similar structure. The functions $N_l^{\pm0}(\xi)$, $M_l^{\pm0}(\xi)$, $H_l^{(N,M)\pm0}$, and $G_l^{(N,M)\pm0}$ exponentially stabilize to zero as $|\xi_1|\to+\infty$ (in the sense of (2.3.28)).

Here

$$H_l^{N\pm0}=L_{\xi\xi}N_l^{\pm0}, \qquad G_l^{N\pm0}(\xi)=\sum_{m,j=1}^{s}A_{mj}\frac{\partial N_l^{\pm0}}{\partial\xi_j}n_m,$$

and $\bar{H}_l^M$, $\bar{G}_l^M$, M_l, $H_l^{M\pm0}$, $G_l^{M\pm0}$, $M_l^{\pm0}$, $G_l^{M\pm0}$ are obtained by replacing N with M. We require that the conditions

$$H_l^{(N,M)\pm0}(\xi)=0, \quad H_l^{(N,M)\pm}(\xi)=0,$$

$$G_l^{(N,M)\pm0}(\xi)=0, \qquad G_l^{(N,M)\pm}(\xi)=0, \quad \xi'\in\partial\beta,$$

$$\bar{N}_l\Big(-\frac{b_2}{\mu},\xi'\Big)=\widetilde{\Phi}(0,\xi')h_l^{N-}+O(e^{-c/\mu}), \quad \bar{M}_l\Big(-\frac{b_2}{\mu},\xi'\Big)=\widetilde{\Phi}(0,\xi')h_l^{M-}+O(e^{-c/\mu}),$$

$$\bar{N}_l\Big(\frac{b_1}{\mu},\xi'\Big)=\widetilde{\Phi}(0,\xi')h_l^{N+}+O(e^{-c/\mu}), \quad \bar{M}_l\Big(\frac{b_1}{\mu},\xi'\Big)=\widetilde{\Phi}(0,\xi')h_l^{M+}+O(e^{-c/\mu}),$$

$$N_l(+0,\xi')+[N_l^{+0}]\big|_{\xi_1=0}=0, \qquad -N_l(-0,\xi')+[N_l^{-0}]\big|_{\xi_1=0}=0,$$

$$M_l(+0,\xi')+[M_l^{+0}]\big|_{\xi_1=0}=0, \qquad -M_l(-0,\xi')+[M_l^{-0}]\big|_{\xi_1=0}=0,$$

$$G_l^{(N,M)}(+0,\xi')+[G_l^{(N,M)+0}]\big|_{\xi_1=0}=0,$$

$$-G_l^{(N,M)}(-0,\xi') + [G_l^{(N,M)-0}]\big|_{\xi_1=0} = 0 \tag{2.3.56}$$

and the conditions

$$[N_l^{\pm 0}] = 0, \qquad [N_l^{\pm}] = 0, \quad \xi \in \Sigma,$$

$$[G_l^{\pm 0}] = 0, \qquad [G_l^{\pm}] = 0, \quad \xi \in \Sigma \tag{2.3.57}$$

hold. For N_l and M_l we obtain analogs of problems (2.3.11)-(2.3.14) (specific for $\xi_1 < 0$ and $\xi_1 > 0$) and analogs of problems (2.3.30), (2.3.31) for the functions $N_l^{\pm}$, $M_l^{\pm}$. Finally, for $N_l^{\pm}$ we obtain the following problems in the cylinder $R \times \beta$:

$$L_{\xi\xi} N_l^{\pm 0} = 0, \quad \xi \in R \times \beta,$$

$$\frac{\partial}{\partial \nu_\xi} N_l^{\pm 0} = 0, \quad \xi' \in \partial\beta,$$

$$[N_l^{\pm 0}]\big|_{\xi_1=0} = \mp N_l(\pm 0,\xi') + \widetilde{\Phi}(0,\xi')\bar{h}_l^{N\pm 0},$$

$$\Big[\frac{\partial N_l^{\pm 0}}{\partial \nu_\xi}\Big]\Big|_{\xi_1=0} = \mp\Big(\frac{\partial N_l}{\partial \nu_\xi}(\pm 0,\xi') + A_{11}N_{l-1}(\pm 0,\xi')n_1\Big) + \widetilde{\Phi}(0,\xi')\bar{\bar{h}}_l^{N\pm 0}, \tag{2.3.58}$$

where $n_1 > 0$, the $\bar{d} \times d$ constant matrices $\bar{h}_l^{N\pm 0}$ and $\bar{\bar{h}}_l^{N\pm 0}$ are chosen from the condition of existence of a solution exponentially stabilizing to zero. Here

$$\bar{\bar{h}}_l^{N\pm 0} =$$

$$= \langle \widetilde{\Phi}^*(0,\xi')\widetilde{\Phi}(0,\xi')\rangle^{-1}\Big\langle \widetilde{\Phi}^*(0,\xi')\Big(\pm\Big(\frac{\partial N_l}{\partial \nu_\xi}(\pm 0,\xi') + A_{11}N_{l-1}(\pm 0,\xi')n_1\Big)\Big)\Big\rangle. \tag{2.3.59}$$

Equation (2.3.59) follows from Theorem 2.3.5.

Theorem 2.3.5. *Let $A_{ij}(\xi)$ satisfy the conditions stated at the beginning of the section for $\xi_1 < 0$ and $\xi_1 > 0$, let the s-dimensional vectors $F_j(\xi) \in L^2(I\!R \times \beta)$, $j = 0,\dots,s$ satisfy estimates (2.3.32), let s-dimensional vectors $u_0(\xi') \in H^{1/2}(\{0\} \times \beta)$, $u_1(\xi') \in L^2(\beta)$, and let the d-dimensional constant vector be given by*

$$\bar{\bar{h}} = -\langle \widetilde{\Phi}^*(0,\xi')\widetilde{\Phi}(0,\xi')\rangle^{-1}\Big(\frac{1}{mes\ \beta}\int_{I\!R\times\beta} \widetilde{\Phi}^*(\xi)F_0(\xi)\,d\xi + \langle \widetilde{\Phi}^*(0,\xi')u_1(\xi')\rangle\Big). \tag{2.3.60}$$

Then there exists a solution $u(\xi)$ to the problem

$$L_{\xi\xi} u = F_0(\xi) + \sum_{j=1}^{s} \frac{\partial F_j}{\partial \xi_j}, \quad \xi_1 > 0, \quad \xi' \in \beta, \tag{2.3.61}$$

$$\frac{\partial u}{\partial \nu_\xi} = \sum_{j=2}^{s} F_j n_j, \quad \xi' \in \partial\beta, \tag{2.3.62}$$

$$[u] = u_0(\xi'), \quad \xi_1 = 0, \tag{2.3.63}$$

$$\Big[\sum_{j=1}^{s} A_{1j}\frac{\partial u}{\partial \xi_j}\Big] = [F_1] + u_1(\xi') + \widetilde{\Phi}(0,\xi')\bar{\bar{h}}, \quad \xi_1 = 0 \tag{2.3.64}$$

such that for some rigid displacement $w(\xi) = \widetilde{\Phi}(\xi)\bar{h}$, where $\bar{h}$ is a constant $\bar{d}$-dimensional vector, inequality (2.3.36) holds, and, in addition,

$$\int_{(\sigma,\sigma+1)\times\beta} |u|\, d\xi \le c_1 e^{-c_2\sigma}, \quad \sigma < -1.$$

Here the normal to the discontinuity interface $\xi_1 = 0$ is co-directed with the $O\xi_1$-axis. The solution to problem (2.3.61)–(2.3.64) is thought of as an s-vector function $u(\xi) \in H^{1\,\mathrm{loc}}$ satisfying the following integral identity for any vector-function $\Psi(\xi) \in C^\infty(I\!R \times \bar{\beta})$ equal to zero for ξ_1 sufficiently large:

$$-\sum_{i,j=1}^{s}\int_{I\!R\times\beta}\Big(A_{ij}\frac{\partial u}{\partial \xi_j}, \frac{\partial \Psi}{\partial \xi_i}\Big)\, d\xi = \int_{I\!R\times\beta}(F_0,\Psi)\, d\xi - \sum_{j=1}^{s}\int_{I\!R\times\beta}\Big(F_j, \frac{\partial \Psi}{\partial \xi}\Big)\, d\xi$$

$$+ \int_\beta (\widetilde{\Phi}(0,\xi')\bar{\bar{h}} + u_1, \Psi)\, d\xi'.$$

The choice of the constant $\bar{h}$ is carried out as that of the constant h^{N0} in problem (2.3.30).

By analogy with the problems (2.3.1)–(2.3.4) and (2.3.48), from (2.3.51)–(2.3.55) we obtain the homogenized problem: equations (2.3.17) with coefficients h_l^N and h_l^M constant on each half-line $R_- = (-\infty,0)$ and $R_+ = (0,+\infty)$, boundary conditions (2.3.38), (2.3.39) at $x_1 = -b_2$ and $x_1 = b_1$, and the interface conditions at $x_1 = 0$,

$$[\omega] + \sum_{l=1}^{\infty}\sum_{\pm}\mu^l \bar{h}_l^{N\pm)(\mathrm{cut}\, d)}\frac{d^l\omega}{dx_1^l}\Big|_{x_1=\pm 0} +$$

$$+\sum_{l=0}^{\infty}\sum_{\pm}\mu^{l+2}\bar{h}^{M\pm 0(\mathrm{cut}\, d)}\frac{d^l\psi}{dx_1^l}\Big|_{x_1=\pm 0} = 0, \tag{2.3.65}$$

$$\mu\Big(\Big[\frac{d}{dx}(\omega^2,\dots,\omega^s)^*\Big] + \sum_{l=2}^{\infty}\sum_{\pm}\mu^l \bar{h}_l^{N\pm O(\mathrm{oth})}\frac{d^l\omega}{dx_1^l}\Big|_{x_1=\pm 0}$$

$$+\sum_{l=0}^{\infty}\mu^{l+2}\bar{h}_l^{M\pm O(\mathrm{oth})}\frac{d^l\psi}{dx_1^l}\Big|_{x_1=\pm 0} = 0, \tag{2.3.66}$$

$$\sum_{l=0}^{\infty}\sum_{\pm}\mu^{l-1}\bar{h}^{N\pm 0}\frac{d^l\omega}{dx_1^l}\Big|_{x_1=\pm 0} + \sum_{l=0}^{\infty}\sum_{\pm}\mu^{l+1}\bar{h}^{M\pm 0}\frac{d^l\psi}{dx_1^l}\Big|_{x_1=\pm 0} = 0. \tag{2.3.67}$$

Theorem 2.3.6. *The matrices* $\bar{\bar{h}}^{N\pm O(\text{cut } d)}$ *are diagonal, and* $\bar{\bar{h}}_0^{N\pm 0} = 0$. *For* $s = 3$ *we have*

$$\bar{\bar{h}}_1^{N\pm 0} = \pm \begin{pmatrix} \bar{E}_{\pm}^{(3)} & 0 & 0 & 0 \\ 0 & 0 & 0 & 0 \\ 0 & 0 & 0 & 0 \\ 0 & 0 & 0 & \bar{M}_{\pm}^{(3)} \\ 0 & 0 & 0 & 0 \\ 0 & 0 & 0 & 0 \end{pmatrix},$$

$$\bar{\bar{h}}_2^{N\pm 0} = \pm \begin{pmatrix} C_1 & 0 & 0 & 0 \\ 0 & 0 & 0 & 0 \\ 0 & 0 & 0 & 0 \\ 0 & 0 & 0 & C_2 \\ 0 & \frac{\bar{J}_{2\pm}^{(3)}}{\langle \xi_2^2 \rangle} & 0 & 0 \\ 0 & 0 & \frac{\bar{J}_{2\pm}^{(3)}}{\langle \xi_3^2 \rangle} & 0 \end{pmatrix},$$

$$\bar{\bar{h}}_3^{N\pm 0(\text{cut } d)} = \pm \begin{pmatrix} C_3 & 0 & 0 & 0 \\ 0 & -\bar{J}_{2\pm}^{(3)} & 0 & 0 \\ 0 & 0 & -\bar{J}_{3\pm}^{(3)} & 0 \\ 0 & 0 & 0 & C_4 \end{pmatrix}$$

and for $s = 2$ *we have*

$$\bar{\bar{h}}_1^{N\pm 0} = \pm \begin{pmatrix} \bar{E}_{\pm}^{(2)} & 0 \\ 0 & 0 \\ 0 & 0 \end{pmatrix}, \quad \bar{\bar{h}}_2^{N\pm 0} = \pm \begin{pmatrix} C_5 & 0 \\ 0 & 0 \\ 0 & \bar{J}_{2\pm}^{(2)} \end{pmatrix},$$

$$\bar{\bar{h}}_3^{N\pm 0(\text{cut } d)} = \pm \begin{pmatrix} C_6 & 0 \\ 0 & -\bar{J}_{2\pm}^{(2)} \end{pmatrix},$$

where $\bar{E}_{\pm}^{(s)}$, $\bar{M}_{\pm}^{(s)}$, $\bar{J}_{2\pm}^{(s)}$, $\bar{J}_{3\pm}^{(s)} > 0$ *are the constants from Theorem 1 (the sign + or - denotes the half-line* R_+ *or* R_-, *where the coefficients* $\bar{\bar{h}}_l^{N\pm 0}$ *are calculated by* $A_{ij}^{\pm}(\xi)$), C_1, C_2, C_3, C_4, C_5, C_6 *are certain constants.*

As before, we seek the f.a.s. of a homogenized problem of infinite order represented as the series (2.3.18) and obtain a recurrent chain of problems for the function ω_j: equations (2.3.19) for $(-b_2, 0)$ and $(0, b_1$ (with their specific coefficients on each half-line), boundary conditions (2.3.40) at $x_1 = -b_2$ and $x_1 = b_1$ and interface conditions at $x_1 = 0$ of the form

$$[\omega_j] = g_{j0}^1,$$

$$\left[\frac{d\omega_j^q}{dx_1}\right] = g_{j0}^{2q}, \quad q = 2, \ldots, s,$$

$$\left[\bar{E}^{(s)} \frac{d\omega_j^1}{dx_1}\right] = g_{j0}^3,$$

$$\Big[-\bar{J}_2^{(s)}\frac{d^3\omega_j^2}{dx_1^3}\Big]=g_{j0}^4,$$

$$\Big[\frac{\bar{J}_2^{(s)}}{\langle\xi_2^2\rangle}\frac{d^2\omega_j^2}{dx_1^2}\Big]=g_{j0}^5,$$

When $s=3$ we add also

$$\Big[-\bar{J}_3^{(s)}\frac{d^3\omega_j^3}{dx_1^3}\Big]=g_{j0}^6,$$

$$\Big[\bar{M}^{(s)}\frac{d^4\omega_j}{dx_1}\Big]=g_{j0}^7,$$

$$\Big[\frac{\bar{J}_3^{(s)}}{\langle\xi_3^2\rangle}\frac{d^2\omega_j^3}{dx_1^2}\Big]=g_{j0}^8. \tag{2.3.68}$$

Here we take the values $\bar{E}^{(s)}$, $\bar{M}^{(s)}$, $\bar{J}_2^{(s)}$ and $\bar{J}_3^{(s)}$ for R_+ at the point $x_1=+0$ and for R_- at the point $x_1=-0$, g_{j0}^k depend on ω_{j_1} for $j_1<j$.

2.3.6 Homogenized problem of zero order

The leading term of the homogenized problem of infinite order as $\mu\to 0$ will be called a *homogenized problem of zero order*.

From (2.3.17) we obtain

$$\Re_\mu v=\bar{F}\psi, \tag{2.3.69}$$

where

$$\Re_\mu=\begin{pmatrix}\bar{E}^{(3)}\partial^2 & 0 & 0 & 0\\ 0 & -\mu^2\bar{J}_2^{(3)}\partial^4 & 0 & 0\\ 0 & 0 & -\mu^2\bar{J}_3^{(3)}\partial^4 & 0\\ 0 & 0 & 0 & \bar{M}^{(3)}\partial^2\end{pmatrix}$$

for $s=3$ and for $s=2$;

$$\Re_\mu=\begin{pmatrix}\bar{E}^{(2)}\partial^2 & 0\\ 0 & -\mu^2\bar{J}_2^{(2)}\partial^4\end{pmatrix},\qquad \partial=\frac{d}{dx_1}$$

v is the unknown vector function in (2.3.69).

From (2.3.39) we obtain

$$V=0,\quad x_1=0,\ b, \tag{2.3.70}$$

where for $s=3$ we define

$$V=(v^1,v^2,v^3,v^4,\mu\partial v^2,\mu\partial v^3)^*,$$

and for $s=2$ we define

$$V=(v^1,v^2,\mu\partial v^2)^*.$$

and v^q are the components of the vector v.

Thus, the homogenized problem of zero order for (2.3.1)–(2.3.4) is problem (2.3.69),(2.3.70).

The boundary conditions of second kind give (2.3.46), whence follows

$$-\Lambda_\mu v = 0, \tag{2.3.71}$$

where

$$\Lambda_\mu = \begin{pmatrix} \bar{E}^{(3)}\partial & 0 & 0 & 0 \\ 0 & -\mu^2 \bar{J}_2^{(3)}\partial^3 & 0 & 0 \\ 0 & 0 & -\mu^2 \bar{J}_3^{(3)}\partial^3 & 0 \\ 0 & 0 & 0 & \bar{M}^{(3)}\partial \\ 0 & \mu\frac{\bar{J}_2^{(3)}}{\langle\xi_2^2\rangle}\partial^2 & 0 & 0 \\ 0 & 0 & \mu\frac{\bar{J}_3^{(3)}}{\langle\xi_3^2\rangle}\partial^2 & 0 \end{pmatrix}$$

for $s = 3$ and

$$\Lambda_\mu = \begin{pmatrix} \bar{E}^{(2)}\partial & 0 \\ 0 & -\mu^2 \bar{J}_2^{(2)}\partial^3 \\ 0 & \mu\frac{\bar{J}_2^{(2)}\partial^2}{\langle\xi_2^2\rangle} \end{pmatrix}$$

for $s = 2$.

Thus, the homogenized problem of zero order for (2.3.1)–(2.3.4), (2.3.3′) has the representation (2.3.69), (2.3.70) if $x_1 = b$ and (2.3.71) if $x_1 = 0$.

The interface conditions at $x_1 = 0$ yield the homogenized interface conditions of zero order

$$[V] = 0, \qquad x_1 = 0, \Lambda_\mu^+ v\big|_{x_1=+0} - \Lambda_\mu^- v\big|_{x_1=-0} = 0, \tag{2.3.72}$$

where Λ_μ^+ and Λ_μ^- are obtained by replacing $\bar{E}^{(s)}$, $\bar{J}_r^{(s)}$, $\bar{M}^{(s)}$ with $\bar{E}_+^{(s)}$, $\bar{J}_{r+}^{(s)}$, $\bar{M}_+^{(s)}$ or $\bar{E}_-^{(s)}$, $\bar{J}_{r-}^{(s)}$, $\bar{M}_-^{(s)}$, respectively. We define the operators $\Re_\mu^+$ and $\Re_\mu^-$ in a similar way. Then problem (2.3.48) gives the homogenized problem of zero order

$$\begin{gathered} \Re_\mu^+ v = \bar{F}\psi, \qquad x_1 \in (0, b_1), \\ \Re_\mu^- v = \bar{F}\psi, \qquad x_1 \in (-b_2, 0), \\ V = 0 \quad \text{when} \quad x_1 = b_1, \quad x_1 - b_2, \\ [V] = 0 \quad \text{when} \quad x_1 = 0, \Lambda_\mu^+ v\big|_{x_1=+0} - \Lambda_\mu^- v\big|_{x_1=-0} = 0. \end{gathered} \tag{2.3.73}$$

To calculate the coefficients of the operator $\Re_\mu$, we must solve the cell problems

$$\sum_{m,j=1}^{s} \frac{\partial}{\partial\xi_m}\Big(A_{mj}(\xi)\frac{\partial}{\partial\xi_j}(N_1^{(1)} + \mathbf{e}\xi_1)\Big) = 0, \quad \xi' \in \beta,$$

$$\frac{\partial}{\partial\nu_\xi}(N_1^{(1)} + \mathbf{e}\xi_1) = 0, \quad \xi' \in \partial\beta, \tag{2.3.74}$$

$$\sum_{m,j=1}^{s} \frac{\partial}{\partial \xi_m}\Big(A_{mj}(\xi)\frac{\partial}{\partial \xi_j}(N_1^{(4)} + \eta\xi_1)\Big) = 0, \quad \xi' \in \beta,$$

$$\frac{\partial}{\partial \nu_\xi}(N_1^{(4)} + \eta\xi_1) = 0, \quad \xi' \in \partial\beta, \tag{2.3.75}$$

where $N_1^{(1)}$ and $N_1^{(4)}$ are 1-periodic in ξ_1 s-vector functions with components n_1^{k1} and n_1^{k4}, $\mathbf{e} = (1,0,0)^*$ for $s=3$, $\mathbf{e} = (1,0)^*$ for $s=2$ and $\eta = (0, -\xi_3, \xi_2)^*$. On solving these equations, we obtain the longitudinal stiffness

$$\bar{E}^{(s)} = \sum_{j,k=1}^{s} \Big\langle a_{1j}^{1k}\frac{\partial}{\partial \xi_j}(n_1^{k1} + \delta_{k1}\xi_1)\Big\rangle \tag{2.3.76}$$

and the torsional stiffness

$$\bar{M}^{(s)} = \sum_{j,k=1}^{s} \Big\langle a_{1j}^{1k}\frac{\partial}{\partial \xi_j}(n_1^{k4} + \eta^k\xi_1)\Big\rangle. \tag{2.3.77}$$

Further, if $N_2^{(2)}, \dots, N_2^{(s)}$ are s-dimensional vector-functions one-periodic in ξ_1 with components $n_2^{k2}, \dots, n_2^{ks}$, which are solutions to the problem

$$\sum_{m,j=1}^{s} \frac{\partial}{\partial \xi_m}\Big(A_{mj}(\xi)\frac{\partial}{\partial \xi_j}(N_2^{(q)} - A_{m1}(\xi)\mathbf{e}\xi_q)\Big) = 0, \quad q = 2,\dots,s, \quad \xi' \in \beta,$$

$$\sum_{m,j=1}^{s} n_m\Big(A_{mj}(\xi)\frac{\partial}{\partial \xi_j}N_2^{(q)} - A_{m1}(\xi)\mathbf{e}\xi_q)\Big) = 0, \quad \xi' \in \partial\beta, \tag{2.3.78}$$

then the bending stiffness is calculated by means of the formula

$$\bar{J}_q^{(s)} = -\Big\langle\Big(\sum_{j,k=1}^{s} a_{1j}^{1k}\frac{\partial}{\partial \xi_j}n_2^{kq} - a_{11}^{11}\xi_q\Big)\xi_q\Big\rangle. \tag{2.3.79}$$

Indeed,

$$\begin{aligned} -\bar{J}_q^{(s)} &= h_4^{qq} = \Big\langle a_{1j}^{qk}\frac{\partial n_3^{kq}}{\partial \xi_j} + a_{11}^{qk}n_2^{kq}\Big\rangle = \Big\langle a_{qj}^{1k}\frac{\partial n_3^{kq}}{\partial \xi_j} + a_{q1}^{1k}n_2^{kq}\Big\rangle \\ &= \Big\langle\Big(a_{mj}^{1k}\frac{\partial n_3^{kq}}{\partial \xi_j} + a_{m1}^{1k}n_2^{kq}\Big)\frac{\partial \xi_q}{\partial \xi_m}\Big\rangle = \Big\langle\frac{\partial}{\partial \xi_m}\Big(a_{mj}^{1k}\frac{\partial n_3^{kq}}{\partial \xi_j} + a_{m1}^{1k}n_2^{kq}\Big)\xi_q\Big\rangle \\ &= \Big\langle\Big(a_{1j}^{1k}\frac{\partial n_2^{kq}}{\partial \xi_j} + a_{11}^{1k}n_1^{kq}\Big)\xi_q\Big\rangle = \Big\langle\Big(a_{1j}^{1k}\frac{\partial n_2^{kq}}{\partial \xi_j} + a_{11}^{11}\xi_q\Big)\xi_q\Big\rangle. \end{aligned}$$

Here n_l^{kq} are the components of the matrix N_l.

Consider the isotropic case, in which

$$a_{ij}^{kl}(\xi) = (\delta_{ij}\delta_{kl} + \delta_{il}\delta_{jk})M(\xi) + \lambda(\xi)\delta_{ik}\delta_{jl},$$

where $\lambda(\xi)$, $M(\xi)$ are scalar 1-periodic in ξ_1 functions (Lamé coefficients).

If the coefficients λ and M are constant, then the following theorem holds true:

Theorem 2.3.7. *The following equalities hold:*

$$\bar{E}^{(3)} = \frac{M(3\lambda + 2M)}{\lambda + M}, \tag{2.3.80}$$

$$\bar{J}_r^{(3)} = \bar{E}^{(3)} \langle \xi_r^2 \rangle, \quad r = 2, 3, \tag{2.3.81}$$

$$\bar{M}^{(3)} = M\Big(1 - \Big\langle \Big(\frac{\partial Y}{\partial \xi_2}\Big)^2 + \Big(\frac{\partial Y}{\partial \xi_3}\Big)^2 \Big\rangle\Big), \tag{2.3.82}$$

where $Y(\xi')$ is the solution to the Laplace equation $\Delta Y = 0$ in β with the boundary condition

$$\frac{\partial Y}{\partial n} - \xi_3 n_2 a + \xi_2 n_3 a = 0$$

on $\partial\beta$, and

$$\bar{E}^{(2)} = \frac{(\lambda + 2M)^2 - \lambda^2}{\lambda + 2M}, \tag{2.3.80$'$}$$

$$\bar{J}_2^{(2)} = \bar{E}^{(2)} \langle \xi_2^2 \rangle. \tag{2.3.81$'$}$$

The proof of Theorem 2.3.7 will be given below.

Remark 2.3.3. Equations (2.3.69) with constant coefficients are well known (e.g., see [1],[189]), but these equations with variable coefficients were apparently first obtained in [83],[84]. Homogenized problems of infinite order specify these equations.

2.3.7 The justification of the asymptotic expansion.

First we prove Lemmas 2.3.2-2.3.5 justifying theorem 2.3.1.

Lemma 2.3.2. *Let condition A hold and let the matrices $N_l(\xi)$ satisfy the relations $\langle \Phi^* N_l \rangle = 0$, $l > 0$. Then the elements n_l^{kp} of the matrices N_l satisfy the following relations:*

$$n_l^{kp}(\xi) = (-1)^{\delta_{\mathcal{A}k} + \tilde{\delta}_{\mathcal{A}p}} n_l^{kp}(S_{\mathcal{A}}\xi), \quad \widetilde{\delta}_{\mathcal{A}p} = \delta_{\mathcal{A}p} + \delta_{p4}, \quad \mathcal{A} = 2, \dots, s, \tag{2.3.83}$$

where $k = 1, 2, 3$ and $p = 1, 2, 3, 4$ for $s = 3$ and $k = 1, 2$ and $p = 1, 2$ for $s = 2$; moreover, the matrices h_l^N are diagonal.

If the sum $l + \delta_{1k} + \delta_{1p}$ is odd and condition B is satisfied, then the relations $n_l^{kp} = 0$ and

$$\varphi^{mk} \left(a_{ij}^{mq} \frac{\partial n_l^{qp}}{\partial \xi_j} + a_{i1}^{mq} n_{l-1}^{qp} \right) = 0, \quad i = 2, 3, \tag{2.3.84}$$

hold (here and below summation over repeated indices from 1 to s is implied).

If the sum $\delta_{2i}+\tilde{\delta}_{2k}+\tilde{\delta}_{2p}$ or the sum $\delta_{3i}+\tilde{\delta}_{3k}+\tilde{\delta}_{3p}$ is odd, then for the averages we have

$$\left\langle \varphi^{mk}\left(a_{ij}^{mq}\frac{\partial n_l^{qp}}{\partial \xi_j}+a_{i1}^{mq}n_{l-1}^{qp}\right)\right\rangle=0,\quad i=2,3. \tag{2.3.85}$$

The proof of Lemma 2.3.2 is similar to that of [16], Theorems 6.3.3 and 6.3.4: applying the identity

$$a_{ij}^{kl}(S_{\mathcal{A}}\xi)=(-1)^{\delta_{\mathcal{A}i}+\delta_{\mathcal{A}j}+\delta_{\mathcal{A}k}+\delta_{\mathcal{A}l}}a_{ij}^{kl}(\xi),\quad \mathcal{A}=2,3,$$

and substituting into (2.3.11)-(2.3.14) (by induction on l), we see that $(-1)^{\delta_{\mathcal{A}k}+\tilde{\delta}_{\mathcal{A}p}}$ $n_l^{kp}(S_{\mathcal{A}}\xi)$ is a solution. Thus,

$$\frac{\partial}{\partial \xi_i}\left(a_{ij}^{kp}(\xi)\frac{\partial n_l^{pr}(S_{\mathcal{A}}\xi)}{\partial \xi_j}(-1)^{\delta_{\mathcal{A}p}+\tilde{\delta}_{\mathcal{A}r}}\right)=(-1)^{\delta_{\mathcal{A}i}+\delta_{\mathcal{A}j}+\delta_{\mathcal{A}p}+\tilde{\delta}_{\mathcal{A}r}+\delta_{\mathcal{A}i}+\delta_{\mathcal{A}j}+\delta_{\mathcal{A}k}+\delta_{\mathcal{A}p}}$$

$$\times\frac{\partial}{\partial \eta_i}\left(a_{ij}^{kp}(\eta)\frac{\partial n_l^{pr}(\eta)}{\partial \eta_j}\right)\Bigg|_{\eta=S_{\mathcal{A}}\xi}=(-1)^{\delta_{\mathcal{A}k}+\tilde{\delta}_{\mathcal{A}r}}\frac{\partial}{\partial \eta_i}\left(a_{ij}^{kp}(\eta)\frac{\partial n_l^{pr}}{\partial \eta_j}\right)\Bigg|_{\eta=S_{\mathcal{A}}\xi},$$

$$\frac{\partial}{\partial \xi_i}\left(a_{i1}^{kp}(\xi)n_{l-1}^{pr}(S_{\mathcal{A}}\xi)(-1)^{\delta_{\mathcal{A}p}+\tilde{\delta}_{\mathcal{A}r}}\right)=(-1)^{\delta_{\mathcal{A}i}+\delta_{\mathcal{A}k}+\delta_{\mathcal{A}p}+\delta_{\mathcal{A}p}+\tilde{\delta}_{\mathcal{A}r}+\delta_{\mathcal{A}i}}$$

$$\times\frac{\partial}{\partial \eta_i}\left(a_{i1}^{kp}(\eta)n_{l-1}^{pr}(\eta)\right)\big|_{\eta=S_{\mathcal{A}}\xi}=(-1)^{\delta_{\mathcal{A}k}+\tilde{\delta}_{\mathcal{A}r}}\frac{\partial}{\partial \eta_i}\left(a_{i1}^{kp}(\eta)n_{l-1}^{pr}(\eta)\right)\Big|_{\eta=S_{\mathcal{A}}\xi}, \tag{2.3.86}$$

$$a_{1j}^{kp}(\xi)\frac{\partial}{\partial \xi_j}n_{l-1}^{pr}(S_{\mathcal{A}}\xi)(-1)^{\delta_{\mathcal{A}p}+\tilde{\delta}_{\mathcal{A}r}}=(-1)^{\delta_{\mathcal{A}k}+\delta_{\mathcal{A}p}+\delta_{\mathcal{A}j}+\delta_{\mathcal{A}j}+\delta_{\mathcal{A}p}+\tilde{\delta}_{\mathcal{A}r}}$$

$$\times\left(a_{1j}^{kp}(\eta)\frac{\partial}{\partial \eta_j}n_{l-1}^{pr}(\eta)\right)\Bigg|_{\eta=S_{\mathcal{A}}\xi}=(-1)^{\delta_{\mathcal{A}k}+\tilde{\delta}_{\mathcal{A}r}}\left(a_{1j}^{kp}(\eta)\frac{\partial}{\partial \eta_j}n_{l-1}^{pr}(\eta)\right)\Bigg|_{\eta=S_{\mathcal{A}}\xi}, \tag{2.3.87}$$

$$a_{11}^{kp}(\xi)n_{l-2}^{pr}(S_{\mathcal{A}}\xi)(-1)^{\delta_{\mathcal{A}p}+\tilde{\delta}_{\mathcal{A}r}}=(-1)^{\delta_{\mathcal{A}k}+\tilde{\delta}_{\mathcal{A}r}}a_{11}^{kp}(\eta)n_{l-2}^{pr}(\eta)\big|_{\eta=S_{\mathcal{A}}\xi}. \tag{2.3.88}$$

Finally,

$$\left\langle \varphi^{mq}(\xi)\left(a_{1j}^{mq}\frac{\partial n_{l-1}^{pr}(S_{\mathcal{A}}\xi)}{\partial \xi_j}+a_{11}^{mp}n_{l-2}^{pr}\right)(-1)^{\delta_{\mathcal{A}p}+\tilde{\delta}_{\mathcal{A}r}}\right\rangle$$

$$=\left\langle \varphi^{mq}(S_{\mathcal{A}}\xi)(-1)^{\delta_{\mathcal{A}m}+\tilde{\delta}_{\mathcal{A}q}}(-1)^{\delta_{\mathcal{A}m}+\tilde{\delta}_{\mathcal{A}r}}\left(a_{1j}^{mp}(\eta)\frac{\partial n_{l-1}^{pr}}{\partial \eta_j}+a_{11}^{mp}(\eta)n_{l-2}^{pr}(\eta)\right)\Bigg|_{\eta=S_{\mathcal{A}}\xi}\right\rangle$$

$$=(-1)^{\tilde{\delta}_{\mathcal{A}q}+\tilde{\delta}_{\mathcal{A}r}}h_l^{qr}, \tag{2.3.89}$$

$$\varphi^{kq}(\xi)\left\langle \varphi^{mq}\left(a_{1j}^{mp}\frac{\partial n_{l-1}^{pr}(S_{\mathcal{A}}\xi)}{\partial \xi_j}(-1)^{\delta_{\mathcal{A}p}+\tilde{\delta}_{\mathcal{A}r}}+a_{11}^{mp}n_{l-2}^{pr}(S_{\mathcal{A}}\xi)(-1)^{\delta_{\mathcal{A}p}+\tilde{\delta}_{\mathcal{A}r}}\right)\right\rangle$$

$$=(-1)^{\delta_{\mathcal{A}k}+\tilde{\delta}_{\mathcal{A}q}}\varphi^{kq}(S_{\mathcal{A}}\xi)(-1)^{\tilde{\delta}_{\mathcal{A}q}+\tilde{\delta}_{\mathcal{A}r}}h_l^{qr}=(-1)^{\delta_{\mathcal{A}k}+\tilde{\tilde{\delta}}_{\mathcal{A}r}}\varphi(S_{\mathcal{A}}\xi)h_l^{qr}, \tag{2.3.90}$$

where h_l^{qr} are entries of the matrices h_l^N and φ^{mq} are entries of the matrix Φ.

By induction on l, from relations (2.3.86)–(2.3.90) we can readily derive that if N_l are solutions to problems (2.3.11)–(2.3.14), then matrices with entries $(-1)^{\delta_{\mathcal{A}k}+\tilde{\delta}_{\mathcal{A}p}} n_l^{kp}(S_{\mathcal{A}}\xi)$ are also solutions and for $l > 0$ both matrices satisfy the relations

$$\langle \Phi^* N_l \rangle = 0, \quad \langle \varphi^{mk}(\xi)(-1)^{\delta_{\mathcal{A}m}+\tilde{\delta}_{\mathcal{A}r}} n_l(S_{\mathcal{A}}\xi) \rangle = (-1)^{\tilde{\delta}_{\mathcal{A}k}+\tilde{\delta}_{\mathcal{A}r}} \langle \varphi^{mk}(\eta) n_l(\eta) \rangle = 0.$$

Since a solution of problem (2.3.11)-(2.3.14) orthogonal to rigid-body displacements is unique, we have (2.3.83). This and (2.3.89) imply $h_l^{qr} = (-1)^{\tilde{\delta}_{\mathcal{A}q}+\tilde{\delta}_{\mathcal{A}r}}$ h_l^{qr}, i.e., the matrices h_l^N are diagonal.

By analogy with (2.3.89), we have

$$\left\langle \varphi^{mk} \left(a_{ij}^{mq} \frac{\partial n_l^{qp}}{\partial \xi_j} + a_{i1}^{mq} n_{l-1}^{qp} \right) \right\rangle = (-1)^{\delta_{\mathcal{A}i}+\tilde{\delta}_{\mathcal{A}k}+\tilde{\delta}_{\mathcal{A}p}} \left\langle \varphi^{mk} \left(a_{ij}^{mq} \frac{\partial n_l^{qp}}{\partial \xi_j} + a_{i1}^{mq} n_{l-1}^{qp} \right) \right\rangle;$$

this implies (2.3.85).

Similarly, by induction on l, we can prove that if condition B is satisfied, then the matrix $(-1)^{l+\delta_{1p}+\delta_{1r}} n_l^{pr}(\xi)$ is a solution to problem (2.3.11)–(2.3.14), and this implies relations (2.3.84), (2.3.85) and $n_l^{pr}(\xi) = (-1)^{l+\delta_{1p}+\delta_{1r}} n_l^{pr}(\xi)$. Hence, $h_l^{qq} = (-1)^l h_l^{qq}$, $h_l^{qq} = 0$ for all odd l. This completes the proof of the lemma. □

Lemma 2.3.3. *Let condition A be satisfied. Then $h_2^{11} > 0$, $h_2^{qq} = 0$, $q = 2, \ldots, s$, and $h_2^{44} > 0$ for $s = 3$; moreover, $h_3^{qq} = 0$, $q = 2, \ldots, s$.*

Proof.

Let us obtain a new representation for the matrix

$$h_2^N = \left\langle \Phi^* \left(\sum_{j=1}^{s} A_{1j} \frac{\partial N_1}{\partial \xi_j} + A_{11} \Phi \right) \right\rangle.$$

We have

$$h_2^N = \left\langle \Phi^* \sum_{j=1}^{s} A_{1j} \left(\frac{\partial N_1}{\partial \xi_j} + \frac{\partial}{\partial \xi_j} (\xi_1 \Phi) \right) \right\rangle = \left\langle \Phi^* \sum_{j=1}^{s} A_{1j} \frac{\partial}{\partial \xi_j} (N_1 + \xi_1 \Phi) \right\rangle.$$

Since

$$\sum_{m=1}^{s} \frac{\partial}{\partial \xi_m} (\Phi^* \xi_1) A_{mj} = \sum_{m=1}^{s} \left(A_{jm} \frac{\partial (\Phi \xi_1)}{\partial \xi_m} \right)^* = (A_{j1} \Phi)^* = \Phi^* A_{1j},$$

we obtain

$$h_2^N = \sum_{m,j=1}^{s} \left\langle \frac{\partial}{\partial \xi_m} (\Phi^* \xi_1) A_{mj} \frac{\partial}{\partial \xi_j} (N_1 + \xi_1 \Phi) \right\rangle. \tag{2.3.91}$$

On the other hand, it follows from the integral identity for problem (2.3.11)–(2.3.14) with the test matrix-valued function N_1 that

$$0 = \left\langle \sum_{m,j=1}^{s} \frac{\partial}{\partial \xi_m} N_1^* A_{mj} \frac{\partial}{\partial \xi_j} (N_1 + \xi_1 \Phi) \right\rangle. \tag{2.3.92}$$

Summing (2.3.91) and (2.3.92), we obtain the following representation for h_2^N:

$$h_2^N = \Big\langle \sum_{m,j=1}^{s} \frac{\partial}{\partial \xi_m}(N_1 + \xi_1 \Phi)^* A_{mj} \frac{\partial}{\partial \xi_j}(N_1 + \xi_1 \Phi)\Big\rangle.$$

Now let us prove that the elements h_2^{kl} of the matrix h_2^N satisfy the relations $h_2^{11} > 0$ and $h_2^{22} = 0$ and if $s = 3$, then we also have $h_2^{33} = 0$ and $h_2^{44} > 0$. Denote the matrix elements by means of a pair of superscripts. Then we have

$$\begin{aligned} h_2^{ll} &= \Big\langle \frac{\partial}{\partial \xi_m}(N_1 + \xi_1 \Phi)^{kl} a_{mj}^{kq} \frac{\partial}{\partial \xi_j}(N_1 + \xi_1 \Phi)^{ql} \Big\rangle = \Big\langle \frac{1}{2}\Big(\frac{\partial}{\partial \xi_m}(N_1 + \xi_1 \Phi)^{kl} \\ &\quad + \frac{\partial}{\partial \xi_k}(N_1 + \xi_1 \Phi)^{ml}\Big) \\ &\quad \times a_{mj}^{kq} \frac{1}{2}\Big(\frac{\partial}{\partial \xi_j}(N_1 + \xi_1 \Phi)^{ql} + \frac{\partial}{\partial \xi_q}(N_1 + \xi_1 \Phi)^{jl}\Big)\Big\rangle \\ &\geq \frac{\kappa}{4}\Big\langle \sum_{j,q=1}^{s} \Big(\frac{\partial}{\partial \xi_j}(N_1 + \xi_1 \Phi)^{ql} + \frac{\partial}{\partial \xi_q}(N_1 + \xi_1 \Phi)^{jl}\Big)^2 \Big\rangle. \end{aligned}$$

Moreover,

$$\begin{aligned} h_2^{11} &\geq \kappa \Big\langle \Big(\frac{\partial}{\partial \xi_1}(N_1 + \xi_1 \Phi)^{11}\Big)^2 \Big\rangle \\ &\geq \kappa \Big\langle \frac{\partial}{\partial \xi_1}(N_1 + \xi_1 \Phi)^{11}\Big\rangle^2 = \kappa > 0, \end{aligned}$$

because $\langle (\partial/\partial \xi_1) N_1 \rangle = 0$ and N_1 is 1-periodic with respect to ξ_1. Moreover,

$$\begin{aligned} h_2^{44} &\geq \frac{\kappa}{4} \sum_{r=2}^{s} \Big\langle \Big(\frac{\partial}{\partial \xi_1}(N_1 + \xi_1 \Phi)^{r4} + \frac{\partial}{\partial \xi_r}(N_1 + \xi_1 \Phi)^{14}\Big)^2 \Big\rangle + \kappa \Big\langle \Big(\frac{\partial N_1^{14}}{\partial \xi_1}\Big)^2 \Big\rangle \\ &= \frac{\kappa}{4} \sum_{r=2}^{s} \Big\langle \Big(\frac{\partial}{\partial \xi_1} n_1^{r4} + \varphi^{r4} + \frac{\partial}{\partial \xi_r} n_1^{14}\Big)^2 \Big\rangle + \kappa \Big\langle \Big(\frac{\partial n_1^{14}}{\partial \xi_1}\Big)^2 \Big\rangle \geq 0. \end{aligned}$$

Let us prove that $h_2^{44} \neq 0$. Assume the contrary, i.e., let $h_2^{44} = 0$. Then the last estimate implies $\partial n_1^{14}/\partial \xi_1 = 0$, and n_1^{14} does not depend on ξ_1, moreover,

$$\Big\langle \Big(\frac{\partial}{\partial \xi_1} n_1^{r4} + \varphi^{r4} + \frac{\partial}{\partial \xi_r} n_1^{14}\Big)^2 \Big\rangle = 0.$$

Hence,

$$\Big\langle \Big(\frac{\partial}{\partial \xi_1} n_1^{r4}\Big)^2 \Big\rangle + 2\Big\langle \frac{\partial n_1^{r4}}{\partial \xi_1} \varphi^{r4} \Big\rangle + 2\Big\langle \frac{\partial n_1^{r4}}{\partial \xi_1} \frac{\partial n_1^{14}}{\partial \xi_r} \Big\rangle + \Big\langle \Big(\varphi^{r4} + \frac{\partial n_1^{14}}{\partial \xi_r}\Big)^2 \Big\rangle = 0.$$

The second integral and the third integral are equal to zero, because n_1^{14} does not depend on ξ_1; hence, n_1^{14} satisfies the system of equations

$$-a\xi_3 + \frac{\partial n_1^{14}}{\partial \xi_2} = 0, \quad a\xi_2 + \frac{\partial n_1^{14}}{\partial \xi_3} = 0,$$

whence

$$\frac{\partial^2 n_1^{14}}{\partial\xi_3\partial\xi_2} = a, \qquad \frac{\partial^2 n_1^{14}}{\partial\xi_2\partial\xi_3} = -a.$$

This contradiction proves that $h_2^{44} > 0$.

The constants h_2^{qq} vanish for $q = 2, \ldots, s$, because the direct substitution into (2.3.11)–(2.3.13) for $l = 2$ proves that the second (the third, …, the s–th) column, $(N_2^{(2)}, \ldots, N_2^{(s)})$, of the matrix N_2 satisfies $N_2^q = -(\xi_q, 0, \ldots, 0)^*$; hence, the corresponding columns of the matrix $N_2 + \xi_1\Phi$ represent the rigid-body displacements.

Now let us prove that $h_3^{qq} = 0$ for $q = 2, \ldots, s$. We have

$$h_3^{22} = \left\langle a_{1j}^{2p}\frac{\partial n_2^{p2}}{\partial\xi_j} + a_{11}^{2p}n_1^{p2}\right\rangle = \left\langle a_{2j}^{1p}\frac{\partial n_2^{p2}}{\partial\xi_j} + a_{21}^{1p}n_1^{p2}\right\rangle = \left\langle\left(a_{mj}^{1p}\frac{\partial n_2^{p2}}{\partial\xi_j} + a_{m1}^{1p}n_1^{p2}\right)\frac{\partial\xi_2}{\partial\xi_m}\right\rangle.$$

Applying the integral identity for problem (2.3.11)–(2.3.14) with $l = 2$, we transform the last integral to the following form ($h_2^{12} = 0$):

$$-\left\langle\frac{\partial}{\partial\xi_m}\left(a_{mj}^{1p}\frac{\partial n_2^{p2}}{\partial\xi_j} + a_{m1}^{1p}n_1^{p2}\right)\xi_2\right\rangle = -\left\langle\left(h_2^{12} - a_{1j}^{1p}\frac{\partial n_1^{p2}}{\partial\xi_j} - a_{11}^{1p}n_0^{p2}\right)\xi_2\right\rangle$$

$$= \left\langle\left(a_{1j}^{1p}\frac{\partial n_1^{p2}}{\partial\xi_j} + a_{11}^{1p}n_0^{p2}\right)\xi_2\right\rangle = \left\langle\left(a_{1j}^{1p}\frac{\partial}{\partial\xi_j}(n_1^{p2} + \xi_1\delta_{p2})\right)\xi_2\right\rangle = 0,$$

since the second column of the matrix $N_1 + \xi_1\Phi$ is the rigid-body displacement. For the case $s = 3$, we similarly obtain $h_3^{33} = 0$. This proves the lemma. □

Let the elasticity operator

$$-\sum_{i,j=1}^{s}\frac{\partial}{\partial x_i}\left(A_{ij}\left(\frac{x}{\mu}\right)\frac{\partial u}{\partial x_j}\right), \quad x \in U_\mu, \tag{2.3.93}$$

and the derivation with respect to the co-normal

$$-\frac{\partial u}{\partial\nu} = -\sum_{i,j=1}^{s} n_i A_{ij}\left(\frac{x}{\mu}\right)\frac{\partial u}{\partial x_j}, \quad x \in \partial U_\mu, \tag{2.3.94}$$

be given and let u be a vector function from $H^1_{b-per}(U_\mu)$; this space is the completion (by the $H^1(C_\mu)$–norm) of the space of b-periodic in x_1 differentiable in $\bar{U}_\mu$ vector-valued functions. Assume that $\int_{C_\mu}\Phi^* u\,dx = 0$.

Lemma 2.3.4. *The function u satisfies the following inequality (for sufficiently small μ):*

$$\int_{C_\mu}\left(A_{ij}\frac{\partial u}{\partial x_j}, \frac{\partial u}{\partial x_i}\right)dx \geq c(\mu^4\|\nabla u\|^2_{L^2(C_\mu)} + \mu^2\|u\|^2_{L^2(C_\mu)}), \quad c > 0. \tag{2.3.95}$$

Proof.

Let us perform the following changes of variables in formulas (2.3.93) and (2.3.94):

1) $x_1 = \xi_1,\ x_2 = \xi_2\mu, \dots, x_s = \xi_s\mu$,

2) $u_1 = w_1,\ u_2 = w_2/\mu, \dots, u_s = w_s/\mu$,

3) multiply the components of the vector (2.3.93), from the second to the $s-$th one, and the first component of the vector (2.3.94) by μ^{-1} and multiply the components of the vector (2.3.94) from the second to the $s-$th one, by μ^{-2}.

Then formulas (2.3.93) and (2.3.94) take the form

$$-\sum_{i,j=1}^{s} \frac{\partial}{\partial \xi_i}\Big(B_{ij}(\xi)\frac{\partial w}{\partial \xi_j}\Big), \quad \xi \in Q_0 = \{\xi_1 \in (0,b),\ \xi' \in \beta\}, \tag{2.3.96}$$

$$-\sum_{i,j=1}^{s} n_i B_{ij}(\xi)\frac{\partial w}{\partial \xi_j}, \quad \xi' \in \partial\beta, \tag{2.3.97}$$

where the entries b_{ij}^{kl} of the matrices B_{ij} are

$$b_{ij}^{kl}(\xi) = a_{ij}^{kl}\Big(\frac{\xi_1}{\mu}, \xi'\Big)\Big(\frac{1}{\mu}\Big)^{\delta_{i2}+\dots+\delta_{is}+\delta_{j2}+\dots+\delta_{js}+\delta_{k2}+\dots+\delta_{ks}+\delta_{l2}+\dots+\delta_{ls}}.$$

Clearly,

$$b_{ij}^{kl} = b_{kj}^{il} = b_{ji}^{lk}$$

and for any symmetric matrix with elements η_i^k ($\eta_i^k = \eta_k^i$) the following inequality holds:

$$\sum_{i,j,k,l=1}^{s} b_{ij}^{kl}\eta_i^k\eta_j^l =$$

$$= \sum_{i,j,k,l=1}^{s} a_{ij}^{kl}\eta_j^l\mu^{-(\delta_{j2}+\dots+\delta_{js}+\delta_{l2}+\dots+\delta_{ls})}\eta_i^k\mu^{-(\delta_{i2}+\dots+\delta_{is}+\delta_{k2}+\dots+\delta_{ks})}$$

$$\geq \kappa\sum_{j,l=1}^{s}\big(\eta_j^l\mu^{-(\delta_{j2}+\dots+\delta_{js}+\delta_{l2}+\dots+\delta_{ls})}\big)^2 \geq \kappa\sum_{j,l=1}^{s}(\eta_j^l)^2.$$

Moreover, we have

$$\int_{Q_0}\Phi^* w\, d\xi = 0. \tag{2.3.98}$$

The following inequality holds for any function from $H^1(Q_0)$ that satisfies relation (2.3.98) (see [55])

$$\sum_{i,j=1}^{s}\int_{Q_0}\Big(B_{ij}\frac{\partial w}{\partial \xi_j}, \frac{\partial w}{\partial \xi_i}\Big)\,d\xi \geq \kappa c\|w\|^2_{W_2^1(Q_0)},$$

$c > 0$, where c is the constant of Korn's inequality for Q_0. Since we have

$$\sum_{i,j=1}^{s}\int_{Q_0}\Big(B_{ij}\frac{\partial w}{\partial \xi_j}, \frac{\partial w}{\partial \xi_i}\Big)\,d\xi = \mu^{-(s-1)}\sum_{i,j,k,l=1}^{s}\int_{C_\mu} a_{ij}^{kl}e_j^l(u)e_i^k(u)\,dx,$$

where

$$e_j^l(u) = \frac{1}{2}\Big(\frac{\partial u^j}{\partial x_l} + \frac{\partial u^l}{\partial x_j}\Big),$$

and since

$$\|w\|^2_{H^1(Q_0)} \geq \mu^{-s+5}\|\nabla u\|^2_{L^2(C_\mu)} + \mu^{-s+3}\|u\|^2_{L^2(C_\mu)},$$

we obtain the inequality of Lemma 2.3.4. □

Lemma 2.3.5. *The diagonal elements of the matrix h_4^N, from the second to the $s-$th one, are negative.*

Proof.

Let the function

$$u_\mu^{(K)}(x) = \sum_{l=0}^{K+1} \mu^l N_l\Big(\frac{x}{\mu}\Big)\frac{\partial^l V(x_1)}{\partial x_1^l}$$

be specified in the space U_μ, where $N_l(\xi)$ are the matrices constructed above, $V(x_1) \in C^\infty(I\!R)$ is a b-periodic d-dimensional vector that does not depend on μ, and K is sufficiently large. One can verify by direct substitution that

$$\sum_{i,j=1}^{s} \frac{\partial}{\partial x_i}\Big(A_{ij}\Big(\frac{x}{\mu}\Big)\frac{\partial u_\mu^{(K)}}{\partial x_j}\Big) = \Phi\sum_{l=2}^{K+1}\mu^{l-2}h_l^N\frac{\partial^l V}{\partial x_1^l} + \theta_k,$$

where

$$\theta_k = \bar\theta_k + \bar{\bar\theta}_k, \qquad \bar\theta_k = \mu^K\sum_{i=2}^{s}\frac{\partial}{\partial \xi_i}(A_{i1}N_{K+1})\frac{\partial^{K+2}V}{\partial x_1^{K+2}},$$

$$\bar{\bar\theta}_k = \mu^K\Big(\frac{\partial}{\partial \xi_1}(A_{11}N_{K+1}) + \sum_{j=1}^{s}A_{1j}\frac{\partial N_{K+1}}{\partial \xi_j} + A_{11}N_K\Big)\frac{\partial^{K+2}V}{\partial x_1^{K+2}}$$

$$+\mu^{K+1}A_{11}N_{K+1}\frac{\partial^{K+3}V}{\partial x_1^{K+3}};$$

$$\sum_{i,j=1}^{s} n_iA_{ij}\frac{\partial u_\mu^{(K)}}{\partial x_j} = \sum_{i=2}^{s}\mu^{K+1}n_iA_{i1}N_{K+1}\frac{\partial^{K+2}V}{\partial x_1^{K+2}}, \quad x \in \partial U_\mu.$$

A similar discrepancy arises on the discontinuity surfaces for the coefficients Σ_μ.

On the other hand

$$\int_{C_\mu}\Big(\sum_{i,j=1}^{s}\frac{\partial}{\partial x_i}\Big(A_{ij}\frac{\partial u_\mu^{(K)}}{\partial x_j}\Big), u_\mu^{(K)}\Big)\,dx = -\int_{C_\mu}\sum_{i,j=1}^{s}\Big(A_{ij}\frac{\partial u_\mu^{(K)}}{\partial x_j}, \frac{\partial u_\mu^{(K)}}{\partial x_i}\Big)\,dx$$

$$+\int_{\partial C_\mu}\Big(\sum_{i,j=1}^{s}n_iA_{ij}\frac{\partial u_\mu^{(K)}}{\partial x_j}, u_\mu^{(K)}\Big)\,ds + \int_{\Sigma_\mu}\Big(\sum_{i,j=1}^{s}n_i\Big[A_{ij}\frac{\partial u_\mu^{(K)}}{\partial x_j}\Big], u_\mu^{(K)}\Big)\,ds,$$

whence the integral I has the form

$$I = -\int_{C_\mu}\Big(\Phi\sum_{l=2}^{K+1}\mu^{l-2}h_l^N\frac{\partial^l V}{\partial x_1^l}, u^{(K)}\Big)\,dx$$

$$= \int_{C_\mu} \sum_{i,j=1}^{s} \left(A_{ij} \frac{\partial u_\mu^{(K)}}{\partial x_j}, \frac{\partial u_\mu^{(K)}}{\partial x_i} \right) dx + O(\mu^{K+s-1}). \qquad (2.3.99)$$

Since $\langle \Phi^* N_l \rangle = \delta_{l0}$, by analogy with the proof of [16], Lemma 4.2.1, we obtain

$$I = -\int_{C_\mu} \left(\Phi \sum_{l=2}^{K+1} \mu^{l-2} h_l^N \frac{\partial^l V}{\partial x_1^l}, V \right) dx + O(\mu^K)$$

$$= -\mu^{s-1} \sum_{l=2}^{K+1} \mu^{l-2} \operatorname{mes} \beta \int_0^b \left(h_l^N \frac{\partial^l V}{\partial x_1^l}, V \right) dx_1 + O(\mu^{K+s-1}).$$

Let the first and the fourth (for $s = 3$) components of the vector V be zero. Then ($K > 3$)

$$I = -\mu^{s+1} \operatorname{mes} \beta \int_0^b \left(h_4^N \frac{\partial^4 V}{\partial x_1^4}, V \right) dx_1 + O(\mu^{s+2})$$

and it follows from (2.3.99) that

$$-\mu^{s+1} \operatorname{mes} \beta \int_0^b \left(h_4^N \frac{\partial^4 V}{\partial x_1^4}, V \right) dx_1 =$$

$$\int_{C_\mu} \left(\sum_{i,j=1}^{s} A_{ij} \frac{\partial u_\mu^{(K)}}{\partial x_j}, \frac{\partial u_\mu^{(K)}}{\partial x_i} \right) dx + O(\mu^{s+2}). \qquad (2.3.100)$$

It follows from Lemma 2.3.4 that the right-hand side of this equality is bounded from below by $c_1 \mu^2 \| u_\mu^{(K)} \|^2_{L^2(U_\mu)}$; in its turn, for small μ, this expression is bounded from below by

$$\frac{c_1 \mu^2}{2} \| \Phi V \|^2_{L^2(U_\mu)} = \frac{c_1 \mu^{s+1}}{2} \operatorname{mes} \beta \, \| V \|^2_{L^2(0,b)}.$$

Here constant c_1 is independent of μ. By taking the second component of the vector V, for example, in the form $\sin(2\pi x_1/b)$ and by setting the other components equal to zero, we see from equation (2.3.100) that $h_4^{22} < 0$. Similarly for $s = 3$ we can prove that $h_4^{33} < 0$.

The statements of Theorem 2.3.1 follow from Lemmas 2.3.2–2.3.5.

Lemma 2.3.6. *If condition A for problem* (1)–(4) *is satisfied, then the matrices* h_0^{Nr} *and* h_1^{Nr} *have the following structure: if* $s = 2$*, then*

$$h_0^{Nr} = \begin{pmatrix} 1 & 0 \\ 0 & 1 \\ 0 & 0 \end{pmatrix}$$

and the third row of h_1^{Nr} *is equal to* $(0, 1)$*; if* $s = 3$*, then*

$$h_0^{Nr} = \begin{pmatrix} 1 & 0 & 0 & 0 \\ 0 & 1 & 0 & 0 \\ 0 & 0 & 1 & 0 \\ 0 & 0 & 0 & 1 \\ 0 & 0 & 0 & 0 \\ 0 & 0 & 0 & 0 \end{pmatrix}$$

and the fifth row and the sixth row of the matrix h_1^{Nr} *are equal to* $(0,1,0,0)$ *and to* $(0,0,1,0)$*, respectively.*

Proof.

Since $N_0 = \Phi$, the trivial matrices $N_0^r = 0$ are solutions of boundary layer problems (2.3.30) and (2.3.31), and h_0^{Nr} indeed has the above structure; so,

$$\widetilde{\Phi}(0,\xi')h_0^{Nr} = \Phi(\xi').$$

The columns $N_1^{(j)}$ of the matrix N_1, from the second to the sth one, are equal to

$$N_1^{(j)} = (-\xi_j, 0, \ldots, 0)^*, \quad j = 2, \ldots, s$$

(see the proof of Lemma 2.3.3). Then the columns $N_1^{r(j)}$ of the matrices N_1^r, from the second to the $s-$th one, are solutions of the homogeneous equations $L_{\xi\xi}N_1^{r(j)} = 0$ with boundary conditions

$$\frac{\partial N_1^{r(j)}}{\partial \nu_\xi} = 0 \ \text{ for } \ \xi' \in \partial\beta \quad \text{and} \quad N_1^{r(j)}(0,\xi') = -N_1^{(j)}(0,\xi') + \widetilde{\Phi}(0,\xi')h_1^{Nr(j)},$$

where $h_1^{Nr(j)}$ is the $j-$th column of the matrix h_1^{Nr}. Setting the third element of $h_1^{Nr(j)}$ equal to 1 for $s = 2$ and the $(j+3)-$th element equal to 1 for $s = 3$ and setting the other elements equal to zero, we see that $N_1^{r(j)} = 0$ is a solution of the boundary layer problem.

To complete the proof, it remains to establish that the first (for $s = 3$, the last) element of the last (for $s = 3$, the next to the last) row of the matrix h_1^{Nr} is equal to zero. To do this, it suffices to prove that for $s = 3$, the first and the fourth columns of a solution to problem (2.3.37) stabilize to the rigid-body displacement orthogonal to the vectors $(-\xi_2, \xi_1, 0)$ and $(-\xi_3, 0, \xi_1)$ and for $s = 2$, the first column of solution to problem (2.3.37) stabilizes to a constant. In its turn, this assertion follows from the fact that the solution $\widetilde{N}_1^0$ to problem (2.3.37) satisfies (under condition A) relations similar to equations (2.3.83):

$$\widetilde{n}_1^{kp}(\xi) = (-1)^{\delta_{\mathcal{A}k}+\tilde{\delta}_{\mathcal{A}p}}\widetilde{n}_1^{kp}(S_{\mathcal{A}}\xi), \quad \mathcal{A} = 2, \ldots, s. \tag{2.3.101}$$

Indeed, it follows from (2.3.101) that $\widetilde{n}_1^{11}(\xi)$ is an even function (and $\widetilde{n}_1^{14}(\xi)$ is an odd function for $s = 3$) of $\xi_2, \ldots, \xi_s$, hence, they cannot stabilize to a rigid-body displacement with the first component of the form $c_0 + c_2\xi_2 + \cdots + c_s\xi_s$ whose coefficients $c_2, \ldots, c_s$ satisfy $|c_2| + \cdots + |c_s| \neq 0$. This means that $\widetilde{N}_1^0$ stabilizes to the rigid-body displacement $(-\widetilde{\Phi}h_1^{N0})$, where the matrix h_1^{N0} has the zero first element in the last row for $s = 2$ and the first and the last zero elements in the two last rows for $s = 3$.

The proof of relations (2.3.101) is perfectly similar to that of Lemma 2.3.2.

This completes the proof of Lemma 2.3.6. □

The proof of Theorems 2.3.2, 2.3.3, and 2.3.5 is perfectly similar to the proof of Theorems 4 and 5 in [130] and of the theorems (of Phrägmen-Lindelöf type) 8.1 and 8.3 of Chapter 1 in [129].

Lemma 2.3.7. *Let condition A be satisfied. Then the solutions of problems (30′) satisfy the following relations*

$$ {}^0n_l^{pq}(\xi) = (-1)^{\delta_{\mathcal{A}p}+\tilde{\delta}_{\mathcal{A}q}}\, {}^0n_l^{pq}(S_{\mathcal{A}}\xi), \quad \mathcal{A} = 2,\dots,s, \tag{2.3.102} $$

$$ \chi_l^{qr} = \int_{(0,+\infty)\times\beta} \widetilde{\varphi}^{mq}\Big(a_{1j}^{kp}\frac{\partial^0 n_{l-1}^{pr}}{\partial \xi_j} + a_{11}^{kp}\, {}^0n_{l-2}^{pr}\Big)\, d\xi $$

$$ = (-1)^{\tilde{\tilde{\delta}}_{\mathcal{A}q}+\tilde{\delta}_{\mathcal{A}r}}\chi_l^{qr}, \quad \mathcal{A} = 2,\dots,s. \tag{2.3.103} $$

$$ \tilde{\tilde{\delta}}_{\mathcal{A}q} = \begin{cases} \widetilde{\delta}_{\mathcal{A}q} & \text{for} \quad q = 1,\dots,d, \\ \delta_{\mathcal{A},q-3} & \text{for} \quad q = 5,6, \ s = 3, \\ 1 \ \text{for} & q = 3, \ s = 2. \end{cases} $$

Here ${}^0n_l^{pq}$ are the entries of the matrices $N_l^0(\xi)$.

The proof of relations (2.3.102) is similar to that of Lemma 2.3.2 (by substitution). Relation (2.3.103) follows from equations (2.3.102):

$$ \chi_l^{qr} = \int_{(0,+\infty)\times\beta} (-1)^{\delta_{\mathcal{A}k}+\tilde{\tilde{\delta}}_{\mathcal{A}q}}\widetilde{\varphi}^{kq}(\eta)(-1)^{\delta_{\mathcal{A}k}+\delta_{\mathcal{A}p}+\delta_{\mathcal{A}p}+\tilde{\delta}_{\mathcal{A}r}} $$

$$ \times\Big(a_{1j}^{kp}(\eta)\frac{\partial^0 n_{l-1}^{pr}}{\partial \eta_j} + a_{11}^{kp}(\eta)^0 n_{l-2}^{pr}\Big)\Big|_{\eta=S_{\mathcal{A}}\xi} d\xi = (-1)^{\tilde{\tilde{\delta}}_{\mathcal{A}q}+\delta_{\mathcal{A}r}}\chi_l^{qr}. $$

It follows from Lemmas 2.3.2 and 2.3.7 that nonzero elements of h_l^{N0} can occupy only the following positions marked by stars: for $s = 3$ we have

$$ \begin{pmatrix} * & 0 & 0 & 0 \\ 0 & * & 0 & 0 \\ 0 & 0 & * & 0 \\ 0 & 0 & 0 & * \\ 0 & * & 0 & 0 \\ 0 & 0 & * & 0 \end{pmatrix}, $$

for $s = 2$ we have

$$ \begin{pmatrix} * & 0 \\ 0 & * \\ 0 & * \end{pmatrix}. $$

Lemma 2.3.8. *For problem (2.3.1)-(2.3.4), (2.3.3') we have*

$$ {}^0h_1^{qq} = -h_2^{qq} \ \textit{for} \ q = 1,\dots,d, \qquad {}^0h_2^{q+3,q} = \langle\xi_q^2\rangle^{-1}h_4^{qq} \ \textit{for} \ q = 2,3, \ s = 3, $$

$$ {}^0h_2^{32} = \langle\xi_2^2\rangle^{-1}h_4^{22} \ \textit{for} \ s = 2, \qquad {}^0h_3^{qq} = -h_4^{qq} \ \textit{for} \ q = 2,\dots,s. $$

Here ${}^0h_l^{pq}$ are the entries of the matrices h_l^{N0}.

Proof.

Since $T_1^{N0} = 0$ and the second column (and the third column for $s = 3$) of the matrix T_2^{N0} is zero, the integral in formulas (2.3.45) for $\bar{h}_1^{qq}$ and $\bar{h}_2^{q+3,q}$, over the set $(0,+\infty)\times\beta$, vanishes. Let us prove that this integral also vanishes in formulas (2.3.45) for $\bar{h}_3^{qq}$. Let $q=2$ and $s=3$ (the proof for $q=3$ and $s=2$ is similar). We have:

$$\int_{(0,+\infty)\times\beta}\left(a_{1j}^{2p}\frac{\partial\,{}^0n_2^{p2}}{\partial\xi_j}+a_{11}^{2p}\,{}^0n_1^{p2}\right)d\xi=\int_{(0,+\infty)\times\beta}\left(a_{mj}^{1p}\frac{\partial\,{}^0n_2^{p2}}{\partial\xi_j}+a_{m1}^{1p}\,{}^0n_1^{p2}\right)\frac{\partial\xi_2}{\partial\xi_m}\,d\xi$$

$$=-\left\langle\left(a_{1j}^{1p}\frac{\partial\,{}^0n_2^{p2}}{\partial\xi_j}+a_{11}^{1p}\,{}^0n_1^{p2}\right)\bigg|_{\xi_1=0}\xi_2\right\rangle-\int_{(0,+\infty)\times\beta}\frac{\partial}{\partial\xi_m}\left(a_{mj}^{1p}\frac{\partial^0n_2^{p2}}{\partial\xi_j}+a_{m1}^{1p}\,{}^0n_1^{p2}\right)\xi_2\,d\xi$$

$$=\left\langle\left(a_{1j}^{1p}\frac{\partial n_2^{p2}}{\partial\xi_j}+a_{11}^{1p}n_1^{p2}\right)\bigg|_{\xi_1=0}\xi_2\right\rangle+\sum_{p=1}^{\bar{d}}\langle\tilde{\varphi}^{1p}(0,\xi')\,{}^0h_2^{p2}\xi_2\rangle$$

$$+\int_{(0,+\infty)\times\beta}\left(a_{1j}^{1p}\frac{\partial\,{}^0n_1^{p2}}{\partial\xi_j}+a_{11}^{1p}\,{}^0n_0^{p2}\right)\xi_2\,d\xi.$$

The last integral vanishes because ${}^0n_1^{p2}={}^0n_0^{p2}=0$. By equations (2.3.45), we have

$$\sum_{p=1}^{\bar{d}}\langle\tilde{\varphi}^{1p}(0,\xi')\,{}^0h_2^{p2}\xi_2\rangle=-\langle\xi_2^2\rangle{}^0h_2^{52}=$$

$$=\langle\xi_2^2\rangle\langle\xi_2^2\rangle^{-1}\left\langle\varphi^{l5}(0,\xi')\left(a_{1j}^{lq}\frac{\partial n_2^{q2}}{\partial\xi_j}+a_{11}^{lq}n_1^{q2}\right)\right\rangle=-\left\langle\left(a_{1j}^{1q}\frac{\partial n_2^{q2}}{\partial\xi_j}+a_{11}^{1q}n_1^{q2}\right)\xi_2\right\rangle,$$

whence the above integral is zero.

Now the equations ${}^0h_1^{qq}=-h_2^{qq}$ for $q=1,\dots,d$, and ${}^0h_3^{qq}=-h_4^{qq}$ follow from the fact that relations (2.3.45) and (2.3.14) give the same expressions for these quantities.

Finally,

$$-{}^0h_2^{52}=-\langle\xi_2^2\rangle^{-1}\left\langle\left(a_{1j}^{1q}\frac{\partial n_2^{q2}}{\partial\xi_j}+a_{11}^{1q}n_1^{q2}\right)\xi_2\right\rangle=\langle\xi_2^2\rangle^{-1}\left\langle\frac{\partial}{\partial\xi_m}\left(a_{mj}^{1q}\frac{\partial n_3^{q2}}{\partial\xi_j}+a_{m1}^{1q}n_2^{q2}\right)\xi_2\right\rangle$$

$$=-\langle\xi_2^2\rangle^{-1}\left\langle a_{2j}^{1q}\frac{\partial n_3^{q2}}{\partial\xi_j}+a_{21}^{1q}n_2^{q2}\right\rangle=-\langle\xi_2^2\rangle^{-1}\left\langle a_{1j}^{2q}\frac{\partial n_3^{q2}}{\partial\xi_j}+a_{11}^{2q}n_2^{q2}\right\rangle=-\langle\xi_2^2\rangle^{-1}h_4^{22}.$$

The statements concerning ${}^0h_2^{63}$ and ${}^0h_2^{32}$ for $s=2$ can be proved similarly.

This completes the proof of Lemma 2.3.8. □

The statement of Theorem 2.3.4 follows from the statements of Lemmas 2.3.7 and 2.3.8, taking into account the homogeneity of equations (2.3.58). The proof of Theorem 2.3.6 can be given similarly to that of Lemmas 2.3.6–2.3.8.

Lemma 2.3.9. *A solution of the equation $Pu=f$ in the domain C_μ with boundary conditions (2.3.2),(2.3.3) and with the right-hand side $f\in L^2(C_\mu)$ satisfies the a priori estimate*

$$\|u\|_{H^1(C_\mu)}\le c\mu^{-2}\|f\|_{L^2(C_\mu)}.$$

Proof.

By analogy with Lemma 2.3.4, we obtain estimate (2.3.95). Furthermore, it follows from the integral identity that

$$\int_{C_\mu}\Big(A_{ij}\frac{\partial u}{\partial x_j},\frac{\partial u}{\partial x_i}\Big)\,dx=-\int_{C_\mu}(f,u)\,dx\le$$

$$\le\|f\|_{L^2(C_\mu)}\|u\|_{L^2(C_\mu)}\le\|f\|_{L^2(C_\mu)}\|u\|_{H^1(C_\mu)}.$$

This, together with inequality (2.3.95), implies the desired estimate. □

Lemma 2.3.9 is used to obtain the estimate for the proximity of the exact solution of problem (2.3.1)–(2.3.4) and the asymptotic expansion $u^{(\infty)}$. Namely, as it was done in section 2.2, substitute the $(K+1)-$th partial sum $u^{(K)}$ of the series $u^{(\infty)}$ into equations (2.3.1)–(2.3.4). Grouping the terms in the same way as in the series itself, we see that the residual on the right-hand side of the equation can be estimated by $O(\mu^{K-1})$. A similar residual in the boundary conditions is transferred to the right-hand side by substitution (see section 2.2). Then we can apply the *a priori* estimate of Lemma 2.3.9 and obtain

$$\|u-u^{(K)}\|_{H^1(C_\mu)}=O(\mu^K)\sqrt{mes\ C_\mu}.$$

The solvability of the ordinary differential equations for ω_j is clear.

Thus we obtain

Theorem 2.3.8. *Let* $K\in\{0,1,2,...\}$. *Denote* χ *a function from* $C^{(K+2)}([0,b])$ *such that* $\chi(t)=1$ *when* $t\in[0,b/3]$, $\chi(t)=0$ *when* $t\in[2b/3,b]$. *Consider a function*

$$u^{(K)}(x)=u_B^{(K)}(x)+u_P^{0(K)}(x)\chi(x_1)+u_P^{1(K)}(x)\chi(b-x_1)+\tilde\Phi(0,\frac{x'}{\mu})\rho(x_1),$$

where

$$u_B^{(K)}=\sum_{l=0}^{K+1}\mu^lN_l\Big(\frac{x}{\mu}\Big)\frac{d^l\omega^{(K)}(x_1)}{dx_1^l}+\sum_{l=0}^{K-1}\mu^{l+2}M_l\Big(\frac{x}{\mu}\Big)\frac{d^l\psi(x_1)}{dx_1^l},$$

$$u_P^{0(K)}=\Big(\sum_{l=0}^{K+1}\mu^lN_l^0(\xi)\frac{d^l\omega^{(K)}}{dx_1^l}+\sum_{l=0}^{K-1}\mu^{l+2}M_l^0(\xi)\frac{d^l\psi}{dx_1^l}\Big)\Big|_{\xi=x/\mu},$$

$$u_P^{1(K)}=\Big(\sum_{l=0}^{K+1}\mu^lN_l^1(\xi)\frac{d^l\omega^{(K)}}{dx_1^l}+\sum_{l=0}^{K-1}\mu^{l+2}M_l^1(\xi)\frac{d^l\psi}{dx_1^l}\Big)|_{\xi_1=(x_1-b)/\mu\xi'=x'/\mu},$$

$$\omega^{(K)}=\sum_{j=-2}^{K}\mu^j\omega_j(x_1);$$

$$\rho(x_1)=(1-x_1/b)q_0+(x_1/b)q_1,$$

$$q_r = \Big(\sum_{l=K+1}^{2K+1} \mu^l \sum_{j+p=l,-2\le j\le K, 0\le p\le K+1} h_p^{N_r} \frac{d^p \omega_j}{dx_1^p} + \mu^{K+1} h_{K-1}^{M_r} \frac{d^{K-1}\psi}{dx_1^{K-1}} \Big)|_{x_1=rb}, r=1,2,$$

$N_l, N_l^0, N_l^1, M_l, M_l^0, M_l^1, \omega_j$ are defined in subsection 2.3. Then the estimate holds

$$\|u^{(K)} - u\|_{(H^1(C_\mu))^s} = O(\mu^K)\sqrt{mes\ C_\mu}.$$

Proof repeats the proof of Theorem 2.2.3 with the following corrections.

1. The products of type

$$A_{mj}\Big(\frac{x}{\mu}\Big)\frac{\partial u_B^{(K)}}{\partial x_j}\frac{\partial \phi}{\partial x_m}$$

are replaced by inner products

$$\Big(A_{mj}\Big(\frac{x}{\mu}\Big)\frac{\partial u_B^{(K)}}{\partial x_j}, \frac{\partial \phi}{\partial x_m}\Big).$$

2. At the second stage of the proof, the integral $\tilde{J}_B^N(\phi)$ is presented in the form

$$\tilde{J}_B^N(\phi) = \sum_{l=0}^{K+1} \int_{\mu^{-1}C_\mu} \mu^{s+l-2} \sum_{m=1}^{s} A_{ml}^{BN}(\xi) : \frac{\partial}{\partial \xi_m}\Big(\Big(\phi \frac{d^l \omega^{(K)*}(x_1)}{dx_1^l}\Big)|_{x=\mu\xi} -$$

$$- A_{1,l-1}^{BN}(\xi) : \Big(\phi \frac{d^l \omega^{(K)*}(x_1)}{dx_1^l}\Big)|_{x=\mu\xi}\ d\xi -$$

$$- \mu^{s+K} \int_{\mu^{-1}C_\mu} A_{1,K+1}^{BN}(\xi) : \Big(\phi \frac{d^{K+2}\omega^{(K)*}(x_1)}{dx_1^{K+2}}\Big)|_{x=\mu\xi}\ d\xi,$$

where : denotes the element-wise multiplication of matrices, i.e. if A and B are two matrices of the same dimension with the elements a_{ij} and b_{ij} respectively then $A : B$ is the matrix with the elements $a_{ij}b_{ij}$ (no summation!).

3. At the fifth stage of the proof,

$$I(\phi) = \int_{C_\mu} \Big(F\psi - \Phi(\frac{x'}{\mu}) \sum_{l=0}^{K+1} \mu^{l-2} h_l^N \frac{d^l \omega^{(K)}}{dx_1^l} -$$

$$- \tilde{F}\psi - \Phi(\frac{x'}{\mu}) \sum_{l=1}^{K-1} \mu^l h_l^M \frac{d^l \psi}{dx_1^l}\Big)\phi(x)\ dx + \Delta_3(\phi).$$

4. The *a priori* estimate constant of lemma 2.3.9 depends on μ and therefore (2.2.40) transforms into

$$\mu^4 \kappa \|u - u^{(K)}\|_{H^1(C_\mu)}^2 \le I(u - u^{(K)}) =$$

$$= \int_{C_\mu} \sum_{m,j=1}^{s} A_{mj}\left(\frac{x}{\mu}\right) \frac{\partial(u-u^{(K)})}{\partial x_j} \frac{\partial(u-u^{(K)})}{\partial x_m}\, dx \;\le$$

$$\le\; c_{10}\mu^K \|u-u^{(K)}\|_{H^1(C_\mu)} \sqrt{mes\ C_\mu}, \qquad (2.3.40')$$

with a constant $c_{10} > 0$ independent of μ.

So,

$$\|u-u^{(K)}\|_{H^1(C_\mu)} \;\le\; c_{10}/\kappa\mu^{K-4}\sqrt{mes\ C_\mu};$$

and therefore

$$\|u-u^{(K+4)}\|_{H^1(C_\mu)} \;\le\; c_{10}/\kappa\mu^{K}\sqrt{mes\ C_\mu}.$$

By definition,

$$\|u^{(K)}-u^{(K+4)}\|_{H^1(C_\mu)} \;=\; O(\mu^K\sqrt{mes\ C_\mu}).$$

Thus, by triangle inequality,

$$\|u-u^{(K)}\|_{H^1(C_\mu)} \;=\; O(\mu^K\sqrt{mes\ C_\mu}).$$

This completes the proof of the theorem.

Similarly we can prove the analogous theorems for problems (2.3.1), (2.3.2), (2.3.3') and (2.3.48),(2.3.4)).

Proof of Theorem 2.3.7.

Substituting

$$n_1^{p1} = (1-\delta_{p1})\frac{(-\lambda)}{2M+\lambda}\xi_p \text{ for } p=2 \quad\text{and}\quad n_1^{p1} = (1-\delta_{p1})\frac{(-\lambda)}{2(M+\lambda)}\xi_p$$

for $s=3$ into (2.3.11)–(2.3.13), we directly verify that these functions satisfy problem (2.3.11)–(2.3.13) for $l=1$. Then

$$h_2^{11} = \sum_{j,p=2}^{s}\left\langle a_{1j}^{1p}\frac{\partial n_1^{p1}}{\partial \xi_j} + a_{11}^{11}\right\rangle = \bar{E}^{(s)} = \begin{cases} \frac{(\lambda+2M)^2-\lambda^2}{\lambda+2M} & \text{for} \quad s=2,\\ \frac{M(3\lambda+2M)}{\lambda+M} & \text{for} \quad s=3.\end{cases}$$

The function n_1^{14} is a solution $Y(\xi')$ of the Laplace equation in the domain β with the boundary condition $\partial Y/\partial n - a\xi_3 n_2 + a\xi_2 n_3 = 0$ on the boundary $\partial\beta$. Then

$$h_2^{44} = Ma\left\langle -\xi_3\frac{\partial n_1^{14}}{\partial \xi_2} + \xi_2\frac{\partial n_1^{14}}{\partial \xi_3} + \xi_2^2 a + \xi_3^2 a\right\rangle$$

$$= \frac{Ma}{\text{mes}\,\beta}\left(a\int_\beta \xi_2^2+\xi_3^2\,d\xi' + \int_{\partial\beta}(-\xi_3 n_2+\xi_2 n_3)n_1^{14}\,ds\right). \qquad (2.3.104)$$

On the other hand, for each $z \in H^1(\beta)$, the function n_1^{14} satisfies the integral identity

$$\int_\beta \left(\frac{\partial n_1^{14}}{\partial \xi_2} \frac{\partial z}{\partial \xi_2} + \frac{\partial n_1^{14}}{\partial \xi_3} \frac{\partial z}{\partial \xi_3} \right) d\xi' = -a \int_{\partial\beta} (-\xi_3 n_2 + \xi_2 n_3) z \, ds. \tag{2.3.105}$$

By setting $z = n_1^{14}$ and replacing the last integral in (104) according to (2.3.105), we obtain

$$h_2^{44} = M \left\langle a^2(\xi_2^2 + \xi_3^2) - \left(\frac{\partial n_1^{14}}{\partial \xi_2} \right)^2 - \left(\frac{\partial n_1^{14}}{\partial \xi_3} \right)^2 \right\rangle =$$

$$= M\left(1 - \left\langle \left(\frac{\partial n_1^{14}}{\partial \xi_2} \right)^2 + \left(\frac{\partial n_1^{14}}{\partial \xi_3} \right)^2 \right\rangle \right) \leq M.$$

Consider problems (2.3.11)–(2.3.14) for the second column (for $s = 3$, for the third column) of N_l: $N_l^{(2)}, N_l^{(3)}$. For $s = 2$ we obtain

$$N_0^{(2)} = (0,1)^*, \quad N_1^{(2)} = (-\xi_2, 0)^*, \quad N_2^{(2)} = \frac{\lambda}{2(\lambda + 2M)} (0, \xi_2^2 - \langle \xi_2^2 \rangle)^*,$$

$$A_{12} \frac{\partial N_3^{(2)}}{\partial \xi_2} + A_{11} N_2^{(2)} = (0, \bar{E}^{(2)}(\xi_2^2 - 1/4)/2)^*;$$

and for $s = 3$ we have

$$N_0^{(2)} = (0,1,0)^*, \quad N_1^{(2)} = (-\xi_2, 0, 0)^*,$$

$$N_2^{(2)} = \left(0, \frac{\nu}{2}(\xi_2^2 - \xi_3^2 - \langle \xi_2^2 - \xi_3^2 \rangle), \nu \xi_2 \xi_3 \right)^*, \quad \nu = \frac{\lambda}{2(M + \lambda)};$$

$$\left\langle a_{1j}^{ip} \frac{\partial n_3^{p2}}{\partial \xi_j} + a_{11}^{ip} n_2^{p2} \right\rangle = \left\langle a_{ij}^{1p} \frac{\partial n_3^{p2}}{\partial \xi_j} + a_{i1}^{1p} n_2^{p2} \right\rangle$$

$$= \left\langle \left(a_{mj}^{1p} \frac{\partial n_3^{p2}}{\partial \xi_j} + a_{m1}^{1p} n_2^{p2} \right) \frac{\partial \xi_i}{\partial \xi_m} \right\rangle = \left\langle - \frac{\partial}{\partial \xi_m} \left(a_{mj}^{1p} \frac{\partial n_3^{p2}}{\partial \xi_j} + a_{m1}^{1p} n_2^{p2} \right) \xi_i \right\rangle$$

$$= \left\langle \left(a_{1j}^{1p} \frac{\partial n_2^{p2}}{\partial \xi_j} + a_{11}^{1p} n_1^{p2} - h_3^{12} \right) \xi_i \right\rangle = -\bar{E}^{(3)} \langle \xi_2^2 \rangle \delta_{i2},$$

where the next to last equation follows from the fact that $N_3^{(2)}$ is a solution to problem (11)–(14), and the last equation follows from the explicit form of $N_2^{(2)}$ and $N_1^{(2)}$. Here summation over repeated indices j and m from 2 to 3 and over the index p from 1 to 3 is implied. Similarly one can prove that

$$\left\langle a_{1j}^{ip} \frac{\partial n_3^{p3}}{\partial \xi_j} + a_{11}^{ip} n_2^{p3} \right\rangle = -\bar{E}^{(3)} \langle \xi_3^2 \rangle \delta_{i3}.$$

Hence $h_4^{rr} = -\bar{E}^{(s)} \langle \xi_r^2 \rangle < 0$, $r = 2, 3$, and $h_2^{rr} = 0$, $r = 2, 3$. This completes the proof of Theorem 2.3.7. □

Remark 2.3.2. Applying the explicit formulae for components of N_l we can prove that in isotropic case any sufficiently smooth $d-$dimensional vector-valued function $\omega(x_1)$ satisfies the asymptotic equality

$$\sum_{l=1}^{\infty} \mu^{l-1} \Big\langle \tilde{\Phi}^*(\xi) \Big(\sum_{j=1}^{s} A_{1j} \frac{\partial N_l}{\partial \xi_j} + A_{11} N_{l-1} \Big) \Big\rangle \frac{d^l \omega}{dx_1^l} =$$

$$= \begin{cases} \begin{pmatrix} \bar{E}\frac{d\omega^1}{dx_1} + O(\mu) \\ -\mu^2 \bar{E}\langle \xi_2^2 \rangle \frac{d^3\omega^2}{dx_1^3} + O(\mu^3) \\ \mu \bar{E}\langle \xi_2^2 \rangle \frac{d^2\omega^2}{dx_1^2} + O(\mu^2) \end{pmatrix} & for \quad s=2, \\ \begin{pmatrix} \bar{E}\frac{d\omega^1}{dx_1} + O(\mu) \\ -\mu^2 \bar{E}\langle \xi_2^2 \rangle \frac{d^3\omega^2}{dx_1^3} + O(\mu^3) \\ -\mu^2 \bar{E}\langle \xi_3^2 \rangle \frac{d^3\omega^3}{dx_1^3} + O(\mu^3) \\ \bar{\mathcal{M}}\frac{d\omega^4}{dx_1} + O(\mu) \\ \mu \bar{E}\langle \xi_2^2 \rangle \frac{d^2\omega^2}{dx_1^2} + O(\mu^2) \\ \mu \bar{E}\langle \xi_3^2 \rangle \frac{d^2\omega^3}{dx_1^2} + O(\mu^2) \end{pmatrix} & for \quad s=3. \end{cases}$$

Denote the main term of this asymptotic expression as $\Lambda_\mu \omega$, i.e.

$$\Lambda_\mu \omega = \begin{cases} \begin{pmatrix} \bar{E}\frac{d\omega^1}{dx_1} \\ -\mu^2 \bar{E}\langle \xi_2^2 \rangle \frac{d^3\omega^2}{dx_1^3} \\ \mu \bar{E}\langle \xi_2^2 \rangle \frac{d^2\omega^2}{dx_1^2} \end{pmatrix} & for \quad s=2, \\ \begin{pmatrix} \bar{E}\frac{d\omega^1}{dx_1} \\ -\mu^2 \bar{E}\langle \xi_2^2 \rangle \frac{d^3\omega^2}{dx_1^3} \\ -\mu^2 \bar{E}\langle \xi_3^2 \rangle \frac{d^3\omega^3}{dx_1^3} \\ \bar{\mathcal{M}}\frac{d\omega^4}{dx_1} \\ \mu \bar{E}\langle \xi_2^2 \rangle \frac{d^2\omega^2}{dx_1^2} \\ \mu \bar{E}\langle \xi_3^2 \rangle \frac{d^2\omega^3}{dx_1^2} \end{pmatrix} & for \quad s=3. \end{cases}$$

Remark 2.3.3. We have studied in this section the case when the rod satisfies some symmetry conditions (Condition A). The general case was studied by different techniques in [117,119] and by the techniques of this section in [101] by A. Majd. The main idea is in multiplication of components of the ansatz by some normalizing factors.

2.4 Non steady-state conductivity of a rod

2.4.1 Statement of the problem

Consider a non steady-state conductivity equation ($s = 3!$)

$$-\rho(x/\mu)\frac{\partial u}{\partial t} + \sum_{m,j=1}^{3} \frac{\partial}{\partial x_m}\Big(A_{mj}(x/\mu)\frac{\partial u}{\partial x_j}\Big) = F_\mu(x,t), \quad x \in C_\mu, \tag{2.4.1}$$

with the boundary conditions

$$\frac{\partial u}{\partial \nu} \equiv \sum_{i,j=1}^{s} A_{ij}\Big(\frac{x}{\mu}\Big)\frac{\partial u}{\partial x_j}n_i = 0, \quad x \in \partial U_\mu, \tag{2.4.2}$$

$$u = 0, \quad x_1 = 0 \quad \text{or} \quad x_1 = b. \tag{2.4.3}$$

and with junction conditions

$$[u] = 0; \;\; \Big[\sum_{i,j=1}^{s} A_{ij}\partial u/\partial x_j n_i\Big] = 0, \tag{2.5.4}$$

(as in section 2.3) and with initial conditions

$$u\Big|_{t=0} = 0. \tag{2.4.5}$$

Here $A_{ij}(\xi)$ are functions from section 2.2 and the function $\rho(\xi)$ is 1-periodic with respect to ξ_1, $\rho(\xi) \geq \kappa > 0$, and $\rho(\xi)$ is piecewise smooth in the sense of section 2.2. Here κ is the same as in section 2. Besides, $\rho(S_\mathcal{A}\xi) = \rho(\xi)$ for $\mathcal{A} = 2, 3$ and

$$F_\mu(x,t) = F\Big(\frac{x}{\mu}\Big)\psi(x_1,t),$$

where F is the same as in section 2.2 and

$$\psi(x_1,t) \in C^\infty, \quad \psi(x_1,t) = 0 \quad \text{for} \quad t \leq \theta_0, \quad \theta_0 > 0.$$

The existence and uniqueness of a solution follows from [85],[54]. This solution belongs to the space $H^{1,0}(C_\mu \times (0,T))$ of functions with a norm

$$\|u\|_{H^{1,0}(C_\mu\times(0,T))} = \sqrt{\sup_{t\in(0,T)} \|u(x,t)\|^2_{L^2(C_\mu)} + \int_0^T \|u(x,t)\|^2_{H^1(C_\mu)}dt}.$$

2.4.2 Inner expansion

An inner f.a.s. of problem (2.4.1)-(2.4.5) is sought as a series

$$u^{(\infty)} = \sum_{l+q=0}^{\infty} \mu^{l+q}N_{lq}(x/\mu)\frac{\partial^{l+q}\omega(x_1,t)}{\partial x_1^l \partial t^q} +$$

$$+\sum_{l+q=0}^{\infty} \mu^{l+q+2} M_{lq}(x/\mu) \frac{\partial^{l+q}\psi(x_1,t)}{\partial x_1^l \partial t^q}, \tag{2.4.6}$$

where ω is a C^∞−function and N_{lq} and M_{lq} are analogs of the functions N_l and M_l from section 2; N_{lq} are (1-periodic with respect to ξ_1) solutions of the problems

$$L_{\xi\xi} N_{lq} = -T_{lq}^N(\xi) + h_{lq}^N, \quad \xi' \in \beta, \tag{2.4.7}$$

$$\frac{\partial N_{lq}}{\partial \nu_\xi} = -\sum_{m=2}^{3} A_{m1} N_{l-1,q} n_m, \quad \xi' \in \partial\beta, \tag{2.4.8}$$

$$[N_{lq}]\big|_\Sigma = 0, \quad \left[\frac{\partial N_{lq}}{\partial \nu_\xi}\right]\Big|_\Sigma = -\Big[\sum_{m=1}^{3} A_{m1} N_{l-1,q} n_m\Big]\Big|_\Sigma, \tag{2.4.9}$$

$$T_{lq}^N(\xi) = \sum_{j=1}^{3} \frac{\partial}{\partial \xi_j}(A_{j1} N_{l-1,q}) + \sum_{j=1}^{3} A_{1j} \frac{\partial N_{l-1,q}}{\partial \xi_j} + A_{11} N_{l-2,q} - \rho N_{l-1,q-1}, \tag{2.4.10}$$

$$h_{lq}^N = \left\langle\left(\sum_{j=1}^{3} A_{1j} \frac{\partial N_{l-1,q}}{\partial \xi_j} + A_{11} N_{l-2,q} - \rho N_{l-1,q-1}\right)\right\rangle, \tag{2.4.11}$$

M_{lq} are solutions of the same problems with N_{lq} replaced by M_{lq}. However, M_{00} is a solution of problem (2.2.15).

As in section 2.2, we obtain a homogenized equation

$$-\langle\rho\rangle \frac{\partial \omega}{\partial t} + h_2^N \frac{\partial^2 \omega}{\partial x_1^2} +$$

$$+\sum_{l+q=3}^{\infty} \mu^{l+q-2} h_{lq}^N \frac{\partial^{l+q}\omega}{\partial x_1^l \partial t^q} +$$

$$+\sum_{l+q=0}^{\infty} \mu^{l+q} h_{lq}^M \frac{\partial^{l+q}\psi}{\partial x_1^l \partial t^q} - \bar{F}\psi_\mu = 0, \tag{2.4.13}$$

where the positive coefficient h_2^N is the same as in section 2.

Let us seek f.a.s. ω of (2.4.13) in the form of a regular series in powers of μ:

$$\omega = \sum_{j=0}^{\infty} \mu^j \omega_j. \tag{2.4.14}$$

Substituting (2.4.14) into (2.4.13) we obtain a chain of parabolic equations

$$-\langle\rho\rangle \frac{\partial \omega_j}{\partial t} + h_2^N \frac{\partial^2 \omega_j}{\partial x_1^2} = f_j(x_1,t), \tag{2.4.15}$$

where right hand sides f_j are some linear combinations of derivatives of functions ω_r with $r < j$ and $f_0 = \bar{F}\psi$.

2.4.3 Boundary layer corrector

The boundary layer corrector can be constructed similarly to that in section 2.2. The f.a.s. to problem (2.4.1)–(2.4.5) is sought in the form,

$$u^{(\infty)} = u_B + u_P^0 + u_P^1, \tag{2.4.16}$$

where u_B is defined by equation (2.4.6) and

$$u_P^0 = \Big(\sum_{l+q=0}^{\infty} \mu^{l+q} N_{lq}^0(x/\mu) \frac{\partial^{l+q}\omega(x_1,t)}{\partial x_1^l \partial t^q} +$$

$$+ \sum_{l+q=0}^{\infty} \mu^{l+q+2} M_{lq}^0(x/\mu) \frac{\partial^{l+q}\psi(x_1,t)}{\partial x_1^l \partial t^q} \Big)\Big|_{\xi=x/\mu},$$

$$u_P^1 = \Big(\sum_{l+q=0}^{\infty} \mu^{l+q} N_{lq}^1(x/\mu) \frac{\partial^{l+q}\omega(x_1,t)}{\partial x_1^l \partial t^q} +$$

$$+ \sum_{l+q=0}^{\infty} \mu^{l+q+2} M_{lq}^1(x/\mu) \frac{\partial^{l+q}\psi(x_1,t)}{\partial x_1^l \partial t^q} \Big)|_{\xi_1=(x_1-b)/\mu \xi'=x'/\mu \cdot}, \tag{2.5.17}$$

and N_{lq}^r, M_{lq}^r are exponentially decaying in ξ_1, 1-periodic in all variables except of ξ_1 functions. They are found from the chain of boundary value problems similar to that of section 2.2.4, but right hand sides of the boundary layer problems contain the supplementary term $-\rho N_{l-1,q-1}^r$, $r = 0, 1$ (or respectively $-\rho M_{l-1,q-1}^r$, $r = 0, 1$).

2.4.4 Justification

Theorem 2.4.1. *Let $K \in \{0, 1, 2, ...\}$. Let χ be a function from section 3. Consider a function*

$$u^{(K)}(x,t) = u_B^{(K)}(x,t) + u_P^{0(K)}(x,t)\chi(x_1) + u_P^{1(K)}(x,t)\chi(b - x_1) + \tilde{\rho}(x_1,t),$$

where

$$u_B^{(K)} = \sum_{l+q=0}^{K+1} \mu^{l+q} N_{lq}\Big(\frac{x}{\mu}\Big) \frac{\partial^{l+q}\omega(x_1,t)}{\partial x_1^l \partial t^q} +$$

$$+ \sum_{l+q=0}^{K-1} \mu^{l+q+2} M_{lq}\Big(\frac{x}{\mu}\Big) \frac{\partial^{l+q}\psi(x_1,t)}{\partial x_1^l \partial t^q},$$

$$u_P^0 = \Big(\sum_{l+q=0}^{K+1} \mu^{l+q} N_{lq}^0(x/\mu) \frac{\partial^{l+q}\omega(x_1,t)}{\partial x_1^l \partial t^q} +$$

$$+\sum_{l+q=0}^{K-1}\mu^{l+q+2}M^0_{lq}(x/\mu)\frac{\partial^{l+q}\psi(x_1,t)}{\partial x_1^l\partial t^q}\Big)\Big|_{\xi=x/\mu},$$

$$u_P^1 = \Big(\sum_{l+q=0}^{K+1}\mu^{l+q}N^1_{lq}(x/\mu)\frac{\partial^{l+q}\omega(x_1,t)}{\partial x_1^l\partial t^q}+$$

$$+\sum_{l+q=0}^{K-1}\mu^{l+q+2}M^1_{lq}(x/\mu)\frac{\partial^{l+q}\psi(x_1,t)}{\partial x_1^l\partial t^q}\Big)|_{\xi_1=(x_1-b)/\mu\xi'=x'/\mu},$$

$$\omega^{(K)} = \sum_{j=0}^{K}\mu^j\omega_j(x_1);$$

$$\tilde{\rho}(x_1,t) = (1-x_1/b)q_0(t)+(x_1/b)q_1(t),$$

$$q_r(t) = \Big(\sum_{l+q=K+1}^{2K+1}\mu^{l+q-2}\sum_{j+p+\alpha=l,0\le j\le K,0\le p+\alpha\le K+1}h^{Nr}_{p\alpha}\frac{\partial^{p+\alpha}\omega_j}{\partial x_1^p\partial t^\alpha}+$$

$$+\mu^{K+1}\sum_{l+q=K-1}h^{Mr}_{lq}\frac{\partial^{K-1}\psi}{\partial x_1^l\partial t^q}\Big)|_{x_1=rb},\ r=1,2,$$

$N_{lq}, N^0_{lq}, N^1_{lq}, M_{lq}, M^0_{lq}, M^1_{lq}, \omega_j$ are defined in subsections 2.2.2, 2.2.3. Then the estimate holds

$$\|u^{(K)}-u\|_{H^{1,0}(C_\mu\times(0,T))} = O(\mu^K)\sqrt{mes\ C_\mu}. \tag{2.4.18}$$

Justification of the asymptotic is also similar to that in section 2.2. Namely, it is well known that applying the improving smoothness technique [85],[86],[54] it can be shown that $\frac{\partial(u-u^{(K)})}{\partial t}$ belongs to $H^1(C_\mu)$ for all $t\in(0,T)$. So as in section 2.2 we estimate the discrepancy functional

$$I_t(\phi) = \int_{C_\mu}\rho\Big(\frac{x}{\mu}\Big)\frac{\partial(u-u^{(K)})}{\partial t}\phi dx +$$

$$+\int_{C_\mu}\sum_{m,j=1}^{s}A_{mj}\Big(\frac{x}{\mu}\Big)\frac{\partial(u-u^{(K)})}{\partial x_j}\frac{\partial\phi}{\partial x_m}\,dx, \tag{2.4.19}$$

where for all $t\in(0,T)$, $\phi\in H^1(C_\mu)$, $\phi=0$ for $x_1=0$ or b.

The second integral is developed in the same way as in section 2.2, while the contribution of the first integral corresponds to the supplementary terms $-\rho N_{l-1,q-1}, -\rho N^r_{l-1,q-1}$, $r=0,1$ and $-\rho M_{l-1,q-1}, -\rho M^r_{l-1,q-1}$, $r=0,1$ in cell problems (2.4.7) and boundary layer problems.

We obtain thus for all $t \in (0,T)$, the same representation for $I_t(\phi)$ as in Remark 2.2.4, i.e.

$$I_t(\phi) = \int_{C_\mu} \Psi_{0,\mu}(x,t)\phi \, dx + \sum_{m=1}^{s} \int_{C_\mu} \Psi_{m,\mu}(x,t) \frac{\partial \phi}{\partial x_m} \, dx, \tag{2.4.20}$$

where functions $\Psi_{0,\mu}, \Psi_{m,\mu}, m = 1, ..., s$, are estimated as

$$\sup_{t\in(0,T)} \|\Psi_{0,\mu}(x,t)\|_{L^2(C_\mu)}, \sup_{t\in(0,T)} \|\Psi_{m,\mu}(x,t)\|_{L^2(C_\mu)} \leq c\mu^K \sqrt{mes\ C_\mu} \tag{2.4.21}$$

with a constant $c > 0$ independent of μ.

Taking $\phi = u - u^{(K)}$, and integrating (2.4.20) from 0 to t in time we obtain the identity

$$\int_0^t I_t(\frac{\partial(u-u^{(K)})}{\partial t})dt =$$

$$= \frac{1}{2}\int_0^t \int_{C_\mu} \frac{\partial}{\partial t}\Big(\rho\Big(\frac{x}{\mu}\Big)\Big(u - u^{(K)}\Big)^2\Big)dxdt +$$

$$+ \int_0^t \int_{C_\mu} \sum_{m,j=1}^{s} \Big(A_{mj}\Big(\frac{x}{\mu}\Big) \frac{\partial(u-u^{(K)})}{\partial x_j} \frac{\partial(u-u^{(K)})}{\partial x_m}\Big) \, dxdt, =$$

$$= \int_0^t \Big(\int_{C_\mu} \Psi_{0,\mu}(x,t)(u-u^{(K)}) \, dx +$$

$$+ \sum_{m=1}^{s} \int_{C_\mu} \Psi_{m,\mu}(x,t) \frac{\partial(u-u^{(K)})}{\partial x_m} \, dx\Big)dt. \tag{2.4.22}$$

It is possible to show by induction that

$$\omega_j(x_1,t) = 0 \quad \text{for} \quad t \leq \theta_0,$$

and therefore

$$u^{(K)}\big|_{t=0} = 0.$$

Taking into account estimate (2.5.21) we estimate the right hand side of identity (2.5.22) by

$$c(s+1)\mu^K \sqrt{T mes\ C_\mu \int_0^T \|u-u^{(K)}\|^2_{H^1(C_\mu)}dt}.$$

Therefore for all $t \in (0,T)$, the integral

$$\frac{1}{2}\int_{C_\mu} \Big(\rho\Big(\frac{x}{\mu}\Big)\Big(u - u^{(K)}\Big)^2\Big)dx +$$

$$+\int_0^t \int_{C_\mu} \sum_{m,j=1}^{s} \left(A_{mj}\left(\frac{x}{\mu}\right) \frac{\partial (u-u^{(K)})}{\partial x_j} \frac{\partial (u-u^{(K)})}{\partial x_m} \right) dxdt$$

is estimated by

$$c(s+1)\mu^K \sqrt{Tmes\ C_\mu \int_0^T \|u-u^{(K)}\|^2_{H^1(C_\mu)} dt}.$$

Passing to $sup_{t\in(0,T)}$ in this last estimate we obtain the following inequality

$$\frac{\kappa}{2} \sup_{t\in(0,T)} \|u-u^{(K)}\|^2_{L^2(C_\mu)} \leq$$

$$\leq c(s+1)\mu^K \sqrt{Tmes\ C_\mu \int_0^T \|u-u^{(K)}\|^2_{H^1(C_\mu)} dt};$$

taking $t = T$, we have

$$\int_0^T \int_{C_\mu} \sum_{m,j=1}^{s} \left(A_{mj}\left(\frac{x}{\mu}\right) \frac{\partial (u-u^{(K)})}{\partial x_j} \frac{\partial (u-u^{(K)})}{\partial x_m} \right) dxdt \leq$$

$$\leq c(s+1)\mu^K \sqrt{Tmes\ C_\mu \int_0^T \|u-u^{(K)}\|^2_{H^1(C_\mu)} dt}.$$

So,

$$\|u-u^{(K)}\|_{H^{1,0}(C_\mu\times(0,T))} =$$

$$= O(\mu^K)\sqrt{mes\ C_\mu}.$$

Theorem 2.4.1 is proved.

2.5 Non steady-state elasticity of a rod

2.5.1 Statement of the problem

Consider a non steady-state elasticity theory system of equations

$$-\rho(x/\mu)\frac{\partial^2 u}{\partial t^2} + \sum_{m,j=1}^{3} \frac{\partial}{\partial x_m}\left(A_{mj}(x/\mu)\frac{\partial u}{\partial x_j} \right) = F_\mu(x,t), \quad x \in C_\mu, \qquad (2.5.1)$$

with the boundary conditions

$$\frac{\partial u}{\partial \nu} \equiv \sum_{i,j=1}^{s} A_{ij}\left(\frac{x}{\mu}\right)\frac{\partial u}{\partial x_j} n_i = 0, \quad x \in \partial U_\mu, \qquad (2.5.2)$$

$$u = 0, \quad x_1 = 0 \quad \text{or} \quad x_1 = b. \tag{2.5.3}$$

and with junction conditions

$$[u] = 0; \quad [\sum_{i,j=1}^{s} A_{ij} \partial u / \partial x_j n_i] = 0, \tag{2.5.4}$$

(as in section 2.3) and with initial conditions

$$u\Big|_{t=0} = 0, \quad \frac{\partial u}{\partial t}\Big|_{t=0} = 0. \tag{2.5.5}$$

Here $A_{ij}(\xi)$ are matrix functions from section 2.3 and the function $\rho(\xi)$ is 1-periodic with respect to ξ_1, $\rho(\xi) \geq \kappa > 0$, and $\rho(\xi)$ is piecewise smooth in the sense of section 2.3. Here κ is the same as in section 3. Besides, $\rho(S_{\mathcal{A}}\xi) = \rho(\xi)$ for $\mathcal{A} = 2, 3$ and

$$F_\mu(x,t) = \Phi\left(\frac{x'}{\mu}\right) F\left(\frac{x}{\mu}\right) \psi(x_1, t),$$

where Φ and F are the same as in section 2.3 (F is diagonal), and the four-dimensional vector ψ_μ is

$$\psi_\mu = (\psi^1(x_1,t), \mu^2\psi^2(x_1,\mu t), \mu^2\psi^3(x_1,\mu t), \psi^4(x_1,t))^*,$$

$$\psi^k(x_1,\theta) \in C^\infty, \quad \psi^k(x_1,\theta) = 0 \quad \text{for} \quad \theta \leq \theta_0, \quad \theta_0 > 0, \ k = 1,2,3,4.$$

It means that the right hand side has two different time-scalings: for the longitudinal and for the torsional components time scale is of order of unity while for two bending components a change $\theta = \mu t$ is done.

2.5.2 Inner expansion

An inner f.a.s. of problem (2.5.1)-(2.5.5) is sought as a series

$$u^{(\infty)} = \sum_{l+q=0}^{\infty} \mu^{l+q} N_{lq}(x/\mu) \frac{\partial^{l+q}\omega(x_1,t)}{\partial x_1^l \partial t^q} +$$

$$+ \sum_{l+q=0}^{\infty} \mu^{l+q+2} M_{lq}(x/\mu) \frac{\partial^{l+q}\psi_\mu(x_1,t)}{\partial x_1^l \partial t^q}, \tag{2.5.6}$$

where ω is a four-dimensional vector of the form

$$(\omega^1(x_1,t), \omega^2(x_1,\mu t), \omega^3(x_1,\mu t), \omega^4(x_1,t))^T,$$

N_{lq} and M_{lq} are analogs of the functions N_l and M_l from section 2; N_{lq} are (1-periodic with respect to ξ_1) solutions of the problems

$$L_{\xi\xi} N_{lq} = -T_{lq}^N(\xi) + \Phi h_{lq}^N, \quad \xi' \in \beta, \tag{2.5.7}$$

$$\frac{\partial N_{lq}}{\partial \nu_\xi} = -\sum_{m=2}^{3} A_{m1} N_{l-1,q} n_m, \quad \xi' \in \partial\beta, \tag{2.5.8}$$

$$[N_{lq}]\big|_\Sigma = 0, \quad \left[\frac{\partial N_{lq}}{\partial \nu_\xi}\right]\Big|_\Sigma = -\Big[\sum_{m=1}^{3} A_{m1} N_{l-1,q} n_m\Big]\Big|_\Sigma, \tag{2.5.9}$$

$$T_{lq}^N(\xi) = \sum_{j=1}^{3} \frac{\partial}{\partial \xi_j}(A_{j1} N_{l-1,q}) + \sum_{j=1}^{3} A_{1j} \frac{\partial N_{l-1,q}}{\partial \xi_j} + A_{11} N_{l-2,q} - \rho N_{l,q-2}, \tag{2.5.10}$$

$$h_{lq}^N = \left\langle \Phi^* \Big(\sum_{j=1}^{3} A_{1j} \frac{\partial N_{l-1,q}}{\partial \xi_j} + A_{11} N_{l-2,q} - \rho N_{l,q-2} \Big) \right\rangle, \tag{2.5.11}$$

M_{lq} are solutions of the same problems with N_{lq} replaced by M_{lq}. However, M_{00} is a solution of problem (2.3.15).

As in section 2.3, we obtain a homogenized equation

$$\sum_{l+q=2}^{\infty} \mu^{l+q-2} h_{lq}^N \frac{\partial^{l+q}\omega}{\partial x_1^l \partial t^q} + \sum_{l+q=0}^{\infty} \mu^{l+q} h_{lq}^M \frac{\partial^{l+q}\psi_\mu}{\partial x_1^l \partial t^q} - \bar{F}\psi_\mu = 0, \tag{2.5.13}$$

whose leading term has the form

$$-\langle \rho \rangle \frac{\partial^2 \omega^1}{\partial t^2} + \bar{E}^{(3)} \frac{\partial^2 \omega^1}{\partial x_1^2} - \bar{F}_{11}\psi^1(x_1, t) + \cdots = 0,$$

$$\mu^2 \Big(-\langle \rho \rangle \frac{\partial^2 \omega^2}{\partial \theta^2} - \bar{J}_2^{(3)} \frac{\partial^4 \omega^2}{\partial x_1^4} - \bar{F}_{22}\psi^2(x_1, \theta) + \cdots \Big) = 0,$$

$$\mu^2 \Big(-\langle \rho \rangle \frac{\partial^2 \omega^3}{\partial \theta^2} - \bar{J}_3^{(3)} \frac{\partial^4 \omega^3}{\partial x_1^4} - \bar{F}_{33}\psi^3(x_1, \theta) + \cdots \Big) = 0,$$

$$-\frac{\langle \rho(\xi_2^2 + \xi_3^2) \rangle}{\langle \xi_2^2 + \xi_3^2 \rangle} \frac{\partial^2 \omega^4}{\partial t^2} + \overline{M}^{(3)} \frac{\partial^2 \omega^4}{\partial x_1^2} - \overline{F}_{44}\psi^4(x_1, t) + \cdots = 0, \tag{2.5.14}$$

where $\theta = \mu t$. Let us seek f.a.s. ω of (2.5.13) in the form of a regular series in powers of μ:

$$\omega = \sum_{j=0}^{\infty} \mu^j \omega_j. \tag{2.5.15}$$

Substituting (2.5.15) into (2.5.13) we obtain as in section 2.3 a chain of hyperbolic equations

$$-\langle \rho \rangle \frac{\partial^2 \omega_j^1}{\partial t^2} + \bar{E}^{(3)} \frac{\partial^2 \omega_j^1}{\partial x_1^2} = f_j^1(x_1, t),$$

$$-\langle \rho \rangle \frac{\partial^2 \omega_j^2}{\partial \theta^2} - \bar{J}_2^{(3)} \frac{\partial^4 \omega_j^2}{\partial x_1^4} = f_j^2(x_1, \theta),$$

$$-\langle\rho\rangle\frac{\partial^2\omega_j^3}{\partial\theta^2}-\bar{J}_3^{(3)}\frac{\partial^4\omega_j^3}{\partial x_1^4} = f_j^3(x_1,\theta),$$

$$-\frac{\langle\rho(\xi_2^2+\xi_3^2)\rangle}{\langle\xi_2^2+\xi_3^2\rangle}\frac{\partial^2\omega_j^4}{\partial t^2}+\overline{M}^{(3)}\frac{\partial^2\omega_j^4}{\partial x_1^2} = f_j^4(x_1,t), \tag{2.5.16}$$

where right hand sides f_j^l are some linear combinations of derivatives of functions ω_r^l with $r<j$ and $f_0^l=\bar{F}_{ll}\psi^l$.

2.5.3 Boundary layer corrector

The boundary layer corrector can be constructed similarly to that in section 2.3. The f.a.s. to problem (2.5.1)–(2.5.5) is sought in the form,

$$u^{(\infty)}=u_B+u_P^0+u_P^1, \tag{2.5.17}$$

where u_B is defined by equation (2.5.6) and

$$u_P^0 = \Big(\sum_{l+q=0}^{\infty}\mu^{l+q}N_{lq}^0(x/\mu)\frac{\partial^{l+q}\omega(x_1,t)}{\partial x_1^l\partial t^q}+$$

$$+\sum_{l+q=0}^{\infty}\mu^{l+q+2}M_{lq}^0(x/\mu)\frac{\partial^{l+q}\psi_\mu(x_1,t)}{\partial x_1^l\partial t^q}\Big)\Big|_{\xi=x/\mu},$$

$$u_P^1 = \Big(\sum_{l+q=0}^{\infty}\mu^{l+q}N_{lq}^1(x/\mu)\frac{\partial^{l+q}\omega(x_1,t)}{\partial x_1^l\partial t^q}+$$

$$+\sum_{l+q=0}^{\infty}\mu^{l+q+2}M_{lq}^1(x/\mu)\frac{\partial^{l+q}\psi_\mu(x_1,t)}{\partial x_1^l\partial t^q}\Big)|_{\xi_1=(x_1-b)/\mu\xi'=x'/\mu}., \tag{2.5.18}$$

and N_{lq}^r, M_{lq}^r are exponentially decaying in ξ_1, 1-periodic in all variables except of ξ_1 functions. They are found from the chain of boundary value problems similar to that of section 3, but right hand sides of the boundary layer problems contain the supplementary term $-\rho N_{l,q-2}^r$, $r=0,1$ (or respectively $-\rho M_{l,q-2}^r$, $r=0,1$).

2.5.4 Justification

Theorem 2.5.1. *Let $K\in\{0,1,2,...\}$. Let χ be a function from section 2.3. Consider a function*

$$u^{(K)}(x,t)=u_B^{(K)}(x,t)+u_P^{0(K)}(x,t)\chi(x_1)+u_P^{1(K)}(x,t)\chi(b-x_1)+\tilde{\rho}(x_1,t),$$

where

$$u_B^{(K)} = \sum_{l+q=0}^{K+1}\mu^{l+q}N_{lq}\Big(\frac{x}{\mu}\Big)\frac{\partial^{l+q}\omega(x_1,t)}{\partial x_1^l\partial t^q}+$$

$$+\sum_{l+q=0}^{K-1}\mu^{l+q+2}M_{lq}\Big(\frac{x}{\mu}\Big)\frac{\partial^{l+q}\psi_\mu(x_1,t)}{\partial x_1^l\partial t^q},$$

$$u_P^0 = \Big(\sum_{l+q=0}^{K+1}\mu^{l+q}N_{lq}^0(x/\mu)\frac{\partial^{l+q}\omega(x_1,t)}{\partial x_1^l\partial t^q}+$$

$$+\sum_{l+q=0}^{K-1}\mu^{l+q+2}M_{lq}^0(x/\mu)\frac{\partial^{l+q}\psi_\mu(x_1,t)}{\partial x_1^l\partial t^q}\Big)\Big|_{\xi=x/\mu},$$

$$u_P^1 = \Big(\sum_{l+q=0}^{K+1}\mu^{l+q}N_{lq}^1(x/\mu)\frac{\partial^{l+q}\omega(x_1,t)}{\partial x_1^l\partial t^q}+$$

$$+\sum_{l+q=0}^{K-1}\mu^{l+q+2}M_{lq}^1(x/\mu)\frac{\partial^{l+q}\psi_\mu(x_1,t)}{\partial x_1^l\partial t^q}\Big)|_{\xi_1=(x_1-b)/\mu\xi'=x'/\mu\cdot},$$

$$\omega^{(K)} = \sum_{j=0}^{K}\mu^j\omega_j(x_1);$$

$$\tilde\rho(x_1,t) = (1-x_1/b)q_0(t)+(x_1/b)q_1(t),$$

$$q_r(t) = \Big(\sum_{l+q=K+1}^{2K+1}\mu^{l+q-2}\sum_{j+p+\alpha=l,0\le j\le K,0\le p+\alpha\le K+1}h_{p\alpha}^{Nr}\frac{\partial^{p+\alpha}\omega_j}{\partial x_1^p\partial t^\alpha}+$$

$$+\mu^{K+1}\sum_{l+q=K-1}h_{lq}^{Mr}\frac{\partial^{K-1}\psi_\mu}{\partial x_1^l\partial t^q}\Big)|_{x_1=rb},\ r=1,2,$$

$N_{lq}, N_{lq}^0, N_{lq}^1, M_{lq}, M_{lq}^0, M_{lq}^1, \omega_j$ *are defined in subsections 2.5.2, 2.5.3. Then the estimate holds*

$$\|u^{(K)}-u\|_{H^1(C_\mu\times(0,T))} = O(\mu^K)\sqrt{mes\ C_\mu}. \tag{2.5.19}$$

Justification of the asymptotic is also similar to that in section 2.3. Namely, it is well known that applying the improving smoothness technique [85],[86],[54] it can be shown that $\frac{\partial(u-u^{(K)})}{\partial t}$ and $\frac{\partial^2(u-u^{(K)})}{\partial t^2}$belong to $H^1(C_\mu\times(0,T))$. So as in section 2.3 we estimate the discrepancy functional

$$I_t(\phi) = \int_{C_\mu}\Big(\rho\Big(\frac{x}{\mu}\Big)\frac{\partial^2(u-u^{(K)})}{\partial t^2},\phi\Big)dx+$$

$$+\int_{C_\mu} \sum_{m,j=1}^{s} \left(A_{mj}\left(\frac{x}{\mu}\right) \frac{\partial(u-u^{(K)})}{\partial x_j}, \frac{\partial \phi}{\partial x_m} \right) dx, \qquad (2.5.20)$$

where for all $t \in (0,T)$, $\phi \in H^1(C_\mu)$, $\phi = 0$ for $x_1 = 0$ or b.

The second integral is developed in the same way as in section 2.3, while the contribution of the first integral corresponds to the supplementary terms $-\rho N_{l,q-2}, -\rho N^r_{l,q-2}$, $r = 0,1$ and $-\rho M_{l,q-2}, -\rho M^r_{l,q-2}$, $r = 0,1$ in cell problems (2.5.7) and boundary layer problems.

We obtain thus for all $t \in (0,T)$, the same representation for $I_t(\phi)$ as in Remark 2.2.4, i.e.

$$I_t(\phi) = \int_{C_\mu} \left(\Psi_{0,\mu}(x,t), \phi \right) dx + \sum_{m=1}^{s} \int_{C_\mu} \left(\Psi_{m,\mu}(x,t), \frac{\partial \phi}{\partial x_m} \right) dx, \quad (2.5.21)$$

where vector valued functions $\Psi_{0,\mu}, \Psi_{m,\mu}, m = 1, ..., s$, are estimated as

$$\sup_{t\in(0,T)} \|\Psi_{0,\mu}(x,t)\|_{L^2(C_\mu)}, \sup_{t\in(0,T)} \|\Psi_{m,\mu}(x,t)\|_{L^2(C_\mu)} \le c\mu^K \sqrt{mes\ C_\mu} \quad (2.5.22)$$

with a constant $c > 0$ independent of μ. Moreover, one can easily check that the time derivatives of functions $\Psi_{0,\mu}, \Psi_{m,\mu}, m = 1, ..., s$, satisfy the same estimate, i.e.

$$\sup_{t\in(0,T)} \|\frac{\partial \Psi_{0,\mu}(x,t)}{\partial t}\|_{L^2(C_\mu)}, \sup_{t\in(0,T)} \|\frac{\partial \Psi_{m,\mu}(x,t)}{\partial t}\|_{L^2(C_\mu)} \le \tilde{c}\mu^K \sqrt{mes\ C_\mu} (2.5.23)$$

with a constant $\tilde{c} > 0$ independent of μ.

Taking $\phi = \frac{\partial(u-u^{(K)})}{\partial t}$, and integrating (2.5.21) from 0 to t in time we obtain the identity

$$\int_0^t I_t\left(\frac{\partial(u-u^{(K)})}{\partial t}\right) dt =$$

$$= \frac{1}{2} \int_0^t \int_{C_\mu} \frac{\partial}{\partial t} \left(\rho\left(\frac{x}{\mu}\right) \left(\frac{\partial(u-u^{(K)})}{\partial t} \right)^2 \right) dx dt +$$

$$+ \frac{1}{2} \int_0^t \int_{C_\mu} \sum_{m,j=1}^{s} \frac{\partial}{\partial t} \left(A_{mj}\left(\frac{x}{\mu}\right) \frac{\partial(u-u^{(K)})}{\partial x_j}, \frac{\partial(u-u^{(K)})}{\partial x_m} \right) dx dt =$$

$$= \int_0^t \left(\int_{C_\mu} \left(\Psi_{0,\mu}(x,t), \frac{\partial(u-u^{(K)})}{\partial t} \right) dx + \right.$$

$$+ \sum_{m=1}^{s} \int_{C_\mu} \Big(\Psi_{m,\mu}(x,t), \frac{\partial}{\partial t} \frac{\partial (u - u^{(K)})}{\partial x_m} \Big) \, dx \Big) dt. \tag{2.5.24}$$

It is possible to show by induction that

$$\omega_j^k(x_1, t) = 0 \quad \text{for} \quad k = 1,4 \quad \text{and} \quad t \leq \theta_0,$$

$$\omega_j^k(x_1, \theta) = 0 \quad \text{for} \quad k = 2,3 \quad \text{and} \quad \theta \leq \theta_0,$$

and therefore

$$u^{(K)}\big|_{t=0} = 0 \quad \text{and} \quad \frac{\partial u^{(K)}}{\partial t}\Big|_{t=0} = 0.$$

Therefore the right hand side of identity (2.5.24) can be integrated by parts with respect to t and it will be transformed then into

$$- \int_0^t \Big(\int_{C_\mu} \Big(\frac{\partial}{\partial t}(\Psi_{0,\mu}(x,t)), (u - u^{(K)}) \Big) \, dx \, +$$

$$+ \sum_{m=1}^{s} \int_{C_\mu} \Big(\frac{\partial}{\partial t}(\Psi_{m,\mu}(x,t)), \frac{\partial (u - u^{(K)})}{\partial x_m} \Big) \, dx \Big) dt \, +$$

$$+ \int_{C_\mu} \Big(\Psi_{0,\mu}(x,t), \phi \Big) \, dx \, + \, \sum_{m=1}^{s} \int_{C_\mu} \Big(\Psi_{m,\mu}(x,t), \frac{\partial \phi}{\partial x_m} \Big) \, dx. \tag{2.5.25}$$

Taking into account estimates (2.5.22),(2.5.23) we estimate the right hand side of identity (2.5.24) by

$$(\tilde{c}T + c)(s+1)\mu^K \sqrt{mes\ C_\mu} \sup_{t \in (0,T)} \|u - u^{(K)}\|_{H^1(C_\mu)}.$$

Therefore for all $t \in (0,T)$, the integral

$$\frac{1}{2} \int_{C_\mu} \Big(\rho\Big(\frac{x}{\mu}\Big) \Big(\frac{\partial (u - u^{(K)})}{\partial t} \Big)^2 \Big) dx \, +$$

$$+ \frac{1}{2} \int_{C_\mu} \sum_{m,j=1}^{s} \Big(A_{mj}\Big(\frac{x}{\mu}\Big) \frac{\partial (u - u^{(K)})}{\partial x_j}, \frac{\partial (u - u^{(K)})}{\partial x_m} \Big) \, dx$$

is estimated by

$$(\tilde{c}T + c)(s+1)\mu^K \sqrt{mes\ C_\mu} \sup_{t \in (0,T)} \|u - u^{(K)}\|_{H^1(C_\mu)}.$$

Passing to $sup_{t \in (0,T)}$ in this last estimate we obtain the following inequality

$$\frac{\kappa}{2} \sup_{t \in (0,T)} \|\frac{\partial (u - u^{(K)})}{\partial t}\|^2_{L^2(C_\mu)} \, +$$

$$+\frac{1}{2}\int_{C_\mu}\sum_{m,j=1}^{s}\Big(A_{mj}\Big(\frac{x}{\mu}\Big)\frac{\partial(u-u^{(K)})}{\partial x_j},\frac{\partial(u-u^{(K)})}{\partial x_m}\Big)\,dx\ \le$$

$$\le\ (\tilde{c}T+c)(s+1)\mu^K\sqrt{mes\ C_\mu}\sup_{t\in(0,T)}\|u-u^{(K)}\|_{H^1(C_\mu)}.$$

So, we get

$$\Big(\sup_{t\in(0,T)}\Big(\|\frac{\partial(u-u^{(K)})}{\partial t}\|^2_{L^2(C_\mu)}+$$

$$+\frac{1}{2}\int_{C_\mu}\sum_{m,j=1}^{s}\Big(A_{mj}\Big(\frac{x}{\mu}\Big)\frac{\partial(u-u^{(K)})}{\partial x_j},\frac{\partial(u-u^{(K)})}{\partial x_m}\Big)\,dx\Big)^{1/2}=$$

$$=\ O(\mu^K)\sqrt{mes\ C_\mu}.$$

This norm is stronger than $H^1(C_\mu\times(0,T))$ in (2.5.19), therefore (2.5.19) is the corollary of this estimate. Theorem 2.5.1 is proved.

2.6 Contrasting coefficients

(Multi-component homogenization)

In this section we consider the highly contrasting coefficients in a 2D model for a rod. We will find out that the classical homogenization can be applied here only in some partial case, when the "contrast" is not too high with respect to μ^2. In the opposite case the multi-component homogenization introduced in [146],[147] should be applied. This example shows that the contrast of the coefficients is an important physical parameter which can correlate with the other geometrical parameters. This scale effect (see also [141]) was used for the invention of a method for measuring of the conductivity of highly heterogeneous stratified plate [164].

Consider a model problem of a rod with contrasting coefficients. Assume that

$$K_\omega(y)=\omega K_+\ \ if\ \ y\in(\frac{1}{4},\frac{1}{2}),$$

$$K_\omega(y)=K_0\ \ if\ \ y\in(-\frac{1}{4},\frac{1}{4}),$$

$$K_\omega(y)=\omega K_-\ \ if\ \ y\in(-\frac{1}{2},-\frac{1}{4}),$$

where $K_0,K_+,K_->0,\omega>0$.

Consider the conductivity equation

$$div(K_\omega(\frac{x_2}{\mu})\nabla u_{\mu,\omega})\ =\ \omega f(x_1),\ x_1\in I\!R,\ |x_2|<\frac{\mu}{2}, \tag{2.6.1}$$

with Neumann condition

$$K_\omega(\frac{x_2}{\mu})\frac{\partial u_{\mu,\omega}}{\partial x_2} = 0, \; x_2 = \pm\frac{\mu}{2}, \tag{2.6.2}$$

and the $T-$periodicity condition in x_1; f is assumed to be a smooth $T-$periodic function of $C^\infty(I\!R)$, such that $\int_0^T f(x_1)dx_1 = 0$.

Thus we consider the 2-dimensional rod (or plate) constituted of three layers

$$G_\mu^+ = \{x \in I\!R^2, \; x_2 \in (\frac{\mu}{4}, \frac{\mu}{2})\},$$

$$G_\mu^0 = \{x \in I\!R^2, \; |x_2| < \frac{\mu}{4}\},$$

and

$$G_\mu^- = \{x \in I\!R^2, \; x_2 \in (-\frac{\mu}{2}, -\frac{\mu}{4})\}$$

with coefficients equal to $\omega K_+, K_0$ and ωK_- respectively in G_μ^+, G_μ^0 and G_μ^-. The interface conditions at the surfaces $\{x_2 = \pm\frac{\mu}{4}\}$ are as follows:

$$[u_{\mu,\omega}] = 0; \;\; [K_\omega \frac{\partial u_{\mu,\omega}}{\partial x_2}] = 0. \tag{2.6.3}$$

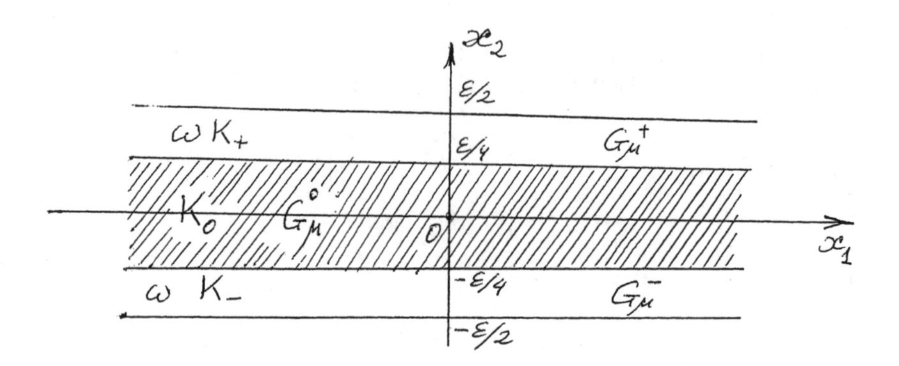

Figure 2.6.1. Three layers plate (rod)

Consider the cases $\mu \to +0$, $\omega \to +\infty$ and $\mu^2\omega = const = \kappa$ or $\mu^2\omega \to +\infty$. In this section we suppose that there exist $\alpha, \beta > 0$, such that $\mu = O(\omega^{-\alpha})$, $\omega^{-1} = O(\mu^{\beta})$.

The asymptotic solution is sought in the form of multi-component homogenization ansatz [146],[147]:

$$u_{\mu,\omega}^{(\infty)} = (\sum_{l=0}^{\infty}\sum_{k=0}^{\infty}\mu^{2l}\omega^{-k}\sum_{\pm}N_{lk}^{\pm}(\xi_2)\frac{d^{2l}v_{\pm}(x_1)}{dx_1^{2l}} +$$

$$+\sum_{l=0}^{\infty}\sum_{k=-1}^{\infty}\mu^{2l}\omega^{-k}M_{lk}(\xi_2)\frac{d^{2l}f(x_1)}{dx_1^{2l}})|_{\xi_2=\frac{x_2}{\mu}}, \tag{2.6.4}$$

where $N_{lk}^{+}, N_{lk}^{-}, M_{lk}, v_+$ and v_- are unknown functions. Ansatz (2.6.4) differs from Bakhvalov's ansatz by the presence of two different functions v_+ and v_-, responsible for the macroscopic description of the upper layer G_μ^+ and of the lower layer G_μ^- respectively.

Substituting (2.6.4) into equation (2.6.1), into Neumann condition (2.6.2) and into interface conditions (2.6.3) we obtain

$$(\sum_{l=0}^{\infty}\sum_{k=0}^{\infty}\mu^{2l-2}\omega^{-k+1}\sum_{\pm}K_{+}(\frac{d^2N_{lk}^{\pm}}{d\xi_2^2}+N_{l-1,k}^{\pm}(\xi_2))\frac{d^{2l}v_{\pm}(x_1)}{dx_1^{2l}} +$$

$$+\sum_{l=0}^{\infty}\sum_{k=-1}^{\infty}\mu^{2l-2}\omega^{-k+1}K_{+}(\frac{d^2M_{lk}}{d\xi_2^2}+M_{l-1,k}(\xi_2))\frac{d^{2l}f(x_1)}{dx_1^{2l}})|_{\xi_2=\frac{x_2}{\mu}} =$$

$$= \omega f(x_1), \quad \xi_2 \in (\frac{1}{4}, \frac{1}{2}), \tag{2.6.5}$$

$$(\sum_{l=0}^{\infty}\sum_{k=0}^{\infty} \mu^{2l-2}\omega^{-k+1} \sum_{\pm} K_-(\frac{d^2 N_{lk}^{\pm}}{d\xi_2^2} + N_{l-1,k}^{\pm}(\xi_2))\frac{d^{2l}v_{\pm}(x_1)}{dx_1^{2l}} +$$

$$+ \sum_{l=0}^{\infty}\sum_{k=-1}^{\infty} \mu^{2l-2}\omega^{-k+1} K_-(\frac{d^2 M_{lk}}{d\xi_2^2} + M_{l-1,k}(\xi_2))\frac{d^{2l}f(x_1)}{dx_1^{2l}})|_{\xi_2=\frac{x_2}{\mu}} =$$

$$= \omega f(x_1), \quad \xi_2 \in (-\frac{1}{2}, -\frac{1}{4}), \tag{2.6.6}$$

$$(\sum_{l=0}^{\infty}\sum_{k=0}^{\infty} \mu^{2l-2}\omega^{-k} \sum_{\pm} K_0(\frac{d^2 N_{lk}^{\pm}}{d\xi_2^2} + N_{l-1,k}^{\pm}(\xi_2))\frac{d^{2l}v_{\pm}(x_1)}{dx_1^{2l}} +$$

$$+ \sum_{l=0}^{\infty}\sum_{k=-1}^{\infty} \mu^{2l-2}\omega^{-k} K_0(\frac{d^2 M_{lk}}{d\xi_2^2} + M_{l-1,k}(\xi_2))\frac{d^{2l}f(x_1)}{dx_1^{2l}})|_{\xi_2=\frac{x_2}{\mu}} =$$

$$= \omega f(x_1), \quad \xi_2 \in (-\frac{1}{4}, \frac{1}{4}), \tag{2.6.7}$$

$$(\sum_{l=0}^{\infty}\sum_{k=0}^{\infty} \mu^{2l-1}\omega^{-k+1} \sum_{\pm} \frac{dN_{lk}^{\pm}}{d\xi_2}\frac{d^{2l}v_{\pm}(x_1)}{dx_1^{2l}} +$$

$$+ \sum_{l=-1}^{\infty}\sum_{k=0}^{\infty} \mu^{2l-1}\omega^{-k+1} \frac{dM_{lk}}{d\xi_2}\frac{d^{2l}f(x_1)}{dx_1^{2l}})|_{\xi_2=\frac{x_2}{\mu}} = 0, \quad \xi_2 = -\frac{1}{2} \text{ or } \frac{1}{2}, \tag{2.6.8}$$

$$(\sum_{l=0}^{\infty}\sum_{k=0}^{\infty} \mu^{2l}\omega^{-k} \sum_{\pm}[N_{lk}^{\pm}]\frac{d^{2l}v_{\pm}(x_1)}{dx_1^{2l}} +$$

$$+ \sum_{l=0}^{\infty}\sum_{k=-1}^{\infty} \mu^{2l}\omega^{-k}[M_{lk}]\frac{d^{2l}f(x_1)}{dx_1^{2l}})|_{\xi_2=\frac{x_2}{\mu}} = 0, \quad \xi_2 = -\frac{1}{4} \text{ or } \frac{1}{4}, \tag{2.6.9}$$

$$\sum_{l=0}^{\infty}\sum_{k=0}^{\infty} \mu^{2l-1}\omega^{-k+1} \sum_{\pm} K_+ \frac{dN_{lk}^{\pm}}{d\xi_2}(\frac{1}{4}+0)\frac{d^{2l}v_{\pm}(x_1)}{dx_1^{2l}} -$$

$$- \sum_{l=0}^{\infty}\sum_{k=0}^{\infty} \mu^{2l-1}\omega^{-k} \sum_{\pm} K_0 \frac{dN_{lk}^{\pm}}{d\xi_2}(\frac{1}{4}-0)\frac{d^{2l}v_{\pm}(x_1)}{dx_1^{2l}} +$$

$$+ \sum_{l=0}^{\infty}\sum_{k=-1}^{\infty} \mu^{2l-1}\omega^{-k+1} K_+ \frac{dM_{lk}}{d\xi_2}(\frac{1}{4}+0)\frac{d^{2l}f(x_1)}{dx_1^{2l}} -$$

$$-\sum_{l=0}^{\infty}\sum_{k=-1}^{\infty}\mu^{2l-1}\omega^{-k}K_0\frac{dM_{lk}}{d\xi_2}(\frac{1}{4}-0)\frac{d^{2l}f(x_1)}{dx_1^{2l}} = 0, \tag{2.6.10}$$

and the analogous condition for $\xi_2 = -\frac{1}{4}$.

Require that $N_{lk}^{\pm}$ and M_{lk} satisfy the following boundary value problems in $G^+ = \{\xi_2 \in (\frac{1}{4}, \frac{1}{2})\}$:

$$K_+(\frac{d^2N_{lk}^{\pm}}{d\xi_2^2} + N_{l-1,k}^{\pm}(\xi_2)) = h_{lk}^{\pm,+},\ \xi_2 \in (\frac{1}{4}, \frac{1}{2}), \tag{2.6.11}$$

$$\frac{dN_{lk}^{\pm}}{d\xi_2}(\frac{1}{2}) = 0, \tag{2.6.12}$$

$$K_+\frac{dN_{lk}^{\pm}}{d\xi_2}(\frac{1}{4}+0) = K_0\frac{dN_{l,k-1}^{\pm}}{d\xi_2}(\frac{1}{4}-0), \tag{2.6.13}$$

and

$$K_+(\frac{d^2M_{lk}}{d\xi_2^2} + M_{l-1,k}(\xi_2)) = h_{lk}^{+},\ \xi_2 \in (\frac{1}{4}, \frac{1}{2}), \tag{2.6.14}$$

$$\frac{dM_{lk}}{d\xi_2}(\frac{1}{2}) = 0, \tag{2.6.15}$$

$$K_+\frac{dM_{lk}}{d\xi_2}(\frac{1}{4}+0) = K_0\frac{dM_{l,k-1}}{d\xi_2}(\frac{1}{4}-0), \tag{2.6.16}$$

analogously, in $G^- = \{\xi_2 \in (-\frac{1}{2}, -\frac{1}{4})\}$:

$$K_-(\frac{d^2N_{lk}^{\pm}}{d\xi_2^2} + N_{l-1,k}^{\pm}(\xi_2)) = h_{lk}^{\pm,-},\ \xi_2 \in (-\frac{1}{2}, -\frac{1}{4}), \tag{2.6.17}$$

$$\frac{dN_{lk}^{\pm}}{d\xi_2}(-\frac{1}{2}) = 0, \tag{2.6.18}$$

$$K_-\frac{dN_{lk}^{\pm}}{d\xi_2}(-\frac{1}{4}-0) = K_0\frac{dN_{l,k-1}^{\pm}}{d\xi_2}(-\frac{1}{4}+0), \tag{2.6.19}$$

and

$$K_-(\frac{d^2M_{lk}}{d\xi_2^2} + M_{l-1,k}(\xi_2)) = h_{lk}^{+},\ \xi_2 \in (-\frac{1}{2}, -\frac{1}{4})\}, \tag{2.6.20}$$

$$\frac{dM_{lk}}{d\xi_2}(-\frac{1}{2}) = 0, \tag{2.6.21}$$

$$K_-\frac{dM_{lk}}{d\xi_2}(-\frac{1}{4}-0) = K_0\frac{dM_{l,k-1}}{d\xi_2}(-\frac{1}{4}+0), \tag{2.6.22}$$

and in $G^0 = \{\xi_2 \in (-\frac{1}{4}, \frac{1}{4})\}$:

$$K_0(\frac{d^2N_{lk}^{\pm}}{d\xi_2^2} + N_{l-1,k}^{\pm}(\xi_2)) = 0, \ \xi_2 \in (-\frac{1}{4}, \frac{1}{4}), \tag{2.6.23}$$

$$N_{lk}^{\pm}(\frac{1}{4} - 0) = N_{lk}^{\pm}(\frac{1}{4} + 0) \ , \tag{2.6.24}$$

$$N_{lk}^{\pm}(-\frac{1}{4} + 0) = N_{lk}^{\pm}(-\frac{1}{4} - 0) \ , \tag{2.6.25}$$

$$K_0(\frac{d^2M_{lk}}{d\xi_2^2} + M_{l-1,k}(\xi_2)) = \delta_{l1}\delta_{k,-1}, \ \xi_2 \in (-\frac{1}{4}, \frac{1}{4}), \tag{2.6.26}$$

$$M_{lk}(\frac{1}{4} - 0) = M_{lk}(\frac{1}{4} + 0) \ , \tag{2.6.27}$$

$$M_{lk}(-\frac{1}{4} + 0) = M_{lk}(-\frac{1}{4} - 0) \ , \tag{2.6.28}$$

Here and below $N_{lk}^{\pm} = 0$ if $l < 0$ or $k < 0$; $M_{lk} = 0$ if $l < 0$ or $k < -1$. Note that $M_{l,-1}(\xi_2) = 0$ if $|\xi_2| > \frac{1}{4}$.

It is clear that $h_{lk}^{\pm,\pm}$ and $h_{lk}^{\pm}$ should be taken in such a way that problems (2.6.11)-(2.6.13), (2.6.14)-(2.6.16), (2.6.17)-(2.6.19), (2.6.20)-(2.6.22) have a solution, ie.,

$$h_{lk}^{\pm,+} = 4(\langle K_+ N_{l-1,k}^{\pm}\rangle_+ - K_0\frac{dN_{l,k-1}^{\pm}}{d\xi_2}(\frac{1}{4} - 0)),$$

$$h_{lk}^{+} = 4(\langle K_+ M_{l-1,k}\rangle_+ - K_0\frac{dM_{l,k-1}}{d\xi_2}(\frac{1}{4} - 0)),$$

$$h_{lk}^{\pm,-} = 4(\langle K_- N_{l-1,k}^{\pm}\rangle_- + K_0\frac{dN_{l,k-1}^{\pm}}{d\xi_2}(-\frac{1}{4} + 0)),$$

$$h_{lk}^{-} = 4(\langle K_- M_{l-1,k}\rangle_- + K_0\frac{dM_{l,k-1}}{d\xi_2}(-\frac{1}{4} + 0)),$$

$$\langle\cdot\rangle_+ = \int_{\frac{1}{4}}^{\frac{1}{2}} .d\xi_2, \ \ \langle\cdot\rangle_- = \int_{-\frac{1}{2}}^{-\frac{1}{4}} .d\xi_2.$$

To initialize the recurrent chain of problems, let us take $N_{00}^{+}(\xi_2) = 1$ in G^+, $N_{00}^{+}(\xi_2) = 2(\xi_2 + \frac{1}{4})$ in G^0, $N_{00}^{+}(\xi_2) = 0$ in G^-, and $N_{00}^{-}(\xi_2) = 0$ in G^+, $N_{00}^{-}(\xi_2) = -2(\xi_2 - \frac{1}{4})$ in G^0, $N_{00}^{-}(\xi_2) = 1$ in G^-, and $M_{00}(\xi_2) = \frac{1}{2K_0}(\xi_2^2 - \frac{1}{16})$ in G^0, $M_{1,-1}(\xi_2) = 0$ in $G^{\pm}$; $M_{1,-1}(\xi_2) = 0$ in $G^{\pm}$; $M_{0,-1}(\xi_2) = 0$ everywhere.

So, we get

$$h_{00}^{\pm,\pm} = 0, \ h_{10}^{++} = 4\langle K_+\rangle_+, \ \ h_{10}^{-,+} = h_{01}^{+-} = 0, \ h_{10}^{--} = 4\langle K_-\rangle_-,$$

$$h_{01}^{++} = -8K_0,\ h_{01}^{-+} = 8K_0,\ h_{01}^{+-} = 8K_0,\ h_{01}^{--} = -8K_0,$$

$$h_{00}^{\pm} = 0,\ h_{10}^{+} = h_{10}^{-} = -1; h_{01}^{+} = h_{01}^{-} = 0.$$

Equations (2.6.5) and (2.6.6) transform into

$$4\omega\langle K_+\rangle_+ \frac{d^2v_+}{dx_1^2} - 8\mu^{-2}K_0(v_+ - v_-) + S^+ = 2\omega f(x_1), \tag{2.6.29}$$

and

$$4\omega\langle K_-\rangle_- \frac{d^2v_-}{dx_1^2} + 8\mu^{-2}K_0(v_+ - v_-) + S^- = 2\omega f(x_1), \tag{2.6.30}$$

where

$$S^+ = \sum_{l,k\geq 0; l+k\geq 2} \mu^{2l-2}\omega^{-k+1}\sum_{\pm} h_{lk}^{\pm+}\frac{d^{2l}v_\pm(x_1)}{dx_1^{2l}} + \\ + \sum_{l,k\geq 0; l+k\geq 2} \mu^{2l-2}\omega^{-k+1}h_{lk}^{+}\frac{d^{2l}f(x_1)}{dx_1^{2l}}$$

and

$$S^- = \sum_{l,k\geq 0; l+k\geq 2} \mu^{2l-2}\omega^{-k+1}\sum_{\pm} h_{lk}^{\pm-}\frac{d^{2l}v_\pm(x_1)}{dx_1^{2l}} + \\ + \sum_{l,k\geq 0; l+k\geq 2} \mu^{2l-2}\omega^{-k+1}h_{lk}^{-}\frac{d^{2l}f(x_1)}{dx_1^{2l}}.$$

and v_+ and v_- are sought as the series

$$v_\pm(x_1) = \sum_{p_1,p_2,p_3=0}^{\infty} \mu^{2p_1}\omega^{-p_2}(\mu^2\omega)^{-p_3}v_{\pm p_1p_2p_3}(x_1) \tag{2.6.31}$$

if $\mu^2\omega \to +\infty$ and

$$v_\pm(x_1) = \sum_{p_1,p_2=0}^{\infty} \mu^{2p_1}\omega^{-p_2}v_{\pm p_1p_2}(x_1) \tag{2.6.32}$$

if $\mu^2\omega = const = \kappa$.

We obtain the recurrent chain of problems for functions $v_{\pm p_1p_2p_3}$ and $v_{\pm p_1p_2}$ of the form

$$4\langle K_+\rangle_+ \frac{d^2v_{+p_1p_2p_3}(x_1)}{dx_1^2} = f_{p_1p_2p_3}^{+}(x_1),$$

$$4\langle K_-\rangle_-\frac{d^2v_{-p_1p_2p_3}(x_1)}{dx_1^2}=f^-_{p_1p_2p_3}(x_1), \tag{2.6.33}$$

and

$$4\langle K_+\rangle_+\frac{d^2v_{+p_1p_2}(x_1)}{dx_1^2}-8\kappa^{-1}K_0(v_{+p_1p_2}(x_1)-v_{-p_1p_2}(x_1))=f^+_{p_1p_2}(x_1),$$

$$4\langle K_-\rangle_-\frac{d^2v_{-p_1p_2}(x_1)}{dx_1^2}+8\kappa^{-1}K_0(v_{+p_1p_2}(x_1)-v_{-p_1p_2}(x_1))=f^-_{p_1p_2}(x_1), \tag{2.6.34}$$

respectively, where $f^{\pm}_{000}(x_1) = 2f(x_1)$, $f^{\pm}_{00}(x_1) = 2f(x_1)$, and $f^{\pm}_{p_1p_2p_3}, f^{\pm}_{p_1p_2}$ depend on $v_{\pm q_1q_1q_2}$ (respectively on $v_{\pm q_1q_2}$) with $q_i \le p_i$ ($i = 1,2$ and eventually 3), $(q_1,q_2,q_3) \ne (p_1,p_2,p_3)$ and $(q_1,q_2) \ne (p_1,p_2)$.

Justification is made as usual by a truncation of the series at the level $l \le J, k \le J, p_i \le J$ ($i = 1,2$ and eventually 3). We obtain in a standard way:

$$\|\nabla(u_{\mu,\omega}-u^{(J)}_{\mu,\omega})\|_{L^2(0,T)\times\beta_\mu)}\le C(\mu^{2J}+\omega^{-J}+(\mu^2\omega)^{-J}) \tag{2.6.35}$$

in the case $\mu^2\omega \to +\infty$ and

$$\|\nabla(u_{\mu,\omega}-u^{(J)}_{\mu,\omega})\|_{L^2(0,T)\times\beta_\mu)}\le C(\mu^{2J}+\omega^{-J}) \tag{2.6.36}$$

in the case $\mu^2\omega = const$, where $u^{(J)}_{\mu,\omega}$ is a truncated sum (2.6.4), (2.6.31), (2.6.32) at $l = J+1, k = J, p_i = J$.

The leading term of the expansion is

$$N^+_{00}(\frac{x_2}{\mu})v_{+0}(x_1)+N^-_{00}(\frac{x_2}{\mu})v_{-0}(x_1)+M_{1,-1}(\frac{x_2}{\mu})f(x_1),$$

where v_{+0}, v_{-0} is a solution of the system

$$2\langle K_+\rangle_+\frac{d^2v_{+0}(x_1)}{dx_1^2}-4\kappa^{-1}K_0(v_{+0}(x_1)-v_{-0}(x_1))=f(x_1),$$

$$2\langle K_-\rangle_-\frac{d^2v_{-0}(x_1)}{dx_1^2}+4\kappa^{-1}K_0(v_{+0}(x_1)-v_{-0}(x_1))=f(x_1),$$

$\kappa = 0$ if $\mu^2\omega \to +\infty$.

It means that when $\mu^2\omega \to +\infty$ the macroscopic fields v_{+0} in G^+_μ and v_{-0} in G^-_μ are independent and different if $K_+ \ne K_-$.

When $\mu^2\omega = const$ there is some interaction between the macroscopic fields of upper and lower layers, nevertheless they are very different if $K_+ \ne K_-$.

Let us consider now the third case when $\mu^2\omega \to +0$. Then the behavior of the solution is close to the case of finite ω (see section 2.2). We seek an asymptotic solution in a form of Bakhvalov's ansatz

$$u_{\mu,\omega}^{(\infty)} = (\sum_{l=0}^{\infty} \mu^{2l} N_l^{(\omega)}(\xi_2) \frac{d^{2l}v(x_1)}{dx_1^{2l}})|_{\xi_2=x_2/\mu}, \quad (2.6.36)$$

where $N_l^{(\omega)}$ are solutions of the cell problems

$$\frac{d}{d\xi_2}(K_\omega(\xi_2)\frac{dN_l^{(\omega)}}{d\xi_2}) + K_\omega(\xi_2)N_{l-1}^{(\omega)}(\xi_2) = h_l, \ \xi_2 \in (-\frac{1}{2}, \frac{1}{2}), \quad (2.6.37)$$

$$\frac{dN_l^{(\omega)}}{d\xi_2}(\pm\frac{1}{2}) = 0, \quad (2.6.38)$$

At the points $\xi_2 = \pm\frac{1}{4}$ the interface conditions hold true

$$[N_l^{(\omega)}] = 0, \ \ [K_\omega(\xi_2)\frac{dN_l^{(\omega)}}{d\xi_2}] = 0, \quad (2.6.39)$$

$$h_l = \langle K_\omega(\xi_2)N_{l-1}^{(\omega)}(\xi_2)\rangle, \quad (2.6.40)$$

here

$$\langle . \rangle = \int_{-\frac{1}{2}}^{\frac{1}{2}} .d\xi_2.$$

We define $N_0 = 1$. We get

$$K_\omega(\xi_2)\frac{dN_l^{(\omega)}}{d\xi_2} = \int_{-\frac{1}{2}}^{\xi_2} (h_l - K_\omega(y)N_{l-1}^{(\omega)}(y))dy$$

and so,

$$\frac{dN_l^{(\omega)}}{d\xi_2} = \frac{1}{K_\omega(\xi_2)}\int_{-\frac{1}{2}}^{\xi_2} (h_l - K_\omega(y)N_{l-1}^{(\omega)}(y))dy.$$

By induction we can prove that

$$h_0 = 0, \ h_l = \sum_{l_1=-l+1}^{l} \omega^{l_1} h_{ll_1}, \ \ l > 0 \quad (2.6.41)$$

$$N_l^{(\omega)}(\xi_2) = \sum_{l_1=-l}^{l} \omega^{l_1} N_{ll_1}^{(\omega)}(\xi_2) \quad (2.6.42)$$

h_{ll_1}, N_{ll_1} do not depend on μ.

Indeed, for $l = 0$, $N_0 = 1$, $h_0 = 0$; let (2.6.41), (2.6.42) be true for some l; then

$$h_{l+1} = \langle K_\omega(\xi_2) N_l^{(\omega)}(\xi_2)) \rangle = \langle K_\omega(\xi_2) \sum_{l_1=-l}^{l} \omega^{l_1} N_{ll_1}^{(\omega)}(\xi_2) \rangle =$$

$$= \sum_{l_1=-l}^{l+1} \omega^{l_1} h_{ll_1};$$

$$\frac{dN_{l+1}^{(\omega)}}{d\xi_2} = \frac{1}{K_\omega(\xi_2)} \int_{-\frac{1}{2}}^{\xi_2} (\sum_{l_1=-l}^{l+1} \omega^{l_1} h_{ll_1} - K_\omega(y) \sum_{l_1=-l}^{l} \omega^{l_1} N_{ll_1}^{(\omega)}(y)) dy =$$

$$= \sum_{l_1=-(l+1)}^{l+1} \omega^{l_1} F_{ll_1}(\xi_2),$$

where F_{ll_1} do not depend on ω.

For v we get the homogenized equation

$$\sum_{l=1}^{\infty} \mu^{2l-2} h_l \frac{d^{2l} v(x_1)}{dx_1^{2l}} = \omega f(x_1), \tag{2.6.43}$$

where $h_1 = \omega(\langle K_+ \rangle_+ + \langle K_- \rangle_-) + K_0/2$; i.e.,

$$\sum_{l=1}^{\infty} \sum_{l_1=-l}^{l} \mu^{2l-2} \omega^{l_1-1} h_{ll_1} \frac{d^{2l} v(x_1)}{dx_1^{2l}} = f(x_1),$$

i.e.

$$\sum_{l=1}^{\infty} \sum_{l_2=0}^{2l} (\mu^2\omega)^{l-1} \omega^{-l_2} h_{l,l-l_2} \frac{d^{2l} v(x_1)}{dx_1^{2l}} = f(x_1), \tag{2.6.44}$$

Then v is sought in a form

$$v = \sum_{j,k=0}^{\infty} (\mu^2\omega)^j \omega^{-k} v_{jk}(x_1), \tag{2.6.45}$$

and we have for v_{jk} the equations

$$(\langle K_+ \rangle_+ + \langle K_- \rangle_-) \frac{d^2 v_{jk}(x_1)}{dx_1^2} = f_{jk}(x_1), \tag{2.6.46}$$

and $f_{jk}(x_1)$ determined by $v_{j_1k_1}$ such that, $j_1 \leq j$, $k_1 \leq k$, $(j_1, k_1) \neq (j, k)$; $f_{00} = f$.

The leading term is v_{00} that is a solution of the equation

$$(\langle K_+ \rangle_+ + \langle K_- \rangle_-) \frac{d^2 v_{00}(x_1)}{dx_1^2} = f(x_1). \tag{2.6.47}$$

For the truncated sum $u^{(J)}_{\mu,\omega}$ (2.6.36), (2.6.45) at $l \leq J+1,\ j \leq J,\ k \leq J$ the following estimate holds true :

$$\|\nabla(u_{\mu,\omega} - u^{(J)}_{\mu,\omega})\|_{L^2((0,T)\times\beta_\mu)} \leq C(\omega^{-J} + (\mu^2\omega)^J) \tag{2.6.48}$$

where C does not depend on the parameters.

So, in this case, when $\mu^2\omega \to +0$, the solution tends to function v_{00} independent of μ and ω.

Thus if $\mu^2\omega$ is a small parameter then problem (2.6.1) - (2.6.3) has a classical homogenization limit v_{00}. If $\mu^2\omega$ is a great or finite parameter then there is no a unique limit function v independent of μ and ω : the solution tends in the upper layer to function v_+ and in the lower layer to function v_-. This is a case of the multi-component homogenization.

2.7 EFMODUL: a code for cell problems

In asymptotic study of partial differential equations auxiliary problems arise - so-called cell problems. The solutions of these problems enable us to define the effective properties of heterogeneous media. These cell problems are solved numerically by the finite-difference method implemented in the code EFMODUL. The numerical results are analyzed and the possibility of an application of the extrapolation method is discussed here below.

Consider the system of equations

$$R\left(\frac{x}{\varepsilon}\right)\frac{\partial^2 u}{\partial t^2} + S\left(\frac{x}{\varepsilon}\right)\frac{\partial u}{\partial t} - \sum_{i,j=1}^{s} \frac{\partial}{\partial x_i}\left(A_{ij}\left(\frac{x}{\varepsilon}\right)\frac{\partial u}{\partial x_j}\right) = f(x,t). \tag{2.7.1}$$

Here $R(\xi), S(\xi)$ and $A_{ij}(\xi)$ are piecewise-smooth 1-periodic square $n{\times}n$ matrices-valued functions of the variable $\xi \in \mathbb{R}^s, s = 1,2,3;\quad u, f$ are $n-$dimensional vectors, $n = 1,2,3;\ \ \varepsilon$ is a small parameter. Equation (2.7.1) describes different processes in composite materials: the wave propagation in elastic non-uniform media, heat transfer ($R = 0$), diffusion, flow in porous media and so on ; if $R = S = 0$ we have a steady state problem. An asymptotic study of equation (1) as $\varepsilon \to 0$ is given in [16] by the homogenization theory. It is proved that in some assumptions on the coefficients the solution to equation (2.7.1) tends to a solution v of the homogenized equation that is,

$$\hat{R}\frac{\partial^2 v}{\partial t^2} + \hat{S}\frac{\partial v}{\partial t} - \sum_{i,j=1}^{s} \frac{\partial}{\partial x_i}\left(\hat{A}_{ij}\frac{\partial v}{\partial x_j}\right) = f(x,t) \tag{2.7.2}$$

with corresponding initial and boundary conditions. The constant $n \times n$ matrix-valued coefficients of this equation are called the effective (or macroscopic, or homogenized) coefficients. One of the fundamental problems of the homogenization theory is the calculation of these coefficients. In [16] it was shown that $\hat{R} = \langle R(\xi)\rangle, \hat{S} = \langle S(\xi)\rangle$, where $\langle.\rangle$ denotes the integral over a periodic cubic cell $\{0 < \xi_i < 1, i = 1,2,\ldots,s\}$. To determine $\hat{A}_{ij}$ we have to solve s, subsidiary cell problems for 1-periodic unknown $n \times n$ matrix-valued functions $N_q(\xi)$ (i.e. for every $q = 1, 2, \dots, s$):

$$\sum_{i,j=1}^{s} \frac{\partial}{\partial \xi_i} \left(A_{ij}(\xi) \frac{\partial (N_q + \xi_q I)}{\partial \xi_j} \right) = 0, \quad \xi \in \mathbb{R}^s, \tag{2.7.3}$$

where I is the unit $n \times n-$ matrix, and put

$$\hat{A}_{ij} = \left\langle \sum_{r=1}^{s} A_{ir}(\xi) \frac{(N_j + \xi_j I)}{\partial \xi_r} \right\rangle. \tag{2.7.4}$$

In the case when the coefficients $A_{ij}(\xi)$ are discontinuous, the solution N_q of systems (2.7.3) is understood in the sense of the variational formulation; in the classical formulation on the surfaces of discontinuity the interface conditions

$$[N_q] = 0, \left[\sum_{i,j=1}^{s} A_{ij}(\xi) \frac{\partial (N_q + \xi_q I)}{\partial \xi_j} n_j(\xi) \right] = 0,$$

are specified, where [.] is the function jump across the surface of discontinuity and $n_j(\xi)$ are the direction cosines of the normal vector. In the composite materials mechanics the quantities $\tilde{A}_{ij}(\xi) = \sum_{r=1}^{s} A_{ir}(\xi) \frac{\partial (N_j + \xi_j I)}{\partial \xi_r}$ are also of interest since in the case of the elasticity problem the local (microscopic) stresses are calculated from the formula

$$\sigma_i^k(x) = \tilde{a}_{ij}^{kl}(x/\varepsilon) e_j^l(v) + O(\epsilon),$$

where $\tilde{a}_{ij}^{kl}(\xi)$ are elements of the matrix $\tilde{A}_{ij}(\xi)$ and $e_j^l(v) = (\partial v^j/\partial x_l + \partial v^l/\partial x_j)/2$ is the mean (macroscopic) strain tensor (see [16] for a more accurate formulation).

For every q, cell problem (2.7.3) for the matrix-valued unknown function can be split into n independent systems relative to the columns N_q^p of the matrix N_q, $p = 1, 2, \dots, n$. We shall rewrite (2.7.3) in the form

$$\sum_{i=1}^{s} \frac{\partial}{\partial \xi_i} A_i(\xi, \nabla_\xi (N_q^p + \xi_q e^p)) = 0, \quad p = 1, 2, \dots, n, \tag{2.7.5}$$

where e^p is the p-th column of the unit matrix, $\nabla_\xi (N_q^p + \xi_q E^p)$ is $n \times s-$matrix constituted of columns $\partial(N_q^p + \xi_q E^p)/\partial \xi_1, \dots, \partial(N_q^p + \xi_q E^p)/\partial \xi_s$ and $A_i(\xi, \nabla_\xi N) = \sum_{j=1}^{s} A_{ij} \partial N / \partial \xi_i$ is an n-dimensional vector. A difference scheme of the second-order approximation (for smooth A_i) is used to solve system (2.7.5)

$$L_h N_q^p = 0 \tag{2.7.6}$$

where

$$L_h N_q^p = \sum_{i=1}^{s} [A_i(\xi_+^{(i)}, (N_q^p + \xi_q e^p)_{\xi_+^{(i)}}) - A_i(\xi_-^{(i)}, (N_q^p + \xi_q e^p)_{\xi_-^{(i)}})]/h_i, \tag{2.7.7}$$

$\xi^{(i)}_{\pm} = \xi \pm e^i h_i/2, e^i = (\delta_{i1} \dots, \delta_{is})^*$ (as above), and for any n-dimensional vector $\varphi(\xi)$ we define the $n \times s-$matrix $\varphi_{\xi^{(i)}_{\pm}}$ in such a way that, its j-th column is equal to $[\pm\varphi(\xi \pm e_i h_i) \mp \varphi(\xi)]/h_i$ if $j = i$, and $[\varphi(\xi \pm e_i h_i + e_j h_j) + \varphi(\xi + e_j h_j) - \varphi(\xi \pm e_i h_i - e_j h_j) - \varphi(\xi - e_j h_j)]/4h_j$ if $j \neq i$ (here there is no summing over the repeated index), and $h_i, \dots, h_s$ are the steps with respect to $\xi_i, \dots, \xi_s$. The mesh is chosen to be uniform : $\xi = (i_1 h_1, \dots, i_s h_s)^t, i_1, \dots, i_s \in \mathbb{Z}$, $*$ is the sign of transposition, and $\mathbb{Z}$ is the set of integers. In order to solve, in turn, the system of algebraic equations (2.7.6), an implicit fix point method was applied with inversion of the difference analogue of the Laplacian (a fast Fourier transform was used, see [173]) :

$$\Lambda N_q^{p(k)} = \Lambda N_q^{p(k-1)} - \tau L_h N_q^{p(k-1)}, \tag{2.7.8}$$

where $N_q^{p(k)}$ is the $k-th$ order iteration approximation, τ is the iteration parameter and

$$\Lambda\varphi = \sum_{i=1}^{s} \varphi_{\bar{\xi}_i \xi_i}$$

is the difference analogue of the Laplace operator [173].

The given method can be used as well in the case of the non-linear system of equations

$$R\left(\frac{x}{\varepsilon}\right)\frac{\partial^2 u}{\partial t^2} + S\left(\frac{x}{\varepsilon}\right)\frac{\partial u}{\partial t} - \sum_{i=1}^{s} \frac{d}{dx_i} A_i\left(\frac{x}{\varepsilon}, \nabla_x u\right) = f(x,t) \tag{2.7.9}$$

where $A_i(\xi, y)$ is 1-periodic in ξ and an n-dimensional vector-valued function that is smooth in y, y is an $n \times s$ matrix, $\nabla_x u = (\partial u/\partial x_1, \dots, \partial u/\partial x_\varepsilon)$, and u is an n-dimensional vector-valued unknown function. An analogue of the cell problems (2.7.3)is the parametric family of systems of equations (y is the matrix parameter)

$$\frac{d}{d\xi_i} A_i(\xi, \nabla_\xi N(\xi, y) + y) = 0, \tag{2.7.10}$$

the solution $N(\xi, y)$ is found in the class of n-dimensional vector functions that are 1-periodic in ξ. Homogenized equation has the form (see [16])

$$\hat{R}\frac{\partial^2 v}{\partial t^2} + \hat{S}\frac{\partial v}{\partial t} - \frac{d}{dx_i}\hat{A}_i(\nabla_x v) = f(x,t), \tag{2.7.11}$$

where

$$\hat{R} = \langle R(\xi)\rangle, \quad \hat{S} = \langle S(\xi)\rangle, \quad \hat{A}_i(y) = \langle A_i(\xi, \nabla_\xi N + y)\rangle$$

The components of the stress tensor are calculated from the formulae
$\sigma_i \approx A_i(x/\varepsilon, \nabla_\xi N(x/\varepsilon, \nabla_x v) + \nabla_x v)$
To solve (2.7.10) the difference scheme

$$L_h N = 0 \tag{2.7.12}$$

with

$$L_h N = \sum_{i=1}^{s} [A_1(\xi_+^{(i)}, N_{\xi_+^{(i)}} + y) - A_i(\xi_-^{(i)}, N_{\xi_-^{(i)}} + y)]/h_i \qquad (2.7.13)$$

is applied and the fixed point method of type (2.7.8) is used :

$$\Lambda N^{(k)} = \Lambda N^{(k-1)} - \tau L_h N^{(k-1)},$$

where Λ is the difference analogue of the Laplace operator.

Consider porous (perforated) media with an ε-periodic in all the x_i pores and denote by A_ε) the domain occupied by pores. Then equation (2.7.9) is set in $x \in R^s \backslash A_\varepsilon$ and it is supplied by the Neumann boundary conditions on pore surface ∂A_ε.

$$A_i(x/\varepsilon, \nabla_x u) n_i(x/\varepsilon) = 0 \quad \text{for} \quad x \in \partial A_\varepsilon.$$

In this case equation (2.7.10) is set out of pores, i.e. for $\xi \in I\!R^s \backslash A_0$, where $A_0 = \{\xi | \ \varepsilon\xi \in A_\varepsilon\}$. On the boundary $\xi \in \partial A_0$ the following boundary condition is set:

$$\sum_{i=1}^{s} A_i(\xi, \nabla_\xi N + y) n_i(\xi) = 0. \qquad (2.7.14)$$

Problem (2.7.10) and (2.7.14) was solved by exactly the same method. In this case in (2.7.8) the quantity A_i was considered to be zero if $\xi_+^{(i)}$ (or $\xi_-^{(i)}$) is in the set A_0

The homogenized equation in this case has the form (2.7.11), where f must be replaced by $f mes(Q \backslash A_0)$, and Q is the unit cube.

The higher order cell problems can be presented in the form

$$\sum_{i=1}^{s} \frac{\partial}{\partial \xi_i} A_i(\xi, \nabla_\xi N + Y(\xi)) = F(\xi) \qquad (2.7.15)$$

where N is a 1-periodic unknown function and $Y(\xi), F(\xi)$ are 1-periodic in ξ (see [16]). These problems can also be solved by the method given above. The cell problems for heterogeneous rods and plates considered in Chapters 2, 3 have also the same form (2.7.15).

In the case when the symmetry of coefficients holds with respect to some coordinate plane, the cell problem can be reduced to the problem in the half-cell by using the properties of evenness and oddness of the components of the solution (see [16], Chapter 5). In this case the periodic problems could be reduced to boundary-value ones.

The above difference schemes were implemented in a form of the computer solver EFMODUL for conductivity and elasticity problems.

Let us consider some results computed by this solver [142].

I. Consider the case of one equation, i.e. $n = 1$. Let

$$A_{ij}(\xi) = \delta_{ij} K(\xi), \quad K(\xi) = \begin{cases} 1 & \text{for} \quad \xi \in G_1 \\ \kappa & \text{for} \quad \xi \in G_2, \end{cases}$$

where G_2 is a sub-domain of the periodic cell $Q = \{\xi \mid \xi_i \in (0,\ 1),\quad i = 1,2,\dots,s\}, G_1 = Q\backslash G_2$.

1. Consider the case $s = 2$, G_2 is a disk of radius 0.25 with its center at the point (0.5,0.5). Then $\hat{A}_{ij} = \delta_{ij}\hat{K}$ (see [16]). Below, values of $\hat{K}$ that depend on κ are computed: if $\kappa = 0, 2, 5, 10, 20$ then the values of $\hat{K}$ are, respectively, 0.66, 1.13, 1.30, 1.39, 1.43. Note that if $\kappa >> 1$ then the iteration parameter $\tau << 1$ and the calculation time increases as κ increases.

On the other hand, in [16] the asymptotic expansion $\hat{K} \sim A_0 + A_i\kappa^{-1} + A_2\kappa^{-2} + \dots$ was obtained as $\kappa \to +\infty$. The second approximation of this expansion gives: $\hat{K} = A_0 + A_1\kappa^{-1} + O(\kappa^{-2})$. If we neglect $O(\kappa^{-2})$ then the calculated values of $\hat{K}$ for $\kappa = 5$ and $\kappa = 10$ can be used to determine A_0 and A_1 and to extrapolate the value $\hat{K}$ for large κ by the approximated formula

$$\hat{K} \approx A_0 + A_1\kappa^{-1} \qquad (2.7.16).$$

Thus , to find A_0 and A_1 the system of equations

$$A_0 + A_1/5 = 1.30, \qquad A_0 + A_1/10 = 1.39$$

must be solved giving $A_0 = 1.48 \quad A_1 = -0.9$. If $\kappa = 20$ the direct calculation of $\hat{K}$ by EFMODUL gives value 1.43, and the calculation by formula (2.7.16) gives value 1.435; the difference between them does not exceed 1 %. This method of undetermined coefficients is the Richardson extrapolation method. The main condition for applicability of the extrapolation method is the existence of an asymptotic expansion for the quantity to be determined.

2. Consider the case $s = 2$, when G_2 is a square with side a and center (0.5, 0.5), $\kappa = 0$ i.e. G_2 is a space inside a periodic cell. Then, as above, $\hat{A}_{ij} = \delta_{ji}\hat{K}$ and for $a = 0.5, 0.75, 0.875$, we get $\hat{K} = 0.619, 0.3, 0.15$ respectively. In [16] it was proved that the asymptotic form $\hat{K} = \mu + A\mu^2 + 0(\mu^3)$ holds if $\mu = (1 - a) \to 0$. By applying the Richardson extrapolation method for $\mu = 0.25$ we obtain $A = 0.8$, i.e. $\hat{K} \approx \mu + 0.8\mu^2$. The somewhat smaller degree of agreement of the extrapolation formula with numerical results is explained by the fact that for small μ the error of the numerical solution increases (because the gradients of the solution increase). That is why for small μ the extrapolation formula is preferable.

It was proved in [16] that in neighborhoods of the vertices Q there is an exponentially fast stabilization of $\tilde{A}_{ij}$ as the distance from the vertex increases, for example $\tilde{A}_{ij} \to \delta_{i1}\delta_{j1}$ as $\xi_1/\mu \to +\infty, \xi_2 \in (0, \mu)$. The graph in Figure 2.7.1 shows how quickly $\tilde{A}_{ii}(\xi_1, 0)$ stabilizes to δ_{i1} as $\xi_i \to +\infty(\overline{A}_{ij} = 0 \text{ if } i \neq j)$; here $\mu = 0.25$.

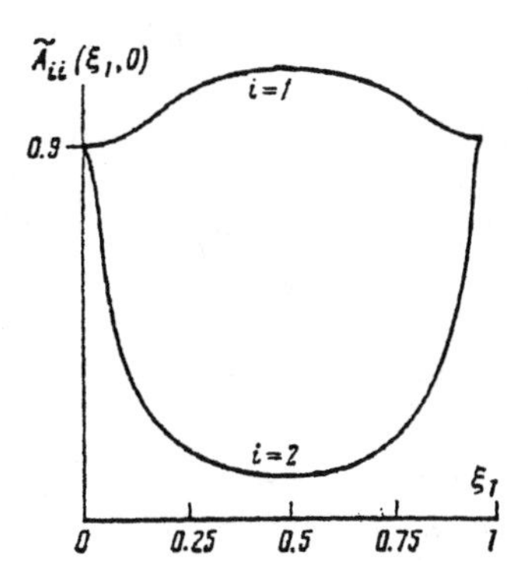

Figure 2.7.1. The $\tilde{A}_{ii}$ micro-stress.

3. Suppose $s = 3, G_2$ is a sphere of radius 0.35 with center at the point (0.5, 0.5, 0.5). We have $\hat{A}_{ij} = \delta_{ij}\hat{K}$. For $\kappa = 0, 2, 5, 10, 20$ we get $\hat{K} = 0.746, 1.15, 1.38, 1.55, 1.69$ respectively. For the values $\kappa = 10$ and $\kappa = 20$ the extrapolation method gives the Richardson extrapolation formula $\hat{K} \approx 1.83 + 2.8\kappa^{-1}$.

4. Assume now that $s = 3, G_2$ is a cube whose volume is equal to the volume of the sphere considered above. Then again $\hat{A}_{ij} = \delta_{ij}\hat{K}$. If $\kappa = 0, 10$ then $\hat{K} = 0.769, 1.48$ respectively.

Comparison of the results of examples 3 and 4 shows that their difference is insignificant.

II. Elasticity equation.

Assume that $n = s = 2$ and that the matrices A_{ij} have the components a_{ij}^{kl}, where

$$a_{ij}^{kl} = \hat{\Lambda}(\xi)\delta_{ik}\delta_{jl} + \mu(\xi)(\delta_{il}\delta_{kj} + \delta_{kl}\delta_{ij}), \Lambda(\xi) = \mu(\xi) = \begin{cases} 1, & \xi \in G_1 \\ \kappa, & \xi \in G_2 \end{cases} . \quad (2.7.17)$$

1. Assume that G_2 is a disk of radius 0.25 with center at the point (0.5, 0.5). Then

$$\hat{a}_{11}^{11} = \hat{a}_{22}^{22} = \alpha, \quad \hat{a}_{11}^{22} = \hat{a}_{21}^{12} = \hat{a}_{12}^{21} = \hat{a}_{22}^{11} = \beta, \quad \hat{a}_{12}^{12} = \hat{a}_{21}^{21} = \gamma \qquad (2.7.18)$$

and the remaining $\hat{a}_{ij}^{kl} = 0$. Below, values of α, β, γ are calculated for different values of κ : if $\kappa = 0, 0.1, 2, 5, 10$ then $\alpha = 2.09, 2.22, 3.41, 3.78, 3.96, \beta = 0.73, 0.77, 1.15, 1.24, 1.28, \gamma = 0.61, 0.67, 1.11, 1.21, 1.25$ respectively. For great values of κ the Richardson extrapolation method can be used to calculate α, β, γ as in I.

2. G_2 is a square with side a and center (0.5 0.5), $\kappa = 0$; then $\hat{a}_{ij}^{kl}$ has structure (2.7.18) ; for $a = 0.5, 0.75, 0.875$, we get $\alpha = 1.94, 0.96, 0.50, \beta = 0.67, 0.34, 0.17$, $\gamma = 0.53, 0.19, 0.076$ respectively. If $\kappa = 2$ then $\alpha = 3.51, 4.33, 4.98, \beta = 1.18, 1.45, 1.67$, $\gamma = 1.14, 1.39, 1.62$.

III. Elasticity equation, $n = s = 3$. Let A_{ij} be matrices with elements a_{ij}^{kl} , and let a_{ij}^{kl} be defined by formula (2.7.17).

Let the cube G_2 have side a and center $(0.5, 0.5, 0.5)$, $\kappa = 0$; then $\hat{a}_{ij}^{kl}$ has the following structure : $\hat{a}_{ii}^{ii} = \alpha, \hat{a}_{ij}^{ij} = \beta, \hat{a}_{ii}^{jj} = \gamma (i \neq j)$; if $a = 0.5, 0.75$ then $\alpha = 2.66, 1.80, \beta = 0.85, 0.51, \gamma = 0.89, 0.62$. If $\kappa = 5, a = 0.5$ then $\alpha = 3.44, \beta = 1.11, \gamma = 1.16$ and if $\kappa = 10, a = 0.5$ then $\alpha = 3.63, \beta = 1.13, \gamma = 1.23$

IV. Non-linear elasticity equation, $n = s = 3$. Assume that $A_i(\xi, y)$ in (2.7.10) has the form (see [63])

$$A_i = (a_i^1, a_i^2, a_i^3)^t,$$

and y is a matrix with components $y_{ml}, \quad m, l = 1, 2, 3,$

$$a_i^r(\xi, y) = \left[\lambda(\xi) + \frac{2}{3}\mu(\xi)\right] \sum_{l=1}^{3} e_{ll}\delta_{ir} + \frac{2}{3}\frac{\Phi(e_u)}{e_u}\left(e_{ir} - \frac{1}{3}\sum_{l=1}^{3} e_{ll}\delta_{ir}\right)$$

$$\lambda(\xi) = \mu(\xi) = \begin{cases} 1 & \text{for} \quad \xi \in G_1 \\ 5 & \text{for} \quad \xi \in G_2 \end{cases},$$

G_2 is a cube with side 0.5 and center $(0.5, 0.5, 0.5)$;

$$e_{ml} = 0.5(y_{ml} + y_{lm}), e_u = \left[\frac{2}{3}\sum_{i,j=1}^{s}\left(e_{ij} - \frac{1}{3}\sum_{l=1}^{s} e_{ll}\delta_{ij}\right)^2\right]^{1/2}$$

$$\Phi(e_u) = 3\mu(\xi)e_u \quad \text{for} \quad e_u < e_{pl} = 0.02,$$

$$\Phi(e_u) = 3\mu(\xi)e_{pl} + (3/7)\mu(\xi)(e_u - e_{pl}) \quad \text{for} \quad e_u > e_{pl}.$$

In this case the values of $\hat{a}_i^r(y)$ are calculated if $y_{11} = \eta, y_{ml} = 0$ (if $m \neq 1$ or $l \neq 1$) for different η: if $\eta = 0.01, 0.02, 0.03$ then $\hat{a}_1^1 = 0.035, 0.0694, 0.0998, \hat{a}_2^2 = 0.11, 0.022, 0.036, \hat{a}_3^3 = \hat{a}_2^2, \hat{a}_i^j = 0$ if $i \neq j$.

2.8 Bibliographical Remark

The homogenization techniques plays an important part in the applied mathematics and mechanics of the twenty's century because it has created the mathematical theory of the heterogeneous media and in particular, composite materials and porous media. We note the pioneer works on the asymptotic analysis of equations with rapidly oscillating coefficients by E. Sanchez-Palencia [177], E. De Giorgi and S. Spagnollo [49], N. Bakhvalov [12]-[14], J. L. Lions, A. Bensoussan, and G. Papanicolaou [22], O. Oleinik [127], V. Berdichevsky [23], I. Babuska [8], F.Murat and L.Tartar [109], [187] and V. Marchenko and E. Khruslov [103]. In particular, in [13] the series permitting the construction of the complete asymptotic expansion appeared for the first time. The form of the series (2.1.2) is close to the asymptotic series of the averaging technique introduced by N. Krylov, N. Bogoljubov, and Yu. Mitropol'sky [29].

Nowadays the homogenization techniques is presented by two branches: one is the construction of asymptotic expansions and the other is the H-convergence (the G-convergence) and its generalization that is the two-scale convergence. The first branch [16,22,128,129] gives more information about the solution (all correctors, boundary layers, high order estimates) but it demands more smoothness of the data. The second approach [3,65,109,187] requires less of smoothness of data but often the results are formulated as some convergence theorems while error estimates are less informative than in the first approach.

The boundary layer techniques appeared in the works by L.Prandtl in the beginning of the twenty's century and were generalized for the partial derivative equations by M.Vishik and L.Lusternik in [193].

The mathematical asymptotic study of a thin homogeneous rod has been developed in [1,53,111,189]. The complete asymptotic expansion of a thin inhomogeneous rod with symmetry condition A was first constructed in [83,84]. The beams with rapidly varying cross-section were studied in [192]. General anisotropic heterogeneous rods were studied by the asymptotic expansion method in [117] and [101] and by the scaling and then application of the H-convergence techniques in [108]. The contact problem of two heterogeneous bars was considered in [149,150,169].

Chapter 3

Heterogeneous Plate

Here we consider the three-dimensional conductivity problem and the three-dimensional elasticity problem in a thin domain (plate) having a heterogeneous structure. We assume that the characteristic size of heterogeneities is of the same order as the "width" of the domain. Assume that it is much less than the "length" of the plate that is a characteristic longitudinal size of the domain. So the ratio of the "width" to the "length" of the plate is the small parameter of the same order as the ratio of the characteristic size of heterogeneity to the "length" of the plate. We assume that the "length" of the plate has a finite value.

The simplest geometrical model of a plate is defined as an intersection of a thin layer $\widetilde{U}_\mu = \{(x_1, x_2) \in I\!R^2,\ x_3/\mu \in (-1/2, 1/2)\}$ orthogonal to the axis x_3 with the "thick" three-dimensional layer $\{x = (x_1, x_2, x_3) \in I\!R^3;\quad x_1 \in (0, b)\}$ orthogonal to the axis x_1. As in Chapter 2, we assume that μ is a small parameter. The microscopical heterogeneity of the plate is simulated by a special dependence of the coefficients of the material on space variable $x = (x_1, x_2, x_3)$; these coefficients are functions of x/μ.

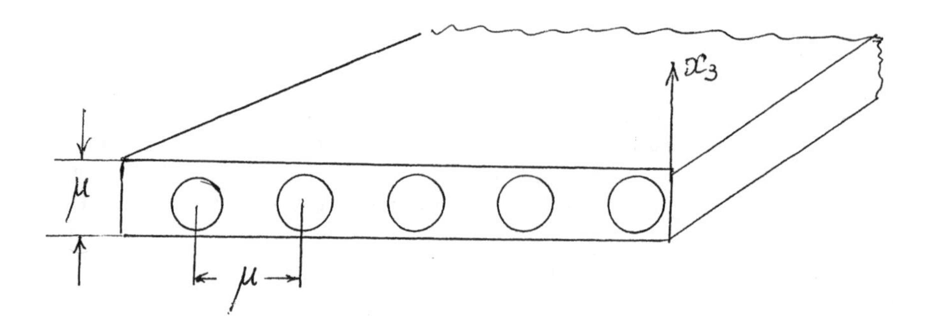

Figure 3.0.1. Heterogeneous plate.

An asymptotic analysis of the mathematical model is developed as μ tends to zero. To this end we apply the homogenization technique and the boundary layer technique (cf. Chapter 2). In section 3.1 we study the conductivity equation; an expansion of a solution is constructed and justified and the two-dimensional homogenized model of the plate is obtained. In section 3.2 the three-dimensional elasticity problem for a plate is studied. The question about existence of an equivalent homogeneous plate is discussed in section 3.3. Non-steady state elasticity equation is studied in section 3.4.

3.1 Conductivity of a plate

3.1.1 Statement of the problem

Definition 1.1 Let $\widetilde{U}_\mu$ be the layer $\{(x_1, x_2) \in I\!R^2,\ x_3/\mu \in (-1/2, 1/2)\}$. The intersection $\widetilde{C}'_\mu = \widetilde{U}_\mu \cap \{x_1 \in (0, b)\}$ will be called the *plate*.

Consider the conductivity equation

$$Pu \equiv \sum_{i,j=1}^{3} \frac{\partial}{\partial x_i}\Big(A_{ij}\Big(\frac{x}{\mu}\Big)\frac{\partial u}{\partial x_j}\Big) = \psi(x_1, x_2), \quad x \in \widetilde{C}'_\mu, \tag{3.1.1}$$

with boundary conditions

$$\frac{\partial u}{\partial \nu} \equiv \pm \sum_{j=1}^{3} A_{3j}\Big(\frac{x_1}{\mu}, \frac{x_2}{\mu}, \pm\frac{1}{2}\Big)\frac{\partial u}{\partial x_j} = 0, \quad x \in \partial\widetilde{U}_\mu, \tag{3.1.2}$$

$$u = 0, \quad x_1 = 0 \quad \text{or} \quad x_1 = b, \tag{3.1.3}$$

and with the condition of T-periodicity with respect to x_2. Here b and T are numbers of the order of 1 that are multiples of μ and $A_{ij}(\xi)$ satisfy the conditions of section 2.2 and are 1-periodic with respect to ξ_1 and ξ_2. These elements are assumed to be piecewise smooth functions (in the sense of section 2.2; for β we take the square $(-1/2, 1/2) \times (-1/2, 1/2)$), while ψ is a C^∞ function that is T-periodic with respect to x_2.

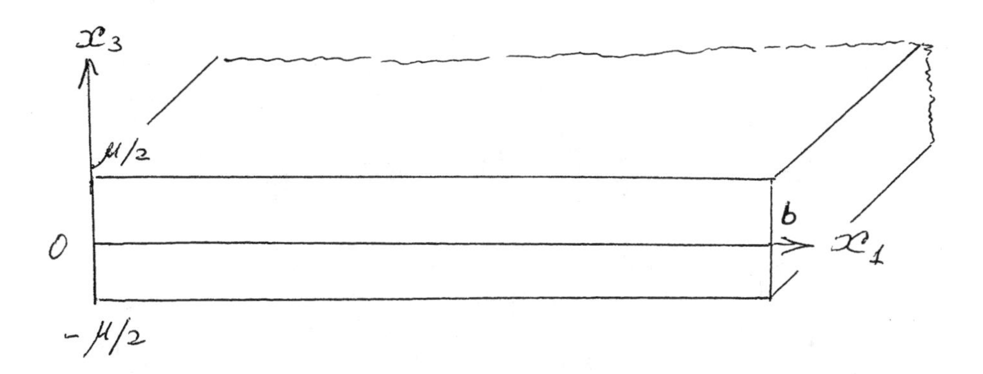

Figure 3.1.1. Plate $\widetilde{C}'_\mu$.

Remark 3.1.1. In this chapter for the reason of simplicity we consider the case when the right side hand function does not depend on the rapid variable x/μ. Of course such dependance could be taken into account in the same way as in the previous chapter.

3.1.2 Inner expansion

We seek a formal asymptotic solution in the form of a series

$$u^{(\infty)} = \sum_{l=0}^{\infty} \mu^l \sum_{|i|=l} N_i\Big(\frac{x}{\mu}\Big) D^i \omega, \tag{3.1.4}$$

where $i = (i_1, \ldots, i_l)$ is a multi-index, $i_j \in \{1, 2\}$, $\omega(x_1, x_2)$ is a smooth T-periodic with respect to x_2 three-dimensional vector function, $N_i(\xi)$ are 3×3 matrix functions that are 1-periodic with respect to ξ_1 and ξ_2.

Substituting series (3.1.4) in (3.1.1), (3.1.2) and (2.2.4) and grouping terms of the same order, we obtain

$$Pu^{(\infty)} - \psi = \sum_{l=0}^{\infty} \mu^{l-2} \sum_{|i|=l} H_i(\xi) D^i \omega - \psi, \tag{3.1.5}$$

$$\frac{\partial u^{(\infty)}}{\partial \nu} = \sum_{l=0}^{\infty} \mu^{l-1} \sum_{|i|=l} G_i(\xi) D^i \omega, \tag{3.1.6}$$

where

$$H_i(\xi) = L_{\xi\xi} N_i + T_i(\xi),$$

$$T_i(\xi) = \sum_{j=1}^{3} \frac{\partial}{\partial \xi_j}(A_{ji_1} N_{i_2 \ldots i_l}) + \sum_{j=1}^{3} A_{i_1 j} \frac{\partial N_{i_2 \ldots i_l}}{\partial \xi_j} + A_{i_1 i_2} N_{i_3 \ldots i_l},$$

$$G_i(\xi) = \sum_{m=1}^{3} \Big(\sum_{j=1}^{3} A_{mj} \frac{\partial N_i}{\partial \xi_j} + A_{mi_1} N_{i_2 \ldots i_l} \Big) n_m.$$

Suppose that

$$H_i(\xi) = h_i, \quad \xi \in I\!R^2 \times \Big(-\frac{1}{2}, \frac{1}{2}\Big); \qquad G_i(\xi) = 0, \quad \xi_3 = \pm\frac{1}{2},$$

where h_i are constants.

Assume that $[N_i] = 0$ and $[G_i] = 0$ on the surfaces of discontinuity.

We obtain the following recurrent chain of problems of the form

$$L_{\xi\xi} N_i = -T_i(\xi) + h_i, \tag{3.1.7}$$

$$\sum_{j=1}^{3} A_{3j}(\xi) \frac{\partial N_i}{\partial \xi_j} = -A_{3i_1} N_{i_2 \ldots i_l}, \quad \xi_3 = \pm\frac{1}{2}, \tag{3.1.8}$$

$$[N_i]\big|_\Sigma = 0, \quad \Big[\frac{\partial N_i}{\partial \nu_\xi}\Big]\Big|_\Sigma = \Big[-\sum_{m=1}^{3} n_m A_{mi_1} N_{i_2 \ldots i_l}\Big]\Big|_\Sigma, \tag{3.1.9}$$

to determine N_i that are 1-periodic functions with respect to ξ_1 and ξ_2; here (n_1, n_2, n_3) is an external normal vector.

The constant matrices h_i are chosen from the solvability conditions for problem (3.1.7)-(3.1.9):

$$h_i = \Big\langle \sum_{j=1}^{3} A_{i_1 j} \frac{\partial N_{i_2 \ldots i_l}}{\partial \xi_j} + A_{i_1 i_2} N_{i_3 \ldots i_l} \Big\rangle, \quad l \geq 2,$$

$$h_\emptyset = 0, \quad h_{i_1} = 0; \tag{3.1.10}$$

here $\langle \cdot \rangle = \int_{(-1/2,1/2)^3} d\xi$ and $N_\emptyset = 1$.

Let us give the variational formulation of (3.1.7)-(3.1.9). Let Q be the unit cube $(0,1)^2 \times (-\frac{1}{2}, \frac{1}{2})$. Let $H^1_{per\ \xi_1,\xi_2}(Q)$ be closure in the norm $H^1(Q)$ of the space of differentiable on $I\!R^2 \times [-1/2, 1/2]$ functions of $\xi = (\xi_1, \xi_2, \xi_3)$, 1-periodic in ξ_1 and ξ_2. Then the variational formulation is: find $N_i \in H^1_{per\ \xi_1,\xi_2}(Q)$, such that

$$\forall \Phi \in H^1_{per\ \xi_1,\xi_2}(Q), \Big\langle - \sum_{m,j=1}^{3} A_{mj}(\xi) \frac{\partial N_i}{\partial \xi_j} \frac{\partial \Phi}{\partial \xi_m} \Big\rangle =$$

$$= \Big\langle \sum_{m=1}^{3} A_{mi_1}(\xi) N_{i_2 \ldots i_l} \frac{\partial \Phi}{\partial \xi_m} \Big\rangle - \Big\langle (\sum_{j=1}^{3} A_{i_1 j}(\xi) \frac{\partial N_{i_2 \ldots i_l}}{\partial \xi_j} + A_{i_1 i_2} N_{i_3 \ldots i_l} - h_i) \Phi \Big\rangle.$$

It could be proved as in section 2.2 (Lemma 2.2.1) that this problem has a solution if and only if (3.1.10) holds true. This solution is defined up to an arbitrary additive constant.

Thus, the algorithm for constructing the functions N_i is recurrent. We formally assume that $N_i = 0$ for $|i| < 0$, $N_\emptyset = 1$, and $N_i(\xi)$ is a solution of problem (3.1.7)–(3.1.10) for $l > 0$. The right-hand side contains N_j with multi-indices j whose length is smaller than $|i|$.

Theorem 3.1.1. *The constant matrix of homogenized coefficients* $\bar{A} = (h_{i_1 i_2})_{1 \le i_1, i_2 \le 2}$ *is positive definite and it is symmetric.*

Proof.

1. It follows from (3.1.10) that

$$h_{i_1 i_2} = \Big\langle \sum_{j=1}^{3} A_{i_1 j} \frac{\partial N_{i_2}}{\partial \xi_j} + A_{i_1 i_2} \Big\rangle = \Big\langle \sum_{j=1}^{3} A_{i_1 j} \frac{\partial}{\partial \xi_j} (N_{i_2} + \xi_{i_2}) \Big\rangle$$

$$= \Big\langle \sum_{m,j=1}^{3} A_{mj} \frac{\partial}{\partial \xi_j} (N_{i_2} + \xi_{i_2}) \frac{\partial}{\partial \xi_m} (N_{i_1} + \xi_{i_1}) \Big\rangle.$$

The last relation arises from the integral identity for problem (3.1.7)–(3.1.9) with $|i| = 2$. The symmetry of the matrix of coefficients $A = (A_{mj})_{1 \le m,j \le 3}$ implies the symmetry of the matrix of homogenized coefficients $\bar{A} = (h_{i_1 i_2})_{1 \le i_1, i_2 \le 2}$. Let us prove that $\bar{A}$ is positive definite. Let $(\eta_1, \eta_2) \in I\!R^2$. Consider the sum

$$\sum_{i_1,i_2=1}^{2} h_{i_1 i_2} \eta_{i_2} \eta_{i_1} = \sum_{m,j=1}^{3} \langle A_{mj} \frac{\partial}{\partial \xi_j} (\sum_{i_2=1}^{2} (N_{i_2} + \xi_{i_2}) \eta_{i_2}) (\sum_{i_1=1}^{2} (N_{i_1} + \xi_{i_1}) \eta_{i_1}) \rangle \ge$$

$$\ge \kappa \sum_{j=1}^{3} \langle (\sum_{k=1}^{2} \frac{\partial}{\partial \xi_j} (N_k + \xi_k) \eta_k)^2 \rangle \ge$$

$$\ge \kappa \sum_{j=1}^{3} \langle (\sum_{k=1}^{2} \frac{\partial}{\partial \xi_j} (N_k + \xi_k) \eta_k) \rangle^2 =$$

$$= \kappa \sum_{k=1}^{2} \eta_k^2,$$

because $\langle \frac{\partial N_k}{\partial \xi_j} \rangle = 0$.
The proof of Theorem 3.1.1 is complete.

Then (3.1.5) takes the form

$$Pu^{(\infty)} - \psi = \sum_{i_1,i_2=1}^{2} h_{i_1 i_2} \frac{\partial^2 \omega}{\partial x_{i_1} \partial x_{i_2}} +$$

$$+ \sum_{l=3}^{\infty} \mu^{l-2} \sum_{|i|=l} h_i \frac{\partial^l \omega}{\partial x_{i_1} \cdots \partial x_{i_l}} - \psi = 0. \tag{3.1.11}$$

Problem (3.1.11) can be regarded as the homogenized equation of infinite order with respect to the three-dimensional vector ω. A formal asymptotic solution (f.a.s.) of this problem is sought in the form of a series

$$\omega = \sum_{j=0}^{\infty} \mu^j \omega_j(x_1, x_2), \tag{3.1.12}$$

where ω_j do not depend on μ. Substituting the series (3.1.12) into (3.1.11), we obtain a recurrent chain of equations for the components ω_j^k of the vectors ω_j in the form

$$\sum_{i_1,i_2=1}^{2} \bar{A}_{i_1 i_2} \frac{\partial^2 \omega_j}{\partial x_{i_1} \partial x_{i_2}} = f_j(x_1, x_2), \tag{3.1.13}$$

where $\bar{A}_{i_1 i_2} = h_{i_1 i_2}$, the functions f_j depend on ω_{j_1}, $j_1 < j$, and on the derivatives of these functions. Thus, the f.a.s. of the problem (3.1.1), (3.1.2), and (2.2.4) is constructed.

The asymptotic analysis of the conductivity equation for inhomogeneous plates was first carried out by D.Caillerie [31], where the homogenized equation of zero order was obtained. Here below we construct the complete asymptotic expansion of a solution of the conductivity equation (taking into account the boundary layers).

3.1.3 Boundary layer corrector

A formal asymptotic solution of problem (3.1.1)–(3.1.3), (2.2.4) will be sought again in the form (2.2.22): $u^{(\infty)} = u_B + u_P^0 + u_P^1$, where u_B is defined by (3.1.4) and

$$u_P^0 = \sum_{l=0}^{\infty} \mu^l \sum_{|i|=l} N_i^0(\xi) D^i \omega \big|_{\xi = x/\mu}, \tag{3.1.14}$$

$$u_P^1 = \sum_{l=0}^{\infty} \mu^l \sum_{|i|=l} N_i^1(\xi) D^i \omega \big|_{\xi=(x_1-b)/\mu,\ \xi'=x'/\mu}. \qquad (3.1.15)$$

Here the functions $N_i^0(\xi)$ and $N_i^1(\xi)$ are 1-periodic with respect to ξ_2 and exponentially stabilize to zero as $|\xi_1| \to +\infty$.

Substituting (3.1.14) and (3.1.15) into (3.1.1)–(3.1.3) and taking into account the fact that u_B is constructed above, we obtain recurrent chains of problems to determine $N_i^0(\xi)$ as in subsection 3.1.2

$$L_{\xi\xi} N_i^0 + T_i^0(\xi) = 0, \quad \xi_1 > 0, \quad \xi' \in I\!R \times \Big(-\frac{1}{2}, \frac{1}{2} \Big),$$

$$\sum_{j=1}^{3} A_{3j} \frac{\partial N_i^0}{\partial \xi_j} = -A_{3i_1} N^0_{i_2 \ldots i_l}, \quad \xi_3 = \pm \frac{1}{2},$$

$$N_i^0(0, \xi') = -N_i(0, \xi') + h_i^0 \qquad (3.1.16)$$

and obtain similar problems for $N_i^1(\xi)$ defined for $\xi_1 < 0$ with the constant h_i^1 instead of h_i^0. h_i^r are constants chosen from the condition that solutions of problems (3.1.16) stabilize to zero as $\xi_1 \to +\infty$ ($\xi_1 \to -\infty$ for $r = 1$), and $N_\emptyset^0 = 0$.

Theorem 3.1.2. *Let $A_{ij}(\xi)$ satisfy the conditions formulated at the beginning of this section, let the three-dimensional vectors $F_j(\xi)$ be 1-periodic with respect to ξ_2 and belong to $L^2([0, +\infty) \times \beta)$, $j = 0, 1, 2, 3$, let $\beta = (-1/2, 1/2)^2$, and let $u_0(\xi')$ be 1-periodic with respect to ξ_2 and belong to $H^{1/2}(\{0\} \times (-1, 1) \times (-1/2, 1/2))$. Then a solution $u(\xi)$ of the problem*

$$L_{\xi\xi} u = F_0(\xi) + \sum_{j=1}^{3} \frac{\partial F_j}{\partial \xi_j}, \quad \xi_1 > 0, \quad \xi' \in I\!R \times \Big(-\frac{1}{2}, \frac{1}{2} \Big),$$

$$\sum_{j=1}^{3} A_{3j} \frac{\partial u}{\partial \xi_j} = F_j, \quad \xi_3 = \pm \frac{1}{2}, \qquad u\big|_{\xi_1=0} = u_0(\xi') \qquad (3.1.17)$$

such that u is 1-periodic with respect to ξ_2 and inequality (2.2.36) holds for constant h, exists and is unique.

Here by a solution of problem (3.1.17) we mean a function $u(\xi)$ satisfying the integral identity from subsection 3.1.2.

Theorem 3.1.2 can be proved similarly to Theorems 4 and 5 in [130] and theorems of the Phragmen-Lindelöf type [89],[90]. The constants h_i^r are defined similarly to the constants h_l^{Nr} from Chapter 2. In this case $h_\emptyset^r = 1$ since $N_\emptyset^r = 0$ and $N_\emptyset = 1$.

After determining all N_i^r and h_i^r by induction on $|i|$, we obtain a homogenized problem for the three-dimensional vector function $\omega(x_1, x_2)$: this problem is formed by equation (3.1.14) with the boundary conditions

$$\Big(\omega + \sum_{l=1}^{\infty} \mu^l \sum_{|i|=l} h_i^r D^i \omega \Big) \Big|_{x_1 = rb} = 0.$$

A f.a.s. of problem (3.1.1)–(3.1.3) is sought in the form of series (3.1.12) with boundary conditions

$$\omega_j\Big|_{x_1=rb} = g_{jr}, \quad r = 0, 1. \tag{3.1.18}$$

where the g_{jr} depend on $\omega_{j_1}^r$ with $j_1 < j$ and on derivatives of these functions.

The homogenized problem of zero order is formed by relations (3.1.13), (3.1.18) with $j = 0$,

$$f_0(x_1, x_2) = \psi(x_1, x_2), \quad g_{0r} = 0. \tag{3.1.19}$$

The estimate for the difference of the exact solution and the asymptotic solution can be obtained as in Chapter 2.

3.1.4 Algorithm for calculating the effective conductivity of a plate

As in section 2.3, we present a calculation algorithm for the coefficients $\bar{A}_{i_1 i_2} = h_{i_1 i_2}$, $i_1, i_2 \in \{1, 2\}$, of the homogenized equation (the effective conductivity tensor in a plane).

We first solve the conductivity equations for $i_1 = 1, 2$:

$$\sum_{m,j=1}^{3} \frac{\partial}{\partial \xi_m}\Big(A_{mj}\frac{\partial}{\partial \xi_j}(N_{i_1} + \xi_{i_1})\Big) = 0, \quad \xi_3 \in \Big(-\frac{1}{2}, \frac{1}{2}\Big),$$

$$\sum_{j=1}^{3} A_{3j}\frac{\partial}{\partial \xi_j}(N_{i_1} + \xi_{i_1}) = 0, \quad \xi_3 = \pm\frac{1}{2}, \tag{3.1.20}$$

where $N_{i_1}(\xi)$ is an unknown function that is 1-periodic with respect to ξ_1 and ξ_2.

The effective conductivity coefficients in a plane can be defined by the following formulas:

$$\bar{A}_{i_1 i_2} = h_{i_1 i_2} = \Big\langle \sum_{j=1}^{3} A_{i_1 j}\frac{\partial}{\partial \xi_j}(N_{i_2} + \xi_{i_2})\Big\rangle. \tag{3.1.21}$$

For constant coefficients in the isotropic case $A_{mj} = \delta_{mj}K$, $K = const$, we have:

$$N_k = 0, \ k = 1, 2, \ \bar{A}_{i_1 i_2} = \delta_{i_1 i_2}K;$$

for constant coefficients in the general case we have

$$\bar{A}_{i_1 i_2} = A_{i_1 i_2} - A_{i_1 3}A_{33}^{-1}A_{3 i_2}. \tag{3.1.22}$$

For coefficients depending on ξ_3 only we have

$$\bar{A}_{i_1 i_2} = \left\langle A_{i_1 i_2} - A_{i_1 3} A_{33}^{-1} A_{3 i_2} \right\rangle,$$

where $\langle \cdot \rangle = \int_{-1/2}^{1/2} d\xi_3$.

Indeed, in this case a solution of problem (3.1.20) is sought as a function of ξ_3 only. We have:

$$A_{33} \frac{\partial N_{i_2}}{\partial \xi_3} + A_{3 i_2} = 0,$$

so

$$\frac{\partial N_{i_2}}{\partial \xi_3} = -A_{33}^{-1} A_{3 i_2},$$

and so (3.1.21) implies

$$\bar{A}_{i_1 i_2} = \left\langle A_{i_1 3} \frac{\partial N_{i_2}}{\partial x_3} + A_{i_1 i_2} \right\rangle = \left\langle A_{i_1 i_2} - A_{i_1 3} A_{33}^{-1} A_{3 i_2} \right\rangle.$$

Remark 3.1.2. In mechanics it is usual to take $\mu \bar{A}_{ij}$ as the effective conductivity coefficients in a plane, where μ is the thickness of a plate.

The homogenized problem of zero order has the form

$$\sum_{m,j=1}^{2} \frac{\partial}{\partial x_m} \left(\bar{A}_{mj} \frac{\partial \omega_0}{\partial x_j} \right) = f(x_1, x_2),$$

$$\omega_0 = 0 \quad \text{for} \quad x_1 = 0, b,$$

i.e.

$$\sum_{m,j=1}^{2} \frac{\partial}{\partial x_m} \left(\mu \bar{A}_{mj} \frac{\partial \omega_0}{\partial x_j} \right) = \mu f(x_1, x_2),$$

$$\omega_0 = 0 \quad \text{for} \quad x_1 = 0, b.$$

Evidently, for any given positive definite symmetric matrix $\mu \bar{A}_{ij},\ i,j = 1,2$ it is possible to find such a constant positive definite symmetric matrix $A_{ij},\ i,j = 1,2,3$ that it is related with the given matrix by equality (3.1.22), for example, $A_{ij} = \bar{A}_{ij},\ i,j = 1,2,\quad A_{33} > 0$, and $A_{i3}, A_{3j} = 0,\ i,j = 1,2,3$. It means that any heterogeneous conductive plate (3.1.1)–(3.1.3) is "equivalent" in the sense of the effective conductivity coefficients in a plane to some homogeneous plate. It is remarkable that the analogous property for elasticity coefficients is not true. This question will be discussed in the section 3.3.

3.1.5 Justification of the asymptotic expansion

Here we prove the following main theorem.

Theorem 3.1.3. *Let $K \in \{0,1,2,...\}$. Denote χ a function from $C^{(K+2)}([0,b])$ such that $\chi(t)=1$ when $t \in [0,b/3]$, $\chi(t)=0$ when $t \in [2b/3,b]$. Consider a function*

$$u^{(K)}(x) = u_B^{(K)}(x) + u_P^{0(K)}(x)\chi(x_1) + u_P^{1(K)}(x)\chi(b-x_1) + \rho(x_1,x_2),$$

where

$$u_B^{(K)} = \sum_{l=0}^{K+1} \mu^l \sum_{|i|=l} N_i\Big(\frac{x}{\mu}\Big) D^i\omega^{(K)}(x_1,x_2),$$

$$u_P^{0(K)} = \Big(\sum_{l=0}^{K+1} \mu^l \sum_{|i|=l} N_i^0(\xi) D^i\omega^{(K)}(x_1,x_2)\Big)\Big|_{\xi=x/\mu},$$

$$u_P^{1(K)} = \Big(\sum_{l=0}^{K+1} \mu^l \sum_{|i|=l} N_i^1(\xi) D^i\omega^{(K)}(x_1,x_2)\Big)|_{\xi_1=(x_1-b)/\mu \xi'=x'/\mu},$$

$$\omega^{(K)} = \sum_{j=0}^{K} \mu^j \omega_j(x_1,x_2);$$

$$\rho(x_1,x_2) = (1-x_1/b)q_0(x_2,\mu) + (x_1/b)q_1(x_2,\mu),$$

$$q_r(x_2,\mu) = \Big(\sum_{l=K+1}^{2K+1} \mu^l \sum_{j+p=l,0\le j\le K,0\le p\le K+1} \sum_{|i|=p} h_i^{Nr} D^i\omega_j\Big)|_{x_1=rb}, \quad r=1,2,$$

$N_l, N_l^0, N_l^1, \omega_j$ are defined in subsection 2.3. Then the estimate holds

$$\|u^{(K)} - u\|_{H^1(\tilde{C}_\mu)} = O(\mu^K)\sqrt{mes\ \tilde{C}_\mu}.$$

Proof.

Let us estimate the discrepancy functional

$$I(\phi) = \int_{\tilde{C}_\mu} \sum_{m,j=1}^{s} A_{mj}\Big(\frac{x}{\mu}\Big)\frac{\partial(u-u^{(K)})}{\partial x_j}\frac{\partial\phi}{\partial x_m}\,dx,$$

where $\phi \in H^1(C_\mu)$, $\phi = 0$ for $x_1 = 0$ or b.

1. At the first stage represent it in the form

$$I(\phi) \;=\; \int_{\tilde{C}_\mu} F\psi\phi \, dx \;-\; J_B(\phi) \;-\; J_0(\phi) \;-\; J_1(\phi) \;-\; J_2(\phi),$$

where

$$J_B(\phi) \;=\; \int_{\tilde{C}_\mu} \sum_{m,j=1}^{s} A_{mj}\Big(\frac{x}{\mu}\Big) \frac{\partial u_B^{(K)}}{\partial x_j} \frac{\partial \phi}{\partial x_m} \, dx,$$

$$J_0(\phi) \;=\; \int_{\tilde{C}_\mu} \sum_{m,j=1}^{s} A_{mj}\Big(\frac{x}{\mu}\Big) \frac{\partial (u_P^{0(K)}(x)\chi(x_1))}{\partial x_j} \frac{\partial \phi}{\partial x_m} \, dx,$$

$$J_1(\phi) \;=\; \int_{\tilde{C}_\mu} \sum_{m,j=1}^{s} A_{mj}\Big(\frac{x}{\mu}\Big) \frac{\partial (u_P^{1(K)}(x)\chi(b-x_1))}{\partial x_j} \frac{\partial \phi}{\partial x_m} \, dx,$$

$$J_2(\phi) \;=\; \int_{\tilde{C}_\mu} \sum_{m,j=1}^{s} A_{mj}\Big(\frac{x}{\mu}\Big) \frac{\partial \rho(x_1)}{\partial x_j} \frac{\partial \phi}{\partial x_m} \, dx \;=$$

$$=\; \int_{\tilde{C}_\mu} \frac{q_1 - q_0}{b} \sum_{m=1}^{s} A_{m1}\Big(\frac{x}{\mu}\Big) \frac{\partial \phi}{\partial x_m} \, dx;$$

so,

$$|J_2(\phi)| \;\le\; c_0 \mu^{K+1} \|\phi\|_{H^1(\tilde{C}_\mu)} \sqrt{mes\, \tilde{C}_\mu}.$$

The boundary layer functions N_l^r decay exponentially as $|\xi_1| \to +\infty$, so (2.2.36) implies the following estimates

$$\Big|\int_{\tilde{C}_\mu \cap \{x_1 \ge b/3\}} \sum_{m,j=1}^{s} A_{mj}\Big(\frac{x}{\mu}\Big) \frac{\partial (u_P^{0(K)}(x)\chi(x_1))}{\partial x_j} \frac{\partial \phi}{\partial x_m} \, dx\Big| \le c_1 e^{-c_2/\mu} \|\phi\|_{H^1(\tilde{C}_\mu)},$$

$$\Big|\int_{\tilde{C}_\mu \cap \{x_1 \le 2b/3\}} \sum_{m,j=1}^{s} A_{mj}\Big(\frac{x}{\mu}\Big) \frac{\partial (u_P^{1(K)}(x)\chi(b-x_1))}{\partial x_j} \frac{\partial \phi}{\partial x_m} \, dx\Big| \le c_1 e^{-c_2/\mu} \|\phi\|_{H^1(\tilde{C}_\mu)}$$

with some constants $c_1, c_2 > 0$ independent of μ.

So

$$I(\phi) \;=\; \int_{\tilde{C}_\mu} F\psi\phi \, dx \;-\; J_B(\phi) \;-\; \hat{J}_0(\phi) \;-\; \hat{J}_1(\phi) \;+$$

$$+\; (\hat{J}_0(\phi) \;-\; J_0(\phi)) \;+\; (\hat{J}_1(\phi) \;-\; J_1(\phi)) \;-\; J_2(\phi),$$

where

$$\hat{J}_r(\phi) \;=\; \int_{\tilde{C}_\mu} \sum_{m,j=1}^{s} A_{mj}\Big(\frac{x}{\mu}\Big) \frac{\partial u_P^{r(K)}(x)}{\partial x_j} \frac{\partial \phi}{\partial x_m} \, dx, \; r = 1, 2,$$

and

$$|\hat{J}_r(\phi) \ - \ J_r(\phi)| \ \leq \ c_1 e^{-c_2/\mu} \|\phi\|_{H^1(\tilde{C}_\mu)}.$$

Consider now

$$J_B(\phi) \ =$$

$$= \ \sum_{l=0}^{K+1} \sum_{m=1}^{s} \int_{\tilde{C}_\mu} \mu^{l-1} \sum_{|i|=l} \Big(\sum_{j=1}^{s} A_{mj}\Big(\frac{x}{\mu}\Big) \frac{\partial N_i(\xi)}{\partial \xi_j}|_{\xi=x/\mu} +$$

$$+ A_{mi_1}\Big(\frac{x}{\mu}\Big) N_{i_2 \dots i_l}\Big(\frac{x}{\mu}\Big)\Big) D^i \omega^{(K)} \frac{\partial \phi}{\partial x_m} \ dx \ +$$

$$+ \ \sum_{m=1}^{s} \int_{\tilde{C}_\mu} \mu^{K+1} \sum_{|i|=K+2} A_{mi_1}\Big(\frac{x}{\mu}\Big) N_{i_2 \dots i_{K+2}} D^i \omega^{(K)} \frac{\partial \phi}{\partial x_m} \ dx.$$

Here $i = (i_1 \dots i_l), \ \ i_j \in \{1, 2\}$.
Denote

$$A^{BN}_{mi}(\xi) \ = \ \sum_{j=1}^{s} A_{mj}(\xi) \frac{\partial N_i(\xi)}{\partial \xi_j} + A_{mi_1}(\xi) N_{i_2 \dots i_l}(\xi),$$

$$\Delta^N_1(\phi) \ = \ \sum_{m=1}^{s} \int_{\tilde{C}_\mu} \mu^{K+1} \sum_{|i|=K+2} A_{mi_1}\Big(\frac{x}{\mu}\Big) N_{i_2 \dots i_{K+2}} D^i \omega^{(K)} \frac{\partial \phi}{\partial x_m} \ dx.$$

If necessary the subscript m at the notation A^{BN}_{mi} can be "included" in the multi-index i, i.e., if $i = (i_1, ..., i_l)$ then A^{BN}_{mi} can be written in the form $A^{BN}_{\tilde{i}}$, where $\tilde{i} = (m, i_1, ..., i_l)$; in this case $|\tilde{i}| = l + 1$. Of course this "inclusion rule" is true if and only if $m \in \{1, 2\}$.

The boundary layer functions $N^r_{i_2 \dots i_{K+2}}$ decay exponentially as $|\xi_1| \to +\infty$, so

$$|\Delta^N_1(\phi)| \ \leq \ c_3 \mu^{K+1} \|\phi\|_{H^1(\tilde{C}_\mu)} \sqrt{mes \ \tilde{C}_\mu},$$

with a constant $c_3 > 0$ independent of μ.

We get

$$J_B(\phi) \ =$$

$$= \ \sum_{l=0}^{K+1} \sum_{m=1}^{s} \int_{\tilde{C}_\mu} \mu^{l-1} \sum_{|i|=l} A^{BN}_{mi}(x/\mu) D^i \omega^{(K)} \frac{\partial \phi}{\partial x_m} \ dx \ +$$

$$+ \ \Delta^N_1(\phi).$$

2. At the second stage consider the sum

$$\tilde{J}_B^N(\phi) =$$

$$= \sum_{l=0}^{K+1}\sum_{m=1}^{s}\int_{\tilde{C}_\mu}\mu^{l-1}\sum_{|i|=l}A_{mi}^{BN}(x/\mu)D^i\omega^{(K)}(x_1,x_2)\frac{\partial\phi}{\partial x_m}\,dx =$$

$$= \sum_{l=0}^{K+1}\sum_{m=1}^{s}\int_{\tilde{C}_\mu}\mu^{l-1}\sum_{|i|=l}A_{mi}^{BN}(x/\mu)D^i\omega^{(K)}\frac{\partial}{\partial x_m}\Big(\phi D^i\omega^{(K)}\Big)\,dx -$$

$$-\sum_{l=0}^{K+1}\int_{\tilde{C}_\mu}\mu^{l-1}\phi\sum_{m=1}^{2}\sum_{|i|=l}A_{mi}^{BN}(x/\mu)\frac{\partial}{\partial x_m}D^i\omega^{(K)}\,dx.$$

Taking into account the mentioned above "inclusion rule" for subscripts we can replace summation $\sum_{m=1}^{2}\sum_{|i|=l}A_{mi}^{BN}\frac{\partial}{\partial x_m}D^i$ by the equivalent summation $\sum_{|i|=l+1}A_i^{BN}D^i$ and therefore

$$\tilde{J}_B^N(\phi) =$$

$$= \sum_{l=0}^{K+1}\sum_{m=1}^{s}\int_{\tilde{C}_\mu}\mu^{l-1}\sum_{|i|=l}A_{mi}^{BN}(x/\mu)D^i\omega^{(K)}\frac{\partial}{\partial x_m}\Big(\phi D^i\omega^{(K)}\Big)\,dx -$$

$$-\sum_{l=0}^{K+1}\int_{\tilde{C}_\mu}\mu^{l-1}\sum_{|i|=l+1}A_i^{BN}(x/\mu)\phi D^i\omega^{(K)}\,dx =$$

$$= \sum_{l=0}^{K+1}\sum_{m=1}^{s}\int_{\mu^{-1}\tilde{C}_\mu}\mu^{s+l-2}\sum_{|i|=l}A_{mi}^{BN}(\xi)\frac{\partial}{\partial\xi_m}\Big(\Big(\phi D^i\omega^{(K)}\Big)|_{x=\mu\xi}\Big)\,d\xi -$$

$$-\sum_{l=0}^{K+1}\int_{\mu^{-1}\tilde{C}_\mu}\mu^{s+l-1}\sum_{|i|=l+1}A_i^{BN}(\xi)\Big(\phi D^i\omega^{(K)}\Big)|_{x=\mu\xi}\,d\xi,$$

where

$$\mu^{-1}\tilde{C}_\mu = \{\xi\in I\!R^s \mid \mu\xi\in\tilde{C}_\mu\} = (0,b/\mu)\times(0,T/\mu)\times(-\frac{1}{2},\frac{1}{2}).$$

Finally

$$\tilde{J}_B^N(\phi) = \sum_{l=0}^{K+1}\int_{\mu^{-1}\tilde{C}_\mu}\mu^{s+l-2}\sum_{|i|=l}\sum_{m=1}^{s}A_{mi}^{BN}(\xi)\frac{\partial}{\partial\xi_m}\Big(\Big(\phi D^i\omega^{(K)}\Big)|_{x=\mu\xi}\Big) -$$

$$- A_i^{BN}(\xi)\Big(\phi D^i\omega^{(K)}\Big)|_{x=\mu\xi}\, d\xi\; -$$

$$- \mu^{s+K}\int_{\mu^{-1}\tilde{C}_\mu}\sum_{|i|=K+2} A_i^{BN}(\xi)\Big(\phi D^i\omega^{(K)}\Big)|_{x=\mu\xi}\, d\xi,$$

where $A_{mi}^{BN} = 0$ when $|i| < 0$ (by convention).

Denote

$$\Delta_2^N(\phi) \;=\; \mu^{s+K}\int_{\mu^{-1}\tilde{C}_\mu}\sum_{|i|=K+2} A_i^{BN}(\xi)\Big(\phi D^i\omega^{(K)}\Big)|_{x=\mu\xi}\, d\xi.$$

3. The variational formulation for problem (3.1.7)-(3.1.9) gives:

$$\forall\tilde{\phi}(\xi) \in H^1_{per\ \xi_1\xi_2}(Q),$$

$$\Big\langle \sum_{m=1}^{s} A_{mi}^{BN}(\xi)\frac{\partial\tilde{\phi}}{\partial\xi_m} - A_i^{BN}(\xi)\tilde{\phi}(\xi)\Big\rangle = \langle h_i^N\tilde{\phi}(\xi)\rangle.$$

Then this identity holds, in particular, for any function $\tilde{\phi} \in H^1_{per\ \xi_1\xi_2}(Q)$, vanishing when $(\xi_1,\xi_2) \in \partial S_Q$, where$S_Q$ is a unit square $(0,1)^2$. So, this identity holds for any function $\tilde{\phi} \in H^1(Q)$, vanishing when $(\xi_1,\xi_2) \in \partial S_Q$, as well as for any function $\tilde{\phi} \in H^1((a_1,a_1+1)\times(a_2,a_2+1)\times(-\frac{1}{2},\frac{1}{2}))$, $a_1,a_2 \in I\!R$ vanishing when $(\xi_1,\xi_2) \in \partial S_{Q_{a_1,a_2}}$, where$S_{Q_{a_1,a_2}}$ is a square $(a_1,a_1+1)\times(a_2,a_2+1)$. In the last case the average $\langle\ \rangle$ is defined as $\int_{(a_1,a_1+1)\times(a_2,a_2+1)\times(-\frac{1}{2},\frac{1}{2})} d\xi$.

Let $\tilde{\phi}$ be a function from $H^1_{per\ \xi_2=T/\mu}(\mu^{-1}\tilde{C}_\mu)$ vanishing when $\xi_1 = 0$ and when $\xi_1 = b/\mu$. (Here $H^1_{per\ \xi_2=T/\mu}(\mu^{-1}\tilde{C}_\mu)$ is the completion with respect to the norm $H^1(\mu^{-1}\tilde{C}_\mu \cap \{0<\xi_2<T/\mu\})$ of the set of infinitely differentiable on the closed layer $\mu^{-1}\tilde{C}_\mu$, $T/\mu-$periodic in ξ_2 functions.)

We represent it in the form

$$\tilde{\phi}(\xi) \;=\; \tilde{\phi}(\xi)\ sin^2(\pi\xi_1)sin^2(\pi\xi_2) \;+\; \tilde{\phi}(\xi)\ cos^2(\pi\xi_1)sin^2(\pi\xi_2)\; +$$

$$+\; \tilde{\phi}(\xi)\ sin^2(\pi\xi_1)cos^2(\pi\xi_2) \;+\; \tilde{\phi}(\xi)\ cos^2(\pi\xi_1)cos^2(\pi\xi_2),$$

where

$\tilde{\phi}(\xi)\ sin^2(\pi\xi_1)sin^2(\pi\xi_2)$ vanishes when $(\xi_1,\xi_2) \in \partial S_{Q_{a_1,a_2}}$, $a_1 \in \{0,1,...,b/\mu\}$, $a_2 \in \{0,1,...,T/\mu,\}$

$\tilde{\phi}(\xi)\ cos^2(\pi\xi_1)sin^2(\pi\xi_2)$vanishes when$(\xi_1,\xi_2) \in \partial S_{Q_{a_1,a_2}}$, $a_1 \in \{0,1/2,3/2,..., b/\mu{-}1/2, b/\mu\}$, $a_2 \in \{0,1,...,T/\mu,\}$

$\tilde{\phi}(\xi)\ sin^2(\pi\xi_1)cos^2(\pi\xi_2)$ vanishes when $(\xi_1,\xi_2) \in \partial S_{Q_{a_1,a_2}}$, $a_1 \in \{0,1,...,b/\mu\}$, $a_2 \in \{-1/2,1/2,3/2,...,T/\mu-1/2,T/\mu+1/2\},\}$

$\tilde{\phi}(\xi)\ cos^2(\pi\xi_1)cos^2(\pi\xi_2)$ vanishes when $(\xi_1,\xi_2) \in \partial S_{Q_{a_1,a_2}}$, $a_1 \in \{0,1/2,3/2,..., b/\mu{-}1/2, b/\mu\}$, $a_2 \in \{-1/2,1/2,3/2,...,T/\mu-1/2,T/\mu+1/2\}.\}$

Therefore, by simple addition the integral identity of the variational formulation for problem (3.1.7)-(3.1.9) can be generalized as

$$\int_{\mu^{-1}\tilde{C}_\mu} \sum_{m=1}^{s} A_{mi}^{BN}(\xi)\frac{\partial\tilde{\phi}}{\partial\xi_m} - A_i^{BN}(\xi)\tilde{\phi}(\xi)\, d\xi = \int_{\mu^{-1}\tilde{C}_\mu} h_i^N\tilde{\phi}(\xi)\, d\xi$$

for any $\tilde{\phi}(\xi) \in H^1_{per\xi_2=T/\mu}(\mu^{-1}\tilde{C}_\mu)$, vanishing when $\xi_1 = 0$ or $\xi_1 = b/\mu$; for instance, it remains valid for $\tilde{\phi}(\xi) = \Big(\phi D^i\omega^{(K)}(x_1,x_2)\Big)|_{x=\mu\xi}$.

Thus,

$$\tilde{J}_B^N(\phi) = \sum_{l=0}^{K+1}\int_{\mu^{-1}\tilde{C}_\mu}\sum_{|i|=l}\mu^{s+l-2}h_i^N\Big(\phi D^i\omega^{(K)}\Big)|_{x=\mu\xi}\, d\xi + \Delta_2^N(\phi) =$$

$$= \sum_{l=0}^{K+1}\int_{\tilde{C}_\mu}\sum_{|i|=l}\mu^{l-2}h_l^N\phi(x)D^i\omega^{(K)}\, dx + \Delta_2^N(\phi),$$

where

$$|\Delta_2^N(\phi)| \le c_4^{(N)}\mu^K\|\phi\|_{H^1(\tilde{C}_\mu)}\sqrt{mes\ C_\mu},$$

with a constant $c_4^{(N)} > 0$ independent of μ.

4. Replacing N_i by N_i^0 in the expressions A_{mi}^{BN} we define A_{mi}^{0N} . Applying the same reasoning as at stages 2 and 3 we obtain the relation

$$\hat{J}_0(\phi) = \sum_{l=0}^{K+1}\int_{\mu^{-1}\tilde{C}_\mu}\sum_{|i|=l}\mu^{s+l-2}\sum_{m=1}^{s}A_{mi}^{0N}(\xi)\frac{\partial}{\partial\xi_m}\Big((\phi D^i\omega^{(K)}\Big)|_{x=\mu\xi} -$$

$$- A_i^{0N}(\xi)\Big((\phi D^i\omega^{(K)}\Big)|_{x=\mu\xi}\, d\xi -$$

$$- \mu^{s+K}\int_{\mu^{-1}\tilde{C}_\mu}\sum_{|i|=K+2}A_i^{0N}(\xi)\Big(\phi\frac{d^{K+2}\omega^{(K)}(x_1)}{dx_1^{K+2}}\Big)|_{x=\mu\xi}\, d\xi.$$

Variational formulation for problem (3.1.16) gives

$$\sum_{|i|=l}\int_{\mu^{-1}\tilde{C}_\mu}\sum_{m=1}^{s}A_{mi}^{0N}(\xi)\frac{\partial\tilde{\phi}}{\partial\xi_m} - A_i^{0N}(\xi)\tilde{\phi}(\xi)\, d\xi = 0$$

for any function $\tilde{\phi} \in H^1_{per(\xi_2,T/\mu)}(\mu^{-1}\tilde{C}_\mu)$ vanishing when $\xi_1 = 0$ or $\xi_1 = b/\mu$.

So,

$$|\hat{J}_0(\phi)| \le c_5\mu^K\|\phi\|_{H^1(\tilde{C}_\mu)}\sqrt{mes\ \tilde{C}_\mu},$$

with a constant $c_5 > 0$ independent of μ.

Similarly,

$$|\hat{J}_1(\phi)| \le c_6\mu^K\|\phi\|_{H^1(\tilde{C}_\mu)}\sqrt{mes\ \tilde{C}_\mu},$$

with a constant $c_6 > 0$ independent of μ.

Thus,

$$I(\phi) = \int_{\tilde{C}_\mu} (\psi - \sum_{l=0}^{K+1} \mu^{l-2} \sum_{|i|=l} h_i^N D^i \omega^{(K)}) \phi(x) \, dx + \Delta_3(\phi),$$

where $\Delta_3(\phi)$ is a linear functional of ϕ such that

$$|\Delta_3(\phi)| \leq c_7 \mu^K \|\phi\|_{H^1(\tilde{C}_\mu)} \sqrt{mes \, \tilde{C}_\mu},$$

with a constant $c_7 > 0$ independent of μ.

Consider the expression

$$B(x_1) = \psi - \sum_{l=0}^{K+1} \mu^{l-2} \sum_{|i|=l} h_i^N D^i \omega^{(K)} =$$

$$= -\Big(\sum_{|i|=2} h_i^N D^i \omega^{(K)} - \psi + \sum_{l=3}^{K+1} \mu^{l-2} \sum_{|i|=l} h_i^N D^i \omega^{(K)} \Big).$$

Substitute $\omega^{(K)}(x_1, x_2) = \sum_{j=0}^K \mu^j \omega_j(x_1, x_2)$ and remind (3.1.13),(3.1.18). Then there exist a constant c_8 independent of μ such that

$$|B(x_1)| \leq c_8 \mu^K.$$

Thus

$$|I(\phi)| \leq |\int_{\tilde{C}_\mu} B(x_1) \phi \, dx| + |\Delta_3(\phi)| \leq$$

$$\leq c_8 \mu^K \|\phi\|_{H^1(\tilde{C}_\mu)} \sqrt{mes \, \tilde{C}_\mu} + |\Delta_3(\phi)| \leq$$

$$\leq (c_8 + c_7) \mu^K \|\phi\|_{H^1(\tilde{C}_\mu)} \sqrt{mes \, \tilde{C}_\mu} \leq$$

$$\leq c_9 \mu^K \|\phi\|_{H^1(\tilde{C}_\mu)} \sqrt{mes \, \tilde{C}_\mu},$$

with a constant $c_9 > 0$ independent of μ.

On the other hand, $u^{(K)} \in H^1(\tilde{C}_\mu)$ and it vanishes when $x_1 = 0$ or $x_1 = b$. Indeed,

$$u^{(K)}|_{x_1=0} = (u_B^{(K)} + u_P^{0(K)})|_{x_1=0} + q_0 =$$

$$= \sum_{l=0}^{K+1} \mu^l \sum_{|i|=l} \Big(N_i\Big(\frac{x}{\mu}\Big) + N_i^0\Big(\frac{x}{\mu}\Big)\Big)|_{x_1=0} D^i \omega^{(K)}|_{x_1=0} + q_0 =$$

$$= \sum_{l=0}^{K+1} \mu^l \sum_{|i|=l} h_i^{N0} D^i \omega^{(K)}|_{x_1=0} + q_0 \ = 0$$

due to relations (3.1.16). Similarly, $u^{(K)}|_{x_1=b} \ = \ 0$.

Taking $\phi = u - u^{(K)}$ we obtain the inequality

$$\kappa \|u - u^{(K)}\|^2_{H^1(\tilde{C}_\mu)} \ \leq \ I(u - u^{(K)}) \ =$$

$$= \int_{\tilde{C}_\mu} \sum_{m,j=1}^{s} A_{mj}\Big(\frac{x}{\mu}\Big) \frac{\partial (u - u^{(K)})}{\partial x_j} \frac{\partial (u - u^{(K)})}{\partial x_m} \, dx \ \leq$$

$$\leq \ c_{10} \mu^K \|u - u^{(K)}\|_{H^1(\tilde{C}_\mu)} \sqrt{mes \ \tilde{C}_\mu}, \qquad (3.1.23)$$

with a constant $c_{10} > 0$ independent of μ.

So,

$$\|u - u^{(K)}\|_{H^1(\tilde{C}_\mu)} \ \leq \ c_{10}/\kappa \mu^K \sqrt{mes \ \tilde{C}_\mu};$$

this completes the proof of the theorem.

Corollary 3.1.1. *The estimate holds*

$$\|u - u_B^{(K-1)} - u_P^{0(K-1)} - u_P^{1(K-1)}\|_{H^1(\tilde{C}_\mu)} \ = O(\mu^K) \sqrt{mes \ \tilde{C}_\mu}, \ K = 1, 2, \ldots.$$

Indeed, the H^1–norm of the differences $u^{0(K)} - u^{0(K)}\chi(x_1)$ and $u^{1(K)} - u^{1(K)}\chi(b - x_1)$ and the H^1–norm of ρ are of order $O(\mu^K)\sqrt{mes \ C_\mu}$. Moreover, $\|u_B^{(K)} - u_B^{(K-1)}\|_{H^1(\tilde{C}_\mu)}$, $\|u_P^{0(K)} - u_P^{0(K-1)}\|_{H^1(\tilde{C}_\mu)}$, $\|u_P^{1(K)} - u_P^{1(K-1)}\|_{H^1\tilde{(C}_\mu)} = O(\mu^K)\sqrt{mes \ \tilde{C}_\mu}$.

So when one replace in the estimate of the theorem $u^{(K)}$ by $u_B^{(K-1)} + u_P^{0(K-1)} + u_P^{1(K-1)}$, the order of this estimate does not change. It proves the corollary.

Remark 3.1.3 The function $u_B^{(K-1)} + u_P^{0(K-1)} + u_P^{1(K-1)}$ does not vanish at the ends of the rod; however, it has the order $O(\mu^K)$ there.

Remark 3.1.4 The plate can be simulated by a thin cylinder $\Omega_\mu = G \times (-\mu/2, \mu/2) \subset R^3$, where G is a two-dimensional domain with a smooth boundary. In this case the the construction of a complete asymptotic expansion is still an open problem. Nevertheless the estimate for $u_B^{(0)}$ can be obtained by means of the technique [16], Chapter 4, section 4.1. Consider equation (3.1.1) set in Ω_μ with condition (3.1.2) on the both bases of cylinder Ω_μ and (3.1.3) on the lateral boundary of the cylinder $\partial G \times (-\mu/2, \mu/2)$. Let ω_0 be a solution of equation (3.1.13) for $j = 0$, supplied with the Dirichlet condition $\omega_0 = 0$ on ∂G. Then the estimate can be proved in the same way as (32)-(33) of Chapter 4, section 4.1 [16]:

$$\|u - u_B^{(0)}\|_{H^1(\Omega_\mu)} = O(\sqrt{\mu mes \Omega_\mu})$$

3.2 Elasticity of a plate.

3.2.1 Statement of the problem.

For $s = 2$, problem (3.1.1)–(3.1.3) can be regarded as the two-dimensional analog of the elasticity theory system of equations in a plate. Below a three-dimensional formulation ($s = 3$) will be considered.

Consider the elasticity theory system of equations

$$Pu \equiv \sum_{i,j=1}^{3} \frac{\partial}{\partial x_i}\Big(A_{ij}\Big(\frac{x}{\mu}\Big)\frac{\partial u}{\partial x_j}\Big) = \psi(x_1, x_2), \quad x \in \widetilde{C}_\mu, \tag{3.2.1}$$

with boundary conditions

$$\frac{\partial u}{\partial \nu} \equiv \pm \sum_{j=1}^{3} A_{3j}\Big(\frac{x_1}{\mu}, \frac{x_2}{\mu}, \pm\frac{1}{2}\Big)\frac{\partial u}{\partial x_j} = 0, \quad x \in \partial\widetilde{U}_\mu, \tag{3.2.2}$$

$$u = 0, \quad x_1 = 0 \quad \text{or} \quad x_1 = b, \tag{3.2.3}$$

and with the condition of T-periodicity with respect to x_2. Here b and T are numbers of the order of 1 that are multiples of μ and $A_{ij}(\xi)$ are 3×3 matrices whose elements satisfy the conditions of subsection 2.2.1 and are 1-periodic with respect to ξ_1 and ξ_2. These elements are assumed to be piecewise smooth functions, while ψ is a C^∞ three-dimensional vector valued function that is T-periodic with respect to x_2.

3.2.2 Inner expansion

We seek a formal asymptotic solution in the form of a series analogous to the series of the previous section

$$u^{(\infty)} = \sum_{l=0}^{\infty} \mu^l \sum_{|i|=l} N_i\Big(\frac{x}{\mu}\Big) D^i \omega(x_1, x_2), \tag{3.2.4}$$

where $i = (i_1, \ldots, i_l)$ is a multi-index, $i_j \in \{1, 2\}$, $\omega(x_1, x_2)$ is a smooth T-periodic with respect to x_2 three-dimensional vector function, $N_i(\xi)$ are 3×3 matrix functions that are 1-periodic with respect to ξ_1 and ξ_2.

Substituting series (3.2.4) in (3.2.1), (3.2.2) and the interface conditions and grouping terms of the same order, we obtain

$$Pu^{(\infty)} - \psi = \sum_{l=0}^{\infty} \mu^{l-2} \sum_{|i|=l} H_i(\xi) D^i \omega - \psi, \tag{3.2.5}$$

$$\frac{\partial u^{(\infty)}}{\partial \nu} = \sum_{l=0}^{\infty} \mu^{l-1} \sum_{|i|=l} G_i(\xi) D^i \omega, \tag{3.2.6}$$

where

$$H_i(\xi) = L_{\xi\xi}N_i + T_i(\xi),$$

$$T_i(\xi) = \sum_{j=1}^{3} \frac{\partial}{\partial \xi_j}(A_{ji_1}N_{i_2\dots i_l}) + \sum_{j=1}^{3} A_{i_1 j}\frac{\partial N_{i_2\dots i_l}}{\partial \xi_j} + A_{i_1 i_2}N_{i_3\dots i_l},$$

$$G_i(\xi) = \sum_{m=1}^{3}\Big(\sum_{j=1}^{3} A_{mj}\frac{\partial N_i}{\partial \xi_j} + A_{mi_1}N_{i_2\dots i_l}\Big)n_m.$$

Suppose that

$$H_i(\xi) = h_i, \quad \xi \in I\!R^2 \times \Big(-\frac{1}{2},\frac{1}{2}\Big); \qquad G_i(\xi) = 0, \quad \xi_3 = \pm\frac{1}{2},$$

where h_i are constant 3×3 matrices.

Assume that $[N_i] = 0$ and $[G_i] = 0$ on the surfaces of discontinuity.

We obtain the following recurrent chain of problems of the form

$$L_{\xi\xi}N_i = -T_i(\xi) + h_i, \tag{3.2.7}$$

$$\sum_{j=1}^{3} A_{3j}(\xi)\frac{\partial N_i}{\partial \xi_j} = -A_{3i_1}N_{i_2\dots i_l}, \quad \xi_3 = \pm\frac{1}{2}, \tag{3.2.8}$$

$$[N_i]\big|_\Sigma = 0, \quad \Big[\frac{\partial N_i}{\partial \nu_\xi}\Big]\Big|_\Sigma = \Big[\sum_{m=1}^{3} A_{mi_1}N_{i_2\dots i_l}\Big]\Big|_\Sigma, \tag{3.2.9}$$

to determine N_i that are 1-periodic functions with respect to ξ_1 and ξ_2. The constant matrices h_i are chosen from the solvability conditions for problem (3.2.7)-(3.2.9):

$$h_i = \Big\langle \sum_{j=1}^{3} A_{i_1 j}\frac{\partial N_{i_2\dots i_l}}{\partial \xi_j} + A_{i_1 i_2}N_{i_3\dots i_l}\Big\rangle, \quad l \geq 2,$$

$$h_\emptyset = 0, \quad h_{i_1} = 0; \tag{3.2.10}$$

here $\langle\cdot\rangle = \int_{(-1/2,1/2)^3} d\xi$ and $N_\emptyset = I$ (the identity matrix).

The sufficiency of condition (3.2.10) can be proved as in Lemma 2.2.1.

Thus, the algorithm for constructing the matrices N_i is recurrent. We formally assume that $N_i = 0$ for $|i| < 0$, $N_\emptyset = I$ (the identity matrix), and $N_i(\xi)$ is a solution to problem (3.2.7)–(3.2.10) for $l > 0$. The right-hand side contains N_j with multi-indices j whose length is smaller than $|i|$.

Let the following condition A' be satisfied:

$$a_{ij}^{kl}(S_3\xi) = (-1)^{\delta_{3i}+\delta_{3k}+\delta_{3j}+\delta_{3l}}a_{ij}^{kl}(\xi).$$

(Remind that $S_3\xi = (\xi_1, \xi_2, -\xi_3)$).

Theorem 3.2.1. *The matrices h_i have the form*

$$h_i = \begin{pmatrix} h_i^{11} & h_i^{12} & 0 \\ h_i^{21} & h_i^{22} & 0 \\ 0 & 0 & h_i^{33} \end{pmatrix}.$$

For $|i| = 2, 3$ we have $h_i^{33} = 0$; for $|i| = 2$ the elements $h_{i_1 i_2}^{kl}$, $k, l \in \{1, 2\}$, of the matrices h_i satisfy the relations

$$\forall e = (e_{i_1 i_2})_{i_1, i_2 \in \{1,2\}},\ e_{i_2}^{l} = e_l^{i_2},$$

the following inequality holds

$$\sum_{i_1, i_2, k, l=1}^{2} h_{i_1 i_2}^{kl} e_{i_2}^{l} e_{i_1}^{k} \geq \kappa \sum_{i_2, l=1}^{2} (e_{i_2}^{l})^2, \tag{3.2.11}$$

$$h_{i_1 i_2}^{kl} = h_{k i_2}^{i_1 l} = h_{i_2 i_1}^{lk}, \tag{3.2.12}$$

$$-\sum_{i_1, i_2, i_3, i_4=1}^{2} h_{i_1 i_2 i_3 i_4}^{33} \eta_{i_1} \eta_{i_2} \eta_{i_3} \eta_{i_4} > 0 \quad \forall (\eta_1, \eta_2) \neq (0, 0). \tag{3.2.13}$$

The proof of theorem 3.2.1 will be given later.
Then (3.2.5) takes the form

$$Pu^{(\infty)} - \psi = \sum_{i_1, i_2=1}^{2} h_{i_1 i_2} \frac{\partial^2 \omega}{\partial x_{i_1} \partial x_{i_2}} + \mu \sum_{i_1, i_2, i_3=1}^{2} h_{i_1 i_2 i_3} \frac{\partial^3 \omega}{\partial x_{i_1} \partial x_{i_2} \partial x_{i_3}}$$

$$+\mu^2 \sum_{i_1, i_2, i_3, i_4=1}^{2} h_{i_1 i_2 i_3 i_4} \frac{\partial^4 \omega}{\partial x_{i_1} \partial x_{i_2} \partial x_{i_3} \partial x_{i_4}} +$$

$$+\sum_{l=5}^{\infty} \mu^{l-2} \sum_{|i|=l} h_i \frac{\partial^l \omega}{\partial x_{i_1} \cdots \partial x_{i_l}} - \psi = 0. \tag{3.2.14}$$

Problem (3.2.14) can be regarded as the homogenized equation of infinite order with respect to the three-dimensional vector ω. A formal asymptotic solution (f.a.s.) of this problem is sought in the form of a series

$$\omega = \sum_{j=-2}^{\infty} \mu^j \omega_j(x_1, x_2), \tag{3.2.15}$$

where ω_j do not depend on μ and $\omega_j^1 = 0$ and $\omega_j^2 = 0$ for $j = -2, -1$. Substituting the series (3.2.15) into (3.2.14), we obtain a recurrent chain of equations for the components ω_j^k of the vectors ω_j in the form

$$\sum_{i_1, i_2=1}^{2} \bar{A}_{i_1 i_2} \frac{\partial^2 \hat{\omega}_j}{\partial x_{i_1} \partial x_{i_2}} = \hat{f}_j(x_1, x_2), \tag{3.2.16}$$

$$\sum_{i_1,i_2,i_3,i_4=1}^{2} h^{33}_{i_1i_2i_3i_4} \frac{\partial^4 \omega^3_j}{\partial x_{i_1}\partial x_{i_2}\partial x_{i_3}\partial x_{i_4}} = f^3_j(x_1,x_2), \quad (3.2.17)$$

where $\hat{\omega}_j$ and $\hat{f}_j$ are the first two components of the vectors ω_j and f_j, $\bar{A}_{i_1i_2}$ are 2×2 matrices with components $h^{kl}_{i_1i_2}$, the functions $\hat{f}_j$ depend on $\hat{\omega}_{j_1}$, $j_1 < j$, and on the derivatives of these functions, while the functions f^3_j depend on $\omega^3_{j_1}$, $j_1 < j$. Thus, the f.a.s. of problem (3.2.1), (3.2.2) is constructed.

3.2.3 Boundary layer corrector

A formal asymptotic solution of problem (3.2.1)–(3.2.3) will be sought again in the form (2.2.22): $u^{(\infty)} = u_B + u^0_P + u^1_P$, where u_B is defined by (3.2.4) and

$$u^0_P = \sum_{l=0}^{\infty}\mu^l \sum_{|i|=l} N^0_i(\xi) D^i\omega\big|_{\xi=x/\mu}, \quad (3.2.18)$$

$$u^1_P = \sum_{l=0}^{\infty}\mu^l \sum_{|i|=l} N^1_i(\xi) D^i\omega\big|_{\xi_1=(x_1-b)/\mu,\ \xi'=x'/\mu}. \quad (3.2.19)$$

Here the matrix functions $N^0_i(\xi)$ and $N^1_i(\xi)$ are 1-periodic with respect to ξ_2 and exponentially stabilize to zero as $\xi_1 \to +\infty$, $\xi' = (\xi_2,\xi_3)$, $x' = (x_2,x_3)$.

Substituting (3.2.18) and (3.2.19) into (3.2.1)–(3.2.3) and taking into account the fact that u_B is constructed above, we obtain recurrent chains of problems to determine $N^0_i(\xi)$ as in subsection 3.1.2

$$L_{\xi\xi}N^0_i + T^0_i(\xi) = 0, \quad \xi_1 > 0, \quad \xi' \in I\!R \times \left(-\frac{1}{2},\frac{1}{2}\right),$$

$$\sum_{j=1}^{3} A_{3j}\frac{\partial N^0_i}{\partial \xi_j} = -A_{3i_1}N^0_{i_2\dots i_l}, \quad \xi_3 = \pm\frac{1}{2},$$

$$N^0_i(0,\xi') = -N_i(0,\xi') + \tilde{\tilde{\Phi}} h^0_i \quad (3.2.20)$$

and obtain similar problems for $N^1_i(\xi)$ defined for $\xi_1 < 0$ with the constant h^1_i instead of h^0_i. Here $\tilde{\tilde{\Phi}}$ is a 3×4 matrix

$$\tilde{\tilde{\Phi}}(\xi) = \begin{pmatrix} 1 & 0 & 0 & -\xi_3 \\ 0 & 1 & 0 & 0 \\ 0 & 0 & 1 & \xi_1 \end{pmatrix},$$

h^r_i are 4×3 constant matrices chosen from the condition that solutions of problems (3.2.20) stabilize to zero as $\xi_1 \to +\infty$ ($\xi_1 \to -\infty$ for $r = 1$), and $N^0_{\emptyset} = 0$.

Theorem 3.2.2. *Let $A_{ij}(\xi)$ satisfy the conditions formulated at the beginning of this section, let the three-dimensional vectors $F_j(\xi)$ be 1-periodic with respect to ξ_2 and belong to $L^2([0,+\infty)\times(-1/2,1/2)^2)$, $j = 0,1,2,3,$ let conditions*

(2.32) be satisfied for $\beta = (-1/2, 1/2)^2$, *and let a three-dimensional vector* $u_0(\xi')$ *be 1-periodic with respect to* ξ_2 *and belong to* $H^{1/2}(\{0\} \times (-1,1) \times (-1/2,1/2))$. *Then a solution* $u(\xi)$ *of the problem*

$$L_{\xi\xi}u = F_0(\xi) + \sum_{j=1}^{3} \frac{\partial F_j}{\partial \xi_j}, \quad \xi_1 > 0, \quad \xi' \in I\!R \times \Big(-\frac{1}{2}, \frac{1}{2}\Big),$$

$$\sum_{j=1}^{3} A_{3j} \frac{\partial u}{\partial \xi_j} = F_j, \quad \xi_3 = \pm\frac{1}{2}, \qquad u\big|_{\xi_1=0} = u_0(\xi') \tag{3.2.21}$$

such that u is 1-periodic with respect to ξ_2 *and inequality* (2.3.36) *holds for some rigid displacement* $w(\xi) = \tilde{\tilde{\Phi}}(\xi)h$, *where h is a constant four-dimensional vector, exists and is unique.*

Here by a solution of problem (3.2.21) we mean a vector function $u(\xi)$ satisfying the integral identity from subsection 3.1.2.

Theorem 3.2.2 can be proved similarly to Theorems 4 and 5 in [130]. The matrices h_i^r are defined similarly to the matrices h_l^{Nr} from Chapter 2. In this case

$$h_\emptyset^r = \begin{pmatrix} 1 & 0 & 0 \\ 0 & 1 & 0 \\ 0 & 0 & 1 \\ 0 & 0 & 0 \end{pmatrix}$$

since $N_\emptyset^r = 0$ and $N_\emptyset = I$. The fourth row of the matrix $h_{i_1}^r$ is equal to $(0,0,1)$. The fact that the first two elements of the row are zeros can be proved, as in Lemma 2.3.6, by using the following relation for the components $n_{i_1}^{kl}(\xi)$ of the matrices $N_{i_1}(\xi)$:

$$n_{i_1}^{kl}(S_3\xi) = (-1)^{\delta_{3k}+\delta_{3l}} n_{i_1}^{kl}(\xi). \tag{3.2.22}$$

It follows from this relation that for $k,l \in \{1,2\}$ these components are even with respect to ξ_3. The number 1 at the end of the row is explained by the fact that the third column of the matrices $N_{i_1}(\xi)$ has the form $(-\xi_3, 0, 0)^*$, and therefore the third column $N_{i_1}^r$ is zero and the coefficient for the rigid displacement $(-\xi_3, 0, \xi_1)$ in boundary condition (3.2.20) is equal to 1.

After determining all N_i^r's and h_i^r's by induction on $|i|$, we obtain a homogenized problem for the three-dimensional vector function $\omega(x_1, x_2)$: this problem is formed by equation (3.2.14) with the boundary conditions

$$\Big(\omega + \sum_{l=1}^{\infty} \mu^l \sum_{|i|=l} h_i^{r(\mathrm{cut}\,3)} D^i \omega\Big)\Big|_{x_1=rb} = 0,$$

$$\Big(\mu \frac{\partial \omega^3}{\partial x_1} + \sum_{l=2}^{\infty} \mu^l \sum_{|i|=l} h_i^{r(\mathrm{oth}\,3)} D^i \omega\Big)\Big|_{x_1=rb} = 0. \tag{3.2.23}$$

Here the superscript 'cut 3' denotes the first three rows of the matrix and 'oth 3' denotes the fourth row.

A f.a.s. of problem (3.2.1)–(3.2.3) is sought in the form of series (3.2.15) with boundary conditions

$$\omega_j\Big|_{x_1=rb} = g_{jr}, \quad r = 0, 1, \qquad \frac{\partial \omega_j^3}{\partial x_1}\Big|_{x_1=rb} = g_{jr}^4, \quad r = 0, 1, \tag{3.2.24}$$

where the g_{jr}^q's depend on $\omega_{j_1}^r$ with $j_1 < j$ and on derivatives of these functions.

The homogenized problem of zero order is formed by relations (3.2.16), (3.2.17), (3.2.24) with $j = 0$,

$$\widehat{f}_0(x_1, x_2) = \widehat{\psi}(x_1, x_2), \quad f_0^3(x_1, x_2) = \psi^3(x_1, x_2), \quad g_{0r} = 0.$$

The estimate for the difference of the exact solution and the asymptotic solution can be obtained as in section 3.1.

3.2.4 Proof of Theorem 3.2.1.

As in Lemma 2.3.2, by substituting into (3.2.7)–(3.2.10), we can prove the relation

$$n_i^{kl}(S_3\xi) = (-1)^{\delta_{3k}+\delta_{3l}} n_i^{kl}(\xi);$$

for the elements h_i^{kl} of the matrices h_i this relation implies the formula $h_i^{kl} = (-1)^{\delta_{3k}+\delta_{3l}} h_i^{kl}$ which yields $h_i^{kl} = 0$ for $k = 3$ and $l \neq 3$ or for $k \neq 3$ and $l = 3$. It follows from (3.2.10) that

$$\begin{aligned} h_{i_1 i_2} &= \Big\langle \sum_{j=1}^{3} A_{i_1 j} \frac{\partial N_{i_2}}{\partial \xi_j} + A_{i_1 i_2} \Big\rangle = \Big\langle \sum_{j=1}^{3} A_{i_1 j} \frac{\partial}{\partial \xi_j}(N_{i_2} + I\xi_{i_2}) \Big\rangle \\ &= \Big\langle \sum_{m,j=1}^{3} A_{mj} \frac{\partial}{\partial \xi_j}(N_{i_2} + I\xi_{i_2}) \frac{\partial}{\partial \xi_m}(N_{i_1} + I\xi_{i_1}) \Big\rangle. \end{aligned} \tag{3.2.25}$$

The last relation arises from the integral identity for problem (3.2.7)–(3.2.9) with $|i| = 2$. Relations (3.2.11) and (3.2.12) can be derived from (3.2.25). Indeed,

$$\begin{aligned} h_{i_1 i_2}^{kl} &= \sum_{m,j,q,p=1}^{3} \Big\langle a_{mj}^{qp} \frac{\partial}{\partial \xi_j}(n_{i_2}^{pl} + \xi_{i_2}\delta_{pl}) \frac{\partial}{\partial \xi_m}(n_{i_1}^{qk} + \xi_{i_1}\delta_{qk}) \Big\rangle \\ &= \sum_{m,j,q,p=1}^{3} \Big\langle a_{mj}^{qp} Z_j^{pl}(N_{i_2} + I\xi_{i_2}) Z_m^{qk}(N_{i_1} + I\xi_{i_1}) \Big\rangle, \end{aligned}$$

where

$$Z_j^{pl}(N_r + I\xi_r) = \frac{1}{2}\Big(\frac{\partial}{\partial \xi_j}(n_r^{pl} + \xi_r\delta_{pl}) + \frac{\partial}{\partial \xi_p}(n_r^{jl} + \xi_r\delta_{jl}) \Big).$$

This implies (3.2.12). For $\eta_{i_1}^k = \eta_k^{i_1}$ we have

$$\sum_{i_1,i_2,k,l=1}^{2} h_{i_1 i_2}^{kl} \eta_{i_2}^l \eta_{i_1}^k = \sum_{m,j,q,p=1}^{3} \sum_{i_1,i_2,k,l=1}^{2} a_{mj}^{qp} Z_j^{pl}(N_{i_2}+I\xi_{i_2})\eta_{i_2}^l Z_m^{qk}(N_{i_1}+I\xi_{i_1})\eta_{i_1}^k \rangle$$

$$\geq \kappa \sum_{j,p=1}^{3} \Big\langle \Big(\sum_{i_2,l=1}^{2} Z_j^{pl}(N_{i_2} + I\xi_{i_2})\eta_{i_2}^p \Big)^2 \Big\rangle$$

$$\geq \kappa \sum_{j,p=1}^{2} \Big\langle \Big(\sum_{i_2,l=1}^{2} Z_j^{pl}(I\xi_{i_2})\eta_{i_2}^p \Big)^2 \Big\rangle$$

$$= \kappa \sum_{j,p=1}^{2} \Big(\frac{1}{2} \sum_{i_2,l=1}^{2} (\delta_{pl}\delta_{ji_2} + \delta_{jl}\delta_{pi_2})\eta_{i_2}^p \Big)^2 = \kappa \sum_{j,p=1}^{2} (\eta_j^p)^2.$$

The third column of the matrices N_{i_1} is $(-\xi_3, 0, 0)^*$, hence, $n_{i_1}^{q3} + \xi_{i_1}\delta_{q3}$ is a rigid displacement and $h_{i_1 i_2}^{33} = 0$; moreover,

$$h_{i_1 i_2 i_3}^{33} = \Big\langle a_{i_1 j}^{3p} \frac{\partial}{\partial \xi_j} n_{i_2 i_3}^{p3} + a_{i_1 i_2}^{3p} n_{i_3}^{p3} \Big\rangle = \Big\langle a_{3j}^{i_1 p} \frac{\partial}{\partial \xi_j} n_{i_2 i_3}^{p3} + a_{3 i_2}^{i_1 p} n_{i_3}^{p3} \Big\rangle$$

$$= \Big\langle \Big(a_{mj}^{i_1 p} \frac{\partial}{\partial \xi_j} n_{i_2 i_3}^{p3} + a_{m i_2}^{i_1 p} n_{i_3}^{p3} \Big) \frac{\partial \xi_3}{\partial \xi_m} \Big\rangle = -\Big\langle \frac{\partial}{\partial \xi_m} \Big(a_{mj}^{i_1 p} \frac{\partial}{\partial \xi_j} n_{i_2 i_3}^{p3} + a_{m i_2}^{i_1 p} n_{i_3}^{p3} \Big) \xi_3 \Big\rangle$$

$$= \Big\langle \Big(a_{i_2 j}^{i_1 p} \frac{\partial n_{i_3}^{p3}}{\partial \xi_j} + a_{i_2 i_3}^{i_1 p} \delta_{p3} \Big) \xi_3 \Big\rangle = \Big\langle \Big(a_{i_2 j}^{i_1 p} \frac{\partial}{\partial \xi_j} (n_{i_3}^{p3} + \delta_{p3}\xi_{i_3}) \Big) \xi_3 \Big\rangle = 0.$$

The proof of the fact that the form (3.2.13) is positive definite is similar to the proof of Lemmas 2.3.4 and 2.3.5.

Namely, the following statement can be proved as in Lemma 2.3.4. Let η_1, η_2 be two positive numbers. Let operators (2.3.93) and (2.3.94) be defined for $x \in \widetilde{U}_\mu$ and $x \in \partial\widetilde{U}_\mu$, respectively. Denote

$$C_\mu^\eta = \Big(-\frac{\pi}{\eta_1}, \frac{\pi}{\eta_1} \Big) \times \Big(-\frac{\pi}{\eta_2}, \frac{\pi}{\eta_2} \Big) \times \Big(-\frac{\mu}{2}, \frac{\mu}{2} \Big).$$

Assume that a three-dimensional vector function $u \in H^1_{\frac{\pi}{\eta_1}, \frac{\pi}{\eta_2} - per}$, where $H^1_{\frac{\pi}{\eta_1}, \frac{\pi}{\eta_2} - per}$ is a completion (by the $H^1(C_\mu^\eta)$–norm) of the space of differentiable in $\mathbb{R}^2 \times [-\frac{\mu}{2}, \frac{\mu}{2}]$ three-dimensional vector functions, $2\pi/\eta_1$-periodic with respect to x_1 and $2\pi/\eta_2$-periodic with respect to x_2. Assume that

$$\int_{C_\mu^\eta} u \, dx = 0.$$

Then the following inequality holds (for sufficiently small μ):

$$\sum_{i,j=1}^{3} \int_{C_\mu^\eta} \Big(A_{ij} \frac{\partial u}{\partial x_j}, \frac{\partial u}{\partial x_i} \Big) dx \geq c \big(\mu^4 \|\nabla u\|^2_{L^2(C_\mu^\eta)} + \mu^2 \|u\|^2_{L^2(C_\mu^\eta)} \big), \quad c > 0, \tag{3.2.26}$$

c_1 does not depend on μ.

Indeed, in (2.3.93) and (2.3.94) let us perform the following changes of variables:

1) $x_1 = \xi_1$, $x_2 = \xi_2$, $x_3 = \xi_3 \mu$,

2) $u_1 = w_1$, $u_2 = w_2$, $u_3 = w_3/\mu$,

3) multiply the third component of (2.3.93) and the first two components of (2.3.94) by μ^{-1} and the last component of (2.3.94) by μ^{-2}; then we obtain for (2.3.93) and (2.3.94)the following presentations:

$$-\sum_{i,j=1}^{3} \frac{\partial}{\partial \xi_i}\Big(B_{ij}(\xi)\frac{\partial w}{\partial \xi_j}\Big), \quad \xi_3 \in \Big(-\frac{1}{2}, \frac{1}{2}\Big), \qquad \sum_{j=1}^{3} B_{1j}(\xi)\frac{\partial w}{\partial \xi_j}, \qquad \xi_3 = \pm\frac{1}{2},$$

where the elements b_{ij}^{kl} of the matrices B_{ij} are

$$b_{ij}^{kl}(\xi) = a_{ij}^{kl}\big(\xi_1/\mu, \xi_2/\mu, \xi_3\big)\big(1/\mu\big)^{\delta_{i3}+\delta_{j3}+\delta_{k3}+\delta_{l3}}.$$

As in Lemma 2.3.6, we see that for any symmetric matrix with elements η_i^k the following inequality holds:

$$\sum_{i,j,k,l=1}^{3} b_{ij}^{kl}\eta_i^k\eta_j^l \geq \kappa \sum_{j,l=1}^{3} (\eta_j^l)^2$$

and we have

$$\int_{Q_0^\eta} w\, d\xi = 0, \quad \text{where} \quad Q_0^\eta = \big(-\pi/\eta_1, \pi/\eta_1\big) \times \big(-\pi/\eta_2, \pi/\eta_2\big) \times \big(-\tfrac{1}{2}, \tfrac{1}{2}\big).$$

Korn's inequality for the function w implies

$$\sum_{i,j=1}^{3} \int_{Q_0^\eta} \Big(B_{ij}\frac{\partial w}{\partial \xi_j}, \frac{\partial w}{\partial \xi_i}\Big)\, d\xi \geq \kappa c_1 \|w\|^2_{H^1(Q_0^\eta)},$$

where c_1 is a positive constant independent of μ. The left-hand side of the last inequality is equal to the left-hand side of inequality (3.2.26) multiplied by μ^{-1} and we have

$$\|w\|^2_{H^1(Q_0^\eta)} \geq \mu^3\|\nabla u\|^2_{L^2(C_\mu^\eta)} + \mu\|u\|^2_{L^2(C_\mu^\eta)},$$

which implies (3.2.26).

Furthermore, in $\widetilde{U}_\mu$ we specify a vector function

$$u_\mu^{(K)}(x) = \sum_{l=0}^{K+1} \mu^l \sum_{|i|=l} N_i\Big(\frac{x}{\mu}\Big) D^i V(x_1, x_2),$$

where $N_i(\xi)$ are the above matrices and $V(x_1, x_2)$ are infinitely differentiable three-dimensional vector functions that are $2\pi/\eta_j$-periodic with respect to x_j, $j = 1, 2$, and the first two components of V are zero: $V = (0, 0, V^3)^*$.

As in Lemma 2.3.5, for sufficiently large K and given η_j, we obtain

$$\sum_{i,j=1}^{3} \int_{C_\mu^\eta} \Big(A_{ij}\frac{\partial u_\mu^{(K)}}{\partial x_j}, \frac{\partial u_\mu^{(K)}}{\partial x_i}\Big)\, dx$$

$$= -\mu^3 \sum_{i_1,i_2,i_3,i_4=1}^{2} h^{33}_{i_1 i_2 i_3 i_4} \int_{(-\pi/\eta_1,\pi/\eta_1)\times(-\pi/\eta_2,\pi/\eta_2)} (D^i V^3) V^3 \, dx_1 \, dx_2 + O(\mu^4)$$

$$\geq \mu^2 \kappa c_1 \| u^{(K)}_\mu \|^2_{L^2(C^\eta_\mu)} \geq \frac{\mu^3}{2} \kappa c_1 \int_{(-\pi/\eta_1,\pi/\eta_1)\times(-\pi/\eta_2,\pi/\eta_2)} (V^3)^2 \, dx_1 \, dx_2,$$

c_1 being constant independent of μ. By setting $V^3 = \sin(\eta_1 x_1 + \eta_2 x_2)$ we see that $D^i V = \eta_{i_1}\eta_{i_2}\eta_{i_3}\eta_{i_4} V^3$; this implies

$$- \sum_{i_1,i_2,i_3,i_4=1}^{2} h^{33}_{i_1 i_2 i_3 i_4} \eta_{i_1}\eta_{i_2}\eta_{i_3}\eta_{i_4} \geq \frac{\kappa c_1}{2} > 0$$

for all μ small enough. Remind that here $\eta_1 > 0$ and $\eta_2 > 0$ are arbitrary positive numbers. Considering $V^3 = \sin(\pm\eta_1 x_1 \pm \eta_2 x_2)$ we can generalize this inequality for any couples $(\eta_1, \eta_2) \neq (0,0)$. (If $\eta_1 = 0$ and $\eta_2 \neq 0$ or $\eta_2 = 0$ and $\eta_1 \neq 0$ we apply the same reasoning with $C^\eta_\mu = (-\frac{1}{2}, \frac{1}{2}) \times (-\frac{\pi}{\eta_2}, \frac{\pi}{\eta_2}) \times (-\frac{\mu}{2}, \frac{\mu}{2})$ or $C^\eta_\mu = (-\frac{\pi}{\eta_1}, \frac{\pi}{\eta_1}) \times (-\frac{1}{2}, \frac{1}{2}) \times (-\frac{\mu}{2}, \frac{\mu}{2})$ respectively).

The proof of Theorem 3.2.1 is complete.

3.2.5 Algorithm for calculating the effective stiffness of a plate

As in section 2.3, we present a calculation algorithm for the coefficients $h^{kl}_{i_1 i_2}$, $i_1, i_2, k, l \in \{1, 2\}$, of the homogenized equation (elasticity modules in a plane) and the coefficients $h^{33}_{i_1 i_2 i_3 i_4}$, $i_1, i_2, i_3, i_4 \in \{1, 2\}$. Below we denote $\bar{a}^{kl}_{i_1 i_2} = h^{kl}_{i_1 i_2}$ and $\bar{c}_{i_1 i_2 i_3 i_4} = -h^{33}_{i_1 i_2 i_3 i_4}$ (bending stiffness coefficients).

We first solve the elasticity theory systems of equations for i_1, $q = 1, 2$:

$$\sum_{m,j=1}^{3} \frac{\partial}{\partial \xi_m} \Big(A_{mj} \frac{\partial}{\partial \xi_j} (N^{(q)}_{i_1} + \mathbf{e}_q \xi_{i_1}) \Big) = 0, \quad \xi_3 \in \Big(-\frac{1}{2}, \frac{1}{2} \Big),$$

$$\sum_{j=1}^{3} A_{3j} \frac{\partial}{\partial \xi_j} (N^{(q)}_{i_1} + \mathbf{e}_q \xi_{i_1}) = 0, \quad \xi_3 = \pm\frac{1}{2},$$

where $\mathbf{e}_q = (\delta_{q1}, \delta_{q2}, \delta_{q3})^*$ and $N^{(q)}_{i_1}$ is an unknown three-dimensional vector function that is 1-periodic with respect to ξ_1 and ξ_2. In this case the elasticity modules in a plane can be defined by the following formulas:

$$\bar{a}^{kq}_{i_1 i_2} = \Big\langle a^{kr}_{i_1 j} \frac{\partial}{\partial \xi_j} (n^{rq}_{i_2} + \delta_{rq} \xi_{i_1}) \Big\rangle.$$

Then we solve the elasticity theory systems of equations for $i_1, i_2 = 1, 2$:

$$\sum_{m,j=1}^{3} \frac{\partial}{\partial \xi_m} \Big(A_{mj} \frac{\partial}{\partial \xi_j} N^{(3)}_{i_1 i_2} - A_{m i_1} \mathbf{e}_{i_2} \xi_3 \Big) = 0, \quad \xi_3 \in \Big(-\frac{1}{2}, \frac{1}{2} \Big),$$

$$\sum_{j=1}^{3} A_{3j}\frac{\partial}{\partial \xi_j}N^{(3)}_{i_1i_2} - A_{3i_1}\mathbf{e}_{i_2}\xi_3 = 0, \quad \xi_3 = \pm\frac{1}{2},$$

related to three-dimensional vector functions $N^{(3)}_{i_1i_2}(\xi)$ that are 1-periodic in ξ_1 and ξ_2. The bending stiffness coefficients $\bar{c}_{i_1i_2i_3i_4}$ are defined by the formula

$$\bar{c}_{i_1i_2i_3i_4} = -\Big\langle \xi_3\Big(\sum_{j,r=1}^{3} a^{i_1r}_{i_2j}\frac{\partial n^{r3}_{i_3i_4}}{\partial \xi_j} - a^{i_1i_4}_{i_2i_3}\xi_3\Big)\Big\rangle.$$

Indeed

$$h^{33}_{i_1i_2i_3i_4} = \Big\langle a^{3p}_{i_1j}\frac{\partial}{\partial \xi_j}n^{p3}_{i_2i_3i_4} + a^{3p}_{i_1i_2}n^{p3}_{i_3i_4}\Big\rangle = \Big\langle a^{i_1p}_{3j}\frac{\partial}{\partial \xi_j}n^{p3}_{i_2i_3i_4} + a^{i_1p}_{3i_2}n^{p3}_{i_3i_4}\Big\rangle$$

$$= \Big\langle\Big(a^{i_1p}_{mj}\frac{\partial}{\partial \xi_j}n^{p3}_{i_2i_3i_4} + a^{i_1p}_{mi_2}n^{p3}_{i_3i_4}\Big)\frac{\partial \xi_3}{\partial \xi_m}\Big\rangle = -\Big\langle\frac{\partial}{\partial \xi_m}\Big(a^{i_1p}_{mj}\frac{\partial}{\partial \xi_j}n^{p3}_{i_2i_3i_4} + a^{i_1p}_{mi_2}n^{p3}_{i_3i_4}\Big)\xi_3\Big\rangle$$

$$= \Big\langle\Big(a^{i_1p}_{i_2j}\frac{\partial n^{p3}_{i_3i_4}}{\partial \xi_j} + a^{i_1p}_{i_2i_3}n^{p3}_{i_4}\Big)\xi_3\Big\rangle = \Big\langle\Big(a^{i_1p}_{i_2j}\frac{\partial n^{p3}_{i_3i_4}}{\partial \xi_j} + a^{i_1p}_{i_2i_3}(-\xi_3\delta_{pi_4})\Big)\xi_3\Big\rangle$$

$$= \Big\langle\Big(a^{i_1p}_{i_2j}\frac{\partial}{\partial \xi_j}n^{p3}_{i_3i_4} - a^{i_1i_4}_{i_2i_3}\xi_3\Big)\xi_3\Big\rangle.$$

For constant coefficients, in the isotropic case we have:

$$\bar{a}^{11}_{11} = \bar{a}^{22}_{22} = \frac{\bar{E}^{(3)}}{1-\nu^2}, \quad \bar{a}^{11}_{22} = M, \quad \bar{a}^{12}_{12} = \frac{\nu\bar{E}^{(3)}}{1-\nu^2},$$

$$\bar{c}_{1111} = \bar{c}_{2222} = \frac{\bar{E}^{(3)}}{12(1-\nu^2)}, \quad \bar{c}_{1212} = M/12, \quad \bar{c}_{1122} = \frac{\nu\bar{E}^{(3)}}{12(1-\nu^2)},$$

where $\nu = \lambda/(2(\lambda+M))$.

Consider an orthotropic case in which for each the tensor $a^{kl}_{ij}(\xi)$ of elastic modules is given by means of nine scalar functions: $E_1(\xi)$, $E_2(\xi)$, $E_3(\xi)$ (Young modules), $\mu_{21}(\xi)$, $\mu_{31}(\xi)$, $\mu_{32}(\xi)$ (shear modules), and $\nu_{21}(\xi)$, $\nu_{31}(\xi)$, $\nu_{32}(\xi)$ (Poisson ratios) as follows:

$$\begin{pmatrix} a^{11}_{11} & a^{12}_{12} & a^{13}_{13} \\ a^{12}_{12} & a^{22}_{22} & a^{23}_{23} \\ a^{13}_{13} & a^{23}_{23} & a^{33}_{33} \end{pmatrix} = \begin{pmatrix} \dfrac{1}{E_1} & -\dfrac{\nu_{21}}{E_2} & -\dfrac{\nu_{31}}{E_3} \\ -\dfrac{\nu_{21}}{E_2} & \dfrac{1}{E_2} & -\dfrac{\nu_{32}}{E_3} \\ -\dfrac{\nu_{31}}{E_3} & -\dfrac{\nu_{32}}{E_3} & \dfrac{1}{E_3} \end{pmatrix}^{-1},$$

$a^{22}_{11} = \mu_{21}$, $a^{33}_{11} = \mu_{31}$, $a^{33}_{22} = \mu_{32}$, and if among the indices i, j, k, l there is an index that differs from the others, then $a^{kl}_{ij} = 0$. Let a^{kl}_{ij} depend on ξ_3 only. Then we have

$$\bar{a}^{11}_{11} = \Big\langle\frac{1}{1/E_1 - \nu_{21}^2/E_2}\Big\rangle, \quad \bar{a}^{22}_{22} = \Big\langle\frac{1}{E_1(1/(E_1E_2) - \nu_{21}^2/E_2^2)}\Big\rangle, \quad \bar{a}^{11}_{22} = \langle\mu_{21}\rangle,$$

$$\bar{a}_{12}^{12} = \left\langle \frac{\nu_{21}}{1/E_1 - \nu_{21}^2/E_2} \right\rangle, \quad \bar{c}_{1111} = \left\langle \frac{\xi_3^2}{1/E_1 - \nu_{21}^2/E_2} \right\rangle,$$

$$\bar{c}_{2222} = \left\langle \frac{\xi_3^2}{E_1(1/(E_1E_2) - \nu_{21}^2/E_2^2)} \right\rangle,$$

$$\bar{c}_{\overline{1122}} = 2\left\langle \left(\frac{\nu_{21}}{1/E_1 - \nu_{21}^2/E_2} + 2\mu_{21} \right) \xi_3^2 \right\rangle, \tag{3.2.27}$$

where $\langle \cdot \rangle = \int_{-1/2}^{1/2} d\xi_3$; the last coefficient is the sum of coefficients with all possible permutations of subscripts.

If among the indices of an element $\bar{a}_{ij}^{kl}$ there is an index that differs from the others, then $\bar{a}_{ij}^{kl} = 0$. The sums of coefficients $c_{i_1i_2i_3i_4}$ over all possible permutations of the sets $(1, 2, 2, 2)$ and $(2, 1, 1, 1)$ are equal to zero.

3.3 Equivalent homogeneous plate problem

Here below we discuss the problem of existence of a homogeneous orthotropic plate that is mechanically equivalent to the given heterogeneous locally orthotropic plate. Besides the nine elasticity modules we can variate the thickness of the plate. Here we will find out that in general case an equivalent homogeneous plate does not exist (remind that it always exists for conductivity equation, compare to section 3.1). We formulate the necessary and sufficient conditions of existence of an equivalent homogeneous orthotropic plate and propose the approximating homogeneous plate when these conditions are not true.

In mechanics it is usual to take $\mu\bar{a}_{ij}^{kl}$ and $\mu^3\bar{c}_{i_1i_2i_3i_4}$ as the bending characteristic and the membrane characteristic of the stiffness of plates, where μ is the thickness of a plate.

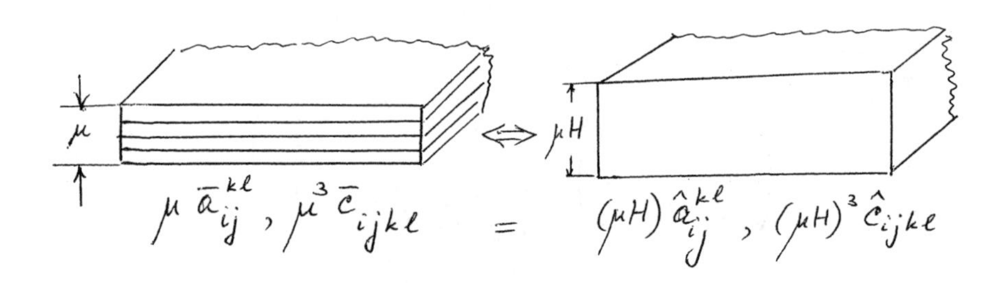

Figure 3.3.1. The heterogeneous plate and an equivalent homogeneous plate

The homogenized problem of zero order for a plate has the form

$$\sum_{i,j,k,l=1}^{2} \frac{\partial}{\partial x_i}\left(\bar{a}_{ij}^{kl}\frac{\partial \omega_0^l}{\partial x_j}\right) = \psi^k(x_1,x_2),$$

$$\bar{c}_{1111}\frac{\partial^4\omega_0^3}{\partial x_1^4} + \bar{c}_{1122}\frac{\partial^4\omega_0^3}{\partial x_1^2\partial x_2^2} + \bar{c}_{2222}\frac{\partial^4\omega_0^3}{\partial x_2^4} = \psi^3(x_1,x_2),$$

$$x_1 \in (0,b), \quad x_2 \in I\!R, \quad k = 1,2,$$

$$\omega_0 = 0, \quad \frac{\partial \omega_0^3}{\partial x_1} = 0 \quad \text{for} \quad x_1 = 0, b.$$

Now we consider the following existence problem for a plate that is homogeneous (with respect to plane and bending characteristics) and equivalent to a stratified plate with the characteristics $\mu\bar{a}_{ij}^{kl}$ and $\mu^3\bar{c}_{ijkl}$ defined by (3.2.27); i.e., for the eight given characteristics (3.2.27) we seek nine constant modules $\hat{E}_1$, $\hat{E}_2$, $\hat{E}_3$, $\hat{\mu}_{21}$, $\hat{\mu}_{31}$, $\hat{\mu}_{32}$, $\hat{\nu}_{21}$, $\hat{\nu}_{31}$, and $\hat{\nu}_{32}$ and a new thickness of the equivalent plate μH (thus, we have ten unknowns in all) such that the plane characteristics $(\mu H)\hat{a}_{ij}^{kl}$ and bending characteristics $(\mu H)^3\hat{c}_{ijkl}$ corresponding to these ten values coincide with the given characteristics $\mu\bar{a}_{ij}^{kl}$ and $\mu^3\bar{c}_{ijkl}$ respectively:

$$(\mu H)\hat{a}_{ij}^{kl} = \mu\bar{a}_{ij}^{kl}, \quad (\mu H)^3\hat{c}_{ijkl} = \mu^3\bar{c}_{ijkl} \tag{3.3.1}$$

Clearly, we can calculate the values $\hat{a}_{ij}^{kl}$ and $\hat{c}_{ijkl}$ from (3.2.27) by replacing the modules E_p, μ_{pq}, and ν_{pq} by unknown constants $\hat{E}_p$, $\hat{\mu}_{pq}$, and $\hat{\nu}_{pq}$ and by replacing the brackets by 1/12 (for example, we set $\langle\xi_3^2\rangle = 1/12$). Note that (3.2.27) does not depend on E_3, μ_{31}, μ_{32}, ν_{31}, and ν_{32}. Thus, from (3.3.1) we pass to the following equivalent problem: to find five constants $\hat{E}_1$, $\hat{E}_2$, $\hat{\mu}_{21}$, $\hat{\nu}_{21}$, and H that satisfy the eight equations

$$\frac{\bar{a}_{11}^{11}}{H} = \frac{1}{1/\hat{E}_1 - \hat{\nu}_{21}^2/\hat{E}_2}, \quad \frac{\bar{a}_{22}^{22}}{H} = \frac{1}{\hat{E}_1(1/(\hat{E}_1\hat{E}_2) - \hat{\nu}_{21}^2/\hat{E}_2^2)},$$

$$\frac{\bar{a}_{22}^{11}}{H} = \mu_{21}, \quad \frac{\bar{a}_{12}^{12}}{H} = \frac{\hat{\nu}_{21}}{1/\hat{E}_1 - \hat{\nu}_{21}^2/\hat{E}_2}, \tag{3.3.2}$$

$$\frac{\bar{c}_{1111}}{H^3} = \frac{1}{12(1/\hat{E}_1 - \hat{\nu}_{21}^2/\hat{E}_2)}, \quad \frac{\bar{c}_{2222}}{H^3} = \frac{1}{12\hat{E}_1(1/(\hat{E}_1\hat{E}_2) - \hat{\nu}_{21}^2/\hat{E}_2^2)},$$

$$\frac{\bar{c}_{1212}}{H^3} = \frac{\hat{\mu}_{21}}{12}, \quad \frac{\bar{c}_{1122}}{H^3} = \frac{\hat{\nu}_{21}}{12(1/\hat{E}_1 - \hat{\nu}_{21}^2/\hat{E}_2)}, \tag{3.3.3}$$

with given $\bar{a}_{11}^{11}, \bar{a}_{22}^{22}, \bar{a}_{22}^{11}, \bar{a}_{12}^{12}$ and $\bar{c}_{1111}, \bar{c}_{2222}, \bar{c}_{1122}, \bar{c}_{1212}$.

Now we can state necessary and sufficient conditions for system (3.3.1) to be solvable (and hence existence conditions for equivalent homogeneous plate):

$$\frac{\bar{c}_{2222}}{\bar{c}_{1111}} = \frac{\bar{a}_{22}^{22}}{\bar{a}_{11}^{11}}, \quad \frac{\bar{c}_{1122}}{\bar{c}_{1111}} = \frac{\bar{a}_{22}^{11}}{\bar{a}_{11}^{11}}, \quad \frac{\bar{c}_{1212}}{\bar{c}_{1111}} = \frac{\bar{a}_{12}^{12}}{\bar{a}_{11}^{11}}. \tag{3.3.4}$$

If conditions (3.3.4) hold, then five constants for equivalent plate can be found from the following relations:

$$H = \sqrt{12\bar{c}_{1111}/\bar{a}_{11}^{11}}, \quad \hat{\mu}_{21} = \bar{a}_{22}^{11}/H, \quad \hat{\nu}_{21} = \bar{a}_{12}^{12}/\bar{a}_{11}^{11},$$

$$\hat{E}_2 = (\bar{a}_{22}^{22} - \bar{a}_{11}^{11}\hat{\nu}_{21}^2)/H, \quad \hat{E}_1 = \bar{a}_{11}^{11}\hat{E}_2/(H\hat{E}_2 + \bar{a}_{11}^{11}\hat{\nu}_{21}^2). \tag{3.3.5}$$

Indeed, relation (3.3.4) holds because the right-hand sides of (3.3.2) and (3.3.3) are proportional. We find H by dividing the first equation of (3.3.2) by the first equation of (3.3.3); then we can find $\hat{\mu}_{21}$ from the third equation of 3.3.2) and determine $\hat{\nu}_{21}$ as the ratio by dividing the fourth equation of (3.3.2) by the first equation of (3.3.2). Finally, we obtain $\hat{E}_1$ and $\hat{E}_2$ from the first two equations of (3.3.2). Relations (3.3.3) follow from (3.3.2) and (3.3.4).

This result was obtained in cooperation with V. E. Grebennikov.

Conditions (3.3.4) hold in particular if the mono-layers are isotropic and symmetric. If conditions (3.3.4) fail, then the equivalent plate does not exist. Nevertheless, one can apply relations (3.3.5) to find an approximation of the five unknowns. The error of this approximation can be evaluated by

$$\Delta = \left(\frac{\bar{c}_{2222}}{\bar{c}_{1111}} - \frac{\bar{a}_{22}^{22}}{\bar{a}_{11}^{11}}\right)^2 + \left(\frac{\bar{c}_{1122}}{\bar{c}_{1111}} - \frac{\bar{a}_{22}^{11}}{\bar{a}_{11}^{11}}\right)^2 + \left(\frac{\bar{c}_{1212}}{\bar{c}_{1111}} - \frac{\bar{a}_{12}^{12}}{\bar{a}_{11}^{11}}\right)^2.$$

This approach was used to adopt a finite element method software based on the concept of a homogeneous plane finite element to the analysis of thin-walled constructions of stratified composite materials (for example, of a stratified car body).

3.4 Time dependent elasticity problem for a plate.

Consider a non-stationary elasticity equation

$$-\rho(x/\mu)\frac{\partial^2 u}{\partial t^2} + \sum_{m,j=1}^{3} \frac{\partial}{\partial x_m}\left(A_{mj}(x/\mu)\frac{\partial u}{\partial x_j}\right) = \psi_\mu(x_1, x_2, t), \quad x \in C_\mu, \tag{3.4.1}$$

with boundary conditions (3.2.2), (3.2.3), junction conditions (2.3.4), and initial conditions

$$u\Big|_{t=0} = 0, \quad \frac{\partial u}{\partial t}\Big|_{t=0} = 0. \tag{3.4.2}$$

Here $A_{ij}(\xi)$ are matrix functions from section 3.2 and the function $\rho(\xi)$ is 1-periodic with respect to ξ_1, $\rho(\xi) \geq \kappa > 0$, and $\rho(\xi)$ is piecewise smooth in the sense of section 2.3. Besides, $\rho(S_{\mathcal{A}}\xi) = \rho(\xi)$ for $\mathcal{A} = 2, 3$ and

$$\psi_\mu(x,t) = (\psi^1(x_1,x_2,t), \psi^2(x_1,x_2,t), \mu^2\psi^3(x_1,x_2,\mu t))^*, \quad \psi^k \in C^\infty, k = 1, 2,$$

$$\psi^3(x_1,x_2,\theta) \in C^\infty, \quad \psi^3(x_1,x_2,\theta) = 0 \quad \text{for} \quad \theta \leq \theta_0, \quad \theta_0 > 0.$$

The inner expansion of solution to problem (3.4.1), (3.4.2), (3.2.2), (3.2.3), (2.3.4) is sought as a series

$$u^{(\infty)} = \sum_{l+q=0}^{\infty} \sum_{|i|=l,\quad i=(i_1,\dots,i_l),i_j\in\{1,2\}} \mu^{l+q} N_{iq}(x/\mu) \frac{\partial^q}{\partial t^q} D^i \omega(x_1,x_2,t) \quad (3.4.3)$$

where ω is a three-dimensional vector of the form

$$(\omega^1(x_1,x_2,t), \omega^2(x_1,x_2,t), \omega^3(x_1,x_2,\mu t))^*,$$

N_{iq} are analogs of the functions N_{iq} from section 3.2; N_{lq} are (1-periodic with respect to ξ_1) solutions of the problems

$$L_{\xi\xi} N_{iq} = -T_{iq}^N(\xi) + h_{iq}^N, \quad \xi \in I\!R^2 \times (-1/2, 1/2), \quad (3.4.4)$$

$$\frac{\partial N_{iq}}{\partial \nu_\xi} = -A_{1i_1} N_{i_2,\dots,i_l,q}, \quad \xi_3 = \pm 1/2, \quad (3.4.5)$$

$$[N_{iq}]\big|_\Sigma = 0, \quad \left[\frac{\partial N_{iq}}{\partial \nu_\xi}\right]\Big|_\Sigma = -\Big[\sum_{m=1}^{3} A_{mi_1} N_{i_2,\dots,i_l,q} n_m\Big]\Big|_\Sigma, \quad (3.4.6)$$

$$T_{iq}^N(\xi) = \sum_{j=1}^{3} \frac{\partial}{\partial \xi_j}(A_{ji_1} N_{i_2,\dots,i_l,q}) + \sum_{j=1}^{3} A_{i_1 j} \frac{\partial N_{i_2,\dots,i_l,q}}{\partial \xi_j} + A_{i_1 i_2} N_{i_3,\dots,i_l,q} - \rho N_{i,q-2}, \quad (3.4.7)$$

$$h_{iq}^N = \left\langle \sum_{j=1}^{3} A_{i_1 j} \frac{\partial N_{i_2,\dots,i_l,q}}{\partial \xi_j} + A_{i_1 i_2} N_{i_3,\dots,i_l,q} - \rho N_{i,q-2} \right\rangle. \quad (3.4.8)$$

As in section 2.3, we obtain a homogenized equation

$$\sum_{l+q=2}^{\infty} \sum_{|i|=l,\quad i_j\in\{1,2\}} \mu^{l+q-2} h_{iq}^N \frac{\partial^q}{\partial x_1^l \partial t^q} D^i \omega \;-\; \psi_\mu = 0, \quad (3.4.9)$$

whose principal part has the form

$$-\langle\rho\rangle \frac{\partial^2 \omega^r}{\partial t^2} + \sum_{q,j,k,r=1}^{2} \frac{\partial}{\partial x_q}(\bar{a}_{qj}^{kr} \frac{\partial \omega^r}{\partial x_j}) - \psi^k(x_1,t) + \cdots = 0, \quad k = 1,2,$$

$$\mu^2\Big(-\langle\rho\rangle \frac{\partial^2 \omega^3}{\partial \theta^2} - \sum_{q,j,p,r=1}^{2} \frac{\partial^2}{\partial x_q \partial x_j}(\bar{c}_{qjpr} \frac{\partial \omega^3}{\partial x_p \partial x_r}) - \psi^3(x_1,\theta) + \cdots\Big) = 0, \quad (3.4.10)$$

where $\theta = \mu t$, and the homogenized coefficients $\bar{a}_{qj}^{kr}$ and $\bar{c}_{qjpr}$ are defined above in section 3.2.

Let us seek f.a.s. ω of (3.4.9) in the form of a regular series in powers of μ:

$$\omega = \sum_{j=0}^{\infty} \mu^j \omega_j.$$

The boundary layer corrector can be constructed similarly to that in section 3.2; the same is true for the justification of the asymptotic expansion: it is similar to that of subsection 2.4.4; the *a priori* estimate follows from [16], [55], [129].

3.5 Bibliographical Remark

The asymptotic analysis of the elasticity equations for homogeneous plates (without strict mathematical justification) was first carried out by A.L.Goldenveizer [59] - [61], and later with mathematical justification by Ph.Ciarlet and P.Destuynder [40], B.Lidsky and S.Nazarov [111]. The review of the results on the asymptotic reduction of dimension from 3D to 2D in the plate theory is given in [38], [39] (see, for example, the assessment in the end of Chapter 3 [38]. For an inhomogeneous plate, the homogenized equation of zero order was obtained by D.Caillerie [31]. As in the general homogenization theory, there are two principal tools in the reduction of dimension from 3D to 2D: the convergence techniques and the asymptotic expansions; as it was stated above, the expansions give more information about the structure of the solution but this technique demands more regularity of the data. The complete asymptotic expansion of a solution of the elasticity theory system of equations for an inhomogeneous plate (taking into account the boundary layers) was constructed by G.Panasenko and M.Reztsov [165] and also in [138],[139],[150]. The plates with rapidly varying thickness were considered in [72],[73]and later in [69]. The singular data models are studied in [191]. We do not discuss in the present book the dimensional reduction for non-linear elasticity, for the shell theory. the review on the state of these questions can be found in [38], [110],[176],[179] and in [97].

Chapter 4

Finite Rod Structures

We consider finite rod structures, i.e. finite connected unions of rods. The definition of the considered geometrical model of such structures is given in section 4.1. We introduce the notions of L-convergence and of FL-convergence for such structures. This tool gives the main term of solution to conductivity problem in a finite rod structure. This approach is implemented then to elasticity equation and it is applied to a shape optimization problem of minimal compliance (section 4.2).

Further we develop a more refined analysis of fields in finite rod structures including asymptotic expansion of elasticity equation based on the expansions of Chapter 2. Namely, these expansions of Chapter 2 are applied to each bar of finite rod structure and then the boundary layers are constructed in neighborhoods of ends. These boundary layers match the inner expansions of each rod and they are similar to that of section 2.3. Finally, the Stokes and the Navier-Stokes equations are studied in a finite system of tubes (or pipes) that is also a variety of finite rod structures.

4.1 Definitions. L-convergence.

Below we consider the finite rod structures; for example , in the two-dimensional case the finite rod structure is a connected union of thin rectangles $B^{\alpha}_{h,j,\mu}$. Their thickness has a magnitude $\mu << 1$. The skeleton is constituted of the limit segment of these rectangles $e^{\alpha}_{h} = \cap_{\mu>0} B^{\alpha}_{h,j,\mu}$. The ends of these segments are called nodes.

Then we consider the boundary value problem for a nonlinear partial derivative equation stated on a finite rod structure.

The asymptotic reduction of the stated problem is done. The main mathematical result of the section is the passage to the limit in the boundary value problem as μ tends to zero. The formal procedure of the reduction of the initial problem to a system of algebraic equations is justified in the sense of so called FL-convergence (i.e. convergence of the normalized Dirichlet's integral). It is the statement of Theorem 4.1.1. In the case of a scalar elliptic linear problem FL-convergence implies L-convergence, i.e. convergence in normalized L^2 norm. The proof of Theorem 4.1.1 uses some auxiliary inequalities (in particular the Poincaré - Friedrichs inequality for a finite rod structure). These inequalities are proved in Appendix A4.1 and Appendix A4.2.

4.1.1 Finite rod structure

First we define some notions [148]: finite rod structure, nodes, skeleton, sections, nodal domain.

Let $\beta_1, ..., \beta_J$ be bounded domains in $I\!R^{s-1}$, $(s = 2, 3)$ with a piecewise smooth boundary, $B_{j,\mu}$ $(j = 1, ..., J)$ the cylinders, defined by

$$B_{j,\mu} = \{x \in R^s \mid (x_2/\mu, ..., x_s/\mu) \in \beta_j,\ x_1 \in I\!R\},$$

and $B^{\alpha}_{h,j,\mu}$ the cylinder obtained from $B_{j,\mu}$ by orthogonal transformation $\tilde{\Pi}$ with the matrix α^T, $\alpha = (\alpha_{il})$ and a translation $h = (h_1, ..., h_s)^T$. Let e^{α}_h be an s-dimensional vector obtained by means of $\tilde{\Pi}$ and h from the vector $i(h, \alpha)$ of the axis Ox_1 with the beginning at the point O .

Definition 4.1.1. *Let B be a union of all segments e^{α}_h when α belongs to a set $\Delta \subset I\!R^{s^2}$ and h to a set $H_{\alpha} \subset I\!R^s$, and these sets are independent of μ. Let B be such that any two segments e^{α}_h can have only one common point which is the end point for both segments. The set B is called skeleton, the end points of e^{α}_h are called nodes.*

If x_0 is an endpoint for a segment e we say that e is an initial segment for x_0.

We associate the cylinder $B^{\alpha}_{h,j,\mu}$ with every e^{α}_h and denote by $\tilde{B}^{\alpha}_{h,j,\mu}$ the part of $B^{\alpha}_{h,j,\mu}$ enclosed between the two planes passing through the ends of segment e^{α}_h and perpendicular to it (we assume that the bases belong to $\tilde{B}^{\alpha}_{h,j,\mu}$). Suppose that Δ and H_{α} are finite sets and B is connected.

Definition 4.1.2. *By a finite rod structure we understand the set of interior points of the union*

$$B_{\mu} = U_{\alpha \in \Delta} \ U_{h \in H_{\alpha}} \ \tilde{B}^{\alpha}_{h,j,\mu}.$$

We assume that B_{μ} is connected, satisfies the strong cone condition [55] and that μ is a small parameter.

Definition 4.1.3. *By sections of a rod structure B_{μ} we understand the cylinders that are the maximal subsets of the cylinders $\tilde{B}^{\alpha}_{h,j,\mu}$ for which any cross-section by the plane perpendicular to the generatrix of $\tilde{B}^{\alpha}_{h,j,\mu}$ is free of points of other cylinders .*

We denote by S_0 the union of sections. Let $c_0\mu \geq d$, where d is the maximal diameter of the connected subsets $B_{\mu} \backslash S_0$, with c_0 being independent of μ.

Let $\tilde{B}^{\alpha,+}_{h,j,\mu}$ (respectively $\tilde{B}^{\alpha,++}_{h,j,\mu}$) be a part of $\tilde{B}^{\alpha}_{h,j,\mu}$, contained between the planes spaced by $c_0\mu$ (respectively by $(c_0+1)\mu$) from the bases of $\tilde{B}^{\alpha}_{h,j,\mu}$. Let B^{+}_{μ} be a union of all $\tilde{B}^{\alpha,+}_{h,j,\mu}$ and let B^{++}_{μ} be a union of all $\tilde{B}^{\alpha,++}_{h,j,\mu}$.

Definition 4.1.4. *The domain $B_\mu \backslash \bar{B}^{++}_{\mu}$ is called the nodal domain.*

Remark 4.1.1. We shall also consider as a finite rod structure the union of the set B_μ with some s-dimensional cubes with the edge

$$d_0\mu \;=\; 2max\{d\ ,\ \mu\ max_{j=1,\dots,J}\ diam\ \beta_j\}$$

and the centers at some nodes. Then we suppose that $c_0 \;>\; 2\sqrt{s}d_0$.

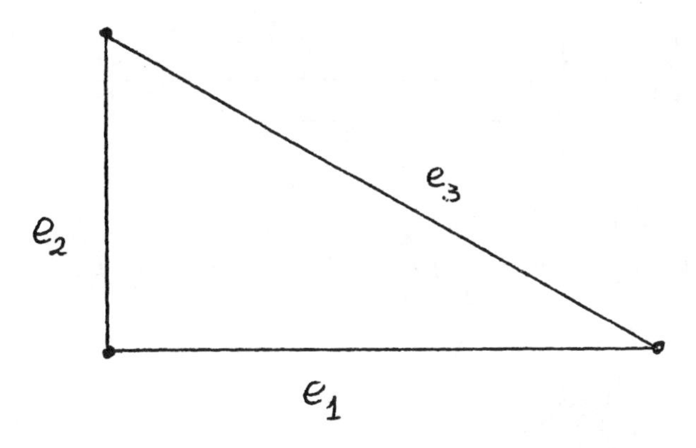

Figure 4.1.1. A skeleton B containing three segments

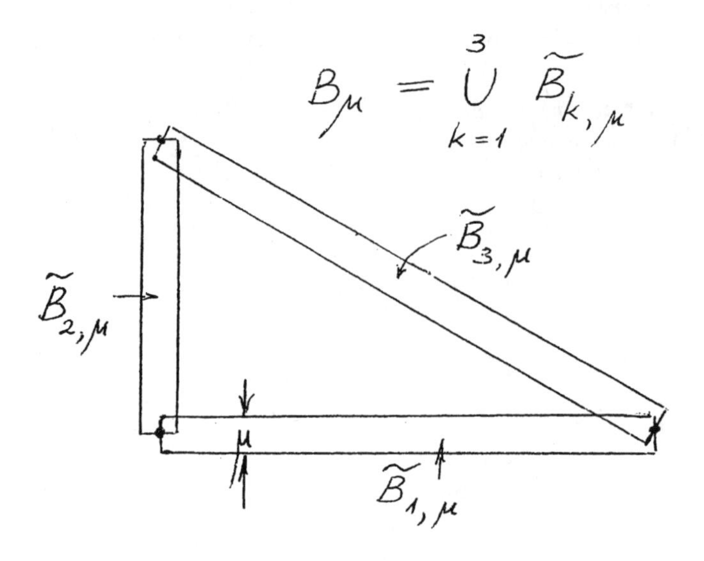

Figure 4.1.2. The corresponding finite rod structure B_μ.

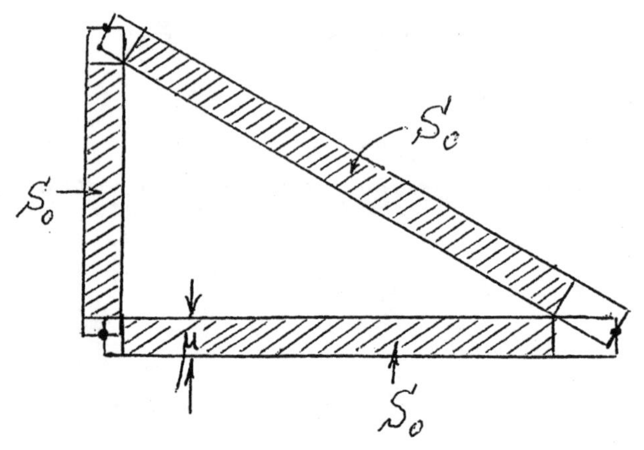

Figure 4.1.3. Sections S_0

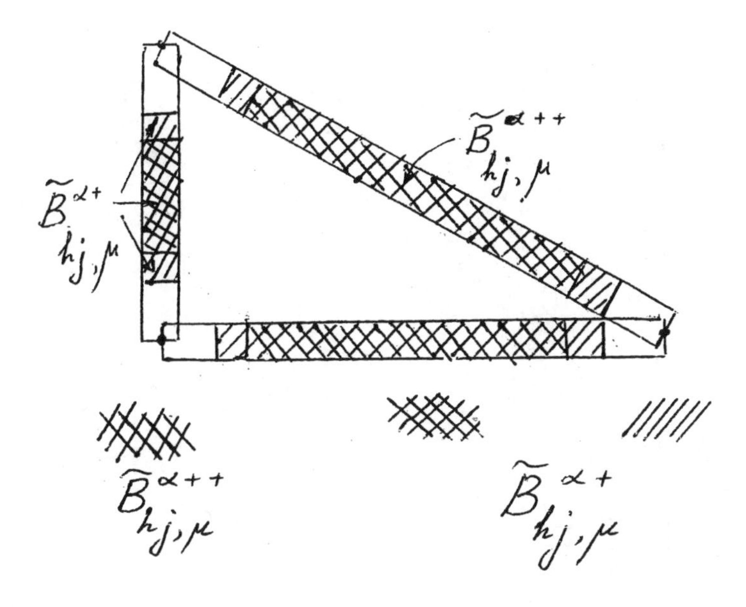

Figure 4.1.4. Sets $\tilde{B}^{\alpha,+}_{h,j,\mu}$ *and* $\tilde{B}^{\alpha,++}_{h,j,\mu}$

4.1.2 L-convergence method for a finite rod structure

Here we consider finite rod structures described in Remark 4.1.1 stretched between the planes $\{x_1 = 0\}$ and $\{x_1 = 1\}$ with the loading at the intersection of the finite rod structure with the plane $\{x_1 = 1\}$ by a proper distribution of the material.

Let G_0, G_1 be two (s-1)-dimensional domains belonging to the planes $\{x_1 = 0\}$ and $\{x_1 = 1\}$ respectively. We consider the set $\mathcal{Q}$ of finite rod structures intersecting the planes $\{x_1 = 0\}$ and $\{x_1 = 1\}$ such that

1) the skeleton B belongs to a cube $[0,1]^s$;

2) the sets of nodes belonging to the domains $G_0 \subset \{x_1 = 0\}$ and $G_1 \subset \{x_1 = 1\}$ are not empty;

3) the rod structure B_μ has additional cubes of Remark 4.1.1

$$C_{x_0} = (x_1^0 - d_0\mu \ , \ x_1^0 + d_0\mu) \times ... \times (x_s^0 - d_0\mu \ , \ x_s^0 + d_0\mu)$$

for all nodes $x^0 = (x_1^0, ..., x_s^0)$ from $G_0 \cup G_1$. We suppose that the constant d_0 is such that the rod structure without these additional cubes does not intersect the planes $\{x_1 = -d_0\mu\}$ and $\{x_1 = 1 + d_0\mu\}$.

These assumptions are not crucial; they are introduced for simplicity of description of the method and its application to the shape design.

Denote Σ_0 the union of the sides of the cubes C_{x_0} belonging to the plane $\{x_1 = -d_0\mu\}$ and Σ_1 the union of the sides of the cubes C_{x_0} belonging to the plane $\{x_1 = 1 + d_0\mu\}$.

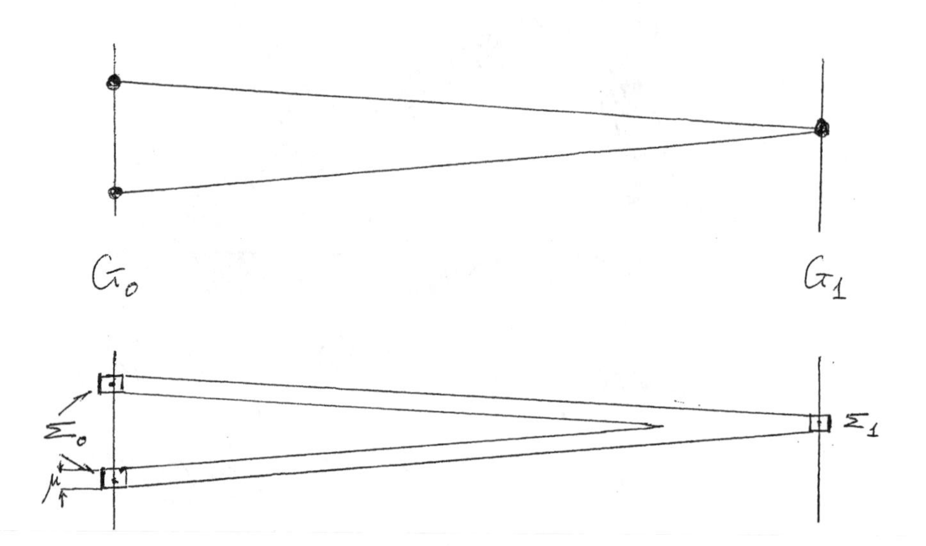

Figure 4.1.5. Rod structure with the supplementary cubes

Consider the problem, set in a finite rod structure $B_\mu \in Q$ for the fixed (given) vector-valued constant T :

$$\sum_{i=1}^{s} \partial/\partial x_i(A_i(\nabla_x u_\mu)) = 0, \quad for\ x \in B_\mu, \tag{4.1.1}$$

$$\partial u_\mu/\partial \nu = \sum_{i=1}^{s} n_i A_i(\nabla_x u_\mu) = 0,\ for\ x \in \Sigma = \partial B_\mu \backslash (\Sigma_0 \cup \Sigma_1), \tag{4.1.2}$$

$$\partial u_\mu/\partial \nu = \sum_{i=1}^{s} n_i A_i(\nabla_x u_\mu) = t = \mu^{s-1} T/meas \Sigma_1,\ for\ x \in \Sigma_1, \tag{4.1.3}$$

$$u_\mu = 0,\ for\ x \in \Sigma_0, \tag{4.1.4}$$

where T, is a constant $s-$dimensional vector $u_\mu(x)$, $A_i(y)$ $(i = 1, ..., s)$ are $n-$dimensional vector-valued functions, $\nabla_x u = (\partial u_k/\partial x_l)$ is an $n \times s$ matrix, $y = (y_{kl})$ is an $n \times s$ matrix-argument, A_i are assumed to be continuously differentiable, and $n_i(x)$ is the cosine of the angle between the axis Ox_i and the exterior normal vector $n(x)$.

We suppose that the problem (4.1.1)-(4.1.4) has the solution, such that for each $n-$dimensional vector-valued function $\varphi(x)$ which belongs to the space $H^1(B_\mu)$ and equal to zero on the surface Σ_0, the integral

$$\int_{B_\mu} \sum_{i=1}^{s} (A_i(\nabla u_\mu) \ , \ \partial\varphi/\partial x_i) \ dx$$

is defined and is equal to

$$\int_{\Sigma_1} (t \ , \ \varphi) \ dx.$$

Remark 4.1.2. In particular, one can consider

$$A_i(\nabla_x u) = \sum_{j=1}^{s} A_{ij} \frac{\partial u}{\partial x_j},$$

where A_{ij} are constants such that the matrix $(A_{ij})_{1 \le i,j \le s}$ is positive definite and symmetric. In this linear case problem (4.1.1)-(4.1.4) surely has a unique solution.

We develop the asymptotic reduction of the problem as μ tends to zero.

Here we introduce two types of "convergence" : the first is L-convergence, i.e. convergence in the normalized $L^2(B_\mu)$ norm. The necessity of normalization is explained by the small measure of the domain B_μ (*meas* B_μ tends to zero). The second type of "convergence" is FL-convergence, i.e. convergence of specially normalized Dirichlet integral.

We describe the formal algorithm of construction of the asymptotic solution $u_\mu^{(1)}$ of the problem (4.1.1) - (4.1.4). The main theorem is proved; it determines the FL-convergence of the solution u_μ of the problem (4.1.1) - (4.1.4) "to" the asymptotic solution $u_\mu^{(1)}$ of the problem (4.1.1) - (4.1.4). In the linear case the FL-convergence implies L-convergence.

Definition 4.1.5. *Let $u_\mu(x)$ be a sequence of functions from $L^2(B_\mu)$, $u_0(x) \in L^2(B)$; (so u_0 is a function defined on the skeleton B). One says that u_μ L - converges to u_0 on B_μ if and only if*

$$\frac{\|u_\mu - \tilde{u}_0\|_{L^2(B_\mu)}}{\sqrt{meas(B_\mu)}} \to 0, \quad (\mu \to 0),$$

where $\tilde{u}_0$ is an extension of $u_0(x)$ onto B_μ such that the values at each point x_0 of the set B_μ^+ is equal to the value of u_0 at the orthogonal projection of x_0 on the corresponding segment of B, and at each connected component of a domain $B_\mu \backslash B_\mu^+$ we pose $\tilde{u}_0$ equal to its value at the node.

The normalization factor $1/\sqrt{meas(B_\mu)}$ is necessary because $\|1\|_{L^2(B_\mu)} = \sqrt{meas(B_\mu)}$. Notice that L-convergence is not a convergence in common sense because the domain depends on small parameters.

Denote $H_{1,0}(B_\mu) = \{\varphi \in H_1(B_\mu), \varphi|_{\Sigma_0} = 0\}$ and

$$\|u_1 \ ; \ u_2\|_{FL(B_\mu)} =$$

$$sup_{\varphi\in H_{1,0}(B_\mu)}\left\{\frac{|\int_{B_\mu}\sum_{i=1}^{s}\ (A_i(\nabla u_1)\ -\ A_i(\nabla u_2)\ ,\ \partial\varphi/\partial x_i)\ dx|}{\|\varphi\|_{H^1(B_\mu)}\sqrt{meas(B_\mu)}}\right\}. \qquad (4.1.5)$$

Definition 4.1.6. *Let $u_{1,\mu}(x)$ and $u_{2,\mu}(x)$ be two sequences of functions from $H_{1,0}(B_\mu)$. One says that the pair $(u_{1,\mu}; u_{2,\mu})$ FL - converges if and only if*

$$\|u_{1,\mu}; u_{2,\mu}\|_{FL(B_\mu)} \to 0, \quad (\mu \to 0). \qquad (4.1.6)$$

Notice that $\|u_{1,\mu}; u_{2,\mu}\|_{FL(B_\mu)}$ is not a norm in a general case. Nevertheless in the linear case for the coercive operator (1) it is a norm of the difference of two functions.

The asymptotic reduction of the problem (4.1.1)-(4.1.4) analogous to that of [134], [136] and to FL-convergence techniques from [148] deduces it to the algebraic system of equations. To each segment $e \subset B$, $(e = e_h^\alpha)$we assign a collection of $n-$dimensional vectors $X_0, ..., X_s$, which satisfies the following three conditions:

1)

$$\sum_{i=1}^{s} A_i(X_1, ..., X_s)\ \nu_i^j \ = \ 0, \qquad (4.1.7)$$

where $j = 1, ..., s-1$, and the vectors

$$\nu^1 \ = \ (\nu_1^1, ..., \nu_s^1)\ ,\ ...\ ,\ \nu^{s-1} \ = \ (\nu_1^{s-1}, ..., \nu_s^{s-1})$$

form the basis in $(s-1)-$dimensional space orthogonal to e.

2) Let the segments $e_1, ..., e_q$ have a common end point; if this end point does not belong to the sets G_0 and G_1 then

$$\sum_{j=1}^{q}\sum_{i=1}^{s} A_i(X_1^{e_j}, ..., X_s^{e_j})\ \eta_i^j meas\tilde{\beta}_j \ = \ 0, \qquad (4.1.8)$$

where $\eta_1^j, ..., \eta_s^j$ are the direction cosines of e_j, (oriented from the common end point) and $\tilde{\beta}_j$ is a cross-section of a cylinder with the axis e_j by the hyperplane orthogonal to this axis, normalized by a factor $\mu^{-(s-1)}$;

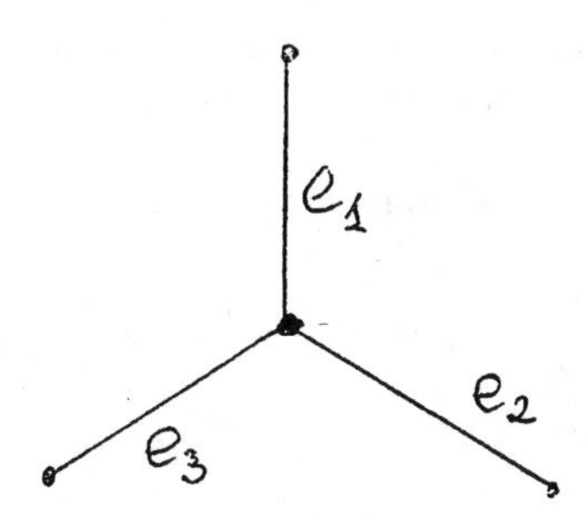

Figure 4.1.6.Equilibrium junction condition

if this end point belongs to the set G_1 then

$$\sum_{j=1}^{q}\sum_{i=1}^{s} A_i(X_1^{e_j},...,X_s^{e_j})\ \eta_i^j meas\tilde{\beta}_j \ = \ -td_0^{s-1}. \tag{4.1.9}$$

Here t is a constant vector equal to $\mu^{s-1}T/meas\Sigma_1$; 3) The vector-valued function defined on each segment $e \subset B$ by

$$u_0(x) \ = \ \sum_{i=1}^{s} X_i^e\ x_i \ + \ X_0^e$$

(for $x \in e$) is a continuous function of $(x_1,...,x_s) \ \in B$, equal to zero for all nodes from G_0.

Let this system (4.1.7)-(4.1.9) have a unique solution.

Remark 4.1.3. The problem (4.1.7)-(4.1.9) does not depend on the form of cross-sections of the cylinders but it depends on their measure.

Denote $\tilde{u}_0$ the extension of $u_0(x)$ onto B_μ according to definition 4.1.4, and denote $\tilde{U}_0$ the constant extension of the value of the function $u_0(x_0)$ for each node x_0 onto the connected part of the nodal domain, containing x_0. Let $\chi(x/\mu)$ be a function equal to zero in $B_\mu \backslash B_\mu^+$ and to 1 in the domain B_μ^{++}, and let

$$|\nabla_z \chi(z)| \ < \ c_0, \quad 0 \le \chi \le 1, \quad \chi \in C^1,$$

where c_0 is a positive constant, independent of μ. Denote

$$u_\mu^{(1)}(x) \ = \ \tilde{u}_0\chi(x/\mu) \ + \ \tilde{U}_0(1-\chi(x/\mu)). \tag{4.1.10}$$

Theorem 4.1.1. *Let u_μ be an exact solution of the problem (4.1.1) - (4.1.4) and let $u_\mu^{(1)}$ be an asymptotic solution satisfying the assumptions (4.1.7) - (4.1.9). Then the estimate holds :*

$$\|u_\mu \ ; \ u_\mu^{(1)}\|_{FL(B_\mu)} \leq C\sqrt{\mu}, \quad (\mu \to 0)$$

where the constant C does not depend on μ, i.e. the pair $(u_\mu \ ; \ u_\mu^{(1)})$ FL-converges and the rate of the FL-convergence is $O(\sqrt{\mu})$.

Proof

For each x from the nodal domain one obtains:

$$|\frac{\partial u_\mu^{(1)}}{\partial x_i}| \ = \ |\frac{\partial \tilde{u}_0}{\partial x_i}\chi(x/\mu) \ + \ \mu^{-1}\frac{\partial \chi}{\partial \xi_i}(\tilde{u}_0 - \tilde{U}_0)| \leq c_1, \tag{4.1.11}$$

where c_1 does not depend on μ.

Estimate the integral

$$I \ = \ \int_{B_\mu} \sum_{i=1}^{s} (A_i(\nabla u_\mu^{(1)}) \ , \ \partial\varphi/\partial x_i) \ dx.$$

Considering the inequality (4.1.11) we obtain

$$I \ = \ \int_{B_\mu^{++}} \sum_{i=1}^{s} (A_i(\nabla u_\mu^{(1)}) \ , \ \partial\varphi/\partial x_i) \ dx + \delta_1,$$

where

$$|\delta_1| \ = \ O(\sqrt{\mu})\|\varphi\|_{H^1(B_\mu)}\sqrt{meas(B_\mu)}$$

As $\varphi = 0$ on S_0 we obtain integrating by parts:

$$I \ = \ -\int_{\partial B_\mu^{++}} \sum_{i=1}^{s} (A_i(\nabla u_\mu^{(1)}) \ , \ \varphi) n_i \ ds \ =$$

$$= \ -\sum_{x^0}\int_{\partial\Pi_{x^0}\cap\partial B_\mu^{++}} \sum_{i=1}^{s} (A_i(\nabla u_\mu^{(1)}) \ , \ \varphi) n_i \ ds \ + \ \delta_2,$$

where the summation is made with respect to all nodal points x^0, which do not belong to G_0, and Π_{x^0} is the connected component of the set $\overline{B_\mu \backslash B_\mu^{++}}$ which contains x^0, and n is an outside normal of the domain Π_{x^0} (and so, n is an inside normal of B_μ^{++}.) The error δ_2 is the same integral for nodes $x^0 \in G_0$ and it could be estimated as

$$|\delta_2| \ \leq \ C_2\|\varphi\|_{L^1(\partial\Pi_{x^0}\cap\partial B_\mu^{++})} \ \leq \ C_3\|\varphi\|_{L^2(\partial\Pi_{x^0}\cap\partial B_\mu^{++})}\sqrt{meas(B_\mu)}.$$

Now applying Lemma 4.A1.1 from the Appendix 4.A1 we obtain

$$|\delta_2| \leq C_3 \|\varphi\|_{L^2(\partial\Pi_{x^0}\cap\partial B_\mu^{++})} \sqrt{meas(B_\mu)} \leq$$

$$\leq C_4 \sqrt{\mu} \|\nabla\varphi\|_{L^2(\Pi_{x^0})} \sqrt{meas(B_\mu)},$$

where C_2, C_3, C_4 do not depend on μ.
For each node x^0 we set

$$<\varphi>_{\Pi_{x^0}} = \frac{1}{meas(\Pi_{x^0})} \int_{\Pi_{x^0}} \varphi dx.$$

We have the presentation

$$\varphi(x) = <\varphi>_{\Pi_{x^0}} + (\varphi - <\varphi>_{\Pi_{x^0}})$$

Lemma 4.A1.2 from the Appendix 4.A1 implies

$$\|\varphi - <\varphi>_{\Pi_{x^0}}\|_{L^2(\partial\Pi_{x^0}\cap\partial B_\mu^{++})} \leq C_7 \sqrt{\mu} \|\nabla\varphi\|_{L^2(\Pi_{x^0})},$$

where C_7 does not depend on μ. Present the integral I in a form

$$I = -\sum_{x^0} \int_{\partial\Pi_{x^0}\cap\partial B_\mu^{++}} \sum_{i=1}^{s} (A_i(\nabla u_\mu^{(1)}) , <\varphi>_{\Pi_{x^0}}) n_i \, ds -$$

$$-\sum_{x^0} \int_{\partial\Pi_{x^0}\cap\partial B_\mu^{++}} \sum_{i=1}^{s} (A_i(\nabla u_\mu^{(1)}) , (\varphi - <\varphi>_{\Pi_{x^0}})) n_i \, ds.$$

Here the first integral is equal to zero for all nodes which do not belong to $G_1 \cup G_2$ (since the properties 1) and 2) of the vectors X_i) and the first integral is equal to $-\mu^{s-1}(t , <\varphi>_{\Pi_{x^0}})\bar{d}_0^{s-1}$ for $x^0 \in G_1$, i.e.

$$-\sum_{x^0} \int_{\partial\Pi_{x^0}\cap\partial B_\mu^{++}} \sum_{i=1}^{s} (A_i(\nabla u_\mu^{(1)}) , <\varphi>_{\Pi_{x^0}}) n_i \, ds = \int_{\Sigma_1} (t , <\varphi>_{\Pi_{x^0}}) ds =$$

$$= \int_{\Sigma_1} (t , \varphi) ds - \int_{\Sigma_1} (t , \varphi - <\varphi>_{\Pi_{x^0}}) ds.$$

Estimating

$$\sum_{x^0} \int_{\partial\Pi_{x^0}\cap\partial B_\mu^{++}} \sum_{i=1}^{s} (A_i(\nabla u_\mu^{(1)}) , (\varphi - <\varphi>_{\Pi_{x^0}})) n_i \, ds$$

and

$$\int_{\Sigma_1} (t , \varphi - <\varphi>_{\Pi_{x^0}}) ds$$

by the sum

$$C_8\Big(\sum_{x^0\notin G_1}\sqrt{meas(B_\mu)}\|\varphi\ -\ <\varphi>_{\Pi_{x^0}}\|_{L^2(\partial\Pi_{x^0}\cap\partial B_\mu^{++})}+$$

$$+\sum_{x^0\in G_1}\sqrt{meas(B_\mu)}\|\varphi\ -\ <\varphi>_{\Pi_{x^0}}\|_{L^2(S_1\cap\partial\Pi_{x^0})}\Big)$$

we can apply Lemma 4.A1.2 of the Appendix 4.A1 to obtain the estimate $O(\sqrt{\mu})\sqrt{meas(B_\mu)}\|\varphi\|_{H^1(B_\mu)}$.

Thus for

$$I_0\ =\ \int_{B_\mu}\sum_{i=1}^{s}(A_i(\nabla u_\mu)\ ,\ \partial\varphi/\partial x_i)\,dx\ =\ \int_{\Sigma_1}(t\ ,\ \varphi)ds$$

we obtain

$$I-I_0\ =\ \int_{B_\mu}\sum_{i=1}^{s}(A_i(\nabla u_\mu^{(1)})\ -\ A_i(\nabla u_\mu)\ ,\ \partial\varphi/\partial x_i)\,dx\ =$$

$$=\ O(\sqrt{\mu})\sqrt{meas(B_\mu)}\|\varphi\|_{H^1(B_\mu)}$$

and therefore the theorem is proved.

Remark 4.1.4. It could be proved using the Poincaré - Friedrichs inequality for rod structures (Appendix 4.2), that L-convergence is a corollary of FL-convergence in the case of a scalar linear elliptic problem.

Indeed, let $A_i(y)$ be $\sum_{j=1}^{s}A_{ij}y_j$ where $(A_{ij})_{1\le i,j\le s}$ is a constant symmetric positive definite matrix. Then using the Poincaré- Friedrichs inequality we estimate (taking $\varphi=u_1-u_2$):

$$\|u_1\ ;\ u_2\|_{FL(B_\mu)}\ge C_9\,\frac{\|\nabla(u_1-u_2)\|^2_{L^2(B_\mu)}}{\|u_1-u_2\|_{H^1(B_\mu)}\sqrt{meas(B_\mu)}}\ge$$

$$\ge C_{10}\,\frac{\|u_1-u_2\|_{H^1(B_\mu)}}{\sqrt{meas(B_\mu)}}\ge$$

$$\ge C_{11}\,\frac{\|u_1-u_2\|_{L^2(B_\mu)}}{\sqrt{meas(B_\mu)}}$$

where the constants C_9, C_{10}, C_{11} do not depend on μ. According to Theorem 4.1.1

$$\|u_1\ ;\ u_2\|_{FL(B_\mu)}\le C\ \sqrt{\mu}.$$

Therefore

$$\frac{\|u_\mu-u_\mu^{(1)}\|_{L^2(B_\mu)}}{\sqrt{meas(B_\mu)}}\le C\ \sqrt{\mu}/C_{11},$$

and thus

$$\frac{\|u_\mu - \tilde{u}_0\|_{L^2(B_\mu)}}{\sqrt{meas(B_\mu)}} = O(\sqrt{\mu}),$$

i.e. u_μ L-converges to u_0 and the rate of the FL-convergence is $O(\sqrt{\mu})$.

The situation changes in the case of elasticity system (see [148],[44],[67]) where the Korn inequality constant depends on μ, nevertheless the discrete model (4.1.7)-(4.1.9) seems to be valid in the case when all rods of the structure work in the tension-compression regime. Thus Theorem 4.1.1 justifies in some sense asymptotic approximation (4.1.10). In case of bending or torsion regime of some rods more general discrete model should be applied. This model was obtained in [148] and (for the junction of two rods) in [168] (see also sections 4.4 and 2.3.5).

Remark 4.1.5. Formulation (4.1.1)-(4.1.4) can be generalized: we can consider right-hand side in the equation (4.1.1) that has a support localized in some $c_0\mu$-vicinities of the nodes and that its magnitude is of order $1/\mu$. Its average appears in the right-hand side of equations (4.1.8). This generalization has been developed in the Ph.D. thesis of R.Chiheb.

4.2 Shape optimization of a finite rod structure.

Below a new algorithm of optimal design based on the asymptotic analysis of rod structures and L-convergence is proposed. It could be applied in case when the given measure of the optimal domain is much less than the square in 2D case or cube in 3D case of the characteristic size of the problem. This case is too often: the modelling computations for various problems of optimal design by existing algorithms show ([3], [4], [34], [186]) that as a rule the optimal structure is a set of connected bars, i.e. rod structure. To obtain this answer these algorithms start from the full 2D or 3D domain, square or cube for example. They iterate then (on the micro-level) the distribution of small anisotropic holes to obtain the optimal domain. Even if the optimal domain is not a rod structure normally the penalization procedure is applied to the optimal domain in order to transform the answer into some sort of a rod structure.

The proposed below algorithm starts from the lattice structure from the very beginning and deals with lattice structure at each step. This principle simplifies the optimization procedure. On the other hand the proposed approach differs from that of the algorithms of optimal topologies of discrete structures (see review [70]) because it starts from the continuous media formulation of the problem containing small parameter, and then obtains the discrete formulation as a result of the asymptotic analysis of the initial problem. This approach allows to estimate the accuracy of thus obtained discrete models and hence of the optimization result of the initial model.

We present as well some numerical experiments. The structure of the section is as follows.

We consider the boundary value problem for non linear partial derivative equations stated on a finite rod structure as in previous section and state the problem of minimization of the stored energy (compliance) of the body with constraint of the given measure of a finite rod structure.

The initial stored energy integral is replaced by the stored energy corresponding to the asymptotic solution obtained in the previous section. Then we introduce the initial configuration [70] as such a skeleton of finite rod structure that contains all admissible skeletons of finite rod structures. We minimize the approximated stored energy by an iterative algorithm of "frozen fluxes"; it is a discrete analog of the iterative algorithm from [35]. At each step of this algorithm we fix the fluxes obtained in the previous iteration for all rods and redistribute then the thicknesses of the rods to minimize the stored energy .

We give some results of numerical experiments (optimal dam, optimal cantilever arm, optimal bridge) calculated by the software OPTIFOR that implements the described method.

4.2.1 Stored energy as the cost

We consider the shape design problem of minimizing the stored energy (compliance) of finite rod structures of section 4.1 under the constraints of given vector-valued constant T and of given total measure of a finite rod structure that is μ^{s-1}.

The stored energy of the body

$$E_{B_\mu} = \int_{B_\mu} \sum_{i=1}^{s} (A_i(\nabla u_\mu) \, , \, \partial u_\mu/\partial x_i) \, dx = \int_{\Sigma_1} (t \, , \, u_\mu) \, dx. \qquad (4.2.1)$$

is a cost. We suppose the existence of the integral (4.2.1). The optimization problem seeks among all finite rod structures $B_\mu \in \mathcal{Q}$ with a fixed measure μ^{s-1} such one B_μ^0 which minimizes the energy (4.2.1):

$$E_{opt} = min_{B_\mu \in \mathcal{Q}} E_{B_\mu}. \qquad (4.2.2)$$

Thus the physical sense of the small parameter μ is the power $(s-1)^{-1}$ of the ratio of given measure of B_μ (given quantity of the material) to the power s of the characteristic size of the problem.

We approximate the energy integral (4.2.1) by the stored energy integral related to the asymptotic solution $u_\mu^{(1)}$

$$\int_{B_\mu} \sum_{i=1}^{s} (A_i(\nabla u_\mu^{(1)}) \, , \, \partial u_\mu^{(1)}/\partial x_i) \, dx.$$

In the linear scalar case under the assumptions of Remark 4.1.4 the difference of these two integrals is $O(\sqrt{\mu}\mu^{s-1})$. Indeed,

$$| \int_{B_\mu} \sum_{i=1}^{s} (A_i(\nabla u_\mu) \, , \, \partial u_\mu/\partial x_i) \, dx - \int_{B_\mu} \sum_{i=1}^{s} (A_i(\nabla u_\mu^{(1)}) \, , \, \partial u_\mu^{(1)}/\partial x_i) \, dx \, | \leq$$

$$\leq | \int_{B_\mu} \sum_{i=1}^{s} ((A_i(\nabla u_\mu) - A_i(\nabla u_\mu^{(1)})) \, , \, \partial u_\mu^{(1)}/\partial x_i) \, dx \, | \, +$$

$$+ | \int_{B_\mu} \sum_{i=1}^{s} (A_i(\nabla u_\mu) \, , \, (\partial u_\mu/\partial x_i - \partial u_\mu^{(1)}/\partial x_i)) \, dx \, | \, \leq$$

$$\leq const \, (\sqrt{\mu}\sqrt{meas(B_\mu)}\|u_\mu^{(1)}\|_{H^1(B_\mu)} +$$

$$+ \|u_\mu\|_{H^1(B_\mu)}\|u_\mu - u_\mu^{(1)}\|_{H^1(B_\mu)}) \; =$$

$$= O(\sqrt{\mu} meas(B_\mu)) \; = \; O(\sqrt{\mu}\mu^{s-1}).$$

Taking into consideration the estimate (4.1.11) we obtain

$$\int_{B_\mu} \sum_{i=1}^{s} (A_i(\nabla u_\mu^{(1)}) \, , \, \partial u_\mu^{(1)}/\partial x_i) \, dx \; =$$

$$= \int_{B_\mu^{++}} \sum_{i=1}^{s} (A_i(\nabla u_\mu^{(1)}) \, , \, \partial u_\mu^{(1)}/\partial x_i) \, dx \; + \; O(\mu^s) \; =$$

$$= \sum_{e_h^\alpha \subset B} \int_{\tilde{B}_{h,j}^\alpha} \sum_{i=1}^{s} (A_i(X_1, ..., X_s) \, , \, X_i) \, dx \; + \; O(\mu^s) \; =$$

$$= \mu^{s-1} \sum_{e_h^\alpha \subset B} |e_h^\alpha| meas\beta_j \sum_{i=1}^{s} (A_i(X_1, ..., X_s) \, , \, X_i) \; + \; O(\mu^s),$$

where $|e_h^\alpha|$ is the length of e_h^α.

Enumerate all segments $e_h^\alpha \subset B$:

$e_1, ..., e_N$ and denote by $\mu^{s-1}m_1, ..., \mu^{s-1}m_N$ the measures of the cross-sections of corresponding cylinders $\tilde{B}_{h,j,\mu}^\alpha$, i.e. $\mu^{s-1}meas\beta_j$ and by $(X_0^k, ..., X_s^k)$ the collection $(X_0^{e_j}, ..., X_s^{e_j})$ for the segment e_k. So the energy integral related to $u_\mu^{(1)}$ is approximated by

$$\int_{B_\mu} \sum_{i=1}^{s} (A_i(\nabla u_\mu^{(1)}) \, , \, \partial u_\mu^{(1)}/\partial x_i) \, dx \; = \; \mu^{s-1}E_B \; + \; O(\mu^s),$$

where

$$E_B \; = \; \sum_{k=1}^{N} |e_k| m_k \sum_{i=1}^{s} (A_i(X_1^k, ..., X_s^k) \, , \, X_i^k), \tag{4.2.3}$$

and $|e_k|$ stand for the lengths of segments e_k.

The approximated constraints are

$$\sum_{k=1}^{N} |e_k| m_k \; = \; 1, \tag{4.2.4}$$

as well as conditions 1)-3) of section 4.1 (i.e. (4.1.7)-(4.1.9)).

Making the change $M_k = |e_k| m_k$, rewrite (4.2.3),(4.2.4) as

$$E_B = \sum_{k=1}^{N} M_k \sum_{i=1}^{s} (A_i(X_1^k, ..., X_s^k) \ , \ X_i^k). \qquad (4.2.3')$$

and

$$\sum_{k=1}^{N} M_k = 1. \qquad (4.2.4')$$

So, we replace the initial shape optimization problem (4.2.2) by the problem of minimization of cost (4.2.3') under constraints (4.2.4'), (4.1.7)-(4.1.9).

4.2.2 Simplification of the set of finite rod structures. Initial configuration

Thus the initial partial derivative equation problem (4.1.1)-(4.1.4) is asymptotically reduced to the algebraic system (4.1.7)-(4.1.9) for a set of unknowns $(X_0^k, ..., X_s^k)$, $k = 1, ..., N$. In the linear case when the matrix of this system is non-singular the problem is stable to the small perturbations of the skeleton B. Indeed the following proposition holds.

Proposition 4.2.1. *Let $B_\mu(\varepsilon)$ be a finite rod structure, such that each segment $e_h^\alpha(\varepsilon)$ of its skeleton is obtained from e_h^α by means of orthogonal transformation with the matrix $Id + O(\varepsilon)$, i.e. identity matrix Id perturbed by a matrix that is $O(\varepsilon)$, by translation on the distance $O(\varepsilon)$, and contraction in $1 + O(\varepsilon)$ times. We suppose that the cross-sections of the cylinders $\tilde{B}_{h,j,\mu}^\alpha$, and their perturbed images $\tilde{B}_{h,j,\mu}^\alpha(\varepsilon)$, are of the same measure. Let the perturbed images of the segments with the end points in G_0 and G_1 have also the end points in these domains and no other segments of the perturbed skeleton have the end points in these domains. Then the solution of perturbed system (4.1.7)-(4.1.9) $(X_0^k(\varepsilon), ..., X_s^k(\varepsilon))$, $k = 1, ..., N$, tends to the solution of non-perturbed system $(X_0^k, ..., X_s^k)$, $k = 1, ..., N$, as ε tends to zero.*

It is the trivial corollary of the regular perturbations theory of linear operators.

This obvious result gives us an opportunity of approximation of the finite rod structure's geometry. We introduce the initial configuration B_I as such skeleton of finite rod structure, that contains all admissible skeletons of finite rod structures, i.e. we consider only such structures that their skeletons belong to the initial configuration. We suppose also that the nodes of the admissible structures belong to the set of nodes of the initial configuration. For example we can choose an $\varepsilon-$periodic lattice-structure of the next Chapter with $\mu << \varepsilon << 1$. We should take such a lattice with sufficiently large set of orientations of segments e_h^α on each periodic cell to approximate "all possible orientations" of segments of the skeleton of finite rod structure.

When the initial configuration is fixed then we can formulate the following approximation to problem (4.2.2). Associate a real positive variable M_k to every segment e_k composing the initial configuration B_I; $(B_I = \cup_{k=1}^{N} e_k)$; every admissible rod structure B_μ corresponds to some ordered set $(M_1, ..., M_N)$, where $M_k = 0$ if the segment e_k is absent in the skeleton of B_μ, and where $M_k = |e_k| m_k$ if e_k is present; the cylinder of B_μ corresponding to e_k has the measure of its cross- section $\mu^{s-1} m_k$ and the length of e_k is $|e_k|$.

The approximate shape optimization problem is thus to find such a control $(M_1, ..., M_N)$ that minimizes cost (4.2.3') under constraints (4.2.4') and 1)-3) of section 4.1 (i.e. (4.1.7)-(4.1.9)). Here the constant vector T of boundary condition (4.1.3) is given.

Further simplifying assumption is that the nodes of non homogeneous boundary condition (4.1.3) on G_1 are fixed in the initial configuration. Instead of this assumption we can apply the below algorithm to all possible combinations of these "nodes of applied fluxes". Normally in the examples we consider the applied fluxes in some given points.

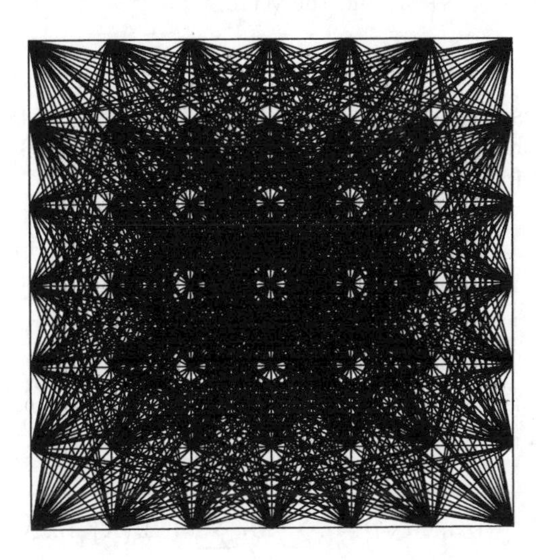

Figure 4.2.1. Initial configuration

4.2.3 An iterative algorithm for the optimal design problem

For the problem of minimization of the approximate energy integral (4.2.3') with constraints (4.2.4') and $B \in B_I$ the numerical algorithm is proposed. It is the iterative optimization algorithm of "frozen fluxes" that is a discrete analog of this from [35]. It starts from the initial configuration B_I with the uniform distribution of the control $M_k = |e_k| m_k$, i.e. $M_1 = ... = M_N = 1/N$.

The n-th step of the algorithm is as follows. We use the finite rod system of the previous $(n-1)-$th step to calculate the solution of the problem (4.1.7)-(4.1.9) and the fluxes for each segment e_k :

$$f_k = (M_k/|e_k|) \sum_{i=1}^{s} (A_i(X_1^k, ..., X_s^k)\, \eta_i^k),$$

where $\eta_1^k, ..., \eta_s^k$ are the direction cosines of e_k. We fix these values of f_k and solve the system of equations for each k for any value of the scalar parameter $\hat{M}_k$ for unknowns $\hat{X}_1^k, ..., \hat{X}_s^k$

$$(\hat{M}_k/|e_k|) \sum_{i=1}^{s} (A_i(\hat{X}_1^k, ..., \hat{X}_s^k)\, \eta_i^k) = f_k,$$

$$\sum_{i=1}^{s} A_i(\hat{X}_1^k, ..., \hat{X}_s^k)\, \nu_i^j = 0,$$

where $j = 1, ..., s-1$, and the vectors

$$\nu^1 = (\nu_1^1, ..., \nu_s^1)\ , \ ...\ ,\ \nu^{s-1} = (\nu_1^{s-1}, ..., \nu_s^{s-1})$$

form the basis in $(s-1)-$dimensional space orthogonal to e.

Suppose that this system has a unique solution for all $\hat{M}_k \in R_+$, that could be presented in a form

$$(\hat{X}_1^k, ..., \hat{X}_s^k) = \hat{M}_k^{-\alpha}(\hat{X}_{01}^k, ..., \hat{X}_{0s}^k)\ ,$$

where $(\hat{X}_{01}^k, ..., \hat{X}_{0s}^k)$ is a solution of the system for $\hat{M}_k = 1, \quad \alpha > 0$. Suppose also that

$$A_i(\hat{X}_1^k, ..., \hat{X}_s^k) = \hat{M}_k^{-1} A_i(\hat{X}_{01}^k, ..., \hat{X}_{0s}^k)\ .$$

Calculate the energy integral of the n-th step

$$E_B^n(\hat{M}_1, ..., \hat{M}_N) = \sum_{k=1}^{N} \hat{M}_k \sum_{i=1}^{s} (\ A_i(\hat{X}_1^k, ..., \hat{X}_s^k),\ \hat{X}_i^k) =$$

$$\sum_{k=1}^{N} \hat{M}_k^{-\alpha} \sum_{i=1}^{s} (A_i(\hat{X}_{01}^k, ..., \hat{X}_{0s}^k)\ ,\ \hat{X}_{0i}^k) \tag{4.2.5}$$

and minimize it with constrains

$$\sum_{k=1}^{N} \hat{M}_k = 1. \tag{4.2.6}$$

Proposition 4.2.2. *Assume that*

$$\sum_{i=1}^{s} (A_i(\hat{X}_{01}^k, ..., \hat{X}_{0s}^k)\ ,\ \hat{X}_{0i}^k) \geq 0, \tag{4.2.7}$$

and

$$\sum_{k=1}^{N}\sum_{i=1}^{s} (A_i(\hat{X}_{01}^k, ..., \hat{X}_{0s}^k) \ , \ \hat{X}_{0i}^k) \ > \ 0. \tag{16}$$

Then the problem of minimization of (4.2.5) with constrains (4.2.6) has a unique solution calculated by the formula

$$\hat{M}_k \ = \ \frac{(\sum_{i=1}^{s} \ (A_i(\hat{X}_{01}^k, ..., \hat{X}_{0s}^k) \ , \ \hat{X}_{0i}^k))^{1/(1+\alpha)}}{\sum_{k=1}^{N}(\sum_{i=1}^{s} \ (A_i(\hat{X}_{01}^k, ..., \hat{X}_{0s}^k) \ , \ \hat{X}_{0i}^k))^{1/(1+\alpha)}}. \tag{4.2.8}$$

Proof.

Using Lagrange multiplier $\lambda > 0$ we obtain the minimization problem of the Lagrange function

$$E_B^n(\hat{M}_1, ..., \hat{M}_N) \ + \ \lambda(\sum_{k=1}^{N} \hat{M}_k - 1).$$

We seek the stationary point where its gradient vanishes and taking in account the relation (4.2.5) in a standard way we get the distribution (4.2.8) of the control parameters $\hat{M}_k$ for the n-th step. The proposition is proved.

Thus we obtain the finite rod structure of the n-th step.

If some of the values of $\hat{M}_k$ are less than the established positive value we can remove such segments e_k from the skeleton (if it does not lead to a lost of the connectivity).

Remark 4.2.1. The asymptotic expansion of the solution of elasticity problem stated in the finite rod system considered in the next section could be used for the correction of the first order discrete model (4.1.7) - (4.1.9).

Remark 4.2.2. The optimal design problem (4.1.1)-(4.2.4), (4.2.1),(4.2.2) can be generalized: we can consider equation (4.1.1) with the given right hand side function $f_\mu(x)$, "concentrated" in the neighborhoods of some fixed interior nodes (end points) $O_1, ..., O_N$, included in all admissible finite rod structures.

Let $T_i, \ i = 1, ..., N$ be vector-valued constants. Let r_0 be a positive value such that all balls with the centers in the nodes and with the radii μr_0 belong to B_μ.

Denote $Y(O_i)$ such ball with the center O_i. Define

$$f_\mu(x) \ = \ \frac{T_i \mu^{s-1}}{meas Y(O_i)}$$

in each ball $Y(O_i)$, and define $f_\mu(x) \ = \ 0$ out of the union of these balls. Consider the problem (4.1.1'), (4.2.2) - (4.2.4) where (4.1.1') is the non-homogeneous equation (4.1.1), i.e.

$$\sum_{i=1}^{s} \partial/\partial x_i(\ A_i(\nabla_x u_\mu)) \ = \ f_\mu(x), \quad for \ x \in B_\mu. \tag{4.1.1'}$$

Then the analogous algorithm can be applied with the replacement of the relation (8) by the relation for nodes $\bar{O}$

$$\sum_{j=1}^{q}\sum_{i=1}^{s} A_i(X_1^{e_j},...,X_s^{e_j})\ \eta_i^j meas\tilde{\beta}_j \ = \ T_{\bar{O}}, \qquad (4.1.8')$$

where $T_{\bar{O}} = 0$ if the node $\bar{O}$ does not belong to the set $\{O_1,...,O_N\}$, and $T_{\bar{O}} = T_r$ if $\bar{O} = O_r$.

The conclusion of Theorem 4.1.1 is valid for this modified problem.

To prove it we present each integral

$$\int_{\Pi_{x^0}} (f_\mu\ ,\ \varphi)\ dx,\ \ x^0\ \in\ \{O_1,...,O_N\},\ x^0 = O_i,$$

in a form

$$\int_{\Pi_{x^0}} (f_\mu\ ,\ <\varphi>_{\Pi_{x^0}})\ dx\ +\ \int_{\Pi_{x^0}} (f_\mu\ ,\ \varphi-<\varphi>_{\Pi_{x^0}})\ dx\ =$$

$$=\ (T_i\ ,\ \mu^{s-1}<\varphi>_{\Pi_{x^0}})\ +\ \int_{\Pi_{x^0}} (f_\mu\ ,\ \varphi-<\varphi>_{\Pi_{x^0}})\ dx$$

and estimate the last integral as

$$\|f_\mu\|_{L^2(\Pi_{x^0})}\ \|\varphi\ -\ <\varphi>_{\Pi_{x^0}}\|_{L^2(\Pi_{x^0})}\ \leq$$

$$\leq\ O(\mu)\ \frac{T_i\mu^{s-1}}{measY(O_i)}\ \sqrt{meas(B_\mu)}\ \|\nabla\varphi\|_{L^2(\Pi_{x^0})}\ =$$

$$=\ O(\mu)\ \sqrt{meas(B_\mu)}\|\varphi\|_{H^1(\Pi_{x^0})}.$$

These modifications complete the proof.

4.2.4 Some results of numerical experiment

Here we present some two-dimensional and three-dimensional numerical examples calculated by the OPTIFOR software which implements the described method. We give below the value of compliance $f^T x$ for every optimal structure.

1. The first example deals with the problem of the reinforcement of a dam. The initial configuration and the distribution of boundary forces are shown in Fig.4.2.2

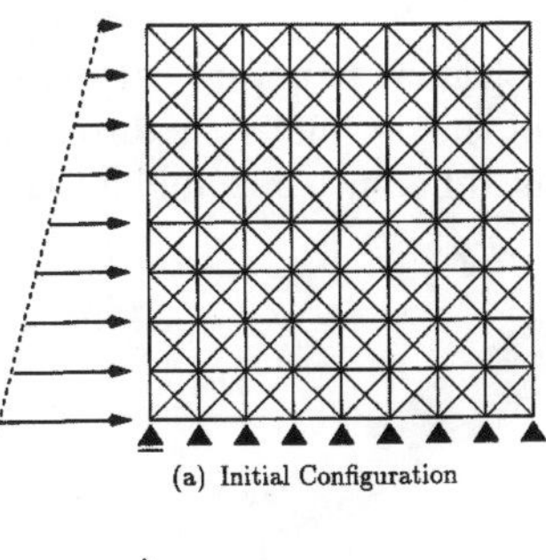

(a) Initial Configuration

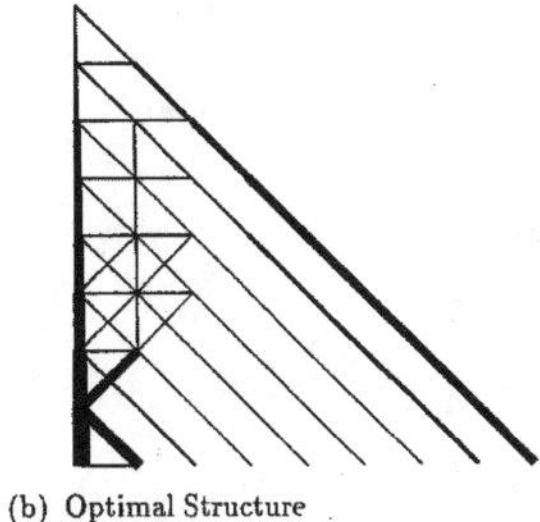

(b) Optimal Structure

Figure 4.2.2. Reinforcement of a dam: the initial configuration and the optimal structure; $f^T x = 12.22$

2. The second example is concerned with the three-dimensional model of a cantilever arm. The left end is fixed while the vertical force is applied to the right end. The optimal result was obtained after 30 iterations.

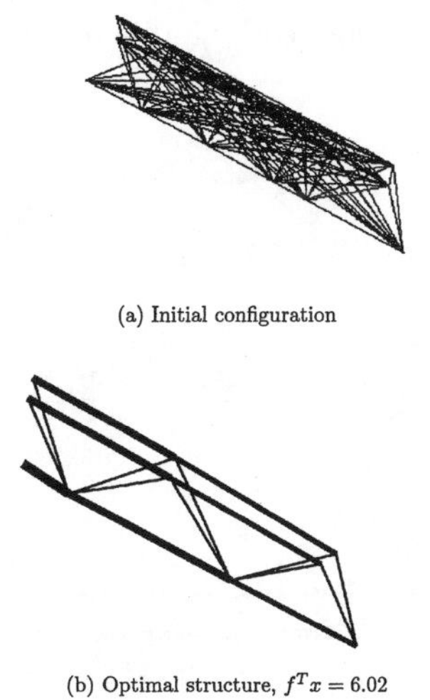

(a) Initial configuration

(b) Optimal structure, $f^T x = 6.02$

Figure 4.2.3. Cantilever arm: the initial configuration and the optimal structure; $f^T x = 6.02$

3. Figure 4.2.4 presents the optimal structure calculated for another initial configuration. The compliance of the optimal structure is here less than for the previous initial configuration. the difference of compliances is about 10 per cent.

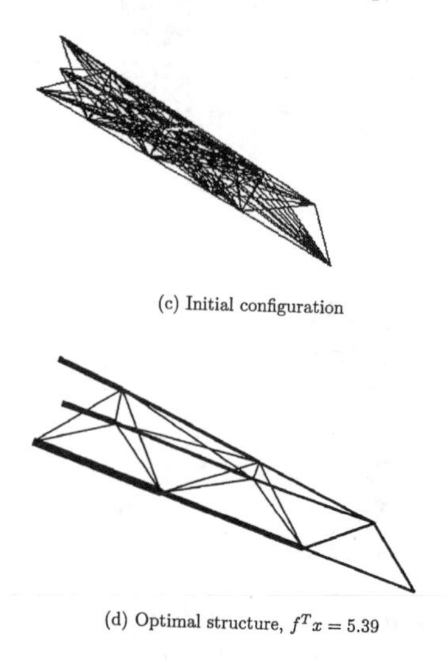

(c) Initial configuration

(d) Optimal structure, $f^T x = 5.39$

Figure 4.2.4. Cantilever arm; another initial configuration: the initial configuration and the optimal structure; $f^T x = 5.39$

Figure 4.2.5 gives the compliance as a function of the number of iterations. The same optimal result was obtained by a different method in [186].

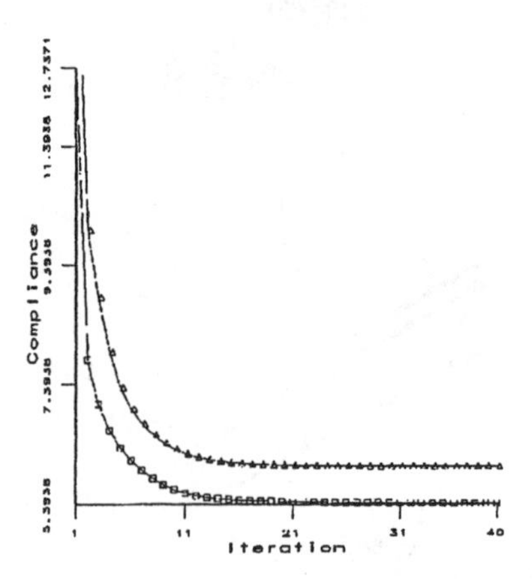

Figure 4.2.5. Compliance histories.

3. The third example is an "optimal bridge". The initial configuration is a square lattice. The structure is supported at the edges of its base. A vertical point force is applied in the middle of the lower side. The initial configuration and the boundary conditions are symmetric with respect to the axis of the force. Therefore the optimal structure is also symmetric, and the computations can be performed on half the initial configuration. The results are presented in the Fig. 4.2.6.

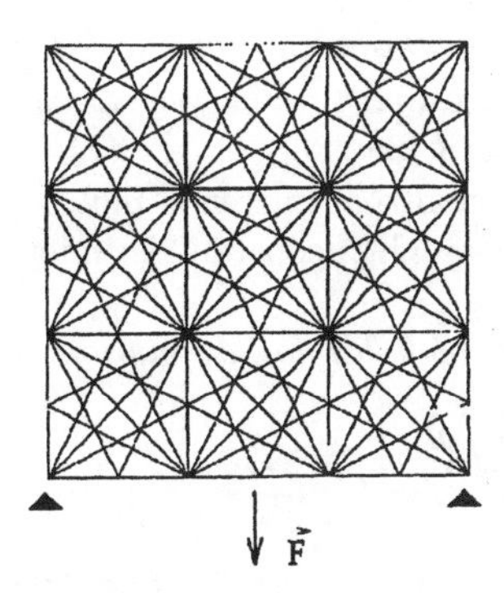

(a) Bridge-type structure with 480 bars

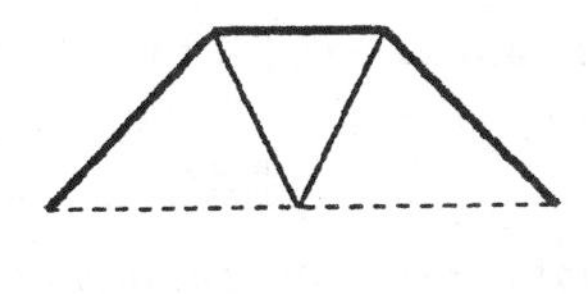

(b) Optimal structure for bridge-type structure, $f^T x = 0.846$

Figure 4.2.6. Initial configuration for a bridge with 480 initial bars and optimal structure of a bridge, $f^T x = 0.846$

4.3 Conductivity: an asymptotic expansion

4.3.1 Construction of asymptotic expansion

In this section we consider finite rod structures B_μ defined in Definition 4.1.2 without the additional cubes.

For any segment $e = e_h^\alpha$, we define local variables $\tilde{x} = \alpha(x - h)$, whose first component is $\tilde{x}_1$.

Consider the conductivity equation in B_μ

$$L_x\, u \;=\; \Delta u \;=\; \begin{cases} f^e(\tilde{x}_1), & x \in S_0, \\ 0, & x \notin S_0, \end{cases} \tag{4.3.1}$$

$$u|_{\partial_1 B_\mu} = 0, \tag{4.3.2}$$

$$\frac{\partial u}{\partial n}|_{\partial_2 B_\mu} = 0, \tag{4.3.3}$$

where $\partial_1 B_\mu$ is a part of the boundary that coincides with the base of one (or several) of the cylinders $\tilde{B}^{\alpha}_{hj}$ (we assume that the nodal point corresponding to this base is an end-point only for one segment $e \subset \mathcal{B}$), $\partial_2 B_\mu = \partial B_\mu \backslash \partial_1 B_\mu$. We call "interior nodes" all nodes not belonging to $\partial_1 B_\mu$.

We assume that $f^e(t) \in C^\infty$. In what follows we omit the superscript e for f.

We seek the solution $u(x)$ in the class of functions $H^1(B_\mu)$.

Below we construct the asymptotic expansion of the solution to problem (4.3.1)-(4.3.3) as $\mu \to 0$.

Theorem 4.3.1. *There exists a unique solution to problem (4.3.1)-(4.3.3).*

This is a well known result on existence and uniqueness of solution to a mixed boundary value conductivity problem ([85],[86]).

The domain B_μ here depends on the small parameter μ but as we see in Appendix A4.2, the Poincaré-Friedrichs inequality constant is independent of μ. This implies

Theorem 4.3.2. *The a priori estimate holds for a solution to problem (4.3.1)-(4.3.3):*

$$\|u\|^2_{H^1(B_\mu)} \leq C\|f\|^2_{L^2(B_\mu)},$$

where constant C does not depend on μ.

We construct the asymptotic solution in two stages. First we construct an asymptotic solution outside neighborhoods of the nodes, and then we build the boundary layers. At the first step for every segment e an asymptotic solution is sought in the form of Bakhvalov's ansatz (section 2.2); in the isotropic case all $N_l = 0$ except of $N_0 = 1$ and so this ansatz is just a function v_e of the longitudinal variable $\tilde{x}_1$, where $\tilde{x} = \alpha_e(x-h)$. This function v_e satisfies the second order differential equation (see section 2.2). For every rod this function is multiplied by a cutting function $\tilde{\rho}$ that is, equal to 1 on the main part of segment e and that vanishes in some *const* $\mu-$vicinities of the ends. Of course, this multiplication perturbs the right hand side and generates some finite support discrepancy in the equation near the nodes.

At the second step we first construct the Taylor expansion of the right hand side in the *const* $\mu-$vicinities of the ends and we obtain the elliptic boundary layer problems with finite right hand sides set in the bundles of half-infinite cylinders related to every node. Such boundary layer problems have solutions stabilizing to constants on every outlet of the bundle if and only if the right hand side is orthogonal to constants. This orthogonality condition generates the flux balance equation (4.3.9) in every node. On the other hand we would like to have the boundary layer solutions exponentially decaying at infinity. It means that the stabilization constants should be equal to zero. It can be seen that these limit constants may be manipulated and changed by varying the values of functions v_e in the ends of segments e. It gives us a possibility to vanish all stabilization limit constants by equating all values of functions v_e in the node (probably with some constant right hand sides δv_e, see (4.3.10)).

Finally we get the second order ordinary differential equations on the skeleton supplied with the junction conditions for the values of v_e' and v_e in the nodes, as well as with the boundary conditions in the nodes of $\partial_1 B_\mu$ and $\partial_2 B_\mu$.

At the third step we seek an asymptotic solution to this problem set on the skeleton in the form of a regular series ansatz (4.3.12) and we obtain the all terms of the asymptotic expansion, and in particular, the leading term v_{l0} satisfying (4.3.23)-(4.3.27).

The justification as usual (see section 2.4.4) is developed as follows: we truncate the ansatz at some level J neglecting the terms of order $O(\mu^J)$ (in H^1-norm) and calculate all discrepancies in the right hand sides of the equation and eventually boundary conditions. We check that these discrepancies are of order $O(\mu^{J-1})$ in L^2-norm. Then we apply the a priori estimate and obtain that

$$\|u - u^{(J)}\|_{H^1(B_\mu)} = O(\mu^{J-1}).$$

Finally we improve this estimate comparing $u^{(J)}$ to $u^{(J-1)}$ and estimating the difference

$$\|u^{(J)} - u^{(J-1)}\|_{H^1(B_\mu)} = O(\mu^{J-1}).$$

Hence, the estimate holds:

$$\|u - u^{(J-1)}\|_{H^1(B_\mu)} = O(\mu^{J-1})$$

and so,

$$\|u - u^{(J)}\|_{H^1(B_\mu)} = O(\mu^J).$$

1. We seek a formal asymptotic solution of problem (4.3.1),(4.3.3) outside neighborhoods of the nodes on any section S_e corresponding to the segment $e = e_h^\alpha$ of the set $\mathcal{B}$, of the form

$$\tilde{u}_e^{(\infty)} \sim \sum_{l=0}^{\infty} (\mu)^l N_l^e\left(\frac{\tilde{x}'}{\mu}\right)\frac{d^l v(\tilde{x}_1)}{d\tilde{x}_1^l}, \tag{4.3.4}$$

where $\tilde{x} = \alpha(x-h)$ are the local co-ordinates introduced in e_h^α, $N_l^e(\tilde{\xi}')$ are the functions constructed in Chapter 2; let us denote $\xi = (\xi_1, \xi')$, and $\xi' = (\xi_2, \ldots, \xi_s)$, v is a smooth function of the variable $\tilde{x}_1$. All these functions N_l^e are equal to zero except of $N_0^e = 1$.

Thus simply

$$\tilde{u}_e^{(\infty)}(x) = v_e(\tilde{x}_1), \quad x \in e,$$

and it is a solution of equation

$$v_e''(\tilde{x}_1) = f(\tilde{x}_1), \quad x \in e.$$

2. Now construct the boundary layers in neighborhoods of the nodes and we extend $\tilde{u}_e^{(\infty)}(x)$ to B_μ. We expand now the constructed above at the first stage $\tilde{u}_e^{(\infty)}$ in a Taylor series in the neighborhood of any node x_0. We obtain

$$\tilde{u}_e^{(\infty)} \sim \sum_{r=0}^{\infty} \mu^r \frac{(\tilde{z}_1)^r}{r!} \frac{d^r v_e(\tilde{x}_1(x_0))}{d\tilde{x}_1^r}, \tag{4.3.5}$$

where $\tilde{z} = \alpha_e z,\ z = (x - x_0)/\mu$,

Here α_e stands for the matrix α of passage from the local (corresponding to the segment e) base to the global one.

Denote by $\Pi_{x_0}$the connected component of the set $B_\mu \backslash \hat{S}_0$ containing the point x_0. Denote by $\Pi_{x_0,z,0}$ the image of Π_{x_0} under the transformation $z = (x - x_0)/\mu$,. Let $\tilde{B}_{hj}^{\alpha}$ be a cylinder whose intersection with Π_{x_0} is nonempty. We extend it to a semi-infinite cylinder behind the basis which has no common points with Π_{x_0}. We denote by $B_{hj\infty}^{\alpha}$ this extended cylinder, and by $\Pi_{x_0\infty}$ the union of all such cylinders $B_{hj\infty}^{\alpha}$ and Π_{x_0}.

Moreover, we denote by $\Pi_{x_0,z,\infty}$ the image of $\Pi_{x_0,\infty}$ under the transformation $z = (x - x_0)/\mu$.

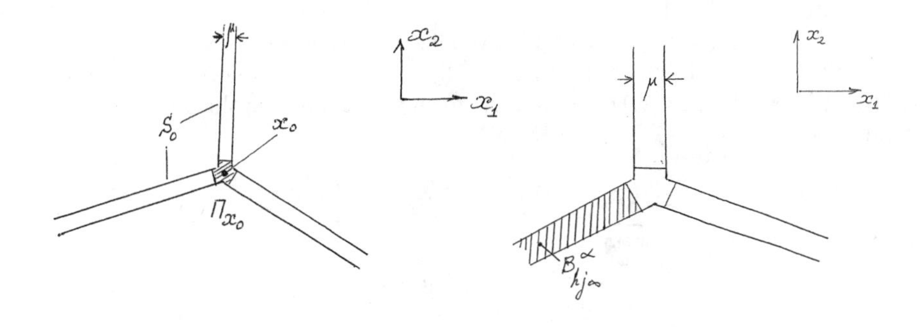

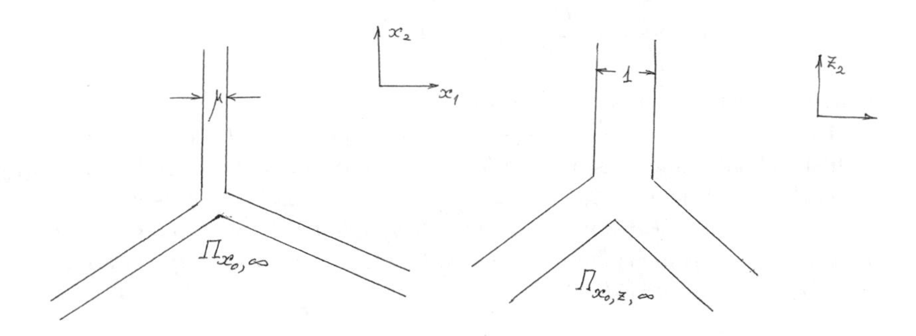

Figure 4.3.1. Sets Π_{x_0}, $B^{\alpha}_{hj\infty}$, $\Pi_{x_0,\infty}$, $\Pi_{x_0,z,\infty}$

Let $\tilde{\chi}_e(z)$ be the characteristic function of $B^{\alpha}_{hj\infty}$ corresponding to the segment e of $\mathcal{B}$.

In the neighborhood of the node x_0 we seek the solution in the form

$$u^{(\infty)}_{\Pi_{x_0}} \sim \sum_{l=0}^{\infty} \mu^l N^0_l(z) + \sum_{e(x_0)} \tilde{\chi}_e(z)\hat{\rho}(\alpha_e z) v_e(\tilde{x}_1(x_0)). \tag{4.3.6}$$

Here the summation of the last sum in (4.3.6) extends over all segments e in $\mathcal{B}$ having x_0 as an end-point, and $\hat{\rho}(t) = \hat{\rho}(t_1)$ (i.e. it depends only on the first component of the variable t in $I\!R^s$) is an infinitely differentiable function that vanishes for $|t_1| \leq c_0$, is equal to unity for $|t_1| \geq c_0 + 1$, and it satisfies the estimate $0 \leq \hat{\rho}(t) \leq 1$.

The functions N_l^0 exponentially stabilize to zero as $|z| \to \infty$.

Substituting (4.3.6) in (4.3.1), (4.3.3), and representing f as the sum of $\hat{\rho}(\alpha_e z)f$ and $(1-\hat{\rho}(\alpha_e z))f$, we find the recurrent chain of problems determining functions N_l^0 :

$$L_z N_l^0 = \sum_{e(x_0)} \bar{\chi}_e(z)(L_z((1-\hat{\rho}(\alpha_e z))\frac{\tilde{z}_1^l}{l!}\frac{d^l v_e(\tilde{x}_1(x_0))}{d\tilde{x}_1^r})$$

$$-(1-\hat{\rho}(\alpha_e z))L_z(\frac{(\tilde{z}_1)^l}{l!}\frac{d^l v_e(\tilde{x}_1(x_0))}{d\tilde{x}_1^l}))$$

$$+\bar{\delta}_{l-2}\frac{(\tilde{z}_1)^r}{r!}(1-\hat{\rho}(\alpha_e z))\frac{d^{l-2}}{d\tilde{x}_1^{l-2}}\tilde{f}(x_0), \quad z \in \Pi_{x_0,z,\infty}, \tag{4.3.7$_1$}$$

$$\frac{\partial N_l^0}{\partial \nu_z} = \sum_{e(x_0)} \bar{\chi}_e(z)(L_z\frac{\partial}{\partial \nu_z}((1-\hat{\rho}(\alpha_e z))\frac{(\tilde{z}_1)^l}{l!})$$

$$-(1-\hat{\rho}(\alpha_e z))\frac{\partial}{\partial \nu_z}(\frac{(\tilde{z}_1)^l}{l!}))\frac{d^l}{d\tilde{x}_1^l}\tilde{v}^e(\tilde{x}_1(x_0)), \quad z \in \partial\Pi_{x_0,z,\infty}, \tag{4.3.7$_2$}$$

where $\bar{\delta}_l = 1$ for $l \geq 0$ and $\bar{\delta}_l = 0$ for $l < 0$.

Here $L_z = \Delta_z$, $\frac{\partial}{\partial \nu_z} = \frac{\partial}{\partial n_z}$. Denote respectively $F_l^{eq}(z)$ and $F_l^b(z)$ the right-hand side functions of equation (4.3.7$_1$) and boundary condition (4.3.7$_2$). Remark that these functions vanish outside of the ball of radius c_0.

Lemma 4.3.1. *Problem* (4.3.7$_1$), (4.3.7$_2$) *is solvable in the class of functions that stabilize exponentially to a constant as $|z|$ tends to infinity (in the sense of [88]-[90]) if and only if*

$$\int_{\Pi_{x_0,z,\infty}} F_l^{eq}(z)dz = \int_{\partial\Pi_{x_0,z,\infty}} F_l^b(z)ds. \tag{4.3.8}$$

This condition implies (in case when $x_0 \notin \partial_1 B_\mu$) that

$$\sum_{e(x_0)} \mu \int_{\Pi_{x_0,z,\infty}} \bar{\chi}_e(z)(1-\hat{\rho}(\alpha_e z))L_x v_e dz$$

$$-\sum_{e(x_0)} \int_{\partial\Pi_{x_0,z,0}\setminus\Pi_{x_0,z,\infty}} \bar{\chi}_e(z)(1-\hat{\rho}(\alpha_e z))\frac{\partial}{\partial \nu_x}v_e dz$$

$$\sim \sum_{e(x_0)} \mu \int_{\Pi_{x_0,z,\infty}} \bar{\chi}_e(z)(1-\hat{\rho}(\alpha_e z))f dz.$$

Here $\sim$ stands for the asymptotic equivalence up to terms of order $O(\mu^K)\forall K$.

Taking into account the fact that

$$L_x v_e \sim v_e'' \sim f,$$
$$\frac{\partial}{\partial \nu_x} v_e \sim v_e',$$

we obtain:

$$\sum_{e(x_0)} meas\ \beta_e\ v_e' \sim 0, \tag{4.3.9}$$

where β_e stands for the cross-section of the cylinder corresponding to e (see the definition of a finite rod structure).

If (4.3.9) is satisfied asymptotically exactly, then problems $(4.3.7_1), (4.3.7_2)$ have solutions that stabilize exponentially to constants on every half-infinite cylinder as $|z| \to \infty$. We keep the notation C_e for these constants

From the representations (4.3.5),(4.3.6) it follows that if the value of $v_e(\tilde{x}_1(x_0))$ changes by a value denoted δv_e, then the value of C_e in the sum $\sum_{l=0}^{\infty} \mu^l N_l^0(z)$ changes also by δv_e. Then we are left with the problem of finding δv_e such that

$$(v_{e_1}(\tilde{x}_1(x_0)) + \delta v_{e_1}) - (v_{e_2}(\tilde{x}_1(x_0)) + \delta v_{e_2}) \sim 0, \qquad \forall e_1(x_0), e_2(x_0), \tag{4.3.10}$$

and the redefined in such manner solutions N_l^0 stabilize to zero as $|z| \to \infty$.

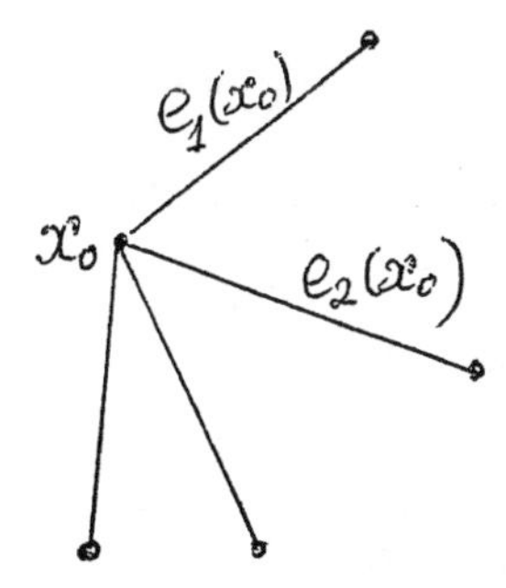

Figure 4.3.2. Segments $e_1(x_0)$, $e_2(x_0)$

If the node x_0 is an end-point for only one segment e and $x_0 \notin \partial_1 B_\mu$, then condition (4.3.10) is cancelled (the constant can be omitted from the solution). If $x_0 \in \partial_1 B_\mu$, then on the part of $\partial \Pi_{x_0,z,\infty}$ corresponding to $\partial_1 B_\mu$, under change of variable $z = (x - x_0)/\mu$ we prescribe the condition $N_l^0 = 0$ instead of boundary condition $(4.3.7_2)$ on this part of the boundary. Then the analogue of Lemma 4.3.1 holds for this problem, but the solvability condition (4.3.8)is no long necessary. The solution stabilizes to a constant C_e. So the condition at the node x_0 takes the form

$$(v_e(\tilde{x}_1(x_0)) + \delta v_e) \sim 0, \tag{4.3.11}$$

where δv_e is chosen so that the solution N_l^0, modified in such a way that it satisfies boundary condition $N_l^0 = -C_e$ instead of $N_l^0 = 0$ on the part of the boundary corresponding to $\partial_1 B_\mu$, stabilizes to zero.

More precisely, we find first the limit constants at infinity C_e^l for each N_l^0 and then prescribe condition (4.3.10) with $\delta v_e \;=\; -\sum_{l=0}^{\infty} \mu^l C_e^l$.

Seeking v_e in the form of a power series in μ,

$$v_e \;=\; \sum_{l=0}^{\infty} \mu^l v_e^l, \tag{4.3.12}$$

we find that , as in Chapter 2, all the v_e^l are determined from the recursive sequence of equations of the form

$$v_e^{l\,"} = f_e(\tilde{x}_1)\delta_{l0}, \quad x \in e, \tag{4.3.12}$$

supplied by the matching conditions at the nodes x_0 that are end-points for at least two segments (and $x_0 \notin \partial_1 B_\mu$)

$$\sum_{e(x_0)} meas \beta_e (v_e^l)'(\tilde{x}_1(x_0)) = 0, \tag{4.3.13}$$

and

$$(v_{e_1}^l(\tilde{x}_1(x_0)) + \delta v_{e_1}^l) = (v_{e_2}^l(\tilde{x}_1(x_0)) + \delta v_{e_2}^l), \qquad \forall e_1(x_0), e_2(x_0), \tag{4.3.14}$$

and boundary conditions at the nodes x_0 in $\partial_1 B_\mu$ or in $\partial_2 B_\mu$, but such that they are end-points for only one segment, of the form

$$(v_e^l)'(\tilde{x}_1(x_0)) = 0, \qquad x_0 \in \partial_2 B_\mu \tag{4.3.15}$$

and

$$(v_e^l(\tilde{x}_1(x_0)) + \delta v_e^l) = 0, \qquad x_0 \in \partial_1 B_\mu, \tag{4.5.16}$$

where $-\delta v_e^l$ are found as the limit constants at infinity for N_l^0.

Indeed, for any l, we solve first problem (4.3.7) for N_l^0 (taking for v_e^l the matching condition $v_{e_1}^l(\tilde{x}_1(x_0)) = v_{e_2}^l(\tilde{x}_1(x_0)), \qquad \forall e_1(x_0), e_2(x_0)$, instead of (4.3.14)) and we obtain different limit constants C_e^l on the outlets (the subscript e means that the outlet corresponds to the segment e of skeleton $\mathcal{B}$); we redefine then $N_l^0(z)$ replacing v_e^l in (4.3.14) by $v_e^l + \delta v_e^l$ with $\delta v_e^l = -C_e^l$; this replacement leads to stabilization of redefined $N_l^0(z)$ to zero. So after such redefinition we have N_l^0 tending to zero on the outlets and we obtain condition (4.3.14) for v_e^l.

Extend δv_e^l on $\mathcal{B}$ in such a way that $(\delta v_e^l)' = 0$ in some neighborhood of each node. Then problem (4.3.12)-(4.3.16) can be reduced to a problem for a sum $w_e = v_e^l + \delta v_e^l$:

equation
$$w_e" = \psi(\tilde{x}_1), \quad x \in e, \tag{4.3.17}$$

supplied by the matching conditions at the nodes x_0 that are end-points for at least two segments (and $x_0 \notin \partial_1 B_\mu$)

$$\sum_{e(x_0)} meas\beta_e w_e'(\tilde{x}_1(x_0)) = 0, \tag{4.3.18}$$

and
$$w_{e_1}(\tilde{x}_1(x_0)) = w_{e_2}(\tilde{x}_1(x_0)), \qquad \forall e_1(x_0), e_2(x_0), \tag{4.3.19}$$

and boundary conditions at the nodes x_0 in $\partial_1 B_\mu$ or in $\partial_2 B_\mu$, but such that they are end-points for only one segment, of the form

$$w_e'(\tilde{x}_1(x_0)) = 0, \qquad x_0 \in \partial_2 B_\mu \tag{4.3.20}$$

and
$$w_e(\tilde{x}_1(x_0)) = 0, \qquad x_0 \in \partial_1 B_\mu. \tag{4.3.21}$$

Let $\mathcal{H}_0^1(\mathcal{B})$ be a space of functions defined on $\mathcal{B}$, having a derivative on each segment e and continuous on $\mathcal{B}$ and vanishing at all nodes of $\partial_1 B_\mu$. Suppose that this space is supplied by the inner product

$$(w, v)_{\mathcal{B}} = \int_{\mathcal{B}} meas\beta_e(w(\tilde{x}_1)v(\tilde{x}_1) + w'(\tilde{x}_1)v'(\tilde{x}_1))d\tilde{x}_1,$$

that is
$$\sum_{e \subset \mathcal{B}} \int_e meas\beta_e(\tilde{w}_e(\tilde{x}_1^e)\tilde{v}_e(\tilde{x}_1^e) + \tilde{w}_e'(\tilde{x}_1^e)\tilde{v}_e'(\tilde{x}_1^e))d\tilde{x}_1^e,$$

where $\tilde{w}_e(\tilde{x}_1^e) = w(\alpha_e(x-h))$ for $x \in e$.

Then the variational formulation for problem (4.3.17)-(4.3.21) is as follows: to find $w \in \mathcal{H}_0^1(\mathcal{B})$ such that, for any $v \in \mathcal{H}_0^1(\mathcal{B})$

$$-\sum_{e \in \mathcal{B}} \int_e w'(\tilde{x}_1)v'(\tilde{x}_1)meas\beta_e d\tilde{x}_1 = \sum_{e \in \mathcal{B}} \int_e \psi(\tilde{x}_1)v(\tilde{x}_1)meas\beta_e d\tilde{x}_1.$$

The existence and uniqueness of such solution is proved by a standard technique of Riesz theorem on representation of a bounded linear functional (or by Lax-Milgramm lemma), using the Poincaré- Friedrichs inequality on $\mathcal{B}$

$$\int_{\mathcal{B}} v^2(\tilde{x}_1)d\tilde{x}_1 \leq C(\mathcal{B}) \int_{\mathcal{B}} (v'(\tilde{x}_1))^2 d\tilde{x}_1,$$

where the constant $C(\mathcal{B})$ depends on $\mathcal{B}$ only. (This inequality is an immediate consequence of the Newton-Leibnitz formula for $\mathcal{B}$: let $e_1, \ldots, e_n$ be segments such, that the initial node of e_i is the final node for e_{i-1} and the beginning of e_1 is a node of $\partial_1 B_\mu$; let $x \in e_j$, then

$$v(x) = \sum_{k=1}^{j-1} \int_{e_k} \frac{\partial \tilde{v}_{e_k}(l)}{\partial l}(l)dl + \int_0^{\tilde{x}_1^{e_j}(x)} \frac{\partial \tilde{v}_{e_j}(l)}{\partial l} dl).$$

We denote by $u^{(J)}$ the truncated (partial) sum of the asymptotic series (4.3.5), obtained by neglecting the terms of order $O(\mu^J)$ (in the $H^1(B_\mu)-$norm. Choosing J sufficiently large, we can show for any K that equation (4.3.1) and conditions (4.3.2) and (4.3.3) are satisfied with remainders of order $O(\mu^K)$ (in the $L^2(B_\mu)-$norm). Hence , using the a priori estimate for problem (4.3.1)-(4.3.3) given by Theorem 4.3.2, we obtain as in section 2.2, the following assertion.

Theorem 4.3.3 *There holds the estimate*

$$\|u - u^{(J)}\|_{H^1(B_\mu)} = O(\mu^J). \tag{4.3.22}$$

This estimate justifies the asymptotic construction.

4.3.2 The leading term of the asymptotic expansion

Thus the limit problem for the leading term v_0 of the asymptotic expansion is as follows:

the equation

$$v_{e\ 0}" = f(\tilde{x}_1), \quad x \in e, \tag{4.3.23}$$

supplied by the matching conditions at the nodes x_0 that are end-points for at least two segments (and $x_0 \notin \partial_1 B_\mu$)

$$\sum_{e(x_0)} v'_{e\ 0} meas \beta_e(\tilde{x}_1(x_0)) = 0, \tag{4.3.24}$$

and

$$v_{e_1\ 0}(\tilde{x}_1(x_0)) = v_{e_2\ 0}(\tilde{x}_1(x_0)), \qquad \forall e_1(x_0), e_2(x_0), \tag{4.3.25}$$

and boundary conditions at the nodes x_0 in $\partial_1 B_\mu$ or in $\partial_2 B_\mu$, but such that they are end-points for only one segment, of the form

$$v'_{e\ 0}(\tilde{x}_1(x_0)) = 0, \qquad x_0 \in \partial_2 B_\mu \tag{4.3.26}$$

and

$$v_{e\ 0}(\tilde{x}_1(x_0)) = 0, \qquad x_0 \in \partial_1 B_\mu. \tag{4.3.27}$$

Extend v_0 on B_μ in a following way. For any connected component of S_0, corresponding to the segment e we define $v(x) = v_{e\ 0}(\tilde{x}_1(x))$, and for any connected component Π_{x_0} of $B_\mu \backslash S_0$ containing x_0 we define $v(x) = v_{e\ 0}(\tilde{x}_1(x_0))$, as a constant. Then the estimate (4.3.22) yields

$$\|u - v\|_{L^2(B_\mu)} / \sqrt{meas\ B_\mu} = O(\sqrt{\mu}). \tag{4.3.28}$$

Remark 4.3.1. The right-hand side function can have more general structure on S_0 :

$$F\left(\frac{\tilde{x}'}{\mu}\right)\psi(\tilde{x}_1), \tag{4.3.29}$$

where

$$\psi \in C^\infty, \quad F \in C^1.$$

In this case F is decomposed as

$$F(\tilde{\xi}') = \bar{F} + \tilde{F}(\tilde{\xi}'), \quad \bar{F} =< F >_\beta .$$

Solution to problem (4.3.1)-(4.3.3) with right-hand side (4.3.29) on a cylinder corresponding to a segment e is sought in a form

$$\tilde{v}_e(\tilde{x}_1) + \sum_{l=0}^{\infty} \mu^l M_l\left(\frac{\tilde{x}'}{\mu}\right)\frac{d^l\psi(\tilde{x}_1)}{d\tilde{x}_1^l},$$

where M_l are solutions to problems

$$\Delta_{\tilde{\xi}'} M_0 = \tilde{F}(\tilde{\xi}'), \quad \tilde{\xi}' \in \beta,$$

$$\frac{\partial M_0}{\partial n_{\tilde{\xi}}} = 0, \quad \tilde{\xi}' \in \partial\beta;$$

$$\Delta_{\tilde{\xi}'} M_l = -M_{l-2} + h_l^M, \quad \tilde{\xi}' \in \beta, \quad l \geq 2,$$

$$\frac{\partial M_l}{\partial n_{\tilde{\xi}}} = 0, \quad \tilde{\xi}' \in \partial\beta;$$

where $h_l^M =< M_{l-2} >_\beta$.

Then for v we have equation

$$\frac{d^2 v_e(\tilde{x}_1)}{d\tilde{x}_1^2} - \bar{F}\psi + \sum_{l=2, l \ even}^{\infty} \mu^l h_l^M \frac{d^l\psi(\tilde{x}_1)}{d\tilde{x}_1^l} = 0.$$

At the second stage we construct boundary layers $u_{\Pi_{x_0}}^{(\infty)}$ in neighborhoods of nodes x_0

$$u_{\Pi_{x_0}}^{(\infty)} \sim \sum_{l=0}^{\infty} \mu^l N_l^0(z) + \sum_{e(x_0)} \bar{\chi}_l(z)\hat{\rho}(\alpha_e z)\left(v^e(\tilde{x}_1) + \sum_{l=0}^{\infty} \mu^{l+2} M_l\left(\frac{\tilde{x}'}{\mu}\right)\frac{d^l\psi(\tilde{x}_1)}{d\tilde{x}_1^l}\right).$$

Applying the same procedure as above we obtain the homogenized problem of the same type as (4.3.12) but with some supplement in the right-hand side. The leading term is the same as in problem (4.3.1)-(4.3.3) with $f = \bar{F}\psi$.

4.4 Elasticity: an asymptotic expansion

4.4.1 Construction of asymptotic expansion

Let as above ξ' be $(s-1)-$dimensional vector of the last $s-1$ components of the vector $\xi = (\xi_1, \ldots, \xi_s)$, i.e. $\xi' = (\xi_2, \ldots, \xi_s)$. Denote by S_α the transformation defined by relations $S_2\xi' = (-\xi_2, \xi_3)$, $S_3\xi' = (\xi_2, -\xi_3)$ for $s = 3$, and by $S_2\xi' = -\xi_2$ for $s = 2$. In what follows we assume that $S_\alpha\beta = \beta$ and $S_\alpha\beta_j = \beta_j$, and we define $S_\alpha\xi = (\xi_1, S_\alpha\xi')$, $\alpha = 2, ..., s-1$. This property means certain symmetry of the cross-sections of the rods of the finite rod structure B_μ.

As in section 4.3, for any segment $e = e_h^\alpha$, we define local variables $\tilde{x} = \alpha(x-h)$, whose first component is $\tilde{x}_1$.

We consider the system of equations of elasticity theory (elasticity equations)in B_μ :

$$L_x\, u_\mu = \sum_{i,j=1}^{s} \frac{\partial}{\partial x_i}\left(A_{ij}\frac{\partial u_\mu}{\partial x_j}\right)$$

$$= \begin{cases} \alpha_e^*(\tilde{f}_1^e(\tilde{x}_1), \mu^2\tilde{f}_2^e(\tilde{x}_1), \ldots, \mu^2\tilde{f}_3^e(\tilde{x}_1))^*, & x \in S_0, \\ 0, & x \notin S_0, \end{cases} \tag{4.4.1}$$

$$u_\mu|_{\partial_1 B_\mu} = 0, \tag{4.4.2}$$

$$\frac{\partial u_\mu}{\partial \nu}|_{\partial_2 B_\mu} = 0, \tag{4.4.3}$$

where $\partial_1 B_\mu$ is a part of the boundary that coincides with the base of one (or several) of the cylinders $\tilde{B}_{hj}^\alpha$ (we assume that the nodal point corresponding to this base is an end-point only for one segment $e \subset \mathcal{B}$), $\partial_2 B_\mu = \partial B_\mu \backslash \partial_1 B_\mu$. We call "interior nodes" all nodes not belonging to $\partial_1 B_\mu$.

Here

$$\frac{\partial u_\mu}{\partial \nu} = \sum_{i,j=1}^{s} A_{ij}\frac{\partial u_\mu}{\partial x_j}n_i,$$

where $A_{ij}$are constant $s \times s$ matrices, the elements a_{ij}^{kl} of the A_{ij} satisfy the relations

$$a_{ij}^{kl} = (\delta_{ij}\delta_{kl} + \delta_{il}\delta_{jk})\mathcal{M} + \lambda\delta_{ik}\delta_{jl},$$

where $\mathcal{M}$ and λ are positive constants, that are the Lamé coefficients, $(n_1, \ldots, n_s)$ is the normal to $\partial B_{\mu\varepsilon}$, $f^e(t) \in C^\infty$. We seek the solution $u_\mu(x)$ in the class of $s-$dimensional vector-valued functions of $H^1(B_\mu)$.

Below we construct the asymptotic expansion of the solution to problem (4.4.1)-(4.4.3) as $\mu \to 0$ under some assumptions PF_1 and PF_2 of geometrical rigidity of the frame. These assumptions will be formulated below for the limit problem.

Theorem 4.4.1. *There exists a unique solution to problem (4.4.1)-(4.4.3).*

This is a well known result on existence and uniqueness of solution to a mixed boundary value problem of elasticity ([55]).

The domain B_μ here depends on the small parameter μ and therefore an a priori estimate *can* also depend on this small parameter. On the other hand as we have seen in section 4.2 the a priori estimate for analogous problem for conductivity equation does not depend on the small parameter because the Poincaré-Friedrichs inequality constant is independent of μ. Is elasticity equation different in this sense?

To answer this question we should study the Korn inequality constant for the domain B_μ. This constant depends on μ. Indeed, the following assertion holds.

Theorem 4.4.2. *For any μ small enough for any finite rod structure B_μ there is a vector valued function $v_\mu \in H^1(B_\mu)$ such that $v_\mu|_{\partial_1 B_\mu} = 0$, $\|\nabla v_\mu\|^2_{L^2(B_\mu)} \geq c_1\mu^{s-1}$ and $E_{B_\mu}(v_\mu) \leq c_2\mu^s$ that is*

$$E_{B_\mu}(v_\mu) \leq (\frac{c_2}{c_1})\mu\|\nabla v_\mu\|^2_{L^2(B_\mu)},$$

where c_1 and c_2 are positive constants independent of μ. Here

$$E_{B_\mu}(\phi) = \int_{B_\mu} \sum_{i,j=1}^{s} (e_i^j(\phi))^2 dx, \quad e_i^j(\phi) = \frac{1}{2}(\frac{\partial \phi_i}{\partial x_j} + \frac{\partial \phi_j}{\partial x_i}).$$

Proof.

Without lost of generality we can prove this assertion for one rod $B^\alpha_{h,j,\mu}$ such that one of its bases belongs to $\partial_1 B_\mu$ (we pass to the general case extending v_μ by zero all over $B_\mu \backslash B^\alpha_{h,j,\mu}$).

We can consider the cylinder $B_{j,\mu}$ instead of $B^\alpha_{h,j,\mu}$ because the orthogonal transformation and the translation do not change the order of the values $\|\nabla v_\mu\|^2_{L^2(B_\mu)}$ and $E_{B_\mu}(v_\mu)$.

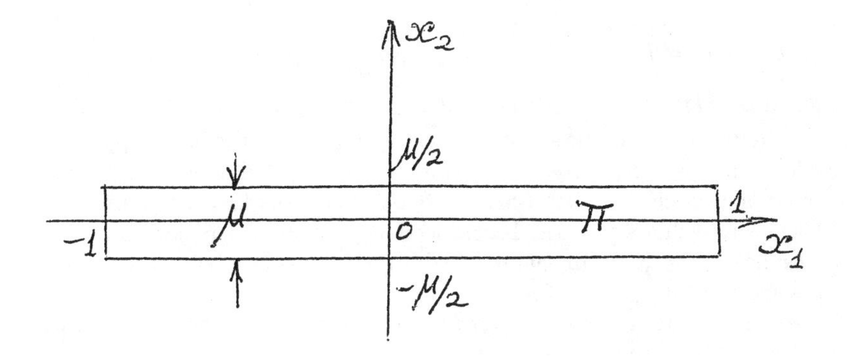

Figure 4.4.1. Domain π.

So, let us consider cylinder $\pi = (-1,1)\times\beta_\mu$, where $\beta_\mu = \{(x'/\mu \in \beta\}$, β is $(s-1)-$dimensional bounded domain with a piecewise-smooth boundary (satisfying the cone condition; in the two-dimensional case β is an interval), i.e. there exist constants $c_1,\ c_2,\ c_3 > 0$, independent of μ, and the s-dimensional vector-valued function $w_\mu(x)$, such that $w_\mu \in H^1(\pi)$ and

$$w_\mu|_{x_1=\pm1} = 0, \quad \|w_\mu\|^2_{L^2(\pi)} \geq c_1\mu^{s-1}, \quad \|\nabla w_\mu\|^2_{L^2(\pi)} \geq c_2\mu^{s-1},$$

$$E_\pi(w_\mu) \leq c_3\mu^s.$$

For example , for $s=2, \mu<1/2$ one can build the following function, satisfying these conditions:

$$w_\mu = \begin{cases} \begin{pmatrix} -x_2\rho(x_1/\mu)\rho((1+x_1)/\mu)\rho((x_1-1)/\mu) \\ 1+x_1 \end{pmatrix} & \text{for } x_1 \leq 0 \\ \begin{pmatrix} x_2\rho(x_1/\mu)\rho((1+x_1)/\mu)\rho((x_1-1)/\mu) \\ 1-x_1 \end{pmatrix} & \text{for } x_1 > 0, \end{cases}$$

where $\rho(t)$ is an even continuously differentiable function, $0 \leq \rho(t) \leq 1$, equal to zero for $|t| \leq 1/3$ and equal to 1 for $|t| \geq 2/3$. (In the case $s=3$ the two first components are the same and the third one is equal to zero.)

Indeed,

$$\|\nabla w_\mu\|^2_{L^2(\pi)} \geq \int_\pi (\frac{\partial w_2}{\partial x_1})^2 dx = 2\ meas\ \beta\ \mu^{s-1},$$

$$\|w_\mu\|^2_{L^2(\pi)} \geq \int_\pi w_2^2 dx = meas\ \ \beta\ \mu^{s-1},$$

and

$$E_\pi(w_\mu) = E_{\pi\cap(\{x_1\leq -1+\mu\}\cup\{|x_1|\leq\mu\}\cup\{x_1\geq 1-\mu\})}(w_\mu) = O(\mu^s),$$

because $\frac{\partial w_1}{\partial x_1} = O(1)$, $\frac{\partial w_1}{\partial x_2} = O(1)$, $\frac{\partial w_2}{\partial x_1} = \pm1$, $\frac{\partial w_2}{\partial x_2} = 0$, and because the strains differ from zero on the set of the measure $O(\mu^s)$. By the linear application we could map this constructed vector-function to any cylinder $B^\alpha_{h,j,\mu}$, and in particular, the cylinder having a base in $\partial_1 B_\mu$. Thus, the theorem is proved.

This theorem shows that the Korn inequality constant depends on μ.

Nevertheless the Korn inequality holds true in this case with the constant that is some negative power of μ.

Theorem 4.4.3. *Any vector-valued function $u \in H^1(B_\mu)$, such that $u|_{\partial_1 B_\mu} = 0$, satisfies the Korn inequality*

$$\|\nabla u\|^2_{L^2(B_\mu)} \leq C_K\mu^{-q}E_{B_\mu}(u),$$

where C_K and q are independent of μ.

Proof.
The proof of this theorem was given first by constructing an extension of u from B_μ to some domain which does not depend on μ (see [145]). We give it here below in Appendix 5.A2.

This theorem with the Poincaré-Friedrichs inequality implies

Theorem 4.4.4. *The a priori estimate holds for a solution to problem (4.4.1)-(4.4.3):*

$$\|u\|^2_{H^1(B_\mu)} \leq C\mu^{-q}\|f\|^2_{L^2(B_\mu)},$$

where constant C does not depend on μ.

Let us remind the definition at section 2.3 of the matrix-valued functions of rigid displacements of an infinite bar $\{ \xi = (\xi_1, \ldots, \xi_s) \in I\!R^s,\ \xi_1 \in I\!R;\ \xi' = (\xi_2, \ldots, \xi_s) \in \beta \}$. For $s = 2$ it is the identity matrix

$$\Phi = I = \begin{pmatrix} 1 & 0 \\ 0 & 1 \end{pmatrix},$$

while for $s = 3$ it depends on the rapid variable ξ and it is equal to

$$\Phi(\xi) = \begin{pmatrix} 1 & 0 & 0 & 0 \\ 0 & 1 & 0 & -\xi_3 a \\ 0 & 0 & 1 & \xi_2 a \end{pmatrix}.$$

Here $a = \left(\frac{\int_\beta (\xi_2^2 + \xi_3^2) d\xi_2 d\xi_3}{meas\ \beta} \right)^{-1/2}$.

Denote d the number of columns of Φ (i.e. $d = 2$ for $s = 2$ and $d = 4$ for $s = 3$).

Moreover, we define as well the extended matrix of rigid displacements in $I\!R^s$: for $s = 2$ it is

$$\tilde{\mathcal{J}}(\xi) = \begin{pmatrix} 1 & 0 & -\xi_2 \\ 0 & 1 & \xi_1 \end{pmatrix},$$

while for $s = 3$ it is

$$\tilde{\mathcal{J}}(\xi) = \begin{pmatrix} 1 & 0 & 0 & 0 & -\xi_2 & -\xi_3 \\ 0 & 1 & 0 & -\xi_3 & \xi_1 & 0 \\ 0 & 0 & 1 & \xi_2 & 0 & \xi_1 \end{pmatrix}.$$

(it differs from $\tilde{\Phi}$ of Chapter 2 only by a normalization factor a of the third for $s = 2$ or the fourth for $s = 3$ column.)

Now we formulate the two assumptions on B_μ.

As in the previous section we will consider segments e which belong to the set of segments e_h^α constituting the skeleton $\mathcal{B}$. Let us define for any segment e of $\mathcal{B}$ the matrix Γ_e of dimension $d \times d$ such that for any $z \in I\!R^s$,

$$\tilde{\mathcal{J}}^*(z)\alpha_e^* = \Gamma_e \tilde{\mathcal{J}}^*(\alpha_e z),$$

where z is a column $(z_1, \ldots, z_s)^*$. It can be proved immediately that

Lemma 4.4.1.

$$\Gamma_e = \begin{pmatrix} & & 0 \\ & \alpha_e^* & 0 \\ 0 & 0 & 1 \end{pmatrix}$$

for $s = 2$, and

$$\Gamma_e = \begin{pmatrix} \alpha_e^* & O \\ O & \bar{\Gamma}_e \end{pmatrix}$$

for $s = 3$, where $\bar{\Gamma}_e$ is an orthogonal 3×3 matrix and O the 3×3 zero matrix.

Proof.

For $s = 2$, let

$$\alpha_e = \begin{pmatrix} cos\ \theta & sin\ \theta \\ -sin\ \theta & cos\ \theta \end{pmatrix}, \quad \theta \in I\!R;$$

let us check the last scalar equality of the relation

$$\tilde{\mathcal{J}}^*(z)\alpha_e^* \ = \ \Gamma_e \tilde{\mathcal{J}}^*(\alpha_e z).$$

The last row of $\tilde{\mathcal{J}}^*(z)\alpha_e^*$ is $(-z_2, z_1)\alpha_e^*$

$$= \begin{pmatrix} -z_2\ cos\ \theta\ +\ z_1\ sin\ \theta \\ z_2\ sin\ \theta\ +\ z_1\ cos\ \theta \end{pmatrix}^* = \begin{pmatrix} -(\alpha_e z)_2 \\ (\alpha_e z)_1 \end{pmatrix}^*$$

because

$$\alpha_e z = \begin{pmatrix} z_1\ cos\ \theta\ +\ z_2\ sin\ \theta \\ -z_1\ sin\ \theta\ +\ z_2\ cos\ \theta \end{pmatrix},$$

and therefore the components $(\alpha_e z)_1$ and $(\alpha_e z)_2$ of this vector satisfy the above relation. The first two scalar equalities of relation

$$\tilde{\mathcal{J}}^*(z)\alpha_e^* \ = \ \Gamma_e \tilde{\mathcal{J}}^*(\alpha_e z)$$

are evident because $I\alpha_e^* = \alpha_e^* I$. So, for $s = 2$ lemma is proved.

If $s = 3$ let us check firstly the last three scalar equalities of the relation

$$\tilde{\mathcal{J}}^*(z)\alpha_e^* \ = \ \Gamma_e \tilde{\mathcal{J}}^*(\alpha_e z).$$

For any $k \in \{1, 2, 3\}$, denote $\mathcal{J}_k(z) = \Pi_k z$, where

$$\Pi_1 = \begin{pmatrix} 0 & 0 & 0 \\ 0 & 0 & -1 \\ 0 & 1 & 0 \end{pmatrix}, \ \Pi_2 = \begin{pmatrix} 0 & -1 & 0 \\ 1 & 0 & 0 \\ 0 & 0 & 0 \end{pmatrix}, \ \Pi_3 = \begin{pmatrix} 0 & 0 & -1 \\ 0 & 0 & 0 \\ 1 & 0 & 0 \end{pmatrix}.$$

Then the matrix $\tilde{\mathcal{J}}(z)$ contains the last three columns $\mathcal{J}_k(z)$, satisfying the following relations:

$$\alpha_e^* \mathcal{J}_k(\alpha_e z) \ = \ \alpha_e^* \Pi_k \alpha_e z \ = \ \sum_{i=1}^{3} \gamma_{ki}(\alpha_e) \Pi_i z,$$

where $\gamma_{ki}(\alpha_e)$ are the components of the 3×3 matrix

$$\gamma(\alpha_e) = \begin{pmatrix} \alpha_{32}\alpha_{23} - \alpha_{22}\alpha_{33} & \alpha_{31}\alpha_{22} - \alpha_{21}\alpha_{32} & \alpha_{31}\alpha_{23} - \alpha_{21}\alpha_{33} \\ \alpha_{22}\alpha_{13} - \alpha_{12}\alpha_{23} & \alpha_{21}\alpha_{12} - \alpha_{11}\alpha_{22} & \alpha_{21}\alpha_{13} - \alpha_{11}\alpha_{23} \\ \alpha_{32}\alpha_{13} - \alpha_{12}\alpha_{33} & \alpha_{31}\alpha_{12} - \alpha_{11}\alpha_{32} & \alpha_{31}\alpha_{13} - \alpha_{11}\alpha_{33} \end{pmatrix},$$

and α_{ij} are the components of the matrix α_e.

One can see that

$$\gamma_{ij}(\alpha_e) = \gamma_{ji}(\alpha_e^*), \quad i, j \in \{1, 2, 3\},$$

so

$$\gamma^*(\alpha_e) = \gamma(\alpha_e^*).$$

On the other hand, the above relation

$$\alpha_e^* \mathcal{J}_k(\alpha_e z) = \sum_{i=1}^{3} \gamma_{ki}(\alpha_e)\Pi_i z$$

can be rewritten in the form

$$\mathcal{J}_k(\alpha_e z) = \sum_{i=1}^{3} \gamma_{ki}(\alpha_e)\alpha_e \mathcal{J}_i(z),$$

and therefore $\mathcal{J}_k(z)$ which is equal to $\mathcal{J}_k(\alpha_e^* \alpha_e z)$ can be presented in the form $\sum_{i=1}^{3} \gamma_{ki}(\alpha_e^*)\alpha_e^* \mathcal{J}_i(\alpha_e z)$, that is

$$\sum_{i,j=1}^{3} \gamma_{ki}(\alpha_e^*)\gamma_{ij}(\alpha_e)\alpha_e^*\alpha_e \mathcal{J}_j(z) = \sum_{i,j=1}^{3} \gamma_{ki}(\alpha_e^*)\gamma_{ij}(\alpha_e)\mathcal{J}_j(z).$$

These vector-valued functions $\mathcal{J}_j(z)$ are linearly independent functions. Therefore, the product $\gamma(\alpha_e^*)\gamma(\alpha_e)$ is the identity matrix I.

On the other hand we have just proved that

$$\gamma^*(\alpha_e) = \gamma(\alpha_e^*).$$

So, $\gamma^*(\alpha_e)\gamma(\alpha_e) = I$, and the matrix $\gamma(\alpha_e)$ is an orthogonal matrix.

Consider now the 3×3–matrix $\tilde{\mathcal{J}}_{last}(z)$ constituted of the three last columns of the matrix $\tilde{\mathcal{J}}(z)$. Its transposition $\tilde{\mathcal{J}}^*_{last}(\alpha_e z)$ consists of three lines:

$$\begin{pmatrix} \mathcal{J}_1^*(\alpha_e z) \\ \mathcal{J}_2^*(\alpha_e z) \\ \mathcal{J}_3^*(\alpha_e z) \end{pmatrix} = \gamma(\alpha_e)(\alpha_e \tilde{\mathcal{J}}_{last}(z))^* = \gamma(\alpha_e)\tilde{\mathcal{J}}^*_{last}(z)\alpha_e^*;$$

so,

$$\gamma^*(\alpha_e)\tilde{\mathcal{J}}^*_{last}(\alpha_e z) = \tilde{\mathcal{J}}^*_{last}(z)\alpha_e^*,$$

and therefore, $\bar{\Gamma}_e = \gamma^*(\alpha_e)$.

The first three scalar equalities of relation

$$\tilde{\mathcal{J}}^*(z)\alpha_e^* = \Gamma_e \tilde{\mathcal{J}}^*(\alpha_e z)$$

are evident because $I\alpha_e^* = \alpha_e^* I$. So, for $s=3$ lemma is proved as well.

Assume now that the two following conditions hold true.

Condition PF_1. Let $\tilde{\Psi}(\tilde{x}_1)$ be any s-dimensional vector-valued function defined on $\mathcal{B}$ vanishing in the nodes belonging to $\partial_1 B_\mu$;

- which has generalized derivative along e for every $e \subset \mathcal{B}$,

- which is such that all its components, except the first, are linear on every segment e,

- which satisfies at all interior nodes x_0 the matching conditions $\alpha_{e_1}^* \tilde{\Psi}^{e_1} = \alpha_{e_2}^* \tilde{\Psi}^{e_2}$ for any two segments $e_1(x_0)$ and $e_2(x_0)$, having the common end-point x_0.

Then *it is assumed* that the following inequality holds:

$$\sum_e \int_e (\tilde{\Psi}_1^e(\tilde{x}_1))^2 d\tilde{x}_1 \leq c \sum_e \int_e (d\tilde{\Psi}_1^e/d\tilde{x}_1(\tilde{x}_1))^2 d\tilde{x}_1$$

with a constant c depending only on $\mathcal{B}$. Here $\tilde{\Psi}_1^e$ is the first component of the vector-valued function $\tilde{\Psi}^e$.

Thus, Condition PF_1 is satisfied for example, if only one node does not belong to $\partial_1 B_\mu$.

Suppose that for $s = 3$ the following Condition PF_2, analogous to the Poincaré- Friedrichs inequality, holds.

Condition PF_2. Let $\tilde{\Psi}(\tilde{x}_1)$ be any three-dimensional 1-periodic vector-valued function defined on $\mathcal{B}$

- whose first component $\tilde{\Psi}_1$ has generalized derivative along e for every segment $e \subset \mathcal{B}$,

- whose second and third components $\tilde{\Psi}_k$, $k = 2, 3$, have two generalized derivatives along e for every segment $e \subset \mathcal{B}$,

- whose second and third components vanish at all nodes, and

- which satisfies at all interior nodes x_0 the matching conditions of the form

$$\bar{\Gamma}_{e_1}^* \begin{pmatrix} \tilde{\Psi}_1^{e_1} \\ d\tilde{\Psi}_2^{e_1}/d\tilde{x}_1 \\ d\tilde{\Psi}_3^{e_1}/d\tilde{x}_1 \end{pmatrix} = \bar{\Gamma}_{e_2}^* \begin{pmatrix} \tilde{\Psi}_1^{e_2} \\ d\tilde{\Psi}_2^{e_2}/d\tilde{x}_1 \\ d\tilde{\Psi}_3^{e_2}/d\tilde{x}_1 \end{pmatrix}$$

for any two segments $e_1(x_0)$ and $e_2(x_0)$, having the common end-point x_0;

- which satisfies at all boundary nodes $x_0 \in \partial_1 B_\mu$ the boundary conditions of the form

$$\tilde{\Psi} = 0; \quad \frac{\partial \tilde{\Psi}_2}{\partial \tilde{x}_1} = 0; \quad \frac{\partial \tilde{\Psi}_3}{\partial \tilde{x}_1} = 0.$$

Then*it is assumed* that the following inequality holds:

$$\int_{\mathcal{B}} \left(\tilde{\Psi}_1^e(\tilde{x}_1))^2 + (\frac{d\tilde{\Psi}_2^e}{d\tilde{x}_1}(\tilde{x}_1))^2 + (\frac{d\tilde{\Psi}_3^e}{d\tilde{x}_1}(\tilde{x}_1))^2 \right) d\tilde{x}_1$$

$$\leq c \int_{\mathcal{B}} \left(\frac{d\tilde{\Psi}_1^e}{d\tilde{x}_1}(\tilde{x}_1))^2 + (\frac{d^2\tilde{\Psi}_2^e}{d\tilde{x}_1^2}(\tilde{x}_1))^2 + (\frac{d^2\tilde{\Psi}_3^e}{d\tilde{x}_1{}^2}(\tilde{x}_1))^2 \right) d\tilde{x}_1$$

with a constant c depending only on $\mathcal{B}$.

We suppose also for simplicity of presentation that for any interior node there are three non-coplanar (if $s=3$), respectively two non-collinear (if $s=2$), segments e having this node as an end-point. This condition will be called the non-coplanarity condition.

Remark 4.4.1 In case $s = 2$ the following conditions analogous to PF_2 is satisfied automatically. Let $\tilde{\Psi}^e$ be any scalar function depending on the variable $\tilde{x}_1$, having two derivatives along segment $e \subset \mathcal{B}$, satisfying the matching conditions at the interior nodes x_0

$$\frac{d\tilde{\Psi}^{e_1(x_0)}}{d\tilde{x}_1} = \frac{d\tilde{\Psi}^{e_2(x_0)}}{d\tilde{x}_1},$$

and vanishing at all nodes (the derivative vanishes as well at the nodes $x_0 \in \partial_1 B_\mu$.) Then the following inequality holds

$$\int_{\mathcal{B}} (\frac{d\tilde{\Psi}^e}{d\tilde{x}_1}(\tilde{x}_1))^2 d\tilde{x}_1$$

$$\leq c \int_{\mathcal{B}} (\frac{d^2\tilde{\Psi}^e}{d\tilde{x}_1^2}(\tilde{x}_1))^2 d\tilde{x}_1$$

with a constant c depending only on $\mathcal{B}$.

We construct the asymptotic solution as usual in two stages. First we construct an asymptotic solution outside neighborhoods of the nodes, and then we build the boundary layers. At the first step for every segment e an asymptotic solution is sought in the form of the ansatz of section 2.3 related to some $d-$dimensional ($d = 2$ if $s = 2$; $d = 4$ if $s = 3$) vector-valued function ω_e depending on the longitudinal variable $\tilde{x}_1$, where $\tilde{x} = \alpha_e(x - h)$. The components of vector-valued function ω_e satisfy some ordinary differential equations (homogenized high order equations). For every segment e this ansatz is multiplied by a cutting function $\tilde{\rho}$ that is, equal to 1 on the main part of segment e and that vanishes in some *const* $\mu-$vicinities of the ends. Of course, this multiplication perturbs the right hand side and generates some finite support discrepancy in the equation near the nodes.

At the second step we first construct the Taylor expansion of the right hand side in the *const* $\mu-$vicinities of the nodes and we obtain the elliptic boundary layer problems with finite right hand sides set in the bundles of half-infinite cylinders related to every node. Such boundary layer problems have solutions stabilizing to some rigid displacement on every outlet of the bundle if and only if the right hand side is orthogonal to all rigid displacements. This orthogonality condition generates the stress balance equation (4.4.11) in every inner node. On the other hand we would like to have the boundary layer solutions exponentially decaying at infinity. It means that the stabilization limit rigid displacements should be equal to zero. It can be seen that these limit constants may be manipulated and changed by varying the values of functions ω_e and its derivatives $\frac{\partial\omega_e^2}{\partial\tilde{x}_1},\ldots,\frac{\partial\omega_e^2}{\partial\tilde{x}_1}$ in the ends of segments e. It gives us a possibility to vanish all stabilization limit rigid displacements by equating all values of functions ω_e and its derivatives $\frac{\partial\omega_e^2}{\partial\tilde{x}_1},\ldots,\frac{\partial\omega_e^2}{\partial\tilde{x}_1}$ in the node multiplied by some passage matrix Γ_e. This matrix is defined as a matrix of orthogonal transform of rigid displacements.

Finally we get the second order ordinary differential equations on the skeleton supplied with the junction conditions for the values and for some derivatives of ω_e in the nodes, as well as with the boundary conditions in the nodes of $\partial_1 B_\mu$ and $\partial_2 B_\mu$.

At the third step we seek an asymptotic solution to this problem set on the skeleton in the form of a regular series ansatz and we obtain the all terms of the asymptotic expansion, and in particular, the leading term satisfying (4.4.55)-(4.4.72).

The justification as usual (see section 2.4.4) is developed as follows: we truncate the ansatz at some level J neglecting the terms of order $O(\mu^J)$ (in H^1-norm) and calculate all discrepancies in the right hand sides of the equation and eventually boundary conditions. We check that these discrepancies are of order $O(\mu^{J-1})$ in L^2-norm. Then we apply the a priori estimate and obtain that

$$\|u-u^{(J)}\|_{H^1(B_\mu)}=O(\mu^{J-1-q}).$$

Factor μ^{-q} appeared here due to the dependency of the Korn inequality constant on μ. This dependency could be made more precise ([75], [118]), but for our study the exact value of q is not too important because we have constructed the complete asymptotic expansion, for *any* J satisfying the equation with accuracy $O(\mu^{J-1})$.

Finally we improve this estimate comparing $u^{(J+q+1)}$ to $u^{(J)}$ and estimating the difference

$$\|u^{(J+q+1)}-u^{(J)}\|_{H^1(B_\mu)}=O(\mu^J).$$

On the other hand,

$$\|u^{(J+q+1)}-u\|_{H^1(B_\mu)}=O(\mu^J).$$

Hence, the estimate holds:

$$\|u-u^{(J)}\|_{H^1(B_\mu)}=O(\mu^J).$$

1. We seek a formal asymptotic solution of problem (4.4.1),(4.4.3) on any section S_e corresponding to the segment $e = e_h^\alpha$ of the set $\mathcal{B}$, of the form

$$\tilde{u}_e^{(\infty)}\sim\sum_{l=0}^{\infty}(\mu)^l N_l^e\Big(\frac{\tilde{x}'}{\mu}\Big)\frac{d^l\tilde{v}(\tilde{x}_1)}{d\tilde{x}_1^l}, \tag{4.4.4}$$

where $N_l^e(\tilde{\xi}')$ are the $s \times d$ matrix valued functions in Chapter 2 and $\tilde{v}(\tilde{x}_1)$ is a $d-$dimensional vector-valued function such that its components have the following expansions:

$$\tilde{v}_1(\tilde{x}_1) \sim \sum_{q=0}^{\infty} \mu^q v_1^q(\tilde{x}_1),$$

$$\tilde{v}_r(\tilde{x}_1) \sim \sum_{q=0}^{\infty} \mu^{q-2} v_r^q(\tilde{x}_1), \quad r = 2, \dots, s,$$

and in case $s = 3$ the fourth component is

$$\tilde{v}_4(\tilde{x}_1) \sim \sum_{q=0}^{\infty} \mu^{q-1} v_1^q(\tilde{x}_1). \tag{4.4.5}$$

Substituting (4.4.4) in (4.4.1) and (4.4.3) we find as in Chapter 2 that (4.4.3) is satisfied asymptotically exactly and that (4.4.1) takes form

$$\sum_{l=2}^{\infty} \mu^{l-2} h_l^N \frac{d^l \tilde{v}(\tilde{x}_1)}{d\tilde{x}_1^l} - \tilde{\mathbf{f}}^{\mathbf{e}} \sim 0, \tag{4.4.6}$$

where $\tilde{\mathbf{f}}^{\mathbf{e}}$ is a $d-$dimensional vector such that $\tilde{\mathbf{f}}^{\mathbf{e}} = f^e$ for $s = 2$ and $\tilde{\mathbf{f}}^{\mathbf{e}} = \begin{pmatrix} f^e \\ 0 \end{pmatrix}$ for $s = 3$. In what follows we omit the superscript e for f.

2. We expand now the constructed above at the first stage $\tilde{u}_e^{(\infty)}$ in a Taylor series in the neighborhood of any node x_0. We obtain

$$\tilde{u}_e^{(\infty)} \sim \sum_{r=0}^{\infty} (\mu)^r \bar{N}_r^e(\tilde{z}) \frac{d^r \tilde{v}(\tilde{x}_1(x_0))}{d\tilde{x}_1^r}, \tag{4.4.7}$$

where $\tilde{z} = \alpha_e z, \ z = (x - x_0)/\mu$,

$$\bar{N}_r^e(\tilde{z}) = \sum_{q=0}^{r} \frac{N_q^e(\tilde{z}')}{(r-q)!} (\tilde{z}_1)^q,$$

and $u_e^{(\infty)} = \alpha_e^* \tilde{u}_e^{(\infty)}$.

Here α_e stands for the matrix α of passage from the local (corresponding to the segment e) base to the global one.

Denote by $\Pi_{x_0}$the connected component of the set $B_\mu \backslash S_0$ containing the point x_0. Denote by $\Pi_{x_0,z,0}$ the image of Π_{x_0} under the transformation $z = (x - x_0)/\mu$. Let $\tilde{B}_{hj}^\alpha$ be a cylinder whose intersection with Π_{x_0} is nonempty. We extend it to a semi-infinite cylinder behind the basis which has no common points with Π_{x_0}. We denote by $B_{hj\infty}^\alpha$ this extended cylinder, and by $\Pi_{x_0\infty}$ the union of all such cylinders $B_{hj\infty}^\alpha$ and Π_{x_0}. Moreover, we denote by $\Pi_{x_0,z,\infty}$ the image of $\Pi_{x_0,\infty}$ under the transformation $z = (x - x_0)/\mu$.

Let $\tilde{\chi}_e(z)$ be the characteristic function of $B^{\alpha}_{hj\infty}$ corresponding to the segment e of $\mathcal{B}$.

In the neighborhood of the node x_0 we seek the solution in the form

$$u^{(\infty)}_{\Pi_{x_0}} \sim \sum_{l=0}^{\infty} \mu^l N^0_r(z) + \sum_{e(x_0)} \tilde{\chi}_e(z)\hat{\rho}(\alpha_e z)u^{(\infty)}_e, \tag{4.4.8}$$

where the summation of the last sum in (4.4.8) extends over all segments e in $\mathcal{B}$ having x_0 as an end-point, and $\hat{\rho}(t) = \hat{\rho}(t_1)$ (i.e. it depends only on the first component of the variable t in $I\!R^s$) is a differentiable function that vanishes for $|t_1| \leq c_0$, is equal to unity for $|t_1| \geq c_0 + 1$, and such that $0 \leq \hat{\rho}(t) \leq 1$.

Substituting (4.4.8) in (4.4.1), (4.4.3), we find the recurrent chain of problems determining functions N^0_l :

$$L_z N^0_l = \sum_{l_1=0}^{l} \sum_{e(x_0)} \bar{\chi}_e(z)(L_z((1-\hat{\rho}(\alpha_e z))\bar{N}^e_{l_1}(z))$$

$$-(1-\hat{\rho}(\alpha_e z))L_z\bar{N}^e_{l_1}(z))\frac{d^{l_1}}{d\tilde{x}_1^{l_1}}\tilde{v}^{e\ l-l_1}(\tilde{x}_1(x_0))$$

$$+\alpha^*_e \begin{pmatrix} \bar{\delta}_{l-2}\frac{\tilde{z}_1^{l-2}}{(l-2)!}(1-\hat{\rho}(\alpha_e z))\frac{d^{l-2}}{d\tilde{x}_1^{l-2}}\tilde{f}_1(x_0) \\ \bar{\delta}_{l-4}\frac{\tilde{z}_1^{l-4}}{(l-4)!}(1-\hat{\rho}(\alpha_e z))\frac{d^{l-4}}{d\tilde{x}_1^{l-4}}(\tilde{f}_2(x_0),\ldots,\tilde{f}_s(x_0))^* \end{pmatrix}, \quad z \in \Pi_{x_0,z,\infty}, \tag{4.4.9$_1$}$$

$$\frac{\partial N^0_l}{\partial \nu_z} = \sum_{l_1=0}^{l} \sum_{e(x_0)} \bar{\chi}_e(z)(L_z\frac{\partial}{\partial \nu_z}((1-\hat{\rho}(\alpha_e z))\bar{N}^e_{l_1}(z))$$

$$-(1-\hat{\rho}(\alpha_e z))\frac{\partial}{\partial \nu_z}\bar{N}^e_{l_1}(z))\frac{d^{l_1}}{d\tilde{x}_1^{l_1}}\tilde{v}^{e\ l-l_1}(\tilde{x}_1(x_0)), \quad z \in \partial\Pi_{x_0,z,\infty}, \tag{4.4.9$_2$}$$

where $\bar{\delta}_l = 1$ for $l \geq 0$ and $\bar{\delta}_l = 0$ for $l < 0$.

Denote respectively $F^{eq}_l(z)$ and $F^b_l(z)$ the right-hand side functions of equation (4.4.9$_1$) and boundary condition (4.4.9$_2$). Remark that these functions vanish outside of the ball of radius c_0. Let $\mathcal{J}$ be the linear span of columns of the matrix $\tilde{\mathcal{J}}$.

Lemma 4.4.2. *Problem* (4.4.9$_1$), (4.4.9$_2$) *is solvable in the class of vector-valued functions that stabilize exponentially as* $|z|$ *tends to infinity (in the sense of [130], [129]) to a vector-valued function in* $\mathcal{J}$ *if and only if*

$$\int_{\Pi_{x_0,z,\infty}} \tilde{\mathcal{J}}^*(z)F^{eq}_l(z)dz = \int_{\partial\Pi_{x_0,z,\infty}} \tilde{\mathcal{J}}^*(z)F^b_l(z)ds. \tag{4.4.10}$$

This stabilization (4.4.10) takes place so that

$$\int_{\Pi_{x_0,z,+\infty}\cap\{\sigma\leq z_1\leq\sigma+1\}} N_l(z) - j(z) \ \ dz \to 0,$$

where $j \in \mathcal{J}$.

Thus condition (4.4.10) implies (in case when $x_0 \notin \partial_1 B_\mu$) that

$$\sum_{e(x_0)} \mu \int_{\Pi_{x_0,z,\infty}} \bar{\chi}_e(z)(1-\hat{\rho}(\alpha_e z))\tilde{\mathcal{J}}^*(z) L_z u_e^{(\infty)} dz$$

$$- \sum_{e(x_0)} \int_{\partial\Pi_{x_0,z,0}\setminus\Pi_{x_0,z,\infty}} \bar{\chi}_e(z)(1-\hat{\rho}(\alpha_e z))\tilde{\mathcal{J}}^*(z) \frac{\partial}{\partial \nu_x} u_e^{(\infty)} dz$$

$$\sim \sum_{e(x_0)} \mu \int_{\Pi_{x_0,z,\infty}} \bar{\chi}_e(z)(1-\hat{\rho}(\alpha_e z))\tilde{\mathcal{J}}^*(z)\alpha_e^* \tilde{\mathbf{f}} dz.$$

Taking into account the fact that

$$L_x u_e^{(\infty)} \sim \alpha^* \sum_{l=2}^{\infty} \mu^{l-2} h_l^N \frac{d^l \tilde{v}(\tilde{x}_1)}{d\tilde{x}_1^l},$$

$$\frac{\partial}{\partial \nu_x} u_e^{(\infty)} \sim \alpha^* \sum_{l=1}^{\infty} \mu^{l-1} \left(\sum_{j=1}^{s} A_{1j} \frac{\partial N_l}{\partial \xi_j} + A_{11} N_{l-1} \right) \frac{d^l \tilde{v}(\tilde{x}_1)}{d\tilde{x}_1^l},$$

as well as the relation of Remark 2.3.2, we obtain

$$\sum_{e(x_0)} \Gamma_e \ \ meas\, \beta_e \ \ \Lambda_{\mu,e} \tilde{v}^e \ + \ \sum_{j=1}^{\infty} \mu^j \sum_{e(x_0)} \ \hat{\Lambda}_{\mu,e,j}(\tilde{v}^e, \tilde{\mathbf{f}}) \ \sim \ 0, \qquad (4.4.11)$$

where $\Lambda_{\mu,e}$ is the operator in (2.3.71) where the constant $\bar{\mathcal{M}}$ is replaced by $\frac{\bar{\mathcal{M}}}{a_e}$, and a_e is given by formula of section 2.3:

$$a = \left(\int_\beta \frac{(\xi_2^2 + \xi_3^2)\, d\xi_2 d\xi_3}{\text{mes}\, \beta} \right)^{-1/2};$$

and $\hat{\Lambda}_{\mu,e,j}(\tilde{v}^e, \tilde{\mathbf{f}})$ are some linear differential operators whose rows have the same order with respect to μ as those of $\Lambda_{\mu,e}$. We remind that Γ_e is the $d \times d$ matrix of the transformation

$$\tilde{\mathcal{J}}^*(z)\alpha_e^* \ = \ \Gamma_e \tilde{\mathcal{J}}^*(\alpha_e z)$$

of the matrix $\tilde{\mathcal{J}}$ and β_e the cross-section of the cylinder corresponding to e (see the definition of a finite rod structure). Denote

$$\hat{\Lambda}\tilde{v}^e \ = \ \sum_{e(x_0)} \Gamma_e \ \ meas\, \beta_e \ \ \Lambda_{\mu,e} \tilde{v}^e,$$

where $\mu = 1$.

If (4.4.11) is satisfied asymptotically exactly, then problems (4.4.8),(4.4.9) have solutions that stabilize exponentially to limit functions of $\mathcal{J}$ on every half-infinite cylinder as $|z| \to \infty$. We keep the notation $\tilde{\mathcal{J}}(z)C_e$ for these limit functions; here C_e is a $\bar{d}-$dimensional vector corresponding to the segment e; we also consider the vectors

$$V_e = \left(\tilde{v}_1^e, \tilde{v}_2^e, \mu\frac{\partial \tilde{v}_2^e}{\partial \tilde{x}_1}\right)^* \tag{4.4.12}$$

for $s = 2$, and

$$V_e = \left(\tilde{v}_1^e, \tilde{v}_2^e, \tilde{v}_3^e, \tilde{v}_4^e, \mu\frac{\partial \tilde{v}_2^e}{\partial \tilde{x}_1}, \mu\frac{\partial \tilde{v}_3^e}{\partial \tilde{x}_1}\right)^* \tag{4.4.13}$$

for $s = 3$.

As in section 4.3, from the representations (4.4.8),(4.4.9) it follows that if the value of V_e changes by δV_e, then the value of $\tilde{\mathcal{J}}(z)C_e$ in the sum $\sum_{l=0}^{\infty} \mu^l N_l^0(z)$ changes by $\alpha_e^* \tilde{\mathcal{J}}(\alpha_e z)C_e \delta V_e$, that is , C_e changes by $\Gamma_e \delta V_e$. Then we are left with the problem of finding δV_e such that

$$\Gamma_{e_1}(V_{e_1} + \delta V_{e_1}) - \Gamma_{e_2}(V_{e_2} + \delta V_{e_2}) \sim 0, \qquad \forall e_1(x_0), e_2(x_0), \tag{4.4.14}$$

and the redefined in such manner solutions N_l^0 stabilize to zero as $|z| \to \infty$.

If the node x_0 is an end-point for only one segment e and $x_0 \notin \partial_1 B_\mu$, then condition (4.4.14) is cancelled (the rigid displacement can be omitted from the solution). If $x_0 \in \partial_1 B_\mu$, then on the part of $\partial \Pi_{x_0,z,\infty}$ corresponding to $\partial_1 B_\mu$, under change of variable $z = (x - x_0)/\mu$ we prescribe the condition $N_l^0 = 0$ instead of boundary condition (4.4.9$_2$) on this part of the boundary. Then the analogue of Lemma 4.4.2 holds for this problem, but the solvability condition (4.4.10)is no long necessary. The solution stabilizes to a rigid displacement $\tilde{\mathcal{J}}C_e$ so the condition at the node x_0 takes the form

$$\Gamma_e(V_e + \delta V_e) \sim 0, \tag{4.4.15}$$

where δV_e is chosen so that the solution stabilizes to zero as in the previous section.

Seeking a formal asymptotic solution of problem (4.4.6), (4.4.11), (4.4.14), (4.4.15) in the form (4.4.5) of a power series in μ, we find that , as in Chapter 2, all the v_e^l are determined from the recursive sequence of equations of the form

$$\mathcal{R}\tilde{v}^{e\ l} = \tilde{f}_l(\tilde{x}_1), \quad x \in e, \tag{4.4.16}$$

where $\mathcal{R}$ is an operator of (2.3.69) with $\mu = 1$; these equations are supplied by the matching conditions at the nodes x_0 that are end-points for at least two segments (and $x_0 \notin \partial_1 B_\mu$)

$$\hat{\Lambda}\tilde{v}^{e\ l} = \Psi_l(x_0), \tag{4.4.17}$$

and

$$\Gamma_{e_1}(V_{e_1}^l + \delta V_{e_1}^l) = \Gamma_{e_2}(V_{e_2}^l + \delta V_{e_2}^l), \qquad \forall e_1(x_0), e_2(x_0), \tag{4.4.18}$$

and boundary conditions at the nodes x_0 in $\partial_1 B_\mu$ or in $\partial_2 B_\mu$, but such that they are end-points for only one segment, of the form

$$\hat{\Lambda}\tilde{v}^{e\ l} = \Psi_l(x_0), \quad x_0 \in \partial_1 B_\mu \tag{4.4.19}$$

$$V_e^l + \delta V_e^l = 0, \quad x_0 \in \partial_2 B_\mu, \tag{4.4.20}$$

where

$$V_e^l = \begin{cases} \left(\tilde{v}_1^{e\ l}, \tilde{v}_2^{e\ l+2}, \dfrac{\partial \tilde{v}_2^{e\ l+1}}{\partial \tilde{x}_1}\right)^*, & s = 2, \\ \left(\tilde{v}_1^{e\ l}, \tilde{v}_2^{e\ l+2}, \tilde{v}_3^{e\ l+2}, \tilde{v}_4^{e\ l+1}, \dfrac{\partial \tilde{v}_2^{e\ l+1}}{\partial \tilde{x}_1}, \dfrac{\partial \tilde{v}_3^{e\ l+1}}{\partial \tilde{x}_1}\right)^*, & s = 3, \end{cases}$$

here $\tilde{f}_l$ and Ψ_l are defined in terms of the functions $\tilde{v}^{e\ l_1}$ with $l_1 < l, \tilde{f}_0 = (\tilde{f}_1, 0, \ldots, 0)$, and $\Psi_0 = 0$. The constant vectors δV_e^l are defined in terms of the solution N_l^0 of problems (4.4.9), found under the assumption that $\delta V_e^l = 0$ in (4.4.18) and (4.4.20); afterwards $N_l^0(z)$ are redefined so that they stabilize to zero as $|z| \to +\infty$ (as above).

Indeed, for any l, we solve first problem (4.4.9) for N_l^0 and we obtain different limit rigid displacements $\mathcal{J}C_e^l$ on the outlets (the subscript e means that the outlet corresponds to the segment e of $\mathcal{B}$); we redefine then $N_l^0(z)$ replacing V_e^l at the end of the segment e in (4.4.9) by $V_e^l + \delta V_e^l$ where δV_e^l is such that $\Gamma_e \delta V_e^l = -C_e^l$; this replacement leads to stabilization of redefined $N_l^0(z)$ to zero. So after such redefinition we have N_l^0 tending to zero on the outlets and we obtain conditions (4.4.18) and (4.4.20) for V_e^l.

Problems (4.4.16)-(4.4.20) split into a pair of problems, for $\tilde{v}_1^l$ and for $(\tilde{v}_2^l, \ldots, \tilde{v}_d^l)^*$:

(i) for any l we solve first a problem for an $s-$dimensional vector $\tilde{w}^l(\tilde{x}_1)$ (denoted $\tilde{w}^{e\ l}(\tilde{x}_1)$ on each segment $e \subset \mathcal{B}$) consisting of

-the equation

$$\bar{E}\frac{d^2\tilde{w}_1^{e\ l}(\tilde{x}_1)}{d\tilde{x}_1^2} = \tilde{f}_{l,1}(\tilde{x}_1), \quad x \in e, \tag{4.4.21}$$

-the conditions

$$\bar{E}\sum_{e(x_0)} meas\ \beta_e \gamma_{\mathbf{e}} \frac{d\tilde{w}_1^{e\ l}(\tilde{x}_1)}{d\tilde{x}_1} = \mathbf{\Psi_1}(x_0) \tag{4.4.22}$$

at all nodes except $x_0 \in \partial_1 B_\mu$, where $\gamma_{\mathbf{e}}$ is a director vector of segment e with initial point x_0;

-the condition

$$\tilde{w}^{e\ l} = -(\delta V_{e1}^l, \delta V_{e2}^{l-2}, ..., \delta V_{es}^{l-2}), \tag{4.4.23}$$

at all nodes $x_0 \in \partial_1 B_\mu$, and

-the condition

$$\alpha^*_{e_1} \tilde{w}^{e_1\ l} = \alpha^*_{e_2} \tilde{w}^{e_2\ l} - \alpha^*_{e_1} \delta V^l_{e_1} + \alpha^*_{e_2} \delta V^l_{e_2} \tag{4.4.24}$$

at all nodes x_0 which are common points for two different segments $e_1(x_0)$ and $e_2(x_0)$; here $\mathbf{\Psi_l}(x_0)$ and δV^l_e are given constants, $\tilde{f}_{l,1}$ is a given right-hand side function; all components of the vector valued function $\tilde{w}^{e\ l}$ (except of $\tilde{w}^{e\ l}_1$) are linear functions;

(ii) secondly we solve a problem for a $(d-1)-$dimensional vector $\tilde{W}^l(\tilde{x}_1)$ (denoted $\tilde{W}^{e\ l}(\tilde{x}_1)$ on each segment $e \subset \mathcal{B}$)

-for $s = 2$ consisting of

-the equation

$$-\bar{E} < \xi_2^2 >_\beta \frac{d^4 \tilde{W}^{e\ l}(\tilde{x}_1)}{d\tilde{x}_1^4} = \tilde{f}_{l-2,2}(\tilde{x}_1), \quad x \in e, \tag{4.4.25}$$

-the matching condition

$$\bar{E} \sum_{e(x_0)} meas\ \beta_e < \xi_2^2 >_\beta \frac{d^2 \tilde{W}^{e\ l}(\tilde{x}_1)}{d\tilde{x}_1^2} = \mathbf{\Psi_l}(x_0) \tag{4.4.26}$$

at all nodes except $x_0 \in \partial_1 B_\mu$,

-the condition

$$\frac{d\tilde{W}^{e_1\ l}(\tilde{x}_1)}{d\tilde{x}_1} = \frac{d\tilde{W}^{e_2\ l}(\tilde{x}_1)}{d\tilde{x}_1} + \eta^l(x_0), \tag{4.4.27}$$

at all nodes x_0 which are common points for two different segments $e_1(x_0)$ and $e_2(x_0)$;

-the condition

$$\tilde{W}^{e_2\ l} = (\tilde{w}^{e_1\ l-2})_{ort\ e_2} + \xi(x_0), \tag{4.4.28}$$

for any $e_2(x_0)$ and for some $e_1(x_0)$ non-collinear to $e_2(x_0)$, here $(\tilde{w}^{e_1\ l-2})_{ort\ e_2}$is the orthogonal projection of the vector $\tilde{w}^{e_1\ l-2}\gamma_{\mathbf{e}}$ on the axis orthogonal to e_2; $\eta^l(x_0), \xi(x_0)$ are given constants;

for $s = 3$, this problem consists of the equations

$$\bar{\mathcal{M}} \frac{d^2 \tilde{W}^{e\ l}(\tilde{x}_1)}{d\tilde{x}_1^2} = 0, \quad x \in e, \tag{4.4.29}$$

$$-\bar{E} < \xi_2^2 >_\beta \frac{d^4 \tilde{W}_2^{e\ l}(\tilde{x}_1)}{d\tilde{x}_1^4} = \tilde{f}_{l-2,2}(\tilde{x}_1), \quad x \in e, \tag{4.4.30}$$

$$-\bar{E} < \xi_3^2 >_\beta \frac{d^4 \tilde{W}_3^{e\ l}(\tilde{x}_1)}{d\tilde{x}_1^4} = \tilde{f}_{l-2,3}(\tilde{x}_1), \quad x \in e, \tag{4.4.31}$$

-the matching conditions

$$\sum_{e(x_0)} \bar{\Gamma}_e \bar{\bar{\Lambda}}_e \tilde{W}^{e\ l} = \Psi^l(x_0), \tag{4.4.32}$$

$$\tilde{W}_2^{e_2\ l} = (\tilde{w}^{e_1\ l-2})_{ort,2} + \xi_2(x_0), \tag{4.4.33}$$

for any $e_2(x_0)$ and for some $e_1(x_0)$ such that its projection $(e_1(x_0))_{ort,2}$ on the local axis $\tilde{x}_2^{e_2}$ is non-zero;

$$\tilde{W}_3^{e_2\ l} = (\tilde{w}^{e_1\ l-2})_{ort,3} + \xi_3(x_0), \tag{4.4.34}$$

for any $e_2(x_0)$ and for some $e_1(x_0)$ such that its projection $(e_1(x_0))_{ort,3}$ on the local axis $\tilde{x}_3^{e_2}$ is non-zero;

$$\bar{\Gamma}_{e_1}(\tilde{W}_1^{e_1\ l}, \frac{d\tilde{W}_2^{e_1\ l}(\tilde{x}_1)}{d\tilde{x}_1}, \frac{d\tilde{W}_3^{e_1\ l}(\tilde{x}_1)}{d\tilde{x}_1})^*$$

$$= \bar{\Gamma}_{e_2}(\tilde{W}_1^{e_2\ l}, \frac{d\tilde{W}_2^{e_2\ l}(\tilde{x}_1)}{d\tilde{x}_1}, \frac{d\tilde{W}_3^{e_2\ l}(\tilde{x}_1)}{d\tilde{x}_1})^* + \eta(x_0),\ \forall e_1(x_0), e_2(x_0), \tag{4.4.35}$$

at nodes x_0 having at least two initial vectors $e_1(x_0), e_2(x_0)$; here

$$\bar{\bar{\Lambda}}_e \tilde{W} = meas\ \beta_e(\frac{\bar{\mathcal{M}}}{a_e}\frac{d\tilde{W}_1^{e\ l}(\tilde{x}_1)}{d\tilde{x}_1}, \bar{E} < \xi_2^2 >_\beta \frac{d^2\tilde{W}_2^{e\ l}(\tilde{x}_1)}{d\tilde{x}_1^2}, \bar{E} < \xi_3^2 >_\beta \frac{d^2\tilde{W}_3^{e\ l}(\tilde{x}_1)}{d\tilde{x}_1^2}), \tag{4.4.36}$$

-and the boundary conditions

$$\tilde{W}^{e\ l}(\tilde{x}_1) = \xi^{\mathbf{l}}(\mathbf{x_0}),\quad \frac{d\tilde{W}_2^e}{d\tilde{x}_1^2} = \eta_2^l(x_0),\quad \frac{d\tilde{W}_3^e}{d\tilde{x}_1} = \eta_3^l(x_0),\quad x_0 \in \partial_1 B_\mu; \tag{4.4.37}$$

-and

$$\bar{\Gamma}_e \bar{\bar{\Lambda}}_e \tilde{W}^{e\ l} = \Psi^0(x_0); \tag{4.4.38}$$

at all nodes having only one initial segment, $x_0 \notin \partial_1 B_\mu$.

Above the functions and constants in right-hand sides are defined by the previous terms of asymptotic expansion with scripts less than l.

We compile then $\tilde{v}^{el} = \tilde{w}^{el} + (0, w^{el})^*$ for $s = 2$ and $\tilde{v}^{el} = (\tilde{w}^{el}, 0)^* + (0, w^{el})^*$ for $s = 3$.

These problems, in turn, could be reduced to the problems with homogeneous matching and boundary conditions (by subtracting of function satisfying to non-homogeneous conditions from the unknown function), i.e. to

1) a problem for an $s-$dimensional vector $\tilde{\nu}^e(\tilde{x}_1)$, consisting of the equation

$$\bar{E}\frac{d^2\tilde{\nu}_1^e(\tilde{x}_1)}{d\tilde{x}_1^2} = \tilde{g}(\tilde{x}_1),\quad x \in e, \tag{4.4.39}$$

the conditions

$$\bar{E} \sum_{e(x_0)} meas\ \beta_e \gamma_{\mathbf{e}} \frac{d\tilde{\nu}^e(\tilde{x}_1)}{d\tilde{x}_1} = 0, \tag{4.4.40}$$

at all nodes except $x_0 \in \partial_1 B_\mu$,
-the condition

$$\tilde{\nu} = 0 \tag{4.4.41}$$

at all nodes $x_0 \in \partial_1 B_\mu$, and
-the condition

$$\alpha^*_{e_1} \tilde{\nu}^{e_1} = \alpha^*_{e_2} \tilde{\nu}^{e_2} \tag{4.4.42}$$

at all nodes x_0 which are common points for two different segments $e_1(x_0)$ and $e_2(x_0)$; all components of the vector valued function $\tilde{\nu}^e$ (except of $\tilde{\nu}^e_1$) are linear functions on every e;

2) a problem for a $(d-1)-$dimensional vector $\tilde{\nu}^e(\tilde{x}_1)$, which for $s = 2$ consists of the equations

$$-\bar{E} < \xi_2^2 >_\beta \frac{d^4\tilde{\nu}^e(\tilde{x}_1)}{d\tilde{x}_1^4} = \tilde{g}(\tilde{x}_1), \quad x \in e, \tag{4.4.43}$$

-the matching condition

$$\bar{E} \sum_{e(x_0)} meas\ \beta_e < \xi_2^2 >_\beta \frac{d^2\tilde{\nu}^e(\tilde{x}_1)}{d\tilde{x}_1^2} = 0, \quad \tilde{\nu} = 0, \tag{4.4.44}$$

at all nodes except $x_0 \in \partial_1 B_\mu$,
-the condition

$$\frac{d\tilde{\nu}^{e_1}(\tilde{x}_1)}{d\tilde{x}_1} = \frac{d\tilde{\nu}^{e_2}(\tilde{x}_1)}{d\tilde{x}_1}, \tag{4.4.45}$$

at all nodes x_0 which are common points for two different segments $e_1(x_0)$ and $e_2(x_0)$;
-the condition

$$\tilde{\nu}^e = 0, \quad \frac{d\tilde{\nu}^e(\tilde{x}_1)}{d\tilde{x}_1} = 0, \tag{4.4.46}$$

at all nodes $x_0 \in \partial_1 B_\mu$, and

$$\bar{E} meas\ \beta_e < \xi_2^2 >_\beta \frac{d^2\tilde{\nu}^e(\tilde{x}_1)}{d\tilde{x}_1^2} = 0, \tag{4.4.47}$$

at all nodes having only one initial segment $e(x_0)$ except $x_0 \in \partial_1 B_\mu$;
for $s = 3$, this problem consists of the equations

$$\bar{\mathcal{M}} \frac{d^2\tilde{\nu}^e_1(\tilde{x}_1)}{d\tilde{x}_1^2} = \tilde{g}_1(\tilde{x}_1), \quad x \in e, \tag{4.4.48}$$

$$-\bar{E} < \xi_2^2 >_\beta \frac{d^4\tilde{\nu}_2^e(\tilde{x}_1)}{d\tilde{x}_1^4} = \tilde{g}_2(\tilde{x}_1), \quad x \in e, \tag{4.4.49}$$

$$-\bar{E} < \xi_3^2 >_\beta \frac{d^4\tilde{\nu}_3^e(\tilde{x}_1)}{d\tilde{x}_1^4} = \tilde{g}_3(\tilde{x}_1), \quad x \in e, \tag{4.4.50}$$

-the matching conditions

$$\sum_{e(x_0)} \bar{\Gamma}_e \bar{\bar{\Lambda}}_e \tilde{\nu}^e = 0, \quad \tilde{\nu}_2^e = 0, \quad \tilde{\nu}_3^e = 0, \tag{4.4.51}$$

$$\bar{\Gamma}_{e_1}(\tilde{\nu}_1^{e_1}, \frac{d\tilde{\nu}_2^{e_1}(\tilde{x}_1)}{d\tilde{x}_1}, \frac{d\tilde{\nu}_3^{e_1}(\tilde{x}_1)}{d\tilde{x}_1})^*$$

$$= \bar{\Gamma}_{e_2}(\tilde{\nu}_1^{e_2}, \frac{d\tilde{\nu}_2^{e_2}(\tilde{x}_1)}{d\tilde{x}_1}, \frac{d\tilde{\nu}_3^{e_2}(\tilde{x}_1)}{d\tilde{x}_1})^*, \ \forall e_1(x_0), e_2(x_0), \tag{4.4.52}$$

at nodes x_0 having at least two initial vectors $e_1(x_0), e_2(x_0)$; and
the boundary conditions

$$\tilde{\nu}^e = 0, \quad \frac{d\tilde{\nu}_2^e(\tilde{x}_1)}{d\tilde{x}_1} = 0, \quad \frac{d\tilde{\nu}_3^e(\tilde{x}_1)}{d\tilde{x}_1} = 0, \tag{4.4.53}$$

at all nodes $x_0 \in \partial_1 B_\mu$, and

$$\bar{\Gamma}_e \bar{\bar{\Lambda}}_e \tilde{\nu}^e = 0, \tag{4.4.54}$$

at all nodes having only one initial segment, $x_0 \notin \partial_1 B_\mu$.

Let conditions PF_1 and PF_2 be satisfied. Denote $\mathcal{H}^1(\mathcal{B})$ and $\mathcal{H}^{1,2,2}(\mathcal{B})$ the spaces of functions described in conditions PF_1 and PF_2 respectively. Denote $\mathcal{H}^2(\mathcal{B})$ the space of functions described in Remark 4.4.1.

Lemma 4.4.2 *If conditions PF_1 and PF_2 hold then problem (4.4.39)-(4.4.42) and problem (4.5.43)-(4.4.47) if $s = 2$, (respectively (4.4.48)-(4.4.54) if $s = 3$) are solvable.*

Proof.

Indeed, problem (4.4.39)-(4.4.42) is equivalent to the following variational formulation:

$$-\int_{\mathcal{B}} meas\ \beta_e \bar{E} \frac{d\tilde{\nu}_1^e(\tilde{x}_1)}{d\tilde{x}_1} \frac{d\tilde{\psi}_1^e(\tilde{x}_1)}{d\tilde{x}_1} d\tilde{x}_1$$

$$= \int_{\mathcal{B}} meas\ \beta_e \tilde{g}_1(\tilde{x}_1) \tilde{\psi}_1^e(\tilde{x}_1) d\tilde{x}_1, \quad \forall \tilde{\psi}_1^e \in \mathcal{H}^1(\mathcal{B}).$$

For $s = 2$, problem (4.4.43)-(4.4.47) is equivalent to the following variational formulation:

$$\int_{\mathcal{B}} meas\ \beta_e < \xi_2^2 >_\beta \bar{E} \frac{d^2\tilde{\nu}^e(\tilde{x}_1)}{d\tilde{x}_1^2} \frac{d^2\tilde{\psi}^e(\tilde{x}_1)}{d\tilde{x}_1^2} d\tilde{x}_1$$

$$= \int_{\mathcal{B}} meas\ \beta_e \tilde{g}(\tilde{x}_1)\tilde{\psi}^e(\tilde{x}_1)d\tilde{x}_1, \quad \forall \tilde{\psi}^e \in \mathcal{H}^2(\mathcal{B}).$$

For $s = 3$ problem (4.4.48)-(4.4.54) is stated as

$$-\int_{\mathcal{B}} meas\ \beta_e \bar{\mathcal{M}} \frac{d\tilde{\nu}_1^e(\tilde{x}_1)}{d\tilde{x}_1}\frac{d\tilde{\psi}_1^e(\tilde{x}_1)}{d\tilde{x}_1}d\tilde{x}_1$$

$$= \int_{\mathcal{B}} meas\ \beta_e \tilde{g}_1(\tilde{x}_1)\tilde{\psi}_1^e(\tilde{x}_1)d\tilde{x}_1,$$

$$\int_{\mathcal{B}} meas\ \beta_e < \xi_2^2 >_\beta \bar{E} \frac{d^2\tilde{\nu}_2^e(\tilde{x}_1)}{d\tilde{x}_1^2}\frac{d^2\tilde{\psi}_2^e(\tilde{x}_1)}{d\tilde{x}_1^2}d\tilde{x}_1$$

$$= \int_{\mathcal{B}} meas\ \beta_e \tilde{g}_2(\tilde{x}_1)\tilde{\psi}^e(\tilde{x}_1)d\tilde{x}_1,$$

$$\int_{\mathcal{B}} meas\ \beta_e < \xi_3^2 >_\beta \bar{E} \frac{d^2\tilde{\nu}_3^e(\tilde{x}_1)}{d\tilde{x}_1^2}\frac{d^2\tilde{\psi}_3^e(\tilde{x}_1)}{d\tilde{x}_1^2}d\tilde{x}_1$$

$$= \int_{\mathcal{B}} meas\ \beta_e \tilde{g}_3(\tilde{x}_1)\tilde{\psi}^e(\tilde{x}_1)d\tilde{x}_1, \quad \forall (\tilde{\psi}_1^e, \tilde{\psi}_2^e, \tilde{\psi}_3^e) \in \mathcal{H}^{1,2,2}(\mathcal{B}).$$

The unique solvability of these variational problems follows from the Riesz functional representation theorem for a bounded (due to PF_1 and PF_2 conditions) linear functional on a Hilbert space.

We denote by $u^{(J)}$ the truncated (partial) sum of the asymptotic series (4.4.4),(4.4.5),(4.4.8) obtained by neglecting the terms of order $O(\mu^J)$ (in the $H^1(B_\mu)$−norm. Choosing J sufficiently large, we can show for any K that equation (4.4.1) and conditions (4.4.2) and (4.4.3) are satisfied with remainders of order $O(\mu^K)$ (in the $L^2(B_\mu)$−norm. Hence , using the a priori estimate for problem (4.4.1)-(4.4.3) given by Theorem 4.4.3, we obtain as in section 2.2, the following assertion.

Theorem 4.4.5 *There holds the estimate*

$$\|u - \tilde{u}^{(J)}\|_{H^1(B_\mu)} = O(\mu^J).$$

More general cases are studied in [119] (non-symmetrical section of a bar, when the conditions of the very beginning of the present section are not respected) and in [121] (the two-dimensional case).

4.4.2 The leading term of asymptotic expansion

Thus the limit problem for the leading term v_0 of the asymptotic expansion is as follows.

For $s = 2$, we define a vector-valued function $v_0^e = (v_{01}^e, v_{02}^e)$ for every segment e; the first component v_{01}^e is a solution to the equation

$$\bar{E}\frac{d^2\tilde{v}^e_{01}(\tilde{x}_1)}{d\tilde{x}_1^2} = \tilde{f}^e_1(\tilde{x}_1), \quad x \in e, \tag{4.4.55}$$

with the matching condition

$$\bar{E}\sum_{e(x_0)} meas\ \beta_e\gamma_{\mathbf{e}}\frac{d\tilde{v}^e_{01}(\tilde{x}_1)}{d\tilde{x}_1} = 0, \quad v^{e_1}_{01} = v^{e_2}_{01} \tag{4.4.56}$$

at all nodes x_0 initial at least for two segments e_1 and e_2, except $x_0 \in \partial_1 B_\mu$, and with the condition

$$\tilde{v}^e_{01} = 0 \tag{4.4.57}$$

at all nodes $x_0 \in \partial_1 B_\mu$, and the condition

$$\bar{E}meas\ \beta_e\gamma_{\mathbf{e}}\frac{d\tilde{v}^e_{01}(\tilde{x}_1)}{d\tilde{x}_1} = 0, \tag{4.4.58}$$

for the nodes initial for the only one segment e;

v^e_{02} is a solution to the equation

$$-\bar{E} < \xi_2^2 >_\beta \frac{d^4\tilde{v}^e_{02}(\tilde{x}_1)}{d\tilde{x}_1^4} = \tilde{f}^e_2(\tilde{x}_1), \quad x \in e, \tag{4.4.59}$$

-the matching condition

$$\bar{E}\sum_{e(x_0)} meas\ \beta_e < \xi_2^2 >_\beta \frac{d^2\tilde{v}^e_{02}(\tilde{x}_1)}{d\tilde{x}_1^2} = 0,$$

$$\frac{d\tilde{v}^{e_1}_{02}(\tilde{x}_1)}{d\tilde{x}_1} = \frac{d\tilde{v}^{e_2}_{02}(\tilde{x}_1)}{d\tilde{x}_1}, \quad \tilde{v}^e_{02} = 0, \tag{4.4.60}$$

at all nodes x_0 which are common points for two different segments $e_1(x_0)$ and $e_2(x_0)$;

-the condition

$$\tilde{v}^e_{02} = 0, \quad \frac{d\tilde{\nu}^e(\tilde{x}_1)}{d\tilde{x}_1} = 0, \tag{4.4.61}$$

at all nodes $x_0 \in \partial_1 B_\mu$, and

$$\bar{E}meas\ \beta_e < \xi_2^2 >_\beta \frac{d^2\tilde{\nu}^e(\tilde{x}_1)}{d\tilde{x}_1^2} = 0, \quad \tilde{v}^e_{02} = 0, \tag{4.4.62}$$

at all nodes having only one initial segment $e(x_0)$ except $x_0 \in \partial_1 B_\mu$;

Extend v^e_0 onto B_μ in a following way: for any connected component of S_0 we define $v(x) = \alpha^*_e v^e_0(\tilde{x}_1(x))$, and for any connected component Π_{x_0} of $B_\mu \backslash S_0$, containing x_0, we define $v(x) = \alpha^*_e v^e_0(\tilde{x}_1(x_0))$, as a constant. Then the estimate of Theorem 4.4.5 yields

$$\|u - v\|_{L^2(B_\mu)}/\sqrt{meas\ B_\mu} = O(\sqrt{\mu}).$$

For $s = 3$, we define a vector-valued function $v_0^e = (v_{01}^e, v_{02}^e, v_{03}^e, v_{04}^e)$, for every segment e; the first component v_{01}^e is a solution to the equation

$$\bar{E}\frac{d^2\tilde{v}_{01}^e(\tilde{x}_1)}{d\tilde{x}_1^2} = \tilde{f}_1^e(\tilde{x}_1), \quad x \in e, \tag{4.4.63}$$

with the matching condition

$$\bar{E}\sum_{e(x_0)} meas\ \beta_e\gamma_{\mathbf{e}}\frac{d\tilde{v}_{01}^e(\tilde{x}_1)}{d\tilde{x}_1} = 0, \tag{4.4.64}$$

at all nodes x_0 initial at least for two segments e_1 and e_2, except $x_0 \in \partial_1 B_\mu$, and with the condition

$$\tilde{v}_{01} = 0 \tag{4.4.64}$$

at all nodes $x_0 \in \partial_1 B_\mu$, and the condition

$$\bar{E}meas\ \beta_e\gamma_{\mathbf{e}}\frac{d\tilde{v}_{01}^e(\tilde{x}_1)}{d\tilde{x}_1} = 0, \tag{4.4.65}$$

for the nodes initial for the only one segment e;
$(v_{04}^e, v_{02}^e, v_{03}^e)$ is a solution to the problem

$$\bar{\mathcal{M}}\frac{d^2\tilde{v}_{04}^e(\tilde{x}_1)}{d\tilde{x}_1^2} = 0, \quad x \in e, \tag{4.4.66}$$

$$-\bar{E} < \xi_2^2 >_\beta \frac{d^4\tilde{v}_{02}^e(\tilde{x}_1)}{d\tilde{x}_1^4} = \tilde{f}_2^e(\tilde{x}_1), \quad x \in e, \tag{4.4.67}$$

$$-\bar{E} < \xi_3^2 >_\beta \frac{d^4\tilde{v}_{03}^e(\tilde{x}_1)}{d\tilde{x}_1^4} = \tilde{f}_3^e(\tilde{x}_1), \quad x \in e, \tag{4.4.68}$$

-the matching conditions

$$\sum_{e(x_0)} \bar{\Gamma}_e\bar{\bar{\Lambda}}_e(v_{04}^e, v_{02}^e, v_{03}^e)^* = 0, \quad \tilde{v}_{02}^e = 0, \quad \tilde{v}_{03}^e = 0, \tag{4.4.69}$$

at all interior nodes x_0, and
-the matching conditions

$$\bar{\Gamma}_{e_1}(\tilde{v}_{04}^{e_1}, \frac{d\tilde{v}_{02}^{e_1}(\tilde{x}_1)}{d\tilde{x}_1}, \frac{d\tilde{v}_{03}^{e_1}(\tilde{x}_1)}{d\tilde{x}_1})^*$$

$$= \bar{\Gamma}_{e_2}(\tilde{v}_{04}^{e_2}, \frac{d\tilde{v}_{02}^{e_2}(\tilde{x}_1)}{d\tilde{x}_1}, \frac{d\tilde{v}_{03}^{e_2}(\tilde{x}_1)}{d\tilde{x}_1})^*, \ \forall e_1(x_0), e_2(x_0), \tag{4.4.70}$$

at nodes x_0 having at least two initial vectors $e_1(x_0), e_2(x_0)$; and

-the boundary conditions

$$\tilde{v}^e_{04} = 0, \quad \tilde{v}^e_{02} = 0, \quad \frac{d\tilde{v}^e_{02}(\tilde{x}_1)}{d\tilde{x}_1} = 0, \quad \tilde{v}^e_{03} = 0, \quad \frac{d\tilde{v}^e_{03}(\tilde{x}_1)}{d\tilde{x}_1} = 0, \tag{4.4.71}$$

at all nodes $x_0 \in \partial_1 B_\mu$, and

$$\tilde{v}^e_{02} = 0, \quad \tilde{v}^e_{03} = 0, \quad \bar{\Gamma}_e \bar{\bar{\Lambda}}_e (v^e_{04}, v^e_{02}, v^e_{03})^* = 0, \tag{4.4.72}$$

at all nodes having only one initial segment, $x_0 \notin \partial_1 B_\mu$.

Extend v^e_0 onto B_μ in a following way: for any connected component of S_0 we define

$$v(x) = \alpha^*_e \{ (v^e_{01}(\tilde{x}_1(x)), v^e_{02}(\tilde{x}_1(x)), v^e_{03}(\tilde{x}_1(x)))^* + \mu^{-1}(0, -\tilde{x}_3(x), \tilde{x}_2(x)) v^e_{04}(\tilde{x}_1(x)) \},$$

and for any connected component Π_{x_0} of $B_\mu \backslash S_0$, containing x_0, we define $v(x) = \alpha^*_e v^e_0(\tilde{x}_1(x_0))$, as a constant. Then the estimate of Theorem 4.5.5 yields

$$\|u - v\|_{L^2(B_\mu)} / \sqrt{meas\ B_\mu} = O(\sqrt{\mu}).$$

4.5 Flows in tube structures

In case of flows in thin domains like finite rod structures these domains are called here below tube structures. The Navier-Stokes problem stated in tube structures (or finite rod structures), i.e. in connected finite unions of the thin cylinders with the ratio of the diameter to the height of the order $\mu << 1$, is considered. Such problems arise in the blood circulation modelling. The asymptotic expansion of the solution is built and justified. Boundary layers are studied. The Navier-Stokes problem in one thin cylindrical domain was considered in [113].

4.5.1 Definitions. One bundle structure

In this section we are going to construct the asymptotic expansion to the solution of the Navier-Stokes problem, stated in a tube structure containing one bundle. We shall justify the error estimate . First we consider the case of right hand side functions concentrated in some neighborhoods of the nodes and then we generalize our construction for the right hand side functions which do not vanish inside of the tubes. Let us define the tube structure containing one bundle. It is the same type of domains as finite rod structure but with more smooth boundary. We consider here two possible dimensions of the space: two and three.

Let $e_1, \dots, e_n$ be n closed segments in $I\!R^s$ $(s = 2, 3)$, which have a single common point O (i.e. the origin of the co-ordinate system), and let it be the common end point of all these segments. Let $\beta_1, ..., \beta_J$ be n bounded (s-1)-dimensional domains in $I\!R^3$, which belong to n hyper-planes containing the point O. Let β_j be orthogonal to e_j. Let β_j^μ be the image of β_j obtained by a homothetic contraction in $1/\mu$ times with the center O. Denote B_j^μ the open cylinders with the bases β_j^μ and with the heights e_j, denote also $\hat{\beta}_j^\mu$ the second base of each cylinder B_j^μ and let O_j be the end of the segment e_j which belongs to the base $\hat{\beta}_j^\mu$. Define the bundle of segments e_j centered in O as

$$B \;=\; \cup_{j=1}^n \; e_j.$$

Denote below $O_0 = O$. Let γ_j^μ, $j = 0, 1, ..., n$, be the images of the bounded domains γ_j, (such that $\bar{\gamma}_j$ contain the ends of the segments O_j and independent of μ) obtained by a homothetic contraction in $1/\mu$ times with the center O_j. Define the tube structure associated with the bundle B as

$$B^\mu \;=\; (\cup_{j=1}^n \; B_j^\mu) \;\cup\; (\cup_{j=0}^n \; \gamma_j^\mu).$$

We suppose it be a domain with C^2-smooth boundary ∂B^μ and we assume that the bases $\hat{\beta}_j^\mu$ of the cylinders B_j^μ $j = 1, ..., n$, are some parts of ∂B^μ. We add the domains γ_j^μ, $j = 0, 1, ..., n$, to make the boundary of the tube structure C^2- smooth surface.

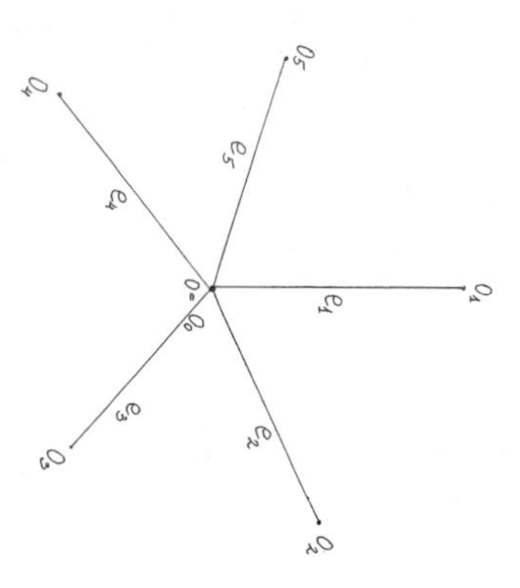

Figure 4.5.1 .a) A bundle of segments .

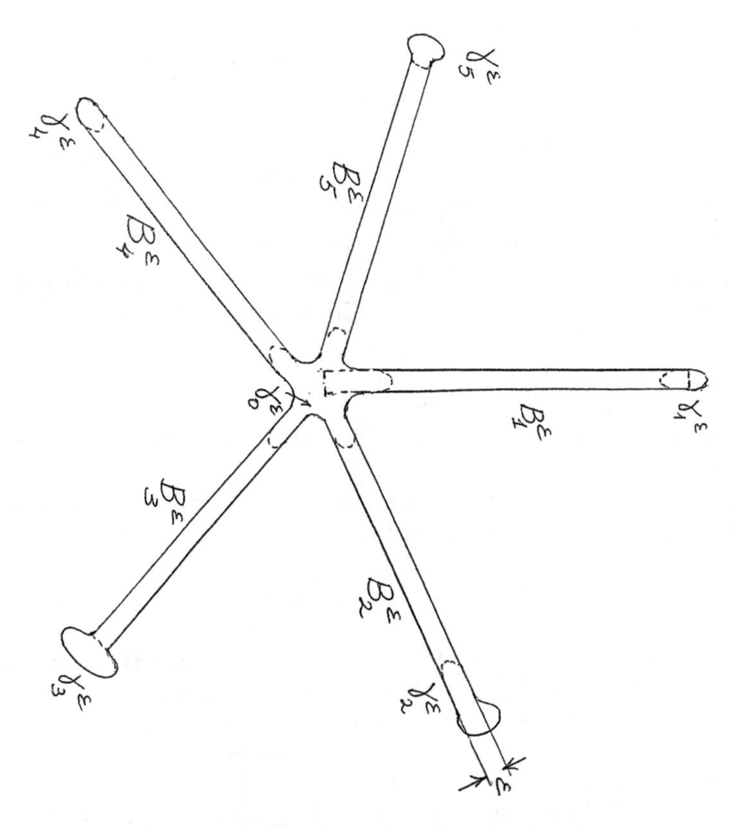

Figure 4.5.1b). A one bundle tube structure.

Consider the Navier-Stokes system of equations

$$\nu \Delta u_\mu - (u_\mu, \nabla) u_\mu - \nabla p_\mu = f, \tag{4.5.1}$$

$$div\ u_\mu = 0, \quad x \in B^\mu \tag{4.5.2}$$

with the Dirichlet condition

$$u_\mu = g \tag{4.5.3}$$

on ∂B^μ.

Here $g = 0$ on the lateral boundary of the cylinders composing B^μ; moreover $g = 0$ everywhere with the exception of the bases $\hat\beta_j^\mu$ of the cylinders B_j^μ (these bases are assumed to belong to the boundary of the tube structure); $g \in C^2(\hat\beta_j^\mu)$, and for each j, $g = \mu^2 g_j(\frac{x-O_j}{\mu})$ on $\hat\beta_j^\mu$, the vector valued function $g_j \in C^2$ do not depend on μ. The solvability condition imposes the relation

$$\int_{\partial B^\mu} gnds = 0. \tag{4.5.4}$$

In this section we will drop the subscript μ for the solution of problem (4.5.1)-(4.5.3).

Consider first the case of a right hand side vector valued function f "concentrated" in some neighborhoods of the nodes O_j, i.e. assume that

$$f = f_j(\frac{x-O_j}{\mu}) \tag{4.5.5}$$

in a vicinity of each O_j, $j = 0, 1, ..., n$, and assume that the vector valued functions $f_j(\xi)$ vanish if $|\xi| > r_0$. Suppose also that these functions (and r_0) do not depend on μ. Thus we have defined f in small domains obtained from *supp* f_j by a contraction in $1/\mu$ times with the centers O_j. We define f as zero in all other points. Let $f \in C^1$, $\nu > 0$.

Let $H_{div=0}(B^\mu)$ be space of vector valued functions from $H^1(B^\mu)$ with vanishing divergence, let $H^0_{div=0}(B^\mu)$ be the subspace of $H_{div=0}(B^\mu)$ of functions vanishing on the boundary. Suppose that g can be continued in B^μ as a vector valued function $\hat g$ of $H_{div=0}(B^\mu)$. The variational formulation is as follows: find $u \in H_{div=0}(B^\mu)$ such that $u - \hat g \in H^0_{div=0}(B^\mu)$, and such that it satisfies to the integral identity

$$-\int_{B^\mu} \nu \sum_{i=1}^{s} (\frac{\partial u}{\partial x_i}, \frac{\partial \varphi}{\partial x_i})\, dx + \int_{B^\mu} \sum_{i=1}^{s} u_i\, (u, \frac{\partial \varphi}{\partial x_i})\, dx = \int_{B^\mu} (f, \varphi)\, dx,$$

for all $\varphi \in H^0_{div=0}(B^\mu)$. Here $(,)$ is the symbol of the scalar product in $I\!R^s$, u_i is the $i-$th component of the vector u, i.e. $\sum_{i=1}^s u_i\, (u, \frac{\partial\varphi}{\partial x_i}) = (u, (u, \nabla)\varphi)$.

The existence and uniqueness of the solution was proved in [87] (for sufficiently small values of μ) .

We construct the asymptotic expansion in a form

$$u^a = \sum_{l=2}^{K} \mu^l \{ \sum_{e=e_j,\ j=1,...,n} u_l^e(\frac{x^{e,L}}{\mu})\chi_\mu(x) + \sum_{i=0}^{n} u_l^{BLO_i}(\frac{x-O_i}{\mu})\}, \tag{4.5.6}$$

$$p^a = \sum_{l=2}^{K+1} \mu^{l-2} \sum_{e=e_j,\ j=1,...,n} p_l^e(x_1^e)\chi_\mu(x) + \sum_{i=0}^{n} p_2^e(O_i)(1-\chi_\mu(x))\theta_i(x)$$

$$+ \sum_{l=2}^{K+1} \mu^{l-1} \sum_{i=0}^{n} p_l^{BLO_i}(\frac{x-O_i}{\mu}), \tag{4.5.7}$$

Here χ_μ is a function equal to zero at the distance not more than $(\hat{d}_0+1)\mu$ from $O_j,\ j=0,1,...,n;\ \hat{d}_0\mu = max\ \{d_0\mu,\ d_1\mu\},\ d_0\mu$ is the infimum of radiuses of all spheres with the center O such that every point of it belongs only to not more than one of the cylinders $B_j\ j=1,...,n;\ d_1$ is the maximal diameter of the domains $\gamma_0, \gamma_1, ..., \gamma_n$, and

$$\theta_j(x) = 0\ if\ |x - O_j|\ >\ min_i\ |e_i|/2,$$

$$\theta_j(x) = 1\ if\ |x - O_j|\ \leq\ min_i\ |e_i|/2.$$

We have introduced here the local system of coordinates $Ox_1^{e_j}...x_s^{e_j}$ associated with a segment e_j such that the direction of the axis $Ox_1^{e_j}$ coincides with the direction of the segment OO_j, i.e. $x_1^{e_j}$ is a longitudinal coordinate. The axes $Ox_1^{e_j}, ..., Ox_s^{e_j}$ form a cartesian coordinate system. We suppose that the function χ_μ is equal to zero on the cylinder B_j^μ if $x_1^{e_j} \leq (d_0+1)\mu$ or if $|x_1^{e_j} - |e_j|| \leq (d_0+1)\mu$ (here $|e_j|$ is the length of the segment e_j), we suppose that the function χ_μ is equal to one on this cylinder if $x_1^{e_j} \geq (d_0+2)\mu$ and $|x_1^{e_j} - |e_j|| \geq (d_0+2)\mu$, and we define χ_μ by relations $\chi_\mu(x) = \chi_j(x_1^{e_j}/\mu)$ if $(d_0+1)\mu \leq x_1^{e_j} \leq (d_0+2)\mu$, and we pose $\chi_\mu(x) = \chi_j((x_1^{e_j} - |e_j|)/\mu)$ if $(d_0+1)\mu \leq |e_j| - x_1^{e_j} \leq (d_0+2)\mu$. Here χ_j is a differentiable on $\mathbb{R}$ function of one variable, it is independent of μ, and it is equal to zero on the segment $[-(d_0+1),\ d_0+1]$ and it is equal to one on the union of the intervals $(-\infty, -(d_0+2)) \cup ((d_0+2), +\infty)$. For any $\gamma_j^\mu,\ \chi_\mu$ is equal to zero on it. The variable $x^{e_j L} = (x_2^{e_j}, ..., x_s^{e_j})$.

Let the relation between the columns (here T is the transposition symbol) x^T and $x^{e_j,T}$ be

$$x^T = \Gamma_j x^{e_j,T} + O,\ j = 1,...,n, \tag{4.5.8}$$

where Γ_i is an orthogonal matrix of passage from the canonic base to the local one (in the previous sections we used for matrices Γ_j the notation α_e^*). Then the vector valued function u_l^e and the scalar functions p_l^e are defined up to the scalar constants c_l^e, d_l^e :

$$u_l^{e_j}(\xi^L) = c_l^{e_j}\Gamma_j(\tilde{u}^{e_j}(\xi^L), 0, ..., 0)^T,\ p_l^{e_j}(x_1^{e_j}) = c_l^{e_j}x_1^{e_j} + d_l^{e_j},\ l = 2, 3, ... \tag{4.5.9}$$

where $\xi^L = (\xi_2, ..., \xi_s)$ and $\tilde{u}^{e_j}$ is the solution of the problem

$$\nu\Delta_{\xi^L}\tilde{u}^{e_j} \;-\; 1 \;=\; 0,\;\; \xi^L\in\beta_j,\quad \tilde{u}^{e_j}|_{\partial\beta_j} \;=\; 0,\; j=1,...,n, \tag{4.5.10}$$

and $d_2^{e_j}=0$.

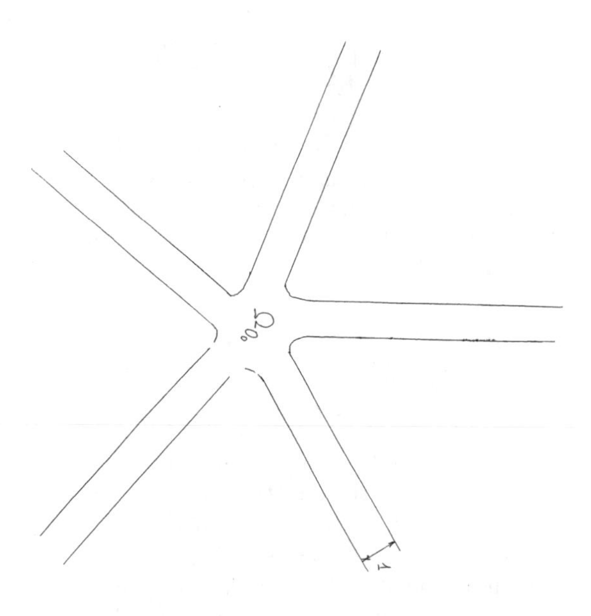

Figure 4.5.2. Dilated domains Ω_{O_j}.

The boundary layer solution is a pair constituted of a vector valued function $u_l^{BLO_j}$ and scalar function $p_l^{BLO_j}$ satisfying to the Stokes system:

$$\nu\Delta_\xi u_l^{BLO_0}-\nabla_\xi p_l^{BLO_0} \;=\; f_0(\xi)\delta_{l,0}\,+$$

$$+\sum_{e=e_j,\; j=1,...,n}\{c_l^{e_j}\{-\nu\Delta_\xi(\chi_j(\xi_1^{e_j})\Gamma_j(\tilde{u}^{e_j}(\xi^L),0,...,0)^T)$$

$$+\,\nabla_\xi(\chi_j(\xi_1^{e_j})\xi_1^{e_j})\}\;+$$

$$+\,(\sum_{p+r=l-1}c_p^{e_j}c_r^{e_j})(\chi_j(\xi_1^{e_j})\tilde{u}^{e_j}(\xi^L)\frac{\partial}{\partial\xi_1^{e_j}}\chi_j(\xi_1^{e_j}))\Gamma_j(\tilde{u}^{e_j}(\xi^L),0,...,0)^T\,+$$

$$+\,d_{l+1}^{e_j}\nabla_\xi(\chi_j(\xi_1^{e_j}))\}, \tag{4.5.11}$$

$$div_\xi u_l^{BLO_0} = -div_\xi\{\sum_{e=e_j,\ j=1,...,n}\{c_l^{e_j}\chi_j(\xi_1^{e_j})\Gamma_j(\tilde{u}^{e_j}(\xi^L),0,...,0)^T\},\quad \xi\in\Omega_{O_0} \tag{4.5.12}$$

with the Dirichlet condition

$$u_l^{BLO_0}|_{\partial\Omega_{O_0}} = 0 \tag{4.5.13}$$

and for $i = 1, ..., n$

$$\nu\Delta_{\hat{\xi}}u_l^{BLO_j} - \nabla_{\hat{\xi}}p_l^{BLO_j} = f_j(\hat{\xi})\delta_{l,0} +$$

$$+\ \hat{c}_l^{e_j}\{-\nu\Delta_{\hat{\xi}}(\chi_j(\hat{\xi}_1^{e_j})\hat{\Gamma}_j(\tilde{u}^{e_j}(\hat{\xi}^L),0,...,0)^T)$$

$$+\ \nabla_{\hat{\xi}}(\chi_j(\hat{\xi}_1^{e_j})\hat{\xi}_1^{e_j})\}\ +$$

$$+\ (\sum_{p+r=l-1}c_p^{e_j}c_r^{e_j})(\chi_j(\xi_1^{e_j})\tilde{u}^{e_j}(\xi^L)\frac{\partial}{\partial\xi_1^{e_j}}\chi_j(\xi_1^{e_j}))\Gamma_j(\tilde{u}^{e_j}(\xi^L),0,...,0)^T\ +$$

$$+\ \hat{d}_{l+1}^{e_j}\nabla_{\hat{\xi}}(\chi_j(\hat{\xi}_1^{e_j})), \tag{4.5.14}$$

$$div_{\hat{\xi}}u_l^{BLO_j} = -div_{\hat{\xi}}\{\hat{c}_l^{e_j}\chi_j(\hat{\xi}_1^{e_j})\hat{\Gamma}_j(\tilde{u}^{e_j}(\hat{\xi}^L),0,...,0)^T\},\quad \hat{\xi}\in\Omega_{O_j} \tag{4.5.15}$$

with the Dirichlet condition

$$u_l^{BLO_j}|_{\partial\Omega_{O_j},\hat{\xi}_1^{e_j}=0} = g_j\delta_{l,2}, \tag{4.5.16}$$

$$u_l^{BLO_j}|_{\partial\Omega_{O_j},\hat{\xi}_1^{e_j}\neq 0} = 0, \tag{4.5.17}$$

where $\Omega_{O_0} = \cup_{j=1}^n\tilde{\Omega}_j \cup \gamma_0$, and $\tilde{\Omega}_i$ are the half-infinite cylinders obtained from B_j^μ by infinite extension behind the base $\hat{\beta}_j^\mu$ and by homothetic dilatation in $1/\mu$ times (with respect to the point O); let Ω_j be obtained from $\tilde{\Omega}_j$ by a symmetric reflection relatively the plain containing β_j^μ and let $\Omega_{O_j} = \Omega_j \cup \gamma_j^t$, where γ_j^t is obtained from γ_j by a translation (such that the point O_j becomes O). The variable $\hat{\xi}_1^{e_j}$ is opposite to $\xi_1^{e_j}$, i.e. to the first component of the vector $\Gamma_j^T\xi^{e_j,T}$. So $\hat{\xi}_1^{e_j} = \hat{\Gamma}_j^T(\xi^{e_j})^T$, where $\hat{\Gamma}_j = \hat{I}d\Gamma_j$ and $\hat{I}d$ is the diagonal matrix with the diagonal elements $-1, 1, ..., 1$. The constants $\hat{c}_l^e, \hat{d}_l^e$ are defined in such a way that the linear functions $p_l^{e_j}(x_1^{e_j}) = c_l^{e_j}x_1^{e_j} + d_l^{e_j}$ and $p_l^{e_j}(x_1^{e_j}) = \hat{c}_l^{e_j}(|e_j| - x_1^{e_j}) + \hat{d}_l^{e_j}$ are equal, i.e.

$$c_l^{e_j} = -\hat{c}_l^{e_j},\ \hat{d}_l^{e_j} = c_l^{e_j}|e_j| + d_l^{e_j}. \tag{4.5.18}$$

We suppose that every term in the sum $\sum_{e=e_j,\ j=1,...,n}$ in (4.5.12) is defined only on the branch of Ω_{O_0}, corresponding to $e=e_j$, and it vanishes in γ_0.

We seek the exponentially decaying at infinity solutions of these boundary layer problems and we choose the constants $c_l^e, \hat{c}_l^{e_j}, d_l^e, \hat{d}_l^{e_j}$ from the conditions of existence of such solutions [120]. We define first $\hat{c}_l^{e_j}$ from the condition of exponential decaying of $u_l^{BLO_j}$ at infinity:

$$\int_{\Omega_{O_j}} div_{\hat{\xi}}(\hat{c}_l^{e_j}\chi_j(\hat{\xi}_1^{e_j})\hat{\Gamma}_i(\tilde{u}^{e_j}(\hat{\xi}^L)),0,...,0)^T)d\hat{\xi} \ = \ \int_{\beta_j}(\hat{\Gamma}_j^T g_j)^1 d\xi,$$

i.e.

$$-\int_{\beta_j}\tilde{u}^{e_j}(\xi^L)d\xi\hat{c}_l^{e_j} \ = \ \int_{\beta_j}(\hat{\Gamma}_j^T g_j)^1 d\xi\delta_{l,2}, \tag{4.5.19}$$

where the upper index 1 corresponds to the first component of the vector.

Note that $\int_{\beta_j}\tilde{u}^{e_j}(\xi^L)d\xi$ is negative due to the principle of maximum for problem (4.5.10). Then we find $c_l^{e_j}=-\hat{c}_l^{e_j}$ and $\hat{d}_l^{e_j}=c_l^{e_j}|e_j|$. Then we determine the constants $d_{l+1}^{e_j}$ from the condition of the exponential decaying of $p_l^{BLO_0}$ at infinity. To this end we consider first the problem (4.5.11)-(4.5.13) without the last term in the equation (4.5.11), i.e.

$$\begin{aligned}\nu\Delta_\xi\bar{u}_l^{BLO_0}-\nabla_\xi\bar{p}_l^{BLO_0} \ &= \ f_0(\xi)\delta_{l,2}\\ &+ \sum_{e=e_j,\ j=1,...,n} c_l^{e_j}\{-\nu\Delta_\xi(\chi_j(\xi_1^{e_j})\Gamma_j(\tilde{u}^{e_j}(\xi^L),0,...,0)^T)\\ &+\nabla_\xi(\chi_j(\xi_1^{e_j})\xi_1^{e_j})\}+\\ +(\sum_{p+r=l-1}c_p^{e_j}c_r^{e_j})(\chi_j(\xi_1^{e_j})\tilde{u}^{e_j}(\xi^L)&\frac{\partial}{\partial\xi_1^{e_j}}\chi_j(\xi_1^{e_j}))\Gamma_j(\tilde{u}^{e_j}(\xi^L),0,...,0)^T\},\end{aligned} \tag{4.5.20}$$

$$div_\xi\bar{u}_l^{BLO_0} \ = \ -div_\xi\{\sum_{e=e_j,\ j=1,...,n}\{c_l^{e_j}\chi_j(\xi_1^{e_j})\Gamma_j(\tilde{u}^{e_j}(\xi^L),0,...,0)^T\}, \ \ \xi\in\Omega_{O_0} \tag{4.5.21}$$

with the Dirichlet condition

$$\bar{u}_l^{BLO_0}|_{\partial\Omega_{O_0}} \ = \ 0. \tag{4.5.22}$$

Here the constants $c_l^{e_j}$ are just defined by (4.5.18), (4.5.19) and satisfy the condition

$$\int_{\Omega_{O_0}} div_\xi\{\sum_{e=e_j,\ j=1,...,n} c_l^e\chi_i(\xi_1^e)\Gamma_j(\tilde{u}^e(\xi^L),0,...,0)^T)\}d\xi \ = \ 0,$$

i.e.

$$\sum_{e=e_j,\ j=1,...,n} \int_{\beta_j} \tilde{u}^{e_j}(\xi^L) d\xi c_l^{e_j} = 0. \tag{4.5.23}$$

Indeed, the choice of the constants $c_l^{e_j} = -\hat{c}_l^{e_j}$ and $\hat{c}_l^{e_j}$ from (4.5.19) and condition (4.5.4) give relation (4.5.23).

It is known (for example [120]) that there exists the unique solution $\{\bar{u}_l^{BLO_0}$, $\bar{p}_l^{BLO_0}\}$ of this problem such that $\bar{u}_l^{BLO_0}$ stabilizes to zero at infinity (on every branch of Ω_{O_0}) and $\bar{p}_l^{BLO_0}$ stabilizes on every branch of Ω_{O_0}, associated with e_j, to its own constant $\bar{p}_l^{BLO_0\infty j}$). These constants are defined uniquely up to one common additional constant, which we fix here by a condition $\bar{p}_l^{BLO_0\infty 1}$ $= 0$. Then we define

$$d_{l+1}^{e_j} = -\bar{p}_l^{BLO_0\infty j}, \tag{4.5.24}$$

$$u_l^{BLO_0} = \bar{u}_l^{BLO_0}, \tag{4.5.25}$$

and

$$p_l^{BLO_0} = \bar{p}_l^{BLO_0} + d_{l+1}^{e_j} \chi_j(\xi_1^{e_j}) \tag{4.5.26}$$

on every branch of Ω_{O_0}, associated with e_j.

Obviously, this pair $\{u_l^{BLO_0}, p_l^{BLO_0}\}$ satisfies equations (4.5.11)-(4.5.13).

The boundary layer functions $u_l^{BLO_j}$ and $p_l^{BLO_j}$, $j = 1, ..., n$, are not defined in the vicinity of O. Therefore we should change a little bit the formulas of u^a and p^a far from the nodes $O_j, j = 0, ..., n$.

Let $\eta_j(x_1^{e_j})$ be a smooth function defined on each segment e_j, let it be one if $|x_1^{e_j} - |e_j|/2| \geq |e_j|/4$ and let it be zero if $|x_1^{e_j} - |e_j|/2| \leq |e_j|/8$. Let $\eta(x) = \eta_j(x_1^{e_j})$ for each cylinder B_j^μ and let $\eta = 1$ on each γ_j^μ. Then we redefine u^a and p^a as

$$u^a = \sum_{l=2}^{K} \mu^l \{ \sum_{e=e_j,\ j=1,...,n} u_l^e(\frac{x^{e,L}}{\mu})\chi_\mu(x) + \sum_{i=0}^{n} u_l^{BLO_i}(\frac{x - O_i}{\mu})\eta(x)\}, \tag{4.5.27}$$

$$p^a = \sum_{l=2}^{K+1} \mu^{l-2} \sum_{e=e_j,\ j=1,...,n} p_l^e(x_1^e)\chi_\mu(x) + \sum_{i=0}^{n} p_2^e(O_i)(1 - \chi_\mu(x))\theta_i(x)$$

$$+ \sum_{l=2}^{K+1} \mu^{l-1} \sum_{i=1}^{N} p_l^{BLO_i}(\frac{x - O_i}{\mu})\eta(x)\}, \tag{4.5.28}$$

The consequence of this redefinition is a small discrepancy in the right hand side of the equations of order $O(exp(-c/\mu))$ with the positive constant c : now we have the relations

$$\nu\Delta\tilde{u}^a - \nabla\tilde{p}^a = f + \Phi, \quad (4.5.29)$$

$$div\ \tilde{u}^a - (\tilde{u}^a\ ,\ -\nabla)\tilde{u}^a = \Psi,\ x \in B^\mu \quad (4.5.30)$$

with the Dirichlet condition

$$\tilde{u}^a = g \quad (4.5.31)$$

on ∂B^μ, where

$$\|\Psi\|_{H^1(B^\mu)} = O(exp(-c/\mu)),\ \ \|\Phi\|_{L^2(B^\mu)} = O(\mu^{K-1+(s-1)/2}) \quad (4.5.32)$$

with a positive constant c and

$$\int_{B^\mu} \Psi dx = 0, \quad (4.5.33)$$

because $\int_{\partial B^\mu}(\tilde{u}^a, n)ds = \int_{\partial B^\mu}(g, n)ds = 0$.

We are going to prove the estimate

$$\|u - \tilde{u}^a\|_{H^1(B^\mu)} = O(\mu^{K-1+(s-1)/2}), \quad (4.5.34)$$

where $c_1 > 0$ does not depend on μ.

First we construct a function φ such that $v = \nabla\varphi$ is a solution of the equation

$$div\ v = \Psi,\ x \in B^\mu, \quad (v, n) = 0,\ x \in \partial B^\mu,$$

i.e. let φ be a solution of the Neumann problem

$$\Delta\varphi = \Psi,\ x \in B^\mu,\ \frac{\partial\varphi}{\partial n} = 0,\ x \in \partial B^\mu.$$

The existence of this solution is provided by the condition $\int_{B^\mu} \Psi dx = 0$. As the boundary belongs to C^2 then $\varphi \in H^2(B^\mu)$. If we consider the solution with vanishing average $\int_{B^\mu} \varphi dx = 0$, then we apply the Poincare inequality for B^μ (cf. Appendix 4.A.2):

$$\forall\varphi \in H^1(B^\mu),\ \ \int_{B^\mu} \varphi^2 dx \leq \frac{1}{mesB^\mu}\ \left(\int_{B^\mu} udx\right)^2 + A\int_{B^\mu} (\nabla\varphi)^2 dx.$$

where the constant A does not depend on μ.

Thus we obtain the estimate $\|\varphi\|_{H^1(B^\mu)} = O(exp(-c/\mu))$ and therefore we can use Agmon, Duglas and Nirenberg [2]theory and obtain the estimate $\|\varphi\|_{H^2(B^\mu)} = O(exp(-c_2/\mu))$, where the positive constant c_2 does not depend on μ.

Therefore $v \in H^1(B^\mu)$ and $\|v\|_{H^1(B^\mu)} = O(exp(-c_2/\mu))$.

The second step is the construction of such a function $w = rot\ \psi$ that $w = -v$ on ∂B^{μ}. This construction is described in [87] and it has a "local nature." We make a partition of unity $1 = \sum_{k=1}^{N} \delta_k$ on B^{μ}, in such a way that all supports of δ_k have the diameters of order μ and satisfy to the condition of the existence of such a change of variables that the procedure of [87] ch.1, sect. 2 can be applied and the corresponding vector valued function ψ_k can be constructed for every support, i.e. $rot\ \psi_k|_{\partial B^{\mu}} = \delta_k v|_{\partial B^{\mu}}$. All these supports have non-empty intersections Δ_k with the boundary of B^{μ}. Moreover we can make homothetic dilatations $\xi = (x - A_k)/\mu$ in $1/\mu$ times with some centers A_k for every support in such a way that its image σ_k does not depend on μ. All these σ_k can be "uniformed", i.e. can be extended up to a finite number (independent of μ) of the domains σ such that all other domains could be obtained from them by rotations and translations. The functions δ_k can be taken satisfying relation $\delta_k(x) = \tilde{\delta}_k((x - A_k)/\mu)$, where the functions $\tilde{\delta}_k$ do not depend on μ. Thus the problem of construction of all vector valued functions ψ_k is reduced to a finite (independent of μ) number of problems : construct such a vector valued function $\tilde{\psi}_\alpha$ that $rot_\xi\ \tilde{\psi}_\alpha|_{\partial B^{\mu}} = \tilde{\delta}_\alpha v|_{\partial B^{\mu}}$ and take $\psi_\alpha(x) = \mu\tilde{\delta}_\alpha((x - A_k)/\mu)$. In this case we can estimate

$$\|rot_\xi\ \tilde{\psi}_\alpha\|_{H^1(\sigma)} \leq C_\alpha \|v(\mu\xi + A_\alpha)\|_{H^1(\sigma)},$$

with the constants C_α uniformly bounded by a constant C independent of μ. Therefore there exists a positive constant c_1 independent of μ such that

$$\|rot\ \psi\|_{H^1(B^{\mu})} = O(exp(-c_1/\mu)),$$

and

$$\|v + w\|_{H^1(B^{\mu})} = O(exp(-c_1/\mu)).$$

Therefore the relation holds true:

$$U^a = \tilde{u}^a - (v + w) \in H_{div=0}(B^{\mu})$$

and

$$-\int_{B^{\mu}} \nu \sum_{i=1}^{s} (\frac{\partial U^a}{\partial x_i}, \frac{\partial \varphi}{\partial x_i})\, dx + \int_{B^{\mu}} (U^a, (U^a, \nabla)\varphi) dx =$$

$$= -\nu \int_{B^{\mu}} \sum_{i=1}^{s} (\frac{\partial(\tilde{u}^a - (v + w))}{\partial x_i}, \frac{\partial \varphi}{\partial x_i})\, dx + \int_{B^{\mu}} (\tilde{u}^a - (v+w), (\tilde{u}^a - (v+w), \nabla)\varphi) dx =$$

$$= \int_{B^{\mu}} (f, \varphi)\, dx + \int_{B^{\mu}} (\Phi, \varphi)\, dx + \nu \int_{B^{\mu}} \sum_{i=1}^{s} (\frac{\partial(v + w)}{\partial x_i}, \frac{\partial \varphi}{\partial x_i})\, dx -$$

$$- \int_{B^{\mu}} ((v + w), (\tilde{u}^a - (v + w), \nabla)\varphi) dx - \int_{B^{\mu}} (\tilde{u}^a, (v + w, \nabla)\varphi) dx,$$

for all $\varphi \in H^0_{div=0}(B^\mu)$.

Now we can apply the Poincaré-Friedrichs estimate (cf. section 4.A.2):

$$\forall \varphi \in H^1_0(B^\mu), \quad \int_{B^\mu} \varphi^2 dx \;\leq\; A \int_{B^\mu} (\nabla \varphi)^2 dx,$$

where the constant A does not depend on μ.

(Moreover, decomposing the domain to some subdomains with diameter of order μ, in such a way that each subdomain contains a part of the boundary ∂B^μ, this estimate can be proved with the factor $\mu^2 A$ instead of A).

Applying this estimate as well as the a priori estimate for Navier-Stokes equation from [87],[188], we obtain:

$$\|U^a - u\|_{H^1(B^\mu)} = O(\mu^{K-1})$$

and

$$\|u^a - u\|_{H^1(B^\mu)} = O(\mu^{K-1}), \tag{4.5.35}$$

where $c_3 > 0$ does not depend on μ. This estimate can be improved in a standard way (see section 2.1) as $\|u^a - u\|_{H^1(B^\mu)} = O(\mu^{K+(s-1)/2})$.

Remark 4.5.1. If the right hand side function is defined on each cylinder B^μ_j as

$$f \;=\; f_j(\frac{x-O_j}{\mu}) \;+\; f_0(\frac{x-O}{\mu}) \;+\; \chi_\mu(x)\Gamma_j(\hat{f}_j(x_1^{e_j}), 0, ..., 0)^T, \tag{4.5.36}$$

where $\hat{f}_j$ are sufficiently smooth functions, and for all $\gamma_j, j = 0, ..., n$, it is defined as earlier

$$f \;=\; f_j(\frac{x-O_j}{\mu}),$$

then this case can be easily reduced to a previous type of right hand side function by a subtraction of a partial solution

$$u_{partial} = 0, \;\; p_{partial} = -\int_0^{x_1^{e_j}} \hat{f}_j(t)dt = -F_j(x_1^{e_j})$$

on each cylinder B^μ_j. Indeed if u and p is the solution of problem (4.5.1)-(4.5.3) with the right hand side (4.5.36) then the pair u, $p - \chi_\mu(x)F_j(x_1^{e_j})$ is the solution of the problem with the right hand side

$$\tilde{f} \;=\; f_j(\frac{x-O_j}{\mu}) + f_0(\frac{x-O}{\mu}) + \chi_\mu(x)\Gamma_j\hat{f}_j(x_1^{e_i}), 0, ..., 0)^T \;-\; \nabla(\chi_\mu(x)F_j(x_1^{e_j})),$$

on each cylinder B^μ_j. This right hand side function has a support concentrated in the vicinities of the nodes O_j, i.e. $supp\ \tilde{f} \subset supp\ (\chi_\mu - 1)$. Now we develop $\hat{f}_j$ by the Taylor's formula in vicinities of the nodes and obtain the set of problems with the right hand side of a form (4.5.5). It can be treated as above.

4.5.2 Tube structure with m bundles of tubes

Consider now the case of m different bundles of segments

$$B_1 = \cup_{j=1}^{n_1} e_{j,1}, ..., B_m = \cup_{j=1}^{n_m} e_{j,m}.$$

We suppose that all common points of these bundles are end points of some segments of these bundles. Let the union of all bundles be connected. Consider the tube structure B_α^μ associated with the bundle B_α. Now we do not require a base $\hat{\beta}_j^\mu$ to be a part of ∂B_α^μ in case if it corresponds to a common point of two bundles. Let

$$B^\mu = \cup_{\alpha=1}^m B_\alpha^\mu$$

be a domain with C^2-smooth boundary.

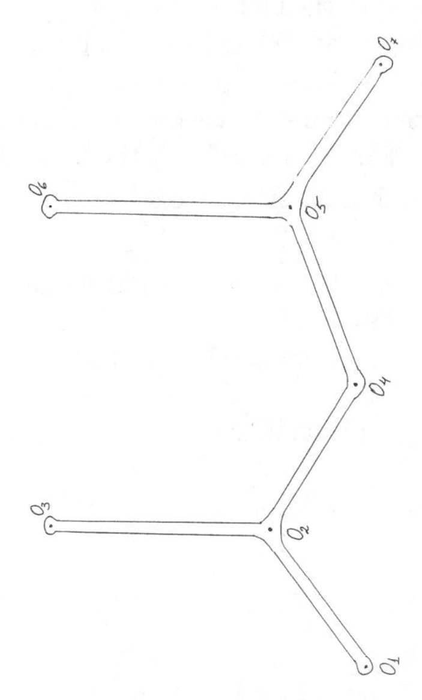

Figure 4.5.3. Tube structure with m (two) bundles of tubes.

Consider the Navier - Stokes system of equations (4.5.1)-(4.5.3) for this B^μ with the right hand side concentrated in the neighborhoods of the nodes (as above).

Let us enumerate all nodes , i.e. all ends $O_1, ..., O_N$ of the segments and all segments $e_1, ..., e_M$. For each e_j introduce local co-ordinates x^{e_j}, related to one of its ends. We seek the solution in a generalized form (4.5.27), (4.5.28), i.e.

$$\tilde{u}^a = \sum_{l=2}^{K} \mu^l \{ \sum_{e=e_j,\ j=1,...,M} u_l^e(\frac{x^{e,L}}{\mu})\chi_\mu(x) + \sum_{i=1}^{N} u_l^{BLO_i}(\frac{x - O_i}{\mu})\eta(x)\}, \tag{4.5.37}$$

$$\tilde{p}^a = \sum_{l=2}^{K+1} \mu^{l-2} \sum_{e=e_j,\ j=1,...,M} p_l^e(x_1^e)\chi_\mu(x) + \sum_{i=1}^{N} p_2(O_i)(1-\chi_\mu(x))\theta_i(x)$$

$$+ \sum_{l=2}^{K+1} \mu^{l-1} \sum_{i=1}^{N} p_l^{BLO_i}(\frac{x-O_i}{\mu})\eta(x), \qquad (4.5.38)$$

where the terms have the same form and sense as above:

$$u_l^{e_j}(\xi^L) = c_l^{e_j}\Gamma_j(\tilde{u}^{e_j}(\xi^L), 0, ..., 0)^T, \quad p_l^{e_j}(x_1^{e_j}) = c_l^{e_j} x_1^{e_j} + d_l^{e_j},$$

here $c_l^{e_j}, d_l^{e_j}$ are scalar constants, $\tilde{u}^{e_j}$ is the solution of the problem (4.5.10) and the boundary value problems are stated for each domain Ω_{O_i}, related to O_i; i.e. let O_i be one of the nodes (one of the ends of $e_{j_1}, ..., e_{j_{q_i}}$) and let the local coordinates for each of these segments are related to O_i; cut all cylinders $B_{j_1}^\mu, ..., B_{j_q}^\mu$, associated with $e_{j_1}, ..., e_{j_q}$ at the distance of $|e_{j_1}|/2, ..., |e_{j_{q_i}}|/2$ respectively and consider the part of B^μ which contains O_i; then we extend this part substituting the deleted parts of the cylinders $B_{j_1}^\mu, ..., B_{j_{q_i}}^\mu$ by the half-infinite cylinders having the same cross-sections and orientations as $B_{j_1}^\mu, ..., B_{j_{q_i}}^\mu$ (these half-cylinders contain only the truncated parts of $B_{j_1}^\mu, ..., B_{j_{q_i}}^\mu$, but not the resting parts); then after the homothetic dilatation of this constructed domain in $1/\mu$ times with respect to O_i we obtain the domain Ω_{O_i}.

Remark 4.5.2. All p_2^e have a common value $p_2^e(O_i)$ in O_i for all e with one of the ends O_i.

The boundary layer problem is similar to (4.5.11)-(4.5.13):

$$\nu\Delta_\xi u_l^{BLO_i} - \nabla_\xi p_l^{BLO_i} = f_i(\xi)\delta_{l2}$$

$$+ \sum_{e=e_j,\ j=j_1,...,j_{q_i}} \{c_l^{e_j}\{-\nu\Delta_\xi(\chi_j(\xi_1^{e_j})\Gamma_j(\tilde{u}^{e_j}(\xi^L)), 0, ..., 0)^T)$$

$$+ \nabla_\xi(\chi_j(\xi_1^{e_j})\xi_1^{e_j})\}+$$

$$+ (\sum_{p+r=l-1} c_p^{e_j} c_r^{e_j})(\chi_j(\xi_1^{e_j})\tilde{u}^{e_j}(\xi^L)\frac{\partial}{\partial \xi_1^{e_j}}\chi_j(\xi_1^{e_j}))\Gamma_j(\tilde{u}^{e_j}(\xi^L), 0, ..., 0)^T +$$

$$d_{l+1}^{e_j}\nabla_\xi(\chi_j(\xi_1^{e_j}))\}, \qquad (4.5.39)$$

$$div_\xi u_2^{BLO_i} = -div_\xi\{ \sum_{e=e_j,\ j=j_1,...,j_{q_i}} \{c_l^{e_j}\chi_j(\xi_1^{e_j})\Gamma_j(\tilde{u}^{e_j}(\xi^L), 0, ..., 0)^T\}, \quad \xi \in \Omega_{O_i}$$

(4.5.40)

with the Dirichlet condition

$$u_l^{BLO_i}|_{\partial\Omega_{O_i}} = 0 \tag{4.5.41}$$

if O_i is not a "boundary node," i.e. if the distance from it to the support of the function g is of order of one, and with the boundary condition

$$u_l^{BLO_i}|_{\partial\Omega_{O_i}} = g_i(\xi) \tag{4.5.42}$$

if O_i is a "boundary node," i.e. if the distance from it to the support of the function g is of order of μ.

We obtain for $c_l^{e_j}$ the equations of the type (4.5.4):

$$\sum_{e=e_j,\ j=j_1,\ldots,j_{q_i}} \int_{\beta_j} \tilde{u}^{e_j}(\xi^L)d\xi c_l^{e_j} = 0 \tag{4.5.43}$$

if O_i is not a "boundary node," or

$$\sum_{e=e_j,\ j=j_1,\ldots,j_{q_i}} \int_{\beta_j} \tilde{u}^{e_j}(\xi^L)d\xi c_l^{e_j} = \int_{\partial\Omega_{O_i}\cap supp\ g_i} (g_i, n)\ d\xi. \tag{4.5.44}$$

if O_i is a "boundary node," n is the outside normal.

Now consider the problem of definition of $c_l^{e_j}$ and $d_{l+1}^{e_j}$ from (4.5.39)-(4.5.44). This problem is equivalent to the problem of definition of piecewise linear function $P_l(x)$ defined on B (linear on every e_j) satisfying conditions (4.5.43),(4.5.44) where $c_l^{e_j} = \partial P_l/\partial x_1^{e_j}$, and satisfying conditions

$$d_l^{e_j} = -\bar{p}_{l-1}^{BLO_i\infty j} + d_l^{O_i},$$

where $\bar{p}_{l-1}^{BLO_i\infty i}$ is a limit of the boundary layer function $\bar{p}_{l-1}^{BLO_i}$ constructed as above in such a way that the pair $\bar{u}_{l-1}^{BLO_i}, \bar{p}_{l-1}^{BLO_i}$ is a solution of the problem (4.5.39)-(4.5.42) for $l-1$ and without the last term of the equation (4.5.39); $d_l^{O_i}$ is a constant independent of j . This condition can be rewritten in a form

$$d_l^{e_j} = -\bar{p}_{l-1}^{BLO_i\infty i} + d_l^{e_{i_1}}. \tag{4.5.45}$$

Let Φ be a function such that it is defined on each segment e, $\partial\Phi/\partial x_1^e = 0$ in the vicinities of all nodes and such that in each node O_i its limit values $\Phi_{e_{j1}}(O_i), \ldots, \Phi_{e_{jq_i}}(O_i)$ for the segments $e_{i_1}, \ldots, e_{i_q}$ respectively satisfy to the relation (4.5.45), i.e.

$$\Phi_{e_j}(O_i) = -\bar{p}_{l-1}^{BLO_i\infty j} + \Phi_{e_{i_1}}(O_i). \tag{4.5.46}$$

We know that $\frac{\partial^2 \hat{P}_l}{\partial x_1^{e\,2}} = 0$ and that the constants $c_l^{e_j} = \frac{\partial \hat{P}_l}{\partial x_1^e}$ satisfy relations (4.5.43),(4.5.44). Then we can reformulate the problem for $\hat{P}_l = P_l - \Phi$ as

$$-\sum_{e=e_j,\ j=1,\dots,M}\int_e \rho_e \frac{\partial \hat{P}_l}{\partial x_1^e}\frac{\partial \hat{\tau}}{\partial x_1^e}dx_1^e =$$

$$-\sum_{e=e_j,\ j=1,\dots,M}\int_e \rho_e \frac{\partial \Phi}{\partial x_1^e}\frac{\partial \hat{\tau}}{\partial x_1^e}dx_1^e$$

$$-\sum_{O=O_i,\ boundary\ nodes}\int_{\partial\Omega_{O_i}}(g_i,n)ds\tau(O_i),$$

where the summation in the last term is developed over the boundary nodes, $\rho_e = \int_{\beta_e}\tilde{u}^e(\xi^L)d\xi$, τ is an arbitrary function of $H^1(B)$.

Now the existence and uniqueness up to a constant of $\hat{P}_l$ is evident (by the Lax - Milgram lemma).

The justification of series (4.5.37), (4.5.38) is the same as in the case of one bundle of segments B : a function $U^a \in H_{div=0}(B^\mu)$ is constructed in the same way as in section 1 and the following estimates are proved

$$\|U^a - u\|_{H^1(B^\mu)} = O(\mu^K), \quad \|u^a - u\|_{H^1(B^\mu)} = O(\mu^K). \tag{4.5.47}$$

4.6 Bibliographical Remark

The asymptotic analysis of the conductivity equation for finite rod structures first appeared as an auxiliary problem in the homogenization of the lattice ("skeletal") structures [136], see also [16], Chapter 8, as well as [44]. The elasticity equations in an L-shaped finite rod structure was first considered in [94], and then in [145, 148] the complete asymptotic expansion of the displacement field in a large class of finite rod structures was constructed. The junction of two elastic rods with contrasting rigidities was studied in [168]. An arbitrary finite rod structure was studied in [123].

The two-dimensional cell problem for the conductivity equation and for the elasticity equations set on a rectangular lattice-like structure was studied in [78]. The complete asymptotic expansion was constructed and the homogenized equation coefficients were calculated explicitly: the effective conductivity (with the first corrector) and the effective shear modulus (which is small).

The various junctions of plates, rods and three dimensional bodies were studied in [38],[80]-[82],[118],[178].

4.7 Appendices

4.A1. Appendix 1: estimates for traces in the pre-nodal domain

Here we prove two estimates that were used in the proof of Theorem 4.1.1.

Lemma 4.A1.1. If the node $x^0 \in G_0$ then the estimate holds :

$$\|\varphi\|_{L^2(\partial\Pi_{x^0}\cap\partial B_\mu^{++})} \leq C\sqrt{\mu}\|\nabla\varphi\|_{L^2(\Pi_{x^0})},$$

where C does not depend on φ and μ.

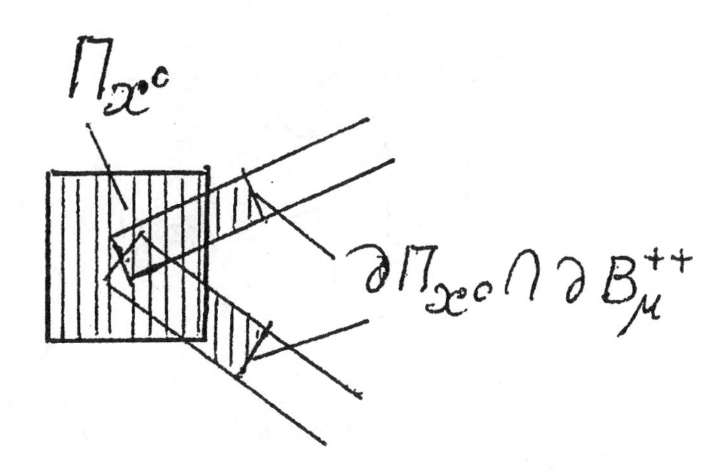

Figure 4.A1.1. Pre-nodal domain Π_{x^0} in Lemma 4.A1.1.

Proof

Make the change $\eta = x/\mu$; one obtains for the images Σ_η of $\partial\Pi_{x^0} \cap \partial B_\mu^{++}$ and Π_η of Π_{x^0} the inequality

$$\|\varphi\|^2_{L^2(\Sigma_\eta)} \leq C_5\|\varphi\|^2_{H^1(\Pi_\eta)},$$

and from the classic Poincaré - Friedrichs inequality we obtain

$$\|\varphi\|^2_{L^2(\Sigma_\eta)} \leq C_6\|\nabla\varphi\|^2_{L^2(\Pi_\eta)}.$$

Making the inverse change we get

$$\mu^{-s+1}\|\varphi\|^2_{L^2(\partial\Pi_{x^0}\cap\partial B_\mu^{++})} \leq C_6\mu^{-s+2}\|\nabla\varphi\|^2_{L^2(\Pi_{x^0})},$$

and it gives the desired estimate.

Lemma 4.A1.2 The estimate holds :

$$\|\varphi - <\varphi>_{\Pi_{x^0}}\|_{L^2(\partial\Pi_{x^0}\cap\partial B_\mu^{++})} \leq C\sqrt{\mu}\|\nabla\varphi\|_{L^2(\Pi_{x^0})},$$

where C does not depend on φ and μ.

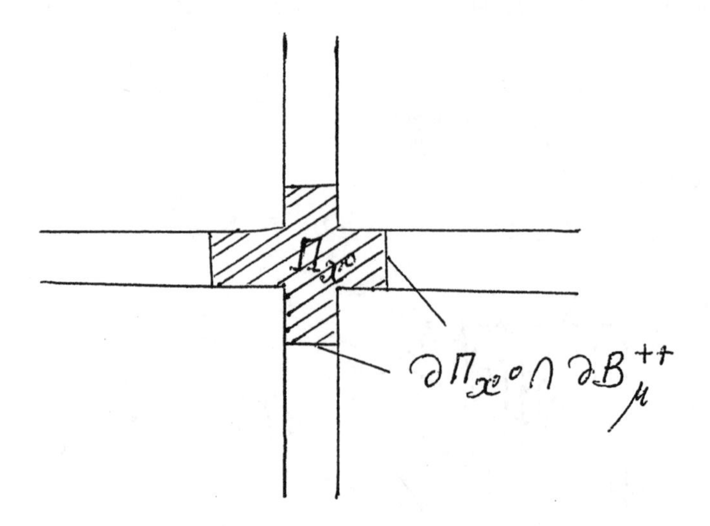

Figure 4.A1.2. Pre-nodal domain in Lemma 4.A1.2.

Proof

In the integral

$$i = \int_{\partial\Pi_{x^0}\cap\partial B_\mu^{++}} |\varphi - <\varphi>_{\Pi_{x^0}}|^2 dx$$

make the change $\eta = x/\mu$; we get

$$i = \mu^{s-1}\int_{\Sigma_\eta} |\varphi - <\varphi>_{\Pi_{x^0}}|^2 d\eta,$$

where Σ_η is the image of the surface $\partial\Pi_{x^0}\cap\partial B_\mu^{++}$ and let Π_η be the image of Π_{x^0} under the transformation $\eta = x/\mu$. Note that Σ_η is independent of μ. The estimate holds true in the dilated domain:

$$\int_{\Sigma_\eta} |\varphi - <\varphi>_{\Pi_{x^0}}|^2 d\eta \leq C\int_{\Pi_\eta}\sum_{j=1}^{s}(\frac{\partial\varphi}{\partial\eta_j})^2 d\eta,$$

where C is independent of μ.

Making the inverse change , we get

$$\int_{\Pi_\eta}\sum_{j=1}^{s}(\frac{\partial\varphi}{\partial\eta_j})^2 d\eta = \mu^{-s+2}\int_{\Pi_{x^0}}\sum_{j=1}^{s}(\frac{\partial\varphi}{\partial x_j})^2 dx.$$

Thus,

$$i \leq C\mu\int_{\Pi_{x^0}}\sum_{j=1}^{s}(\frac{\partial\varphi}{\partial x_j})^2 dx.$$

By taking the square root of the two sides of the inequality, we obtain the assertion of the lemma.

4.A2 Appendix 2: the Poincaré and the Friedrichs inequalities for a finite rod structure

The goal of this appendix is to prove the Poincaré - Friedrichs inequality for a function $u \in H_1(B_\mu)$, $u = 0$ on Σ_0. Below we consider domains which are connected and bounded open sets of $\mathbb{R}^s$ or $\mathbb{R}^{s-1}$ with a piecewise - smooth boundary satisfying the cone condition. We suppose also that μ is sufficiently small.

Lemma 4.A2.1. *Let G_1, G_2 be domains, $G_1 \subset G_2$. Then for each function $u \in H_1(G_2)$ the estimate holds true*

$$(\{u\}_1 - \{u\}_2)^2 \leq c \int_{G_2} (\nabla u)^2 dx,$$

where c is a constant depending on G_1, G_2, and $\{u\}_i = \frac{1}{measG_i} \int_{G_i} u \; dx$.

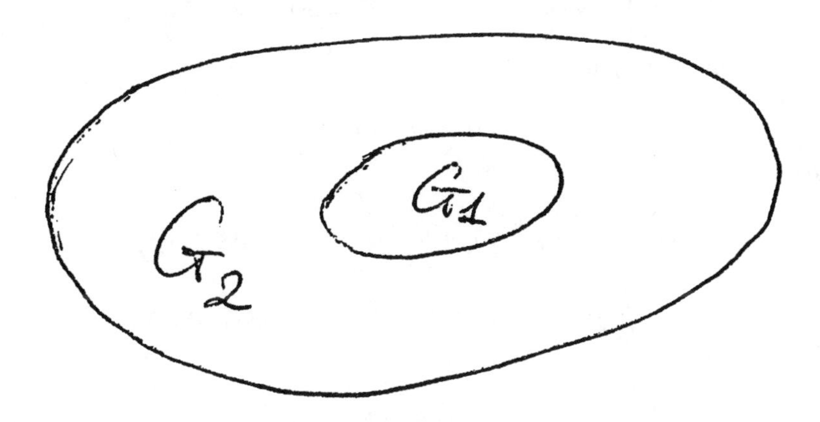

Figure 4.A2.1. Domains of Lemma 4.A2.1.

Proof

$$(\{u\}_1 - \{u\}_2)^2 = \frac{1}{measG_1} \int_{G_1} (\{u\}_1 - \{u\}_2)^2 dx \leq$$

$$\leq \frac{2}{measG_1} \left(\int_{G_1} (\{u\}_1 - u)^2 dx + \int_{G_1} (\{u\}_2 - u)^2 dx \right) \leq$$

$$\leq \frac{2}{measG_1}(\int_{G_1}(\{u\}_1 - u)^2dx + \int_{G_2}(\{u\}_2 - u)^2dx) \leq$$

$$\leq \frac{2c_0}{measG_1}\int_{G_2}(\nabla u)^2dx,$$

c_0 does not depend on u. Lemma is proved.

Lemma 4.A2.2. *Let $G_i^\mu = \{x \in R^s \mid \frac{x}{\mu} \in G_i\}$, where G_1, G_2 are domains, $G_1 \subset G_2$. Then for each $u \in H_1(G_2^\mu)$ the estimate holds*

$$(\{u\}_1^\mu - \{u\}_2^\mu)^2 \leq c\mu^{2-s}\int_{G_2^\mu}(\nabla u)^2dx,$$

where c is independent of μ, but depends on G_1 and G_2,
$\{u\}_i^\mu = \int_{G_i^\mu} udx/measG_i^\mu, \quad i = 1, 2.$

Proof

$$(\{u\}_1^\mu - \{u\}_2^\mu)^2 = \left(\frac{1}{\mu^s measG_1}\int_{G_1^\mu} u \quad dx - \frac{1}{\mu^s measG_2}\int_{G_2^\mu} u \quad dx\right)^2;$$

now the change of variables

$$x' = x/\mu, \quad dx = \mu^s dx'$$

yields

$$(\{u\}_1^\mu - \{u\}_2^\mu)^2 = (\{u(\mu x')\}_1 - \{u(\mu x')\}_2)^2 \leq$$

$$\leq c\int_{G_2}(\nabla_x u(\mu x'))^2dx' = \frac{c\mu^2}{\mu^s}\int_{G_2^\mu}(\nabla_x u(x))^2dx$$

The lemma is proved.

Lemma 4.A2.3. *Let β be an $(s-1)$ dimensional domain, $\beta_{(a,b)}$ be a cylinder $(a,b) \times \beta$; denote*

$$\beta_{(a,b)}^\mu = \{x \in R^s \mid (\frac{x_2}{\mu}, ..., \frac{x_s}{\mu}) \in \beta, \quad x_1 \in (a,b)\}.$$

Let d_1, d_2 be constants independent of μ. Then

$$(\{u\}_{(0,d_1\mu)}^\mu - \{u\}_{(0,d_2)}^\mu)^2 \leq \frac{d_2}{meas\beta} \quad \mu^{1-s}\int_{\beta_{(0,d_2)}^\mu} (\frac{\partial u}{\partial x_1})^2 \quad dx,$$

where

$$u \in H_1(\beta_{(0,d_2)}), \quad \{u\}_{(a,b)}^\mu = \int_{\beta_{(a,b)}^\mu} udx/meas\beta_{(a,b)}^\mu.$$

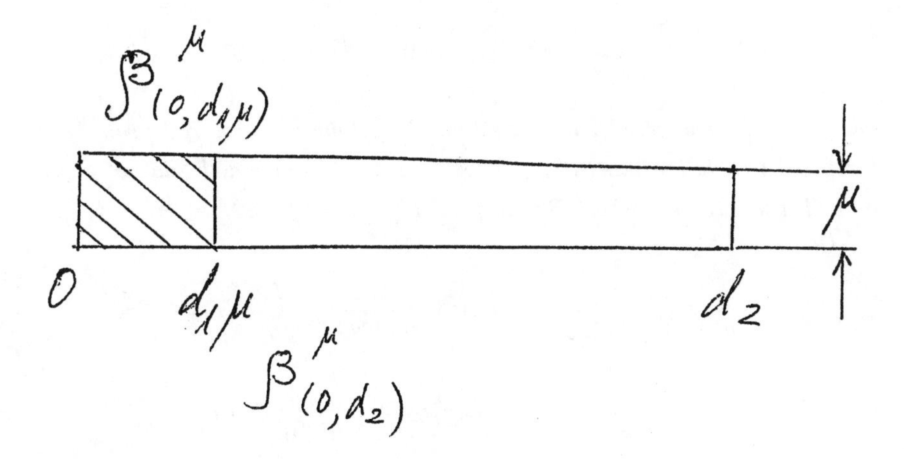

Figure 4.A2.2. Domains of Lemma 4.A2.3.

Proof

We obtain from Newton-Leibnitz formula

$$u(y_1, \tilde{x}_1) - u(y_o, \tilde{x}_1) = \int_{y_0}^{y_1} \frac{\partial u}{\partial x_1} dx_1, \quad \tilde{x}_1 = (x_2, ..., x_s).$$

Integrate this equality in $\tilde{x}_1 \in \beta^{\mu} = \{x \epsilon R^{s-1}, \frac{x}{\mu} \epsilon \beta\}$, in y_1 from 0 to d_2 and in y_0 from 0 to $d_1\mu$.

We obtain then:

$$| \, d_1\mu \int_{\beta^{\mu}_{(0,d_2)}} u dx - d_2 \int_{\beta^{\mu}_{(0,d_1\mu)}} u dx \, |=| \int_0^{d_2} \int_0^{d_1\mu} \int_{y_0}^{y_1} \int_{\beta^{\mu}} \frac{\partial u}{\partial x_1} d\tilde{x}_1 dx_1 dy_1 dy_0 \, |\leq$$

$$d_1 d_2 \mu || \frac{\partial u}{\partial x_1} ||_{L^1(\beta^{\mu}_{(0,d_2)})} \leq$$

$$d_1 d_2 \mu \sqrt{d_2 \mu^{s-1} meas\beta} || \frac{\partial u}{\partial x_1} ||_{L^2(\beta^{\mu}_{(0,d_2)})}.$$

Divide this inequality by $d_1 d_2 \mu^s meas\beta$; we obtain then

$$|\{u\}^{\mu}_{(0,d_2)} - \{u\}^{\mu}_{(0,d_1\mu)}| \leq \frac{\sqrt{d_2}}{\sqrt{\mu^{s-1} meas\beta}} || \frac{\partial u}{\partial x_1} ||_{L^2(\beta^{\mu}_{(0,d_2)})}.$$

Lemma is proved

Lemma 4.A2.4. *Let β_1 and β_2 be two (s - 1) dimensional domains, β_3 be s - dimensional domain. Denote*

$$\beta_i^\mu = \{x \in I\!R^{s-1} \mid \frac{x}{\mu} \in \beta_i\} \quad (i = 1, 2), \quad \beta_3^\mu = \{x \in I\!R^s \mid \frac{x}{\mu} \in \beta_3\},$$

and let $\tilde{\beta}^\mu_{i(a,b)}$ be a cylinder obtained from the cylinder $(a, b) \times \beta_i^\mu$ by some rotations and translations. Let $\tilde{\beta}^\mu_{i(0,d_i)} \cap \beta_3^\mu = \tilde{\beta}^\mu_{i(0,h_i\mu)}$, where d_i, h_i do not depend on μ. Let $u \in H^1(B)$, where $B = (\bigcup_{i=1}^2 \quad \beta^\mu_{i(0,d_i)}) \quad \bigcup \quad \beta_3^\mu$.

Then

$$\frac{1}{meas\tilde{\beta}^\mu_{1(0,d_1)}}(\int_{\tilde{\beta}^\mu_{(0,d_1)}} udx)^2, \quad \frac{1}{\mu meas\beta_3^\mu}(\int_{\beta_3^\mu} udx)^2 \leq$$

$$\frac{c_1}{meas\tilde{\beta}^\mu_{2(0,d_2)}}(\int_{\tilde{\beta}^\mu_{2(0,d_2)}} udx)^2 \quad + \quad c_2 \int_B (\nabla u)^2 dx,$$

where c_1, c_2 do not depend on μ.

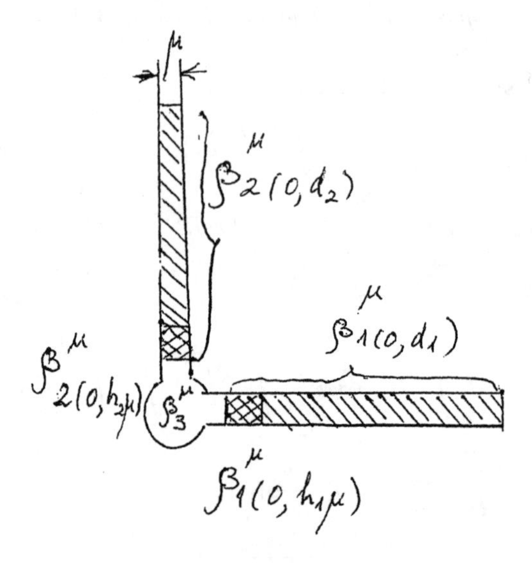

Figure 4.A2.3. Domains of Lemma 4.A2.4.

Proof.

Denote

$\frac{1}{meas\tilde{\beta}^\mu_{1(0,d_1)}}(\int_{\beta^\mu_{1(0,d_1)}} udx)$ as $\{u\}^{d_1}_{1,\mu}$, then

$$meas\tilde{\beta}^\mu_{1(0,d_1)}(\{u\}^{d_1}_{1,\mu})^2 \leq 2d_1\mu^{s-1} meas\beta_1((\{u\}^{h_1\mu}_{1,\mu})^2 \quad + \quad (\{u\}^{d_1}_{1,\mu} \quad - \quad \{u\}^{h_1\mu}_{1,\mu})^2).$$

According to Lemma 4.A2.3 the last expression is not more than

$$2d_1\mu^{s-1}meas\beta_1(\{u\}_{1,\mu}^{h_1\mu})^2 \quad + \quad 2d_1^2K\int_{\tilde{\beta}^{\mu}_{1(0,d_1)}}(\nabla u)^2dx,$$

Here constant K is independent of μ and d_1.

The first term is estimated as follows:

$$2d_1\mu^{s-1} \quad meas\beta_1 \quad (\{u\}_{1,\mu}^{h_1\mu})^2 \le$$

$$\le 4d_1\mu^{s-1} \quad meas\beta_1 \quad (\{u\}_{3,\mu})^2 + 4d_1\mu^{s-1} \quad meas\beta_1 \quad (\{u\}_{1,\mu}^{h_1\mu} - \{u\}_{3,\mu})^2,$$

here $\{u\}_{3,\mu} = \frac{1}{meas\beta_3^{\mu}}\int_{\beta_3^{\mu}} udx$.

Then, applying Lemma 4.A2.2 we obtain

$$2d_1\mu^{s-1} \quad meas\beta_1 \quad (\{u\}_{1,\mu}^{h_1\mu})^2 \le 4d_1\mu^{s-1} \quad meas\beta_1 \quad (\{u\}_{3,\mu})^2 \quad +$$

$$+ \quad 4d_1\mu^{s-1} \quad meas\beta_1 \quad c\mu^{2-s} \quad \int_{\beta_3^{\mu}}(\nabla u)^2dx \le$$

$$\le 4d_1 \quad \frac{meas\beta_1}{meas\beta_3} \quad \frac{1}{\mu} \quad meas\tilde{\beta}_3^{\mu} \quad (\{u\}_{3,\mu})^2 \quad +$$

$$+ \quad (4cd_1meas\beta_1)\mu\int_{\beta_3^{\mu}}(\nabla u)^2dx.$$

Let $\{u\}_{2,\mu}^{b} \quad = \quad \frac{1}{meas\tilde{\beta}^{\mu}_{2(0,b)}}\int_{\tilde{\beta}^{\mu}_{2(0,b)}} udx$. Then

$$meas\tilde{\beta}_3^{\mu}(\{u\}_{3,\mu})^2 \quad = \quad meas\beta_3\mu^s(\{u\}_{3,\mu})^2 \quad \le$$

$$\le 2meas\beta_3\mu^s(\{u\}_{2,\mu}^{h_2\mu})^2 \quad + \quad 2meas\beta_3\mu^s(\{u\}_{2,\mu}^{h_2\mu} \quad - \quad \{u\}_{3,\mu})^2 \quad \le$$

$$\le \quad 4mes\beta_3\mu^s(\{u\}_{2,\mu}^{d_2})^2 \quad + \quad 4meas\beta_3\mu^s(\{u\}_{2,\mu}^{d_2} - \{u\}_{2,\mu}^{h_2\mu})^2+$$

$$+ \quad 2meas\beta_3c\mu^2\int_{\beta_3^{\mu}}(\nabla u)^2dx \quad \le \quad \frac{4mes\beta_3}{d_2mes\beta_2}\mu mes\tilde{\beta}_2^{\mu}(\{u\}_{2,\mu}^{d_2})^2+$$

$$+4meas\beta_3\mu^s\frac{d_2}{meas\beta_2}\mu^{1-s}K\int_{\tilde{\beta}^{\mu}_{2(0,d_2)}}(\nabla u)^2dx \quad + \quad 2\mu cmeas\beta_3\int_{\beta_3^{\mu}}(\nabla u)^2dx \le$$

$$\le \mu(\frac{4meas\beta_3}{d_2meas\beta_2}\frac{1}{meas\tilde{\beta}^{\mu}_{2(0,d_2)}} \quad (\int_{\tilde{\beta}^{\mu}_{2(0,d_2)}} \quad udx)^2+$$

$$+(\frac{4meas\beta_3}{meas\beta_2}d_2K \quad + \quad 2cmeas\beta_3) \quad \int_B(\nabla u)^2dx).$$

Lemma is proved with the following values of constants :

$$c_1 = max\{16\frac{d_1 meas\beta_1}{d_2 meas\beta_2}, \quad \frac{4meas\beta_3}{d_2 meas\beta_2}\},$$

$$c_2 = max\{4d_1 meas\beta_1(\frac{4d_2K}{meas\beta_2}+3c)+2Kd_1^2, \; 4meas\beta_3 d_2K/meas\beta_2+2cmeas\beta_3\},$$

$(\mu < 1)$.

Lemma 4.A2.5. *Let* $u \in H^1(G^\mu)$, $G^\mu = \{x \in I\!R^s | \frac{x}{\mu} \in G\}$ *and* G *is a domain in* $I\!R^s$. *Then the estimate takes place*

$$||u||^2_{L^2(G^\mu)} \le \frac{1}{measG^\mu}(\int_{G^\mu} udx)^2 + \mu^2 c||\nabla u||^2_{L^2(G^\mu)}.$$

Here constant c *depends on* G.

Proof.

Change the variables $\xi = \frac{x}{\mu}, \quad dx = \mu^s d\xi$. The classical Poincare inequality yields

$$||u||^2_{L^2(G^\mu)} = \mu^s \int_G u^2(\mu\xi)d\xi \le \mu^s \frac{1}{measG}(\int_G u(\mu\xi)d\xi)^2+$$

$$+\mu^s c\int_G (\nabla_\xi u(\mu\xi))^2 d\xi = \frac{1}{measG^\mu}(\int_{G^\mu} u(x)dx)^2 + \mu^2 c\int_{G^\mu} (\nabla_x u(x))^2 dx.$$

Lemma is proved.

Lemma 4.A2.6. *Let* $u \in H^1(\beta^\mu_{(0,d)})$ *then*

$$||u||^2_{L^2(\beta^\mu_{(0,d)})} \le \frac{4}{meas\beta^\mu_{(0,d)}}(\int_{\beta^\mu_{(0,d)}} u(x)dx)^2 + 8d^2\int_{\beta^\mu_{(0,d)}} (\nabla u)^2 dx$$

for all sufficiently small μ.

Proof

For any $y_1, y_0 \in R, \quad \tilde{x}_1 \in R^{s-1}$ we obtain

$$(u(y_1,\tilde{x}_1))^2 \le 2(u(y_1,\tilde{x}_1) - u(y_0,\tilde{x}_1))^2 + 2u^2(y_0,\tilde{x}_1) \le$$

$$\le 2(u(y_0,\tilde{x}_1))^2 + 2(\int_{y_o}^{y_1} \frac{\partial u(x)}{\partial x_1}dx_1)^2 \le$$

$$\le 2(u(y_0,\tilde{x}_1))^2 + 2d\int_0^d (\frac{\partial u}{\partial x_1})^2 dx_1$$

Integrate the inequality in $\tilde{x}_1$ over $\beta^\mu_{(0,d)}$, in y_1 from 0 to d, in y_0 from 0 to μ.

$$\mu \int_{\beta^{\mu}_{(0,d)}} u^2 dx \leq 2d \int_{\beta^{\mu}_{(0,\mu)}} u^2 dx + 2d^2 \mu \int_{\beta^{\mu}_{(0,d)}} (\frac{\partial u}{\partial x_1})^2 dx.$$

The Poincaré inequality for $\beta^{\mu}_{(0,\mu)}$ with constant $c_P \mu^2$ gives:

$$\mu \int_{\beta^{\mu}_{(0,d)}} u^2 dx \leq \frac{2d}{\mu^s meas \beta} (\int_{\beta^{\mu}_{(0,\mu)}} u dx)^2 + 2dc_P \mu^2 \int_{\beta^{\mu}_{(0,d)}} (\nabla u)^2 dx +$$

$$+ 2d^2 \mu \int_{\beta^{\mu}_{(0,d)}} (\frac{\partial u}{\partial x_1})^2 dx.$$

So,

$$\mu \int_{\beta^{\mu}_{(0,d)}} u^2 dx \leq 2d(\mu^s meas \beta (\{u\}^{\mu}_{(0,\mu)})^2 +$$

$$+ 2\mu d(d + \mu c_P) \int_{\beta^{\mu}_{(0,d)}} (\nabla u)^2 dx) \leq 4d\mu^s meas \beta ((\{u\}^{\mu}_{(0,d)})^2 + (\{u\}^{\mu}_{(0,\mu)} -$$

$$\{u\}^{\mu}_{(0,d)})^2) + 4d^2 \mu \int_{\beta^{\mu}_{(0,d)}} (\nabla u)^2 dx$$

for all $\mu \leq d/c_P$.

Lemma 4.A2.3 yields

$$\mu \int_{\beta^{\mu}_{(0,d)}} u^2 dx \leq \frac{4\mu}{meas \beta^{\mu}_{(0,d)}} (\int_{\beta^{\mu}_{(0,\mu)}} u dx)^2 + 8d^2 \mu \int_{\beta^{\mu}_{(0,d)}} (\nabla u)^2 dx.$$

Here

$$\{u\}^{\mu}_{(a,b)} = \frac{1}{meas \beta^{\mu}_{(a,b)}} \int_{\beta^{\mu}_{(a,b)}} u dx.$$

Lemma 4.A2.6 is proved.

Lemma 4.A2.6 admits the following generalization:

Lemma 4.A2.7. *Let the domain $\tilde{\beta}^{\mu}_{(a,b)}$ be obtained from $\beta^{\mu}_{(a,b)}$ by means of rotation and translation, then the estimate holds true*

$$\|u\|^2_{L^2(\tilde{\beta}^{\mu}_{(0,d)})} \leq \frac{4}{meas \tilde{\beta}^{\mu}_{(0,d)}} (\int_{\tilde{\beta}^{\mu}_{(0,d)}} u dx)^2 + c_3 \int_{\tilde{\beta}^{\mu}_{(0,d)}} (\nabla u)^2 dx,$$

where c_3 is independent of d and μ, μ is sufficiently small.

Theorem 4.A2.1. (Poincaré inequality for the rod structure) *Let $\beta_1, ..., \beta_r$ be $(s-1)-$dimensional domains, $\beta_{3,1}, ..., \beta_{3,p}$ be $s-$dimensional domains, $\beta^{\mu}_i = \{x \in I\!R^{(s-1)} | x/\mu \in \beta_i\}$, $i = 1, ..., r$, $\beta^{\mu}_{3,j} = \{x \in I\!R^{(s-1)} | x/\mu \in$*

$\beta_{3,j}\}$, $j = 1, ..., p$, $\tilde{\beta}^{\mu}_{i,(a,b)}$ *be cylinders, obtained from the cylinders* $(a,b)\times\beta^{\mu}_i$ *by means of rotations and translations and let* $\tilde{\beta}^{\mu}_{3,j}$ *are obtained from* $\beta^{\mu}_{3,j}$ *by translations only. Let* d_i, h_i *do not depend on* μ*. We assume that the intersection of closures of the sets* $\tilde{\beta}^{\mu}_{i,(0,d_i)}$ *and* $\tilde{\beta}^{\mu}_{j,(0,d_j)}$ *is empty and that the intersection of closures of the sets* $\tilde{\beta}^{\mu}_{3,i}$ *and* $\tilde{\beta}^{\mu}_{3,j}$ *is also empty if* $i \neq j$*. Moreover we assume that for each pair* i, j *either the intersection of closures of the sets* $\tilde{\beta}^{\mu}_{i,(0,d_i)}$ *and* $\tilde{\beta}^{\mu}_{3,j}$ *is empty or* $\tilde{\beta}^{\mu}_{i,(0,d_i)} \cap \tilde{\beta}^{\mu}_{3,j} = \tilde{\beta}^{\mu}_{i,(0,h_i\mu)}$.

Let the set $\tilde{B} = \cup_{i=1}^{r} \tilde{\beta}^{\mu}_{i,(0,d_i)} \cup \cup_{j=1}^{p} \tilde{\beta}^{\mu}_{3,j}$ *be connected. Then the estimate holds true*

$$\int_{\tilde{B}} u^2 dx \leq \frac{1}{meas\tilde{B}} \left(\int_{\tilde{B}} udx\right)^2 + A\int_{\tilde{B}} (\nabla u)^2 dx,$$

where A *does not depend on* μ, μ *is sufficiently small.*

If there exists such a cylinder $\tilde{\beta}^{\mu}_{r,(0,d_r)}$ *that* $\int_{\tilde{\beta}^{\mu}_{r,(0,d_r)}} u\, dx = 0$, *then*

$$\int_{\tilde{B}} u^2 dx \leq A\int_{\tilde{B}} (\nabla u)^2 dx.$$

Proof

Suppose that $\int_{\tilde{\beta}^{\mu}_{r,(0,d_r)}} u\, dx = 0$. We shall prove that

$$\int_{\tilde{B}} u^2 dx \leq A\int_{\tilde{B}} (\nabla u)^2 dx.$$

First we deduce from lemmas 4.A2.6, 4.A2.7 that for all $i = 1, ..., r$

$$\int_{\tilde{\beta}^{\mu}_{i,(0,d_i)}} u^2 dx \leq \frac{4}{meas\tilde{\beta}^{\mu}_{i,(0,d_i)}} \left(\int_{\tilde{\beta}^{\mu}_{i,(0,d_i)}} udx\right)^2 + A_1\int_{\tilde{\beta}^{\mu}_{i,(0,d_i)}} (\nabla u)^2 dx,$$

(when $i = r$ the first term vanishes) and that for all $j = 1, ..., p$

$$\int_{\tilde{\beta}^{\mu}_{3,j}} u^2 dx \leq \frac{1}{meas\tilde{\beta}^{\mu}_{3,j}} \left(\int_{\tilde{\beta}^{\mu}_{3,j}} udx\right)^2 + \mu^2 c\int_{\tilde{\beta}^{\mu}_{3,j}} (\nabla u)^2 dx,$$

where A_1, c are independent of μ. Fix any i (respectively j). Define *the initial set* as the set $\tilde{\beta}^{\mu}_{i,(0,d_i)}$ (respectively the set $\tilde{\beta}^{\mu}_{3,j}$) .

The set $\tilde{B}$ is connected. Therefore there exists the chain c of sets

$$\tilde{\beta}^{\mu}_{3,j_1}, \tilde{\beta}^{\mu}_{i_1,(0,d_{i_1})}, \tilde{\beta}^{\mu}_{3,j_2}, \tilde{\beta}^{\mu}_{i_2,(0,d_{i_2})}, ..., \tilde{\beta}^{\mu}_{3,j_q},$$

such that each two consecutive sets (units) have the nonempty intersection,

such that the intersection of the first of these units with the initial set $\tilde{\beta}^{\mu}_{i,(0,d_i)}$ is also not empty and

such that the intersection of the last of these units $\tilde{\beta}^{\mu}_{3,j_q}$ with the set of the vanishing average $\tilde{\beta}^{\mu}_{r,(0,d_r)}$ is also not empty.

(In the case of the initial set $\tilde{\beta}^{\mu}_{3,j}$ the chain starts from $\tilde{\beta}^{\mu}_{i_1,(0,d_{i_1})}$ directly.)

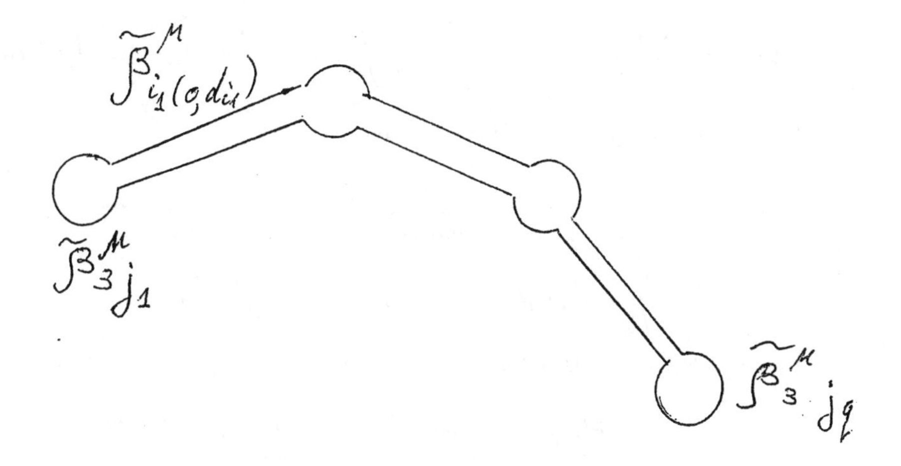

Figure 4.A2.4. A chain of the proof of Theorem 4.A2.1.

Indeed, at the first step we can add to the initial set all units $\tilde{\beta}^{\mu}_{3,l}$ and $\tilde{\beta}^{\mu}_{l,(0,d_l)}$ which have the nonempty intersection with the initial set; then at the second step we add all units which have the nonempty intersection with the units obtained at the first step and so on. After the finite number of steps we shall either obtain the whole set $\tilde{B}$ or come to a contradiction to the connectivity of $\tilde{B}$. We can also suppose that all units are different in the chain C because if there is a unit repeated twice then we can exclude all the units between these two repetitions without loss of connectivity of the chain. Thus there are not more than $r+p$ units in the chain. Applying lemma 4.A2.4 $q+1$ times maximum we obtain the inequality

$$\frac{1}{meas\tilde{\beta}}\left(\int_{\tilde{\beta}} udx\right)^2 \quad \leq \quad \frac{(max\ c_1)^{q+1}}{meas\tilde{\beta}^{\mu}_{r,(0,d_r)}} \quad \left(\int_{\tilde{\beta}^{\mu}_{r,(0,d_r)}} udx\right)^2 \quad +$$

$$max\ c_2 \sum_{i=0}^{q}(max\ c_1)^q \int_{\tilde{B}} (\nabla u)^2 dx,$$

where $\tilde{\beta}$ is the initial set of the chain, $max\ c_1$ and $max\ c_2$ are the maximal values of the constants $c_1 > 1$ and c_2 of lemma 4.A2.4 for all pairs $(\tilde{\beta}^{\mu}_{1,(0,d_1)}, \tilde{\beta}^{\mu}_{2,(0,d_2)})$. Since $\int_{\tilde{\beta}^{\mu}_{r,(0,d_r)}} udx \ = \ 0$ we obtain:

$$\frac{1}{meas\tilde{\beta}}\left(\int_{\tilde{\beta}} udx\right)^2 \quad \leq \quad max\ c_2 \frac{(max\ c_1)^{q+1} \ - \ 1}{max\ c_1 \ - \ 1} \int_{\tilde{B}} (\nabla u)^2 dx.$$

Thus

$$\int_{\tilde{\beta}} u^2 dx \quad \leq \quad A_2 \int_{\tilde{B}} (\nabla u)^2 dx,$$

where A_2 is some constant independent of μ, and, therefore,

$$\int_{\tilde{B}} u^2 dx \leq \sum_{i=1}^{r} \int_{\tilde{\beta}^{\mu}_{i,(0,d_i)}} u^2 dx + \sum_{j=1}^{p} \int_{\tilde{\beta}^{\mu}_{3,j}} u^2 dx \leq (p+r)A_2 \int_{\tilde{B}} (\nabla u)^2 dx.$$

Let now $D_r = \int_{\tilde{\beta}^{\mu}_{r,(0,d_r)}} u dx \neq 0$, then

$$\int_{\tilde{B}} u^2 dx - \frac{1}{meas\tilde{B}} (\int_{\tilde{B}} u dx)^2 \leq$$

$$\int_{\tilde{B}} (u - D_r)^2 dx \leq (p+r)A_2 \int_{\tilde{B}} (\nabla u)^2 dx.$$

Indeed calculate $min_{z\in R} \int_{\tilde{B}} (u-z)^2 dx$. Let $F(z) = \int_{\tilde{B}} (u-z)^2 dx$, then

$$F'(z) = 2\, meas\tilde{B}\, z - 2 \int_{\tilde{B}} u dx$$

and

$$min_{z\in R} F(z) = \int_{\tilde{B}} u^2 dx - \frac{1}{meas\tilde{B}} (\int_{\tilde{B}} u dx)^2.$$

Theorem is proved.

Let now $x^0 \in G_0$, (G_0 be defined in section 4.1). Consider the parallelepiped $\Pi_{x^0} = (-1, -d_0\mu) \times (x_2^0 - d_0\mu, x_2^0 + d_0\mu) \times ... \times (x_s^0 - d_0\mu, x_s^0 + d_0\mu)$. Assume that $\varphi \in H_{1,0}(B_\mu)$. Then we can extend φ as a zero onto $\Pi_{x^0} \setminus B_\mu$. So this extension φ belongs to $H^1(B_\mu \cup \Pi_{x^0})$. Due to Theorem 4.A2.1 the estimate holds

$$\int_{B_\mu \cup \Pi_{x^0}} \varphi^2 dx \leq A \int_{B_\mu \cup \Pi_{x^0}} (\nabla\varphi)^2 dx,$$

and hence

$$\int_{B_\mu} \varphi^2 dx \leq A \int_{B_\mu} (\nabla\varphi)^2 dx.$$

Thus the Poincaré - Friedrichs inequality in B_μ with the constant independent of μ is proved. Formulate it as a theorem.

Theorem 4.A2.2. *Let $\varphi \in H_{1,0}(B_\mu)$. Then*

$$\int_{B_\mu} \varphi^2 dx \leq A \int_{B_\mu} (\nabla\varphi)^2 dx.$$

where the constant A does not depend on μ, μ is sufficiently small.

All results of this Appendix were proved in [144].

4.A3 Appendix 3: the Korn inequality for the finite rod structures.

Here we prove the Korn inequality for the finite rod structures such that every pair of the intersecting constitutive cylinders is star-wise with respect to some ball of the radius of order μ. The proof in a more general case will be given in the appendix to the next chapter.

Lemma 4.A3.1 [129]. *Let Ω be a bounded domain star-like with respect to the ball $Q_{R_1} = \{x : |x| < R_1\}$ let the diameter of Ω be R, $u = (u_1, \dots, u_s) \in (H^1(\Omega))^s$.*

Then

$\|\nabla u\|^2_{L^2(\Omega)} \leq C_1(\frac{R}{R_1})^{s+1}\|e(u)\|^2_{L^2(\Omega)} + C_2(\frac{R}{R_1})^s\|\nabla u\|^2_{L^2(Q_{R_1})}$, *where C_1, C_2 depend on s only,*

$$\|e(u)\|^2_{L^2(\Omega)} = \frac{1}{4}\int_\Omega \sum_{i,j=1}^{s}\left(\frac{\partial u_i}{\partial x_j} + \frac{\partial u_j}{\partial x_i}\right)^2 dx,$$

$e(u)$ is the matrix $s \times s$ with elements $\frac{1}{2}\left(\frac{\partial u_i}{\partial x_j} + \frac{\partial u_j}{\partial x_i}\right)$.

Theorem 4.A3.1. *Let Ω be a domain of Lemma 4.A3.1, let $\widehat{H}$ be a subspace of $(H^1(\Omega))^s$; let Q_{R_1} belongs to a subdomain $\Omega_1 \subset \Omega$ such that there exists a constant C_3 depending on s only such that for any $u \in \widehat{H}$, the following estimate holds :*

$$\|\nabla u\|^2_{L^2(\Omega_1)} \leq C_3\|e(u)\|^2_{L^2(\Omega_1)}.$$

Then the inequality holds for all $u \in \widehat{H}$

$$\|\nabla u\|^2_{L^2(\Omega)} \leq C_4\left(\frac{R}{R_1}\right)^{s+1}\|e(u)\|^2_{L^2(\Omega)} \tag{4.A3.1}$$

where C_4 depends on s only.

Proof

Applying Lemma 4.A3.1 and estimating

$$\|\nabla u\|^2_{L^2(Q_{R_1})} \leq \|\nabla u\|^2_{L^2(\Omega_1)} \leq C_3\|e(u)\|^2_{L^2(\Omega_1)} \leq C_3\|e(u)\|^2_{L^2(\Omega)}$$

we get (4.A3.1) with $C_4 = C_1 + C_2C_3$ because $\frac{R_1}{R} \leq 1$. The theorem is proved.

Consider now a thin rectangle G_μ that is $(0,1) \times \left(-\frac{\mu}{2}, \frac{\mu}{2}\right)$.

Corollary 4.A3.1 (for Theorem 4.A3.1) *There exists $C_6 > 0$ such that for any $\mu > 0$ small enough, for any $u \in (H^1(G_\mu))^s$, such that $u|_{x_1=0} = 0$, the estimate holds true : $\|\nabla u\|^2_{L^2(G_\mu)} \leq C_6\mu^{-3}\|e(u)\|^2_{L^2(G_\mu)}$.*

Proof

Let us consider the square $\Omega_1 = (0, \mu) \times \left(-\frac{\mu}{2}, \frac{\mu}{2}\right)$.

For any $u \in H^1(G_\mu)$ such that $u|_{x_1=0} = 0$, $\|\nabla u\|^2_{L^2(\Omega_1)} \leq C_7\|e(u)\|^2_{L^2(\Omega_1)}$. Applying now theorem Theorem 4.A3.1 with the disk

$$Q_{R_1} = \left\{x_1, x_2 \in \mathbb{R}^2\ ;\ \left(x_1 - \frac{\mu}{2}\right)^2 + x_2^2 < \frac{\mu^2}{4}\right\}$$

we obtain the assertion of the corollary.

Theorem 4.A3.2 *Let Ω be a domain of Lemma 4.A3.1 and let u be a vector-valued function of $H^1(\Omega)$ such that for any rigid displacement η,* $\int_\Omega (\nabla u, \nabla\eta)dx = \sum_{i,j=1} \int_\Omega \frac{\partial u_i}{\partial x_j} \cdot \frac{\partial \eta_i}{\partial x_j} dx = 0.$

Then (4.A3.1) holds with the constant C_4 depending on s only.

Proof

Consider a rigid displacement $\overline{\eta}$, such that $\int_{Q_{R_1}} (\nabla(u-\overline{\eta}), \nabla\eta)dx = 0$, for all rigid displacements η.

Then $\|\nabla(u-\overline{\eta})\|^2_{L^2(\Omega)} = \|\nabla u\|^2_{L^2(\Omega)} - 2\int_\Omega (\nabla u, \nabla\overline{\eta})dx + \|\nabla\overline{\eta}\|^2_{L^2(\Omega)} = \|\nabla u\|^2_{L^2(\Omega)} + \|\nabla\overline{\eta}\|^2_{L^2(\Omega)} \geq$
$\geq \|\nabla u\|^2_{L^2(\Omega)}$.

Applying Lemma 4.A3.1 to $u - \overline{\eta}$, we get

$$\|\nabla(u-\overline{\eta})\|^2_{L^2(\Omega)} \leq C_1 \left(\frac{R}{R_1}\right)^{s+1} \|e(u-\overline{\eta})\|^2_{L^2(\Omega)} + C_2 \left(\frac{R}{R_1}\right)^{s} \|\nabla(u-\overline{\eta})\|^2_{L^2(Q_{R_1})}$$

Applying now the Korn inequality for a function orthogonal to all rigid displacements in a ball (for Q_{R_1}) [129], we have :

$$\|\nabla(u-\overline{\eta})\|^2_{L^2(Q_{R_1})} \leq C_5 \|e(u-\overline{\eta})\|^2_{L^2(Q_{R_1})}$$

with C_5 depending only on s.

So, finally,

$$\|\nabla u\|^2_{L^2(\Omega)} \leq \|\nabla(u-\overline{\eta})\|^2_{L^2(\Omega)} \leq$$

$$\leq C_1 \left(\frac{R}{R_1}\right)^{s+1} \|e(u-\overline{\eta})\|^2_{L^2(\Omega)} + C_2 \left(\frac{R}{R_1}\right)^{s} C_5 \|e(u-\overline{\eta})\|^2_{L^2(Q_{R_1})} \leq$$

$$\leq (C_1 + C_2C_5) \left(\frac{R}{R_1}\right)^{s+1} \|e(u-\overline{\eta})\|^2_{L^2(\Omega)} = (C_1 + C_2C_5) \left(\frac{R}{R_1}\right)^{s+1} \|e(u)\|^2_{L^2(\Omega)}$$

because $e(u-\overline{\eta}) = e(u)$.

The theorem is proved.

Consider the set $\mathcal{A}$ of finite rod structures B_μ such that

- for any pair of cylinders $B^{(1)}_\mu$ and $B^{(2)}_\mu$ constituting B_μ and having common points in one of their bases, there exists a ball $Q_{c_0\mu}$ of the radius $c_0\mu$ (c_0 is independent of μ) such that it belongs to the union $\overline{B}^{(1)}_\mu \cup \overline{B}^{(2)}_\mu$ and such that this union is star-wise with respect to the ball $Q_{c_0\mu}$;

and

- for any cylinder with a base belonging to $\partial_1 B_\mu$ there exists a ball $Q_{c_0\mu}$ of the radius $c_0\mu$ in a μ-vicinity of $\partial_1 B_\mu$ such that this cylinder is a star-wise domain with respect to this ball $Q_{c_0\mu}$ (if such cylinder is associated with an additional cube of Remark 4.1.1, then we require the cylinder with the cube to be star-wise domain with respect to a ball $Q_{c_0\mu}$ belonging to the cube).

It is evident that in the case of dimension 2 all the finite rod structures belong to $\mathcal{A}$, and in the case of dimension 3 if all the bases of cylinders are star-wise with respect to some ball of the radius $c_0\mu$ then B_μ belongs to $\mathcal{A}$. More general finite rod structures will be studied in Appendix 5.A2.

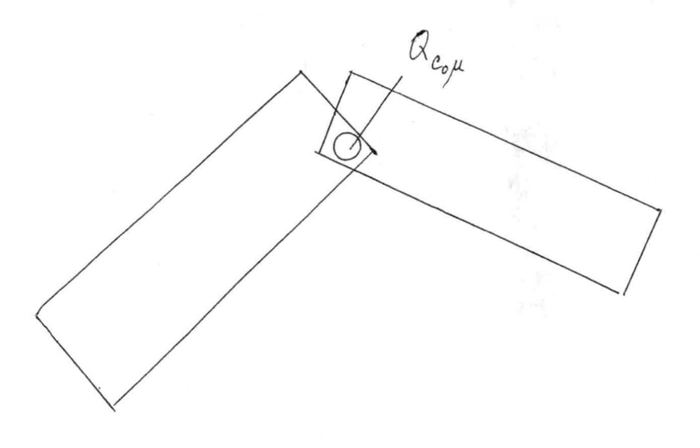

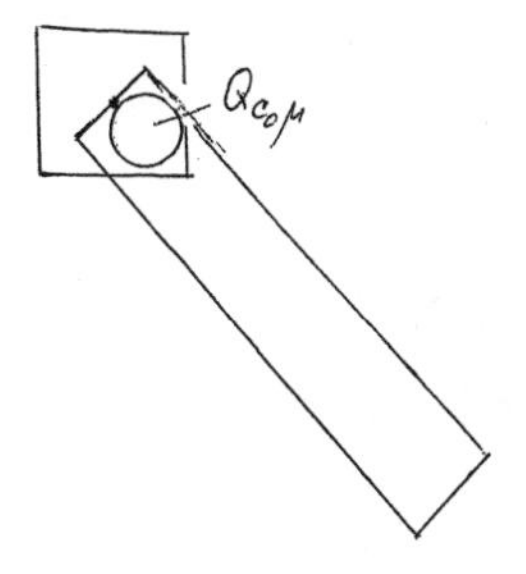

Figure 4.A3.1. To the proof of the Korn inequality

Theorem 4.A3.3. *Let $B_\mu \in \mathcal{A}$. There exists $C_7 > 0$ and $r \in \mathbb{R}$ such that for any positive μ small enough, for any $u \in (H^1(B_\mu))^s, u|_{\partial_1 B_\mu} = 0$ (or $u|_{\sum_0} = 0$), the estimate holds true :*

$$\|\nabla u\|^2_{L^2(B_\mu)} \leq C_7 \mu^{-r} \|e(u)\|^2_{L^2(B_\mu)}.$$

Proof.

Apply lemma 4.A3.1 to all cylinders and their balls $Q_{c_0\mu}$ of the definition of set $\mathcal{A}$. Estimate $\|\nabla u\|^2_{L^2(Q^{(0)}_{c_0\mu})}$ for a ball $Q^{(0)}_{c_0\mu}$ in the μ-vicinity of $\partial_1 B_\mu (i.e. \sum_0)$ by $C_8\|e(u)\|^2_{L^2(Q^{(0)}_{c_0\mu})}$ with a constant $C_8 > 0$ independent of μ as in Theorem 4.A3.1. Let now $Q^{(1)}_{c_0\mu}$ be another ball of the definition of $\mathcal{A}$. Consider a chain of cylinders $B^{(1)}_\mu, \ldots, B^{(N)}_\mu$ relating $Q^{(1)}_{c_0\mu}$ and $Q^{(0)}_{c_0\mu}$ in such a way that $B^{(1)}_\mu$ contains $Q^{(1)}_{c_0\mu}$ and $B^{(N)}_\mu$ contains $Q^{(0)}_{c_0\mu}$ and every two consecutive cylinders $B^{(i)}_\mu$, $B^{(i+1)}_\mu$ contain a ball $Q^{(i+1)}_{c_0\mu}$ such that $Q^{(i+1)}_{c_0\mu} \subset \overline{B}^{(i)}_\mu \cup \overline{B^{(i+1)}_\mu}$ and $\overline{B}^{(i)}_\mu \cup \overline{B}^{(i+1)}_\mu$ is a star-wise domain with respect to $Q^{(i+1)}_{c_0\mu}$.

We have by Lemma 4.A3.1,

$$\|\nabla u\|^2_{L^2(Q^{(1)}_{c_0\mu})} \le \|\nabla u\|^2_{L^2(B^{(1)}_\mu)} \le$$

$$\le C_9\mu^{-(s+1)}(\|e(u)\|^2_{L^2(B^{(1)}_\mu)} + \|\nabla u\|^2_{L^2(Q^{(2)}_{c_0\mu})}) \le$$

$$\le C_9\mu^{-(s+1)}(\|e(u)\|^2_{L^2(B^{(1)}_\mu)} +$$

$$+C_9\mu^{-(s+1)}(\|e(u)\|^2_{L^2(B^{(2)}_\mu)} + \|\nabla u\|^2_{L^2(Q^{(3)}_{c_0\mu})})) \le \ldots$$

$$\le C_{10}\mu^{-N(s+1)}(\|e(u)\|^2_{L^2(B_\mu)} + \|\nabla u\|^2_{L^2(Q^{(0)}_{c_0\mu})}) \le$$

$$\le C_{11}\mu^{-N(s+1)}\|e(u)\|^2_{L^2(B_\mu)},$$

where $C_8, C_9, C_{10}, C_{11} > 0$ do not depend on μ. Theorem is proved.

Chapter 5

Lattice Structures

Here we consider the so called lattice structures, i.e. finite rod structures of Chapter 4 with a great number of elementary rods. Its definition is given at section 5.1.

As in the previous chapter we first study these structures by L-convergence technique for an elliptic equation in section 5.2 and for dynamic problems in section 5.3. We start the study from the simplest (cross-like) rectangular lattice and then (in section 5.3) consider the general case. In section 5.4 the applicability of the L-convergence technique to the elasticity problem is discussed. Then we develop a refined analysis of fields, including asymptotic expansion of elasticity equation based on the expansions of Chapter 2 and Chapter 4. We apply first the analysis of Chapter 4 to obtain some limit differential equations at the axial segments of the elementary rods. Then we apply the homogenization method to find the asymptotic expansion as the period ε tends to zero. Finally, we consider the L-convergence of random lattices. We generalize the L-convergence tools in this case and formulate a result of the convergence for such structures with high probability (section 5.7).

5.1 Definition of lattice structure

The lattice-like domains simulate some widely adopted engineering constructions such as frameworks of houses, trusses of bridges, industrial installations, supports of electric power lines, spaceship grids, etc. as well as some capillary or fissured systems. The methods of solving the system of equations of the strength of materials theory (structural mechanics methods) presently used for calculating the frameworks have the drawback that the number of equations increases with number of nodes of lattice. This greatly impedes the calculation of lattices with large number of nodes, especially when investigating non-stationary and nonlinear processes. The homogenization techniques and splitting principle for the homogenized operator proposed in [134]- [136], [16] allowed essentially simplify the process of calculation of such structures.

Definition 5.1.1 *Let $\beta_1, ..., \beta_J$ be bounded domains in $I\!R^{s-1}$, $(s = 2,3)$ with a piecewise smooth boundary satisfying the strong cone condition, let B_j $(j = 1, ..., J)$ be cylinders, suppose*

$$B_j = \{x \in I\!R^s \mid (x_2/\mu, ..., x_s/\mu \in \beta_j, \ x_1 \in I\!R\},$$

and $B^\alpha_{h,j}$ is the cylinder obtained from B_j by orthogonal transformation $\tilde{\Pi}$ with the matrix α^T, $\alpha = (\alpha_{il})$ and by translation by $h = (h_1, ..., h_s)^T$. Let e^α_h be an s-dimensional vector obtained from the vector $i(h, \alpha)$ of the axis Ox_1 with the beginning at the point O by orthogonal transformation $\tilde{\Pi}$ of $I\!R^s$ with matrix α^T and by translation by $h = (h_1, ..., h_s)$. Let B be a union of all vectors e^α_h when α belongs to a set $\Delta \subset I\!R^{s^2}$ and h to a set $H_\alpha \subset I\!R^s$, and these sets are independent of ε, μ. Let B be such that any two segments e^α_h can have only one common point which is the end point for both segments. The end points of e^α_h are called nodes.

We assume that the system of the segments B is connected and periodic in all the x_i with period 1, with periodicity cube $Q = \{x \in I\!R^s \mid -1/2 \le x_i \le 1/2\}$ intersecting a finite number of segments from B : $e^{\alpha_1}_{h_1}, ..., e^{\alpha_I}_{h_I}$. To each of the segments $e^{\alpha_i}_{h_i}$, $(i = 1, ...I)$, we assign the cylinder $B^{\alpha_i}_{h_i,j_i}$ and denote by $\tilde{B}^{\alpha_i}_{h_i,j_i}$ the part of $B^{\alpha_i}_{h_i,j_i}$ contained between the two planes passing through the ends of segment $e^\alpha_{h\varepsilon}$ and perpendicular to it (we assume that the bases belong to $\tilde{B}^{\alpha_i}_{h_i,j_i}$). Denote $B_{\xi,\mu}$, the set of interior points of the periodic continuation (with the periodicity cube $(0,1)^s$) to $I\!R^s$ of the union of the cylinders $\tilde{B}^{\alpha_i}_{h_i,j_i}$, $i = 1, ..., I$, while assuming that the intersection of the set $B_{\xi,\mu}$ and the cube Q is connected. By lattice structure $B_{\varepsilon,\mu}$ we shall mean the set $\{x \in I\!R^s \mid x/\varepsilon \in B_{\xi,\mu}\}$ produced from $B_{\xi,\mu}$ by homothetic contraction of space in $1/\varepsilon$ times.

We will assume that the lattice structure has a piecewise smooth boundary, satisfying the cone condition.

The maximal subsets of the cylinders $\tilde{B}^\alpha_{h,j,\varepsilon} = \{x \in I\!R^s \mid x/\varepsilon \in \tilde{B}^\alpha_{h,j}\}$ for which any cross-section by the plane perpendicular to the generatrix of $\tilde{B}^\alpha_{h,j,\varepsilon}$ is free of points of other cylinders we call sections. By S_0 we denote the set of sections and let $c_0\varepsilon\mu > d$, where d is the maximal diameter of the connected subsets $B_{\varepsilon,\mu}\backslash S_0$, with c_0 being independent of ε and μ. By $e^\alpha_{h\varepsilon}$ we will mean the segment $\{x \in I\!R^s \mid x/\varepsilon \in e^\alpha_h \}$ produced from e^α_h by homothetic contraction of space in $1/\varepsilon$ times.

Let $\tilde{B}^{\alpha,+}_{h,j,\varepsilon}$ $(\tilde{B}^{\alpha,++}_{h,j,\varepsilon})$ be a part of $\tilde{B}^\alpha_{h,j,\varepsilon}$, contained between the planes spaced by $c_0\varepsilon\mu$ (by $(c_0+1)\varepsilon\mu$) from the bases of $\tilde{B}^\alpha_{h,j,\varepsilon}$. Let $B^+_{\varepsilon\mu}$ be a union of all $\tilde{B}^{\alpha,+}_{h,j,\varepsilon}$ and let $B^{++}_{\varepsilon\mu}$ be a union of all $\tilde{B}^{\alpha,++}_{h,j,\varepsilon}$. The domain $B_{\varepsilon\mu}\backslash\bar{B}^{++}_{\varepsilon\mu}$ is called the nodal domain.

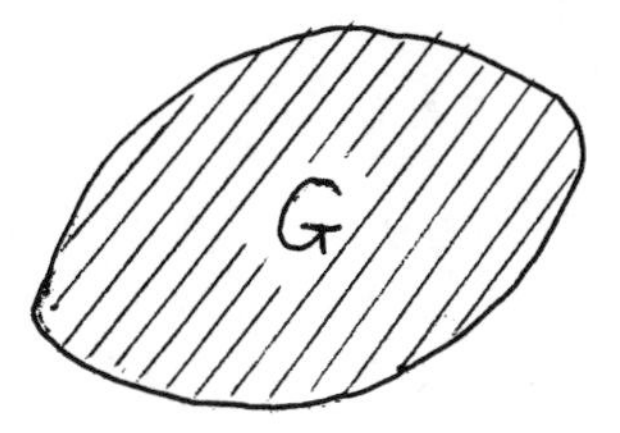

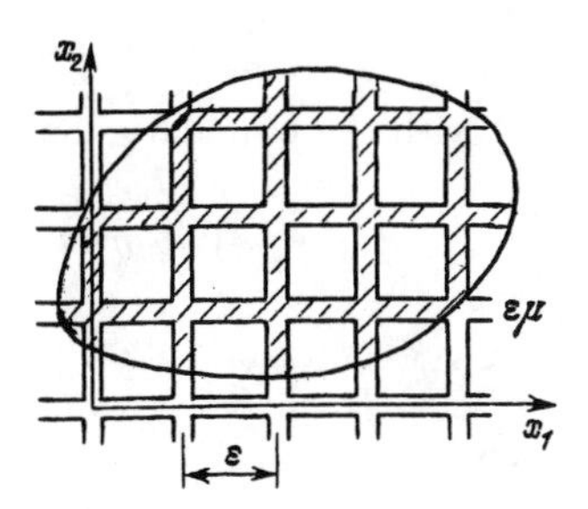

Figure 5.1.1. Domain G independent of small parameters and its intersection with the rectangular lattice.

5.2 L-convergence homogenization of lattices

5.2.1 L-convergence for the simplest lattice

We start from the example of the simplest model of lattice-structures: two-dimensional rectangular lattice. We prove all necessary auxiliary results for this simplest lattice (even if these results are some particular cases of the results of previous chapters); so the first section can be read *absolutely independently* of others. We announce here below some generalizations, proved later.

Definition 5.2.1 *The union*

$$B_{\varepsilon,\mu} \;=\; \cup_{k=-\infty}^{+\infty} \;(\;\{(x_1,x_2)\in I\!R^2 \mid \mid x_2 - k\varepsilon \mid \;<\; \varepsilon\mu/2\;\}$$

$$\cup\;\{(x_1,x_2)\in I\!R^2 \mid \mid x_1 - k\varepsilon \mid \;<\; \varepsilon\mu/2\;\}\;)$$

is called the two-dimensional rectangular lattice.

Thus the rectangular lattice is a union of thin strips of the width $\varepsilon\mu$ stretched in each coordinate direction and forming the $\varepsilon-$ periodic system in each dimension. We also denote

$$B^j_{\varepsilon,\mu} \;=\; \cup_{k=-\infty}^{+\infty}\;\{(x_1,x_2)\in I\!R^2 \mid \mid x_{3-j} - k\varepsilon \mid \;<\; \varepsilon\mu/2\;\}$$

the unions of horizontal ($j=1$) and vertical ($j=2$) strips, so

$$B_{\varepsilon,\mu} = B^1_{\varepsilon,\mu} \cup B^2_{\varepsilon,\mu}.$$

Let G be a domain with the boundary $\partial G \in C^\infty$ which is independent of ε and μ. In the domain $B_{\varepsilon,\mu} \cap G$ we consider the equation

$$-div\ (\ A\ grad\ u_{\varepsilon,\mu}) = f(x)\ , \tag{5.2.1}$$

with the boundary conditions

$$(\ A\ grad\ u_{\varepsilon,\mu}\ ,\ n\ ) = 0,\ for\ x \in \partial B_{\varepsilon,\mu} \cap G, \tag{5.2.2}$$

$$u_{\varepsilon,\mu} = 0,\ for\ x \in \bar{B}_{\varepsilon,\mu} \cap \partial G, \tag{5.2.3}$$

here $x = (x_1, x_2)$, $A = (a_{ij})$ is a constant $(2 \times 2)-$matrix independent of ε and μ , $A = A^T > 0$, i.e. $a_{ij} = a_{ji}$ and A is positive, $f \in C^1(G)$.

Problems (5.2.1)-(5.2.3) simulate a steady state heat field in the lattice structure $B_{\varepsilon,\mu} \cap G$ under the conditions of thermal insulation on the boundary of G with A being the heat conductivity tensor for the material of which the framework is made ("cut out"). For $a_{ij} = \delta_{ij} D$ (5.2.1)-(5.2.3) may be interpreted also as a problem of diffusion of a substance in a fissured rock filled with water or oil, with D being the diffusion coefficient of the substance in the fluid. In this case, the framework $B_{\varepsilon,\mu}$ models a rectangular system of cracks $\varepsilon\mu$ wide, filled with water. This model is very approximate since the system of cracks in a rock has a less regular structure . The model of a random framework (section 5.9) is more suitable for its description.

Numerical solution of problems (5.2.1)-(5.2.3), with $\varepsilon << 1$, $\mu << 1$, is very difficult since the step size of the grid must have an order much less than ε. The implementation of the standard homogenization procedure is also impeded, since the problem on a cell depends on the small parameter μ, and in order to solve it numerically, we must select the step size of the grid to be much less than μ. Hence, an asymptotic investigation of the problem is needed. The result of this investigation can be presented in a form of the following Theorem 5.2.1, proved in [134]-[136]. But first we shall formulate a definition of L-convergence.

Definition 5.2.2 *Let $u_{\varepsilon,\mu}(x)$ is a sequence of functions from $L^2(B_{\varepsilon,\mu} \cap G)$, $u_0(x) \in L^2(G)$. One says that $u_{\varepsilon,\mu}$ L - converges to $u_0(x)$ at $B_{\varepsilon,\mu} \cap G$ if and only if*

$$\frac{\|u_{\varepsilon,\mu} - u_0\|_{L^2(B_{\varepsilon,\mu} \cap G)}}{\sqrt{mes(B_{\varepsilon,\mu} \cap G)}} \to 0, \quad (\varepsilon, \mu \to 0).$$

The normalization factor $1/\sqrt{mes(B_{\varepsilon,\mu} \cap G)}$ is important because $\|1\|_{L^2(B_{\varepsilon,\mu} \cap G)} = \sqrt{mes(B_{\varepsilon,\mu} \cap G)}$. Notice that L-convergence is not a convergence in common sense because the domain depends on small parameters.

Theorem 5.2.1 *Let $\hat{A} = (\hat{a}_{ij})$,*

$$\hat{a}_{11} = 0.5(a_{11} - a_{12}a_{22}^{-1}a_{21}),\ \hat{a}_{12} = 0,$$

$$\hat{a}_{22} = 0.5(a_{22} - a_{21}a_{11}^{-1}a_{12}),\ \hat{a}_{21} = 0.$$

Let $u_0(x)$ is the solution of the homogenized problem

$$-div\ (\ \hat{A}\ grad\ u_0)\ =\ f(x)\ ,\ x \in G, \quad u_0|_{\partial G}\ =\ 0. \qquad (5.2.4)$$

Then $u_{\varepsilon,\mu}$ L-converges to u_0, and

$$\frac{\|u_{\varepsilon,\mu} - u_0\|_{L^2(B_{\varepsilon,\mu}\cap G)}}{\sqrt{mes(B_{\varepsilon,\mu}\cap G)}}\ =\ O(\sqrt{\varepsilon} + \sqrt{\mu}), \quad (\varepsilon, \mu \to 0).$$

Remark 5.2.1 Regularity of ∂G can be reduced up to C^3.

5.2.2 Some auxiliary inequalities

Lemma 5.2.1(The Poincaré - Friedrichs inequality for the simplest lattice) *Let $H_0^1(B_{\varepsilon,\mu}\cap G)\ =\ \{\varphi\ \in\ H^1(B_{\varepsilon,\mu}\cap G), \varphi|_{\partial G}\ =\ 0\}$. Then for each $u\ \in\ H_0^1(B_{\varepsilon,\mu}\cap G)$ there is a constant C independent of ε, μ such that*

$$\|u\|_{L^2(B_{\varepsilon,\mu}\cap G)}\ \le\ C\|\nabla u\|_{L^2(B_{\varepsilon,\mu}\cap G)}.$$

Proof. Extend u onto $H_0^1(B_{\varepsilon,\mu}\cap(-Q,Q)^2)$ by zero, where $(-Q,Q)^2 \supset G$. Then for each strip $B_k^1\cap(-Q,Q)^2,\quad B_k^1 = \{\ |\ x_2 - k\varepsilon\ |\ <\ \varepsilon\mu/2\ , x_1 \in R\}$ we obtain

$$\int_{B_k^1\cap(-Q,Q)^2} u^2 dx = \int_{B_k^1\cap(-Q,Q)^2} (\int_{-Q}^{x_1} \frac{\partial u}{\partial t}(t, x_2)dt)^2 dx\ \le$$

$$(2Q)^2 \int_{B_k^1\cap(-Q,Q)^2} (\frac{\partial u}{\partial x_1})^2\ dx$$

and the same estimate is valid for $B_k^2\cap(-Q,Q)^2$. Then

$$\int_{B_{\varepsilon,\mu}\cap G} u^2 dx\ \le\ (2Q)^2 \int_{B_{\varepsilon,\mu}\cap G} |\nabla u|^2 dx\ .$$

Lemma is proved.

Lemma 5.2.2.*Denote*

$$\|u\|_{FL(B_{\varepsilon,\mu}\cap G)} =$$

$$sup_{\varphi\in H_0^1(B_{\varepsilon,\mu}\cap G)} \left\{ \frac{|\int_{B_{\varepsilon,\mu}\cap G}(A\nabla u\ ,\ \nabla\varphi)\ dx|}{\|\varphi\|_{H^1(B_{\varepsilon,\mu}\cap G)}\sqrt{mes(B_{\varepsilon,\mu}\cap G)}} \right\}. \qquad (5.2.4)$$

Then

$$\frac{\|u\|_{H^1(B_{\varepsilon,\mu}\cap G)}}{\sqrt{mes(B_{\varepsilon,\mu}\cap G)}}\ \le\ C\|u\|_{FL(B_{\varepsilon,\mu}\cap G)},$$

where C does not depend on ε.

Proof. For $\varphi = u$ apply the coercivity and Lemma 5.2.1:

$$\|u\|_{FL(B_{\varepsilon,\mu}\cap G)} \geq \frac{\int_{B_{\varepsilon,\mu}\cap G} |(A\nabla u\ ,\ \nabla u)\ dx|}{\|u\|_{H^1(B_{\varepsilon,\mu}\cap G)}\sqrt{mes(B_{\varepsilon,\mu}\cap G)}} \geq$$

$$\frac{\kappa\|\nabla u\|^2_{L^2(B_{\varepsilon,\mu}\cap G)}}{\|u\|_{H^1(B_{\varepsilon,\mu}\cap G)}\sqrt{mes(B_{\varepsilon,\mu}\cap G)}} \geq \frac{\kappa\|u\|_{H^1(B_{\varepsilon,\mu}\cap G)}}{2(1+2Q)^2\sqrt{mes(B_{\varepsilon,\mu}\cap G)}}.$$

Lemma is proved.

Denote $Q_{a,b} = \{(x_1, x_2) \mid x_1 \in (0,a),\ x_2 \in (-b/2,\ b/2)\}$.

Lemma 5.2.3. *For $u \in H^1(Q_{d,e})$ the estimate holds*

$$\|u\|^2_{L^2(Q_{d,e})} \leq \frac{2d}{e}\|u\|^2_{L^2(Q_{e,e})} + 2d^2\|\frac{\partial u}{\partial x_1}\|^2_{L^2(Q_{d,e})}.$$

Proof. Let $y_1 \in (0,d),\ x_2 \in (-e/2, e/2),\ y_0 \in (0,e)$. We obtain the following inequality:

$$|u(y_1,x_2)|^2 = |u(y_1,x_2) - u(y_0,x_2) + u(y_0,x_2)|^2 =$$

$$|\int_{y_0}^{y_1} \frac{\partial u}{\partial x_1}(x_1,x_2)dx_1 + u(y_0,x_2)|^2 \leq$$

$$2u^2(y_0,x_2) + 2(\int_{y_0}^{y_1} \frac{\partial u}{\partial x_1}(x_1,x_2)dx_1)^2 \leq$$

$$2u^2(y_0,x_2) + 2d\int_{y_0}^{y_1} (\frac{\partial u}{\partial x_1})^2 dx_1.$$

Integrating this inequality over $y_1 \in (0,d),\ x_2 \in (-e/2, e/2),\ y_0 \in (0,e)$ we obtain the estimate of the lemma.

Lemma 5.2.4.(The Poincaré inequality for a cross) *Let $X_{\varepsilon,\mu}$ be a cross*

$$X_{\varepsilon,\mu} = (-\varepsilon/2,\varepsilon/2) \times (-\mu\varepsilon/2,\mu\varepsilon/2) \cup (-\mu\varepsilon/2,\mu\varepsilon/2) \times (-\varepsilon/2,\varepsilon/2),$$

$\varphi \in H^1(X_{\varepsilon,\mu})$, then the Poincaré inequality holds for sufficiently small values of the parameters ε, μ :

$$\|\varphi\|^2_{L^2(X_{\varepsilon,\mu})} \leq \frac{1}{mes(X_{\varepsilon,\mu})}(\int_{X_{\varepsilon,\mu}} \varphi dx)^2 + 9\varepsilon^2\|\nabla\varphi\|^2_{L^2(X_{\varepsilon,\mu})}.$$

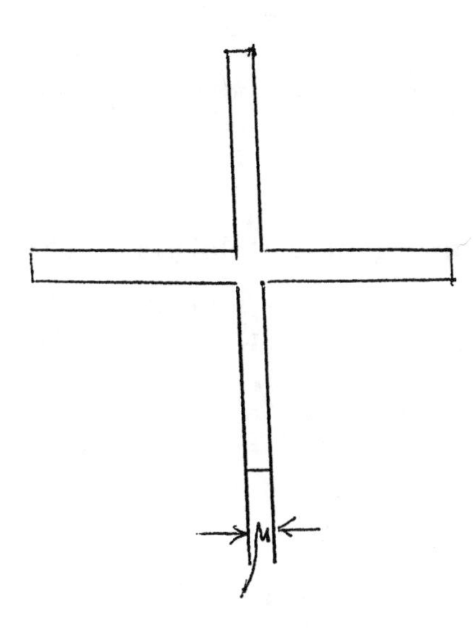

Figure 5.2.1. Periodic cross.

Proof. Consider the function

$$\psi = \varphi - (\varepsilon\mu)^{-2}\int_{(-\mu\varepsilon/2,\mu\varepsilon/2)^2}\varphi dx.$$

Applying Lemma 5.2.3 with $e = \varepsilon\mu$ to each branch of the cross we obtain the estimate

$$\|\psi\|^2_{L^2(X_{\varepsilon,\mu})} \le 4(\frac{2}{\mu}\|\psi\|^2_{L^2((-\mu\varepsilon/2,\mu\varepsilon/2)^2)} + 2\varepsilon^2\|\nabla\varphi\|^2_{L^2(X_{\varepsilon,\mu})}.$$

Since the standard Poincaré inequality one obtains the estimate

$$\|\psi\|^2_{L^2((-\mu\varepsilon/2,\mu\varepsilon/2)^2)} \le (\varepsilon\mu)^2\|\nabla\varphi\|^2_{L^2((-\mu\varepsilon/2,\mu\varepsilon/2)^2)},$$

and therefore

$$\|\psi\|^2_{L^2(X_{\varepsilon,\mu})} \le 9\varepsilon^2\|\nabla\psi\|^2_{L^2(X_{\varepsilon,\mu})} = 9\varepsilon^2\|\nabla\varphi\|^2_{L^2(X_{\varepsilon,\mu})}$$

for sufficiently small values of the parameters.

Note that

$$min_{t\in I\!R}\int_{X_{\varepsilon,\mu}}(\varphi(x)-t)^2dx = \int_{X_{\varepsilon,\mu}}\varphi^2(x)dx - \frac{1}{mes(X_{\varepsilon,\mu})}(\int_{X_{\varepsilon,\mu}}\varphi dx)^2$$

and therefore it is not more than

$$\int_{X_{\varepsilon,\mu}}\psi^2dx,$$

i.e.

$$\int_{X_{\varepsilon,\mu}} \varphi^2(x)dx - \frac{1}{mes(X_{\varepsilon,\mu})}(\int_{X_{\varepsilon,\mu}} \varphi dx)^2 \leq \int_{X_{\varepsilon,\mu}} \psi^2 dx \leq 9\varepsilon^2\|\nabla\varphi\|^2_{L^2(X_{\varepsilon,\mu})},$$

and lemma is proved.

5.2.3 FL-convergence. Relation to the L-convergence

Definition 5.2.3 *Let $u_{\varepsilon,\mu}(x)$ be a sequence of functions from $H^1_0(B_{\varepsilon,\mu} \cap G)$. One says that $u_{\varepsilon,\mu}$ FL - converges to zero if and only if*

$$\|u_{\varepsilon,\mu}\|_{FL(B_{\varepsilon,\mu}\cap G)} \to 0, \quad (\varepsilon, \mu \to 0).$$

Let

$$u^{(1)}_{\varepsilon,\mu}(x) = \begin{cases} u_0 + (-a^{-1}_{22}a_{21}\frac{\partial u_0}{\partial x_1} - \frac{\partial u_0}{\partial x_2})(x_2 - k\varepsilon)\rho(\frac{x_1}{\varepsilon\mu})\psi(\frac{x}{\varepsilon}) & \text{for } x \in B^1_k \cap G, \\ u_0 + (-a^{-1}_{11}a_{12}\frac{\partial u_0}{\partial x_2} - \frac{\partial u_0}{\partial x_1})(x_1 - k\varepsilon)\rho(\frac{x_2}{\varepsilon\mu})\psi(\frac{x}{\varepsilon}) & \text{for } x \in B^2_k \cap G, \end{cases} \tag{5.2.5}$$

where $B^j_k = \{|x_{3-j} - k\varepsilon| < \frac{\varepsilon\mu}{2}\}, \rho(\frac{x_j}{\varepsilon\mu})$ is an $\varepsilon-$ periodic function coinciding on the period $[-\varepsilon/2, \varepsilon/2]$ with the function $\bar{\rho}(\frac{x_j}{\varepsilon\mu})$, and

$$\bar{\rho}(t) = \begin{cases} 0 & \text{for } |t| \leq 3/4, \\ 1 & \text{for } |t| \geq 1, \end{cases}$$

$\bar{\rho}(t) \in C^1$,

$$\psi(x/\varepsilon) = \begin{cases} 0 & \text{in } \varepsilon- \text{ vicinity of } \partial G, \\ 1 & \text{out of } 2\varepsilon- \text{ vicinity of } \partial G, \end{cases}$$

$|\nabla\psi(\xi)| < c_0, \quad 0 \leq \psi \leq 1, \quad \psi \in C^1, \quad c_0$ is a positive constant.

One can obtain from Lemma 5.2.2 that if $u_{\varepsilon,\mu} - u^{(1)}_{\varepsilon,\mu}$ FL-converges to zero ($\|u_{\varepsilon,\mu} - u^{(1)}_{\varepsilon,\mu}\|_{FL(B_{\varepsilon,\mu}\cap G)} = O(\sqrt{\varepsilon} + \sqrt{\mu})$) then

$$\frac{\|u_{\varepsilon,\mu} - u^{(1)}_{\varepsilon,\mu}\|_{H^1(B_{\varepsilon,\mu}\cap G)}}{\sqrt{mes(B_{\varepsilon,\mu} \cap G)}} \to 0.$$

From the estimate

$$\frac{\|u^{(1)}_{\varepsilon,\mu} - u_0\|_{L^2(B_{\varepsilon,\mu}\cap G)}}{\sqrt{mes(B_{\varepsilon,\mu} \cap G)}} = O(\sqrt{\varepsilon} + \sqrt{\mu})$$

and the triangular inequality one obtains

$$\frac{\|u_{\varepsilon,\mu} - u_0\|_{L^2(B_{\varepsilon,\mu}\cap G)}}{\sqrt{mes(B_{\varepsilon,\mu} \cap G)}} \to 0.$$

If

$$\frac{\|u_{\varepsilon,\mu} - u^{(1)}_{\varepsilon,\mu}\|_{H^1(B_{\varepsilon,\mu}\cap G)}}{\sqrt{mes(B_{\varepsilon,\mu}\cap G)}} = O(\sqrt{\varepsilon}+\sqrt{\mu})$$

then

$$\frac{\|u_{\varepsilon,\mu} - u_0\|_{L^2(B_{\varepsilon,\mu}\cap G)}}{\sqrt{mes(B_{\varepsilon,\mu}\cap G)}} = O(\sqrt{\varepsilon}+\sqrt{\mu}).$$

Thus a theorem on relation of L-convergence and FL-convergence is obtained:

Theorem 5.2.2. *If $u_{\varepsilon,\mu} - u^{(1)}_{\varepsilon,\mu}$ FL-converges to zero then $u_{\varepsilon,\mu}$ L-converges to u_0.*

Thus to prove theorem 5.2.1 it is sufficient to prove FL-convergence of $u_{\varepsilon,\mu} - u^{(1)}_{\varepsilon,\mu}$ to zero.

5.2.4 Proof of Theorem 5.2.1

Denote for each $B^j_k = \{ \mid x_{3-j} - k\varepsilon \mid < \varepsilon\mu/2 , x_j \in R\},\ j = 1,2$

$$N_1(x,\xi_{3-j}) = (-a^{-1}_{3-j,3-j}a_{3-j,j}\frac{\partial u_0}{\partial x_j} - \frac{\partial u_0}{\partial x_{3-j}})(\xi_{3-j}-k),$$

then

$$\nabla u^{(1)}_{\varepsilon,\mu}(x) =$$

$$\nabla_x u_0 + \nabla_\xi(N_1(x,\xi_{3-j})\rho(\xi_j/\mu)\psi(\xi)) + \varepsilon\nabla_x(N_1(x,\xi_{3-j})\rho(\xi_j/\mu)\psi(\xi)) =$$

$$\nabla_x u_0 + \nabla_\xi N_1(x,\xi_{3-j}) + R_1,$$

where

$$R_1 = \nabla_\xi(N_1(x,\xi_{3-j})\ (\rho(\xi_j/\mu)\psi(\xi)-1)) + \varepsilon\nabla_x(N_1(x,\xi_{3-j})\rho(\xi_j/\mu)\psi(\xi)),$$

$$\|R_1\|^2_{L^2(B_{\varepsilon,\mu}\cap G)} = O(\varepsilon+\mu)mes(B_{\varepsilon,\mu}\cap G)$$

because the measure of the support of the first term is $O(\varepsilon+\mu)mes(B_{\varepsilon,\mu}\cap G)$, and it is bounded and the second term has a value $O(\varepsilon)$.

Denote $A_i(\nabla u) = \sum_{r=1}^2 a_{i,r}\frac{\partial u}{\partial x_r}$ and estimate the integral

$$I = \int_{B_{\varepsilon,\mu}\cap G} A_i(\nabla_x u^{(1)})\frac{\partial\varphi}{\partial x_i}dx =$$

$$\sum_{j=1}^{2}\int_{B^j_{\varepsilon,\mu}\cap G} A_i(\nabla_x u_0 + \nabla_\xi N_1)\frac{\partial\varphi}{\partial x_i}dx \;+\; \delta_1,$$

where

$$|\delta_1| \;=\; O(\sqrt{\varepsilon}+\sqrt{\mu})\sqrt{mes(B_{\varepsilon,\mu}\cap G)}\|\varphi\|_{H^1(B_{\varepsilon,\mu}\cap G)}.$$

The integral

$$I_j \;=\; \int_{B^j_{\varepsilon,\mu}\cap G}\sum_{i=1}^{2} A_i(\nabla_x u_0 + \nabla_\xi N_1)\frac{\partial\varphi}{\partial x_i}dx \;=$$

$$-\int_{B^j_{\varepsilon,\mu}\cap G}\sum_{i=1}^{2}\frac{\partial}{\partial x_i}A_i(\nabla_x u_0 + \nabla_\xi N_1)|_{\xi=x/\varepsilon}\varphi dx,$$

because $A_{3-j}(\nabla_x u_0 + \nabla_\xi N_1) \;=\; 0$ in $B^j_{\varepsilon,\mu}$.

Therefore

$$I_j \;=\; -\int_{B^j_{\varepsilon,\mu}\cap G}\frac{\partial}{\partial x_j}\hat{A}_j(\nabla_x v)\varphi dx,$$

where

$$\hat{A}_j(\nabla_x u_0) \;=\; A_j(\nabla_x u_0 + \nabla_\xi N_1) \;=\; 2\hat{a}_{jj}\frac{\partial u_0}{\partial x_j}.$$

Consider the intersections of squares $Q_{k_1k_2} \;=\; \{\;|\;x_i - k_i\varepsilon\;|\;<\;\varepsilon/2\;,\;i = 1,2\}$, with the domain $B^j_{\varepsilon,\mu}\cap G$. We obtain the equality

$$I \;=\; -\sum_{j=1}^{2}\int_{B^j_{\varepsilon,\mu}\cap G}\frac{\partial}{\partial x_j}\hat{A}_j(\nabla_x u_0)\varphi dx \;+\; \delta_1 \;= \tag{5.2.6}$$

$$-\sum_{k_1,k_2:Q_{k_1k_2}\subset G}\;\sum_{j=1}^{2}\int_{B^j_{\varepsilon,\mu}\cap Q_{k_1k_2}}\frac{\partial}{\partial x_j}\hat{A}_j(\nabla_x u_0)\varphi dx \;+\; \delta_1 \;+\; \delta_2,$$

where

$$|\delta_2| \;=\; O(\sqrt{\varepsilon}+\sqrt{\mu})\sqrt{mes(B_{\varepsilon,\mu}\cap G)}\|\varphi\|_{H^1(B_{\varepsilon,\mu}\cap G)}.$$

Introduce notation

$$\langle\varphi\rangle_{k_1k_2} \;=\; \frac{\int_{B_{\varepsilon,\mu}\cap Q_{k_1k_2}}\varphi dx}{mes(B_{\varepsilon,\mu}\cap Q_{k_1k_2})},$$

then the following equality is valid:

$$\varphi \;=\; \langle\varphi\rangle_{k_1k_2} \;+\; (\varphi \;-\; \langle\varphi\rangle_{k_1k_2}), \tag{5.2.7}$$

and from lemma 5.2.4 we obtain the estimate

$$\|\varphi \;-\; \langle\varphi\rangle_{k_1k_2}\|_{L^2(B_{\varepsilon,\mu}\cap Q_{k_1k_2})} \;\le\; C\varepsilon\|\nabla\varphi\|_{L^2(B_{\varepsilon,\mu}\cap Q_{k_1k_2})},$$

where C does not depend on ε, μ. Therefore

$$I = - \sum_{k_1,k_2:Q_{k_1k_2}\subset G} \sum_{j=1}^{2} \int_{B^j_{\varepsilon,\mu}\cap Q_{k_1k_2}} \frac{\partial}{\partial x_j} \hat{A}_j(\nabla_x u_0) dx \, \langle\varphi\rangle_{k_1k_2} +$$

$$\delta_1 + \delta_2 + \delta_3,$$

where

$$|\delta_3| = O(\sum_{k_1,k_2:Q_{k_1k_2}\subset G} \sqrt{mes(B_{\varepsilon,\mu}\cap Q_{k_1k_2})} \|\varphi - \langle\varphi\rangle_{k_1k_2}\|_{L^2(B_{\varepsilon,\mu}\cap Q_{k_1k_2})}) =$$

$$O(\varepsilon^2\sqrt{\mu} \sum_{k_1,k_2:Q_{k_1k_2}\subset G} \|\nabla\varphi\|_{L^2(B_{\varepsilon,\mu}\cap Q_{k_1k_2})}.$$

From the Cauchy-Bunyakovskii-Schwartz inequality for sums we obtain then, that

$$|\delta_3| = O(\varepsilon^2\sqrt{\mu} \sqrt{\varepsilon^{-2} \sum_{k_1,k_2:Q_{k_1k_2}\subset G} \|\nabla\varphi\|^2_{L^2(B_{\varepsilon,\mu}\cap Q_{k_1k_2})}}),$$

and therefore

$$|\delta_3| = O(\varepsilon\sqrt{\mu}\|\varphi\|_{H^1(B_{\varepsilon,\mu}\cap G)}) = O(\varepsilon)\sqrt{mes(B_{\varepsilon,\mu}\cap G)}\|\varphi\|_{H^1(B_{\varepsilon,\mu}\cap G)}.$$

Thus

$$I = - \sum_{k_1,k_2:Q_{k_1k_2}\subset G} \int_{B_{\varepsilon,\mu}\cap Q_{k_1k_2}} (1/2) \sum_{j=1}^{2} \frac{\partial}{\partial x_j} \hat{A}_j(\nabla_x u_0)|_{x_1=k_1\varepsilon, x_2=k_2\varepsilon} dx \, \langle\varphi\rangle_{k_1k_2} +$$

$$\delta_1 + \delta_2 + \delta_3 + \delta_4, \tag{5.2.8}$$

where

$$|\delta_4| = O((\varepsilon+\mu) \sum_{k_1,k_2:Q_{k_1k_2}\subset G} mes(B_{\varepsilon,\mu}\cap Q_{k_1k_2})|\langle\varphi\rangle_{k_1k_2}|) =$$

$$O((\varepsilon+\mu)\int_{B_{\varepsilon,\mu}\cap G} |\varphi| dx) = O((\varepsilon+\mu)\sqrt{mes(B_{\varepsilon,\mu}\cap G)}\|\varphi\|_{H^1(B_{\varepsilon,\mu}\cap G)}).$$

Now we use once more the representation (5.2.7) and the Taylor expansion for

$$\frac{\partial}{\partial x_j} \hat{A}_j(\nabla_x u_0)$$

and obtain that

$$I = - \int_{B_{\varepsilon,\mu}\cap G} (1/2) \sum_{j=1}^{2} \frac{\partial}{\partial x_j} \hat{A}_j(\nabla_x u_0) \varphi dx + \delta_5 =,$$

$$-\int_{B_{\varepsilon,\mu}\cap G} div(\hat{A}\nabla_x u_0)\varphi dx \ + \ \delta_5,$$

where

$$|\delta_5| \ = \ O(\sqrt{\varepsilon}+\sqrt{\mu})\sqrt{mes(B_{\varepsilon,\mu}\cap G)}\|\varphi\|_{H^1(B_{\varepsilon,\mu}\cap G)}.$$

From the homogenized equation (5.2.4) we can change this integral by

$$I \ = \ \int_{B_{\varepsilon,\mu}\cap G} f\varphi dx \ + \ \delta_5,$$

and therefore

$$I \ - \ \int_{B_{\varepsilon,\mu}\cap G} f\varphi dx \ =$$

$$\int_{B_{\varepsilon,\mu}\cap G} (A\nabla(u^{(1)}_{\varepsilon\mu} - u_{\varepsilon\mu}) \ , \ \nabla\varphi)dx \ =$$

$$O(\sqrt{\varepsilon}+\sqrt{\mu})\sqrt{mes(B_{\varepsilon,\mu}\cap G)}\|\varphi\|_{H^1(B_{\varepsilon,\mu}\cap G)}.$$

Thus the sequence $u^{(1)}_{\varepsilon\mu} - u_{\varepsilon\mu}$ FL-converges to zero and from theorem 2 $u_{\varepsilon\mu}$ L-converges to v. The theorem is proved.

5.3 Non-stationary problems

5.3.1 Rectangular lattice

Consider the problem

$$R\frac{\partial^2 u}{\partial t^2} + S\frac{\partial u}{\partial t} - div\ (\ A\ grad\ u_{\varepsilon,\mu}) \ = \ f(x,t)\ ,\ for\ x\in B_{\varepsilon,\mu}\cap G,\ t>0, \quad (5.3.1)$$

$$(\ A\ grad\ u_{\varepsilon,\mu}\ ,\ n\ ) \ = \ 0,\ for\ x\in \partial B_{\varepsilon,\mu}\cap G, \quad (5.3.2)$$

$$u_{\varepsilon,\mu} \ = \ 0,\ for\ x\in \bar{B}_{\varepsilon,\mu}\cap\partial G, (5.3.3)$$

$$u_{\varepsilon,\mu}|_{t=0} \ = \ 0,\ \frac{\partial u_{\varepsilon,\mu}}{\partial t}|_{t=0} \ = \ 0, \quad (5.3.3)$$

where R, S ($R>0,\ S\geq 0$) are constants $A=(a_{ij})$ is a constant $(2\times 2)-$matrix independent of ε and μ , $A = A^T > 0$,i.e. $a_{ij} = a_{ji}$ and A is positive, $f\in C^\infty(\bar{G}\times R_+), f(x,0) = \frac{\partial f}{\partial t}|_{t=0} = 0.$

As above ∂G is smooth enough: $\partial G\in C^3$.

Definition 5.3.1 *Let $u_{\varepsilon,\mu}(x,t)$ be a sequence of functions from $L^2((B_{\varepsilon,\mu}\cap G)\times[0,T])$, and $u_0(x)\in L^2(G\times[0,T])$. One says that $u_{\varepsilon,\mu}$ L_t-converges to $u_0(x,t)$ at $(B_{\varepsilon,\mu}\cap G)\times[0,T]$ if and only if*

$$\frac{\|u_{\varepsilon,\mu}-u_0\|_{L^2((B_{\varepsilon,\mu}\cap G)\times[0,T])}}{\sqrt{mes(B_{\varepsilon,\mu}\cap G)}} \to 0, \quad (\varepsilon,\mu\to 0). \quad (5.3.5)$$

For the solution of a homogenized problem

$$R\frac{\partial^2 u_0}{\partial t^2} \ + \ S\frac{\partial u_0}{\partial t} - div\ (\ \hat{A}\ grad\ u_0) \ = \ f(x,t)\ ,\ x\in G,\ t>0, \quad u_0|_{\partial G} \ = \ 0, \tag{5.3.6}$$

$$u_0|_{t=0} \ = \ 0,\ \frac{\partial u_0}{\partial t}|_{t=0} \ = \ 0, \tag{5.3.7}$$

one obtains the following theorem of L_t- convergence,proved in [16]:

Theorem 5.3.1 *The solution $u_{\varepsilon,\mu}$ of the initial problem L_t-converges to the solution u_0 of the homogenized problem and*

$$\frac{\|u_{\varepsilon,\mu} - u_0\|_{L^2((B_{\varepsilon,\mu}\cap G)\times[0,T])}}{\sqrt{mes(B_{\varepsilon,\mu}\cap G)}} \ = \ O(\sqrt{\varepsilon}+\sqrt{\mu}), \quad (\varepsilon,\mu\to 0). \tag{5.3.8}$$

Denote

$$\|u\|_{FLt((B_{\varepsilon,\mu}\cap G)\times[0,T])} =$$

$$sup_{t_0\in[0,T]}sup_{\varphi|_{t=t_0}\in H_0^1(B_{\varepsilon,\mu}\cap G)}\left\{ | \int_{B_{\varepsilon,\mu}\cap G} ((R\frac{\partial^2 u}{\partial t^2} \ + \ S\frac{\partial u}{\partial t})\ ,\ \varphi) + \right.$$

$$\left. (A\nabla u\ ,\ \nabla\varphi)\ dx|/(\|\varphi\|_{H^1(B_{\varepsilon,\mu}\cap G)}\sqrt{mes(B_{\varepsilon,\mu}\cap G)})\right\}.$$

Introduce FLt-convergence.

Definition 5.3.2 *Let $u_{\varepsilon,\mu}(x,t)$ be a sequence of functions from $H_0^1(B_{\varepsilon,\mu}\cap G)$ for almost all $t\in[0,T]$, such that*

$$\frac{\partial^2 u_{\varepsilon,\mu}}{\partial t^2}|_{t=t_0}\ ,\ \frac{\partial u_{\varepsilon,\mu}}{\partial t}|_{t=t_0} \in L^2(B_{\varepsilon,\mu}\cap G)$$

for almost all $t_0\in[0,T]$. One says that $u_{\varepsilon,\mu}$ FLt - converges to zero if and only if

$$\|u_{\varepsilon,\mu}\|_{FLt((B_{\varepsilon,\mu}\cap G)\times[0,T])} \to 0, \quad (\varepsilon,\mu\to 0).$$

The theorems on FLt-convergence of the difference $u_{\varepsilon,\mu} \ - \ u^{(1)}_{\varepsilon,\mu}$ to zero could be formulated and proved as in [16] (see the next subsection).

5.3.2 Lattices: general case

Consider now the general lattices of Definition 5.1.1.

Let G be a bounded domain in $I\!R^s$ with a piece-wise smooth boundary ∂G, (satisfying the cone condition). Consider the problem

$$R(x,t)\frac{\partial^2 u}{\partial t^2} \ + \ S(x,t)\frac{\partial u}{\partial t} - \sum_{i=1}^{s} \partial/\partial x_i(\ A_i(\nabla_x u_{\varepsilon,\mu}, x, t)) \ =$$

$$f(x,t)\ ,\ for\ x \in B_{\varepsilon,\mu} \cap G,\ t > 0, \tag{5.3.9}$$

$$\partial u_{\varepsilon,\mu}/\partial\nu \ =\ n_i A_i(\nabla_x u_{\varepsilon,\mu}, x, t) \ =\ 0,\ for\ x \in \partial B_{\varepsilon,\mu} \cap G, \tag{5.3.10}$$

$$u_{\varepsilon,\mu} \ =\ 0,\ for\ x \in \bar{B}_{\varepsilon,\mu} \cap \partial G, \tag{5.3.11}$$

$$u_{\varepsilon,\mu}|_{t=0} \ =\ 0,\ \frac{\partial u_{\varepsilon,\mu}}{\partial t}|_{t=0} \ =\ 0, \tag{5.3.12}$$

where $R(x,t)$ and $S(x,t)$ are $n \times n$ matrix-valued functions, $u(x,t)$, $f(x,t)$, $A_i(y,x,t)$ $(i = 1,...,s)$ are n -dimensional vector-valued functions, $\nabla_x u = (\partial u_k / \partial x_l)$ is an $n \times s$ matrix, $y = (y_{kl})$ is an $n \times s$ matrix-argument, f, R, S are assumed to be continuously differentiable, and A_i to be twice continuously differentiable, and $n_i(x)$ is the cosine of the angle between the axis Ox_i and the exterior normal vector $n(x)$. We assume that $R > 0$. We can consider a "parabolic" case when $R = 0$ and $S > 0$; if so, the second initial condition (5.3.12) should be cancelled. By formally taking $R = S = 0$, we can consider the steady state problem. Then both conditions (5.3.12) are cancelled.

We suppose that problem (5.3.9)-(5.3.12) has the solution, such that for each $n-$dimensional vector-valued function $\varphi(x,t)$ which belongs to the space $H^1(B_{\varepsilon,\mu} \cap G)$ for each fixed value of t and equal to zero on the boundary ∂G, the integral

$$\int_{B_{\varepsilon,\mu}\cap G} ((R\frac{\partial^2 u_{\varepsilon\mu}}{\partial t^2} \ +\ S\frac{\partial u_{\varepsilon\mu}}{\partial t})\ ,\ \varphi) \ +\ (A\nabla u_{\varepsilon\mu}\ ,\ \nabla\varphi)\ dx$$

is defined and is equal to

$$\int_{B_{\varepsilon,\mu}\cap G} (f(x,t)\ ,\ \varphi)\ dx.$$

We shall describe the procedure of constructing the homogenized operator. It is related to resolving of the cell problem by L-convergence method of section 4.1.2.

To each segment $e \subset B$, we assign a collection of $n-$dimensional vectors (assuming their existence)

$$X_0(y,x,t), ..., X_s(y,x,t) \ \in\ C^2,$$

which satisfies the following four conditions:

1)

$$A_i(X_1(y,x,t), ..., X_s(y,x,t), x, t)\ \nu_i^j \ =\ 0,$$

where $j = 1, ..., s-1$, and the vectors

$$\nu^1 \ =\ (\nu_1^1, ..., \nu_s^1)\ ,\ ...\ ,\ \nu^{s-1} \ =\ (\nu_1^{s-1}, ..., \nu_s^{s-1})$$

form the basis in $(s-1)-$dimensional space orthogonal to e, $y = (y_1, ..., y_s)$ is $n \times s-$matrix-parameter.

2) Let the segments $e_1, ..., e_q$ have a common end point; then

$$\sum_{j=1}^{q} \sum_{i=1}^{s} A_i(X_1^{e_j}(y,x,t), ..., X_s^{e_j}(y,x,t), x, t)\ \eta_i^j mes\tilde{\beta}_j = 0,$$

where $\eta_1^j, ..., \eta_s^j$ are the direction cosines of e_j, and $\tilde{\beta}_j$ is a cross-section of a cylinder with the axis e_j by the hyperplane orthogonal to this axis.

3) The vector-valued function defined on each segment $e \subset B$ by

$$\sum_{i=1}^{s} (X_i^e(y,x,t) - y_i)\ \xi_i + X_0^e(y,x,t)$$

(for $\xi \in e$) is a continuous 1-periodic function of $(\xi_1, ..., \xi_s) \in B$.

4) The functions equal to $X_i^e(y,x,t)$ for $\xi \in e \subset B$, which are piecewise constant in the variables ξ, are 1-periodic.

By the homogenized operator along the e_j, we shall mean the operator $\hat{L}_x^j$:

$$\hat{L}_x^j\ u_0 = M_j \left\{ R(x,t) \frac{\partial^2 u_0}{\partial t^2} + S(x,t) \frac{\partial u_0}{\partial t} - \right.$$

$$\left. \sum_{i=1}^{s} \partial/\partial x_i(\ A_i(X_1(\nabla_x u_0, x, t), ..., X_s(\nabla_x u_0, x, t), x, t)) \right\},$$

where $M_j = mes\tilde{\beta}_j \times length(e_j \cap Q)$. We determine the homogenized operator

$$\hat{L}_x = \sum_{j=1}^{I} \hat{L}_x^j.$$

Let $\chi(x/\varepsilon\mu)$ be a function equal to zero in $B_{\varepsilon\mu} \backslash \bar{B}_{\varepsilon\mu}^{+}$ and to 1 in the domain $B_{\varepsilon\mu}^{++}$, $0 \leq \chi \leq 1$, $\chi \in C^1$, and $|\partial\chi(z)/\partial z_i| < c_1$, where c_1 is independent of ε and μ.

Let $\Psi(x/\varepsilon)$ be a function equal to zero in the $\varepsilon-$neighborhood of ∂G and to 1 outside the $2\varepsilon-$neighborhood of ∂G, $|\partial\Psi(\xi)/\partial\xi_i| < c_1$, $0 \leq \Psi \leq 1$, $\Psi \in C^1$.

We define the vector-valued function

$$u_{\varepsilon\mu}^{(1)} = u_0 + \varepsilon(N^1(\frac{x}{\varepsilon}, \nabla_x u_0, x, t)\chi(x/\varepsilon\mu) + N^2(\frac{x}{\varepsilon}, \nabla_x u_0, x, t)(1-\chi(x/\varepsilon\mu)))\Psi(x/\varepsilon);$$

here

$$N^1(\xi, y, x, t) = \sum_{i=1}^{s} (X_i^e(y,x,t) - y_i)\ \xi_i + X_0^e(y,x,t),$$

if $\varepsilon\xi$ belongs to the section with the axis segment $e_\varepsilon (e_\varepsilon = \{x : \ x/\varepsilon \in e\},)$ $N^1 = 0$ outside the sections, and $N^2(\xi, y, x, t)$ is a function piecewise constant in ξ, equal to

$$\sum_{i=1}^{s} (X_i^e(y,x,t) \;-\; y_i)\, \xi_i^0 \;+\; X_0^e(y,x,t)$$

in the $3c_0\mu\varepsilon-$neighborhood of the nodal point ξ^0, which is the end point for e, and equal to zero in the remaining part of $B_{\varepsilon\mu}$ $(1-\chi=0$ on the discontinuity surfaces of $N^2)$, and let v be a solution of the homogenized problem

$$\hat{L}_x u_0 \;=\; \sum_{j=1}^{I} M_j f\ ,\ x\in G,\ t>0,\quad u_0|_{\partial G} \;=\; 0$$

$$u_0|_{t=0} \;=\; 0,\ \frac{\partial u_0}{\partial t}|_{t=0} \;=\; 0,$$

(we suppose its existence). As above we keep only one of these initial conditions if $R=0,\ S>0$, or no one of them if $R=S=0$.

Theorem 5.3.2 *Let the solution of the homogenized problem be existing and be having all partial derivatives up to the third order which are bounded by a constant independent on* ε,μ. *Then the inequality holds:*

$$\sup_{t_0\in[0,T]}\ \sup_{\varphi|_{t=t_0}\in H_0^1(B_{\varepsilon,\mu}\cap G)} \left\{ |\int_{B_{\varepsilon,\mu}\cap G} ((R\frac{\partial^2(u_{\varepsilon\mu}^{(1)}-u_{\varepsilon\mu})}{\partial t^2} \;+\right.$$

$$S\frac{\partial(u_{\varepsilon\mu}^{(1)}-u_{\varepsilon\mu})}{\partial t}),\varphi) \;+\; (A_i(\nabla_x u_{\varepsilon\mu}^{(1)},x,t) \;-$$

$$\left. A_i(\nabla_x u_{\varepsilon\mu},x,t),\ \frac{\partial\varphi}{\partial x_i})\ dx|/(\|\varphi\|_{H^1(B_{\varepsilon,\mu}\cap G)}\sqrt{mes(B_{\varepsilon,\mu}\cap G)})\right\}|_{t=t_0} \;\le$$

$$C(\sqrt{\varepsilon}+\sqrt{\mu}),\quad (\varepsilon,\mu\to 0),$$

where constant C *is independent of* ε *and* μ, here $\varphi(x,t)$ is an $n-$dimensional vector-valued function which belongs to the space $H^1(B_{\varepsilon,\mu}\cap G)$ for each fixed value of t and equal to zero on the boundary ∂G,

From this theorem (proved in [16]), one could obtain the generalization of L-convergence and Lt-convergence results of the previous sections to the case of lattice-like domains of general type. Thus to prove theorem 5.2.1 it is sufficient to prove FL_t-convergence of $u_{\varepsilon,\mu} \;-\; u_{\varepsilon,\mu}^{(1)}$ to zero.

5.3.3 Proof of Theorem 5.3.2

Proof For every ξ from the nodal domain one obtains:

$$|\frac{\partial}{\partial\xi_i}(N^1\chi+N^2(1-\chi))| \;=\; |\frac{\partial N^1}{\partial\xi_i}\chi \;+\; (N^1-N^2)\frac{\partial\chi}{\partial\xi_i}| =$$

$$|(X_i^e(y,x,t) - y_i)\chi \ + \ (N^1 - N^2)\frac{\partial \chi}{\partial \xi_i}| \leq c_1,$$

with c_1 independent of ε and μ; here χ is considered as a function of ξ/μ.

Estimate the integral

$$I_1 \ = \ \int_{B_{\varepsilon,\mu}\cap G} \left((R\frac{\partial^2 u_{\varepsilon\mu}^{(1)}}{\partial t^2} \, , \, \varphi) + (S\frac{\partial u_{\varepsilon\mu}^{(1)}}{\partial t} \, , \, \varphi) + \sum_{i=1}^{s} \, (A_i(\nabla u_{\varepsilon\mu}^{(1)}, x, t) \, , \, \partial\varphi/\partial x_i) \right) dx.$$

We have $I_1 = I_2 + \delta_1$, where

$$I_2 \ = \ \int_{B_{\varepsilon,\mu}^{++}\cap G} \left((R\frac{\partial^2 u_{\varepsilon\mu}^{(1)}}{\partial t^2} \, , \, \varphi) + (S\frac{\partial u_{\varepsilon\mu}^{(1)}}{\partial t} \, , \, \varphi) + \sum_{i=1}^{s} \, (A_i(\nabla u_{\varepsilon\mu}^{(1)}, x, t) \, , \, \partial\varphi/\partial x_i) \right) dx,$$

$$|\delta_1| \ = \ O\left(\|\varphi\|_{H^1(B_{\varepsilon,\mu}\cap G)} \sqrt{mes((B_{\varepsilon,\mu} \backslash B_{\varepsilon,\mu}^{++}) \cap G)} \right) \ =$$

$$= \ O(\sqrt{\mu}) \|\varphi\|_{H^1(B_{\varepsilon,\mu}\cap G)} \sqrt{mes(B_{\varepsilon,\mu} \cap G)}.$$

The boundedness of the derivatives $\partial^l N_i/\partial t^l \ \ (l, i = 1, 2)$ implies

$$I_2 = \int_{B_{\varepsilon,\mu}^{++}\cap G} \left((R\frac{\partial^2 v}{\partial t^2} \, , \, \varphi) + (S\frac{\partial v}{\partial t} \, , \, \varphi) + \sum_{i=1}^{s} \, (A_i(\nabla u_{\varepsilon\mu}^{(1)}, x, t) \, , \, \partial\varphi/\partial x_i) \right) dx \, +$$

$$+ \ O(\varepsilon) \|\varphi\|_{H^1(B_{\varepsilon,\mu}\cap G)} \sqrt{mes(B_{\varepsilon,\mu} \cap G)}.$$

Denote

$$I_3 = \ \int_{B_{\varepsilon,\mu}^{++}\cap G} \ \sum_{i=1}^{s} \ (A_i(\nabla u_{\varepsilon\mu}^{(1)}, x, t) \ , \ \partial\varphi/\partial x_i) \ dx \ .$$

Considering the smooth dependency of the functions A_i on its arguments, we have

$$I_3 = \ \int_{B_{\varepsilon,\mu}^{++}\cap G} \ \sum_{i=1}^{s} \ (A_i(\nabla_x v + \nabla_\xi (N^1 \Psi), x, t) \ , \ \partial\varphi/\partial x_i) \ dx \ +$$

$$+ \ O(\varepsilon) \|\varphi\|_{H^1(B_{\varepsilon,\mu}\cap G)} \sqrt{mes(B_{\varepsilon,\mu} \cap G)}.$$

From

$$\|\frac{\partial}{\partial \xi_i}((1-\Psi)N^1)\|_{L^2(B_{\varepsilon,\mu}^{++}\cap G)}^2 \ = \ O(\varepsilon)\sqrt{mes(B_{\varepsilon,\mu} \cap G)}$$

we get

$$I_3 = \int_{B^{++}_{\varepsilon,\mu}\cap G} \sum_{i=1}^{s} (A_i(\nabla_x v + \nabla_\xi N^1, x, t) \ , \ \partial\varphi/\partial x_i)\ dx +$$

$$+ O(\sqrt{\varepsilon})\|\varphi\|_{H^1(B_{\varepsilon,\mu}\cap G)}\sqrt{mes(B_{\varepsilon,\mu}\cap G)}.$$

As $\varphi = 0$ on ∂G we obtain integrating by parts:

$$I_3 = -\int_{B^{++}_{\varepsilon,\mu}\cap G} \sum_{i=1}^{s} (\frac{\bar{\partial}}{\partial x_i} A_i(\nabla_x v + \nabla_\xi N^1, x, t) \ , \ \varphi)\ dx -$$

$$-\int_{\partial B^{++}_{\varepsilon,\mu}\cap G} \sum_{i=1}^{s} (n_i A_i(\nabla_x v + \nabla_\xi N^1, x, t) \ , \ \varphi)\ ds +$$

$$+ O(\sqrt{\varepsilon})\|\varphi\|_{H^1(B_{\varepsilon,\mu}\cap G)}\sqrt{mes(B_{\varepsilon,\mu}\cap G)}.$$

Here $\frac{\bar{\partial}}{\partial x_i}$ stands for the full derivative of a function depending on x, t and $\xi = x/\varepsilon$, n is an inside normal vector for $B^{++}_{\varepsilon,\mu}$.

Due to property 1) of the X^e_i, the last integral is

$$I_4 = \int_{\partial B^{++}_{\varepsilon,\mu}\cap G} \sum_{i=1}^{s} (n_i A_i(\nabla_x v + \nabla_\xi N^1, x, t) \ , \ \varphi)\ ds =$$

$$= \sum_{x^0\ :\ \Pi_{x^0}\cap G\neq\emptyset} \int_{\partial\Pi_{x^0}\cap\partial B^{++}_{\varepsilon\mu}\cap G} \sum_{i=1}^{s} (n_i A_i(\nabla_x v + \nabla_\xi N^1, x, t) \ , \ \varphi)\ ds =$$

$$= \sum_{x^0\ :\ \Pi_{x^0}\subset\hat{G}} \int_{\partial\Pi_{x^0}\cap\partial B^{++}_{\varepsilon\mu}\cap G} \sum_{i=1}^{s} (n_i A_i(\nabla_x v + \nabla_\xi N^1, x, t) \ , \ \varphi)\ ds +$$

$$+ O(\sqrt{\varepsilon})\|\varphi\|_{H^1(B_{\varepsilon,\mu}\cap G)}\sqrt{mes(B_{\varepsilon,\mu}\cap G)}.$$

where the summation is made with respect to all nodal points x^0, such that the connected component Π_{x^0} of the set $\overline{B_{\varepsilon\mu}\backslash B^{++}_{\varepsilon\mu}}$ which contains x^0, belongs to $\bar{G}$. Here n is an outside normal of the domain Π_{x^0} (and respectively an inside normal vector for $B^{++}_{\varepsilon,\mu}$).

For each node x^0, we set

$$\langle\varphi\rangle_{\Pi_{x^0}} = \frac{\int_{\Pi_{x^0}} \varphi dx}{mes(\Pi_{x^0})},$$

then the following equality is valid:

$$\varphi = \langle\varphi\rangle_{\Pi_{x^0}} + (\varphi - \langle\varphi\rangle_{\Pi_{x^0}}),$$

and from lemma 4.A1.2 we obtain the estimate

$$\|\varphi - \langle\varphi\rangle_{\Pi_{x^0}}\|_{L^2(\partial\Pi_{x^0}\cap\partial B^{++}_{\varepsilon,\mu})} \leq C\sqrt{\varepsilon\mu}\|\nabla\varphi\|_{L^2(\Pi_{x^0})},$$

where C does not depend on ε, μ. Outside G the function φ is defined on $B_{\varepsilon,\mu}$ as zero.

Therefore

$$\int_{\partial\Pi_{x^0}\cap\partial B_{\varepsilon\mu}^{++}} \sum_{i=1}^{s} (n_i A_i(\nabla_x v + \nabla_\xi N^1, x, t) \ , \ \varphi) \, ds \ =$$

$$= \int_{\partial\Pi_{x^0}\cap\partial B_{\varepsilon\mu}^{++}\cap G} \sum_{i=1}^{s} (n_i A_i(\nabla_x v|_{x=x^0} + \nabla_\xi N^1|_{x=x^0}, x^0, t) \ , \ \varphi) \, ds \ +$$

$$+ \ O(\varepsilon\mu) \int_{\partial\Pi_{x^0}\cap\partial B_{\varepsilon\mu}^{++}} |\varphi| ds \ =$$

$$= \int_{\partial\Pi_{x^0}\cap\partial B_{\varepsilon\mu}^{++}} \sum_{i=1}^{s} (n_i A_i(\nabla_x v|_{x=x^0} + \nabla_\xi N^1|_{x=x^0}, x^0, t) \ , \ \langle\varphi\rangle_{\Pi_{x^0}}) \, ds \ + \ \delta_{x^0},$$

where

$$|\delta_{x^0}| \ \leq \ c_{00} \left(\int_{\partial\Pi_{x^0}\cap\partial B_{\varepsilon\mu}^{++}} |\varphi - \langle\varphi\rangle_{\Pi_{x^0}}| ds \ + \ \varepsilon\mu \int_{\partial\Pi_{x^0}\cap\partial B_{\varepsilon\mu}^{++}} |\varphi| ds \right),$$

with the constant c_{00} being independent of ε, μ, x^0 and φ.

Property 2) of X_i implies that the integral

$$\int_{\partial\Pi_{x^0}\cap\partial B_{\varepsilon\mu}^{++}\cap G} \sum_{i=1}^{s} (n_i A_i(\nabla_x v|_{x=x^0} + \nabla_\xi N^1|_{x=x^0}, x^0, t) \ , \ \langle\varphi\rangle_{\Pi_{x^0}}) \, ds$$

vanishes.

Estimate now δ_{x^0}.

Applying the Cauchy-Bunyakowskii-Schwartz inequality and then Lemma 4.A1.2, we get:

$$\int_{\partial\Pi_{x^0}\cap\partial B_{\varepsilon\mu}^{++}} |\varphi - \langle\varphi\rangle_{\Pi_{x^0}}| ds \ \leq$$

$$\leq \ \sqrt{mes(\partial\Pi_{x^0} \cap \partial B_{\varepsilon\mu}^{++})} \|\varphi - \langle\varphi\rangle_{\Pi_{x^0}}\|_{L^2(\partial\Pi_{x^0}\cap\partial B_{\varepsilon\mu}^{++})} \ \leq$$

$$\leq \ c_{01}\sqrt{\varepsilon\mu} \ \sqrt{mes(\partial\Pi_{x^0} \cap \partial B_{\varepsilon\mu}^{++})} \|\nabla\varphi\|_{L^2(\Pi_{x^0})} \ \leq$$

$$\leq \ c_{02}\sqrt{\varepsilon^s\mu^s} \|\nabla\varphi\|_{L^2(\Pi_{x^0})},$$

with the constants c_{01}, c_{02} being independent of ε, μ, x^0 and φ.

Applying the Cauchy-Bunyakowski-Schwarz inequality and then Lemma 4.A1.1, we get:

$$\int_{\partial\Pi_{x^0}\cap\partial B_{\varepsilon\mu}^{++}} |\varphi| ds \ \leq$$

$$\int_{\partial\Pi_{x^0}\cap\partial B_{\varepsilon\mu}^{++}} |\langle\varphi\rangle_{\Pi_{x^0}}| ds \ +$$

$$+\int_{\partial\Pi_{x^0}\cap\partial B^{++}_{\varepsilon\mu}}|\varphi-\langle\varphi\rangle_{\Pi_{x^0}}|ds.$$

Estimate both of these integrals.
We get

$$|\langle\varphi\rangle_{\Pi_{x^0}}| = \frac{1}{mes\Pi_{x^0}}|\int_{\Pi_{x^0}}\varphi dx| \le \frac{1}{\sqrt{mes\Pi_{x^0}}}\|\varphi\|_{L^2(\Pi_{x^0})} \le$$

$$\le c_{02}\frac{\|\varphi\|_{L^2(\Pi_{x^0})}}{\sqrt{\varepsilon\mu mes(\partial\Pi_{x^0}\cap\partial B^{++}_{\varepsilon\mu})}},$$

where c_{02} is a positive constant being independent of ε,μ,x^0 and φ, such that the following inequality holds true:

$$mes\Pi_{x^0} \ge \frac{1}{c_{02}^2}\varepsilon\mu mes(\partial\Pi_{x^0}\cap\partial B^{++}_{\varepsilon\mu}).$$

So, the integral

$$\int_{\partial\Pi_{x^0}\cap\partial B^{++}_{\varepsilon\mu}}|\langle\varphi\rangle_{\Pi_{x^0}}|ds$$

is estimated by

$$\frac{c_{02}}{\sqrt{\varepsilon\mu}}\|\varphi\|_{L^2(\Pi_{x^0})}\sqrt{mes(\partial\Pi_{x^0}\cap\partial B^{++}_{\varepsilon\mu})}$$

that is $O(\frac{\sqrt{\varepsilon^s\mu^s}}{\varepsilon\mu}\|\varphi\|_{H^1(\Pi_{x^0})})$.

As mentioned above, the same estimate holds true for

$$\int_{\partial\Pi_{x^0}\cap\partial B^{++}_{\varepsilon\mu}}|\varphi-\langle\varphi\rangle_{\Pi_{x^0}}|ds$$

and hence for

$$\int_{\partial\Pi_{x^0}\cap\partial B^{++}_{\varepsilon\mu}}|\varphi|ds.$$

So, $|\delta_{x^0}| \le c_{03}\sqrt{\varepsilon^s\mu^s}\|\varphi\|_{H^1(\Pi_{x^0})}$, where c_{03} is a positive constant being independent of ε,μ,x^0 and φ.

From the Cauchy-Bunyakovskii-Schwartz inequality for sums we obtain then, that

$$\sum_{x^0:\ \Pi_{x^0}\cap G\neq\emptyset}|\delta_{x^0}| \le c_{04}\sqrt{\frac{\varepsilon^s\mu^s}{\varepsilon^s}\sum_{x^0:\ \Pi_{x^0}\cap G\neq\emptyset}\|\varphi\|^2_{H^1(\Pi_{x^0})}} \le$$

$$\le c_{05}\sqrt{\mu}\sqrt{mes(B_{\varepsilon,\mu}\cap G)}\|\varphi\|_{H^1(B_{\varepsilon,\mu}\cap G)}.$$

Thus,

$$I_4 = O(\sqrt{\mu}+\sqrt{\varepsilon})\sqrt{mes(B_{\varepsilon,\mu}\cap G)}\|\varphi\|_{H^1(B_{\varepsilon,\mu}\cap G)},$$

and

$$I_2 = \int_{B^{++}_{\varepsilon,\mu}\cap G} \left((R\frac{\partial^2 v}{\partial t^2}\,,\,\varphi) + (S\frac{\partial v}{\partial t}\,,\,\varphi) - \sum_{i=1}^{s} (\frac{\bar\partial}{\partial x_i} A_i(\nabla_x v + \nabla_\xi N^1, x, t)\,,\,\varphi) \right) dx +$$

$$+ O(\sqrt{\varepsilon} + \sqrt{\mu})\|\varphi\|_{H^1(B_{\varepsilon,\mu}\cap G)}\sqrt{mes(B_{\varepsilon,\mu}\cap G)}.$$

We denote by $Q_{k_1\dots k_s}$ the cube $\{x \in I\!R^s \ |\varepsilon k_i \le \ x_i \ \le \varepsilon(k_i+1)\}$. The number of cubes, such that $Q_{k_1\dots k_s}\cap B_{\varepsilon,\mu}\cap\partial G \neq \emptyset$, does not exceed $const\varepsilon^{1-s}$ and consequently the total measure of these sets is $O(\varepsilon)\ mes\ (B_{\varepsilon,\mu}\cap G)$; therefore

$$I_5 \ = \ - \int_{B^{++}_{\varepsilon,\mu}\cap G} \sum_{i=1}^{s} (\frac{\bar\partial}{\partial x_i} A_i(\nabla_x v + \nabla_\xi N^1, x, t)\ ,\ \varphi)\ dx \ =$$

$$= - \sum_{k_1\dots k_s : Q_{k_1\dots k_s}\cap B_{\varepsilon,\mu}\subset G} \int_{Q_{k_1\dots k_s}\cap B^{++}_{\varepsilon,\mu}} \sum_{i=1}^{s} (\frac{\bar\partial}{\partial x_i} A_i(\nabla_x v + \nabla_\xi N^1, x, t)\,,\,\varphi)\, dx +$$

$$+ O(\sqrt{\varepsilon})\|\varphi\|_{H^1(B_{\varepsilon,\mu}\cap G)}\sqrt{mes(B_{\varepsilon,\mu}\cap G)} =$$

$$= - \sum_{k_1\dots k_s : Q_{k_1\dots k_s}\cap B_{\varepsilon,\mu}\subset G} \int_{Q_{k_1\dots k_s}\cap B^{++}_{\varepsilon,\mu}} \sum_{i=1}^{s} (\frac{\bar\partial}{\partial x_i} A_i(\nabla_x v + \nabla_\xi N^1, x, t)\,,\,\varphi)$$

$$|_{x_i=\varepsilon(k_i+1/2),\ \ i=1,\dots,s}\, dx + O(\sqrt{\varepsilon})\|\varphi\|_{H^1(B_{\varepsilon,\mu}\cap G)}\sqrt{mes(B_{\varepsilon,\mu}\cap G)}.$$

Next, by using on each set $Q_{k_1\dots k_s}\cap B^{++}_{\varepsilon,\mu}$ the representation

$$\varphi \ = \ \langle\varphi\rangle_{k_1\dots k_s} \ + \ (\varphi \ - \ \langle\varphi\rangle_{k_1\dots k_s}),$$

$$\langle\varphi\rangle_{k_1\dots k_s} \ = \ \frac{1}{mes(Q_{k_1\dots k_s}\cap B_{\varepsilon,\mu})} \int_{Q_{k_1\dots k_s}\cap B_{\varepsilon,\mu}} \varphi(x,t)\ dx,$$

the Poincaré-Friedrichs inequality of Appendix 4.A2

$$\|\varphi \ - \ \langle\varphi\rangle_{k_1\dots k_s}\|_{L^2(Q_{k_1\dots k_s}\cap B_{\varepsilon,\mu})} \ \le \ c_{05}\varepsilon\|\nabla\varphi\|_{L^2(Q_{k_1\dots k_s}\cap B_{\varepsilon,\mu})},$$

c_{05} does not depend on ε and μ), and the Cauchy - Bunyakovskii - Schwartz inequality for sums: $\sum_{i=1}^{J} a_i \le \sqrt{J\sum_{i=1}^{J} a_i^2}$, we get

$$\sum_{k_1\dots k_s : Q_{k_1\dots k_s}\cap B_{\varepsilon,\mu}\subset G} \int_{Q_{k_1\dots k_s}\cap B^{++}_{\varepsilon,\mu}} |\varphi - <\varphi>|\ dx \ =$$

$$= \ O\left(\sqrt{\varepsilon^{-s} \sum_{k_1\dots k_s : Q_{k_1\dots k_s}\cap B_{\varepsilon,\mu}\subset G} (\int_{Q_{k_1\dots k_s}\cap B^{++}_{\varepsilon,\mu}} |\varphi - <\varphi>|\ dx)^2}\right) \ =$$

$$= O\left(\sqrt{\varepsilon^{-s}\varepsilon^{s}\mu^{s-1} \sum_{k_1\dots k_s:Q_{k_1\dots k_s}\cap B_{\varepsilon,\mu}\subset G} \|\varphi - <\varphi>\|^2_{L^2(Q_{k_1\dots k_s}\cap B^{++}_{\varepsilon,\mu})}}\right) =$$

$$+\ O(\varepsilon)\|\varphi\|_{H^1(B_{\varepsilon,\mu}\cap G)}\sqrt{mes(B_{\varepsilon,\mu}\cap G)}.$$

$$I_5 \ = \ - \sum_{k_1\dots k_s:Q_{k_1\dots k_s}\cap B_{\varepsilon,\mu}\subset G} \int_{Q_{k_1\dots k_s}\cap B^{++}_{\varepsilon,\mu}} \left(\frac{1}{M}\hat{L}_x v - R\frac{\partial^2 v}{\partial t^2} - \right.$$

$$\left. -S\frac{\partial v}{\partial t}\right)\langle\varphi\rangle_{k_1\dots k_s} dx \ + \ O(\sqrt{\varepsilon}+\sqrt{\mu})\|\varphi\|_{H^1(B_{\varepsilon,\mu}\cap G)}\sqrt{mes(B_{\varepsilon,\mu}\cap G)}.$$

From presentation

$$\varphi \ = \ \langle\varphi\rangle_{k_1\dots k_s} \ + \ (\varphi \ - \ \langle\varphi\rangle_{k_1\dots k_s}),$$

we have again

$$- \int_{B^{++}_{\varepsilon,\mu}\cap G} \sum_{i=1}^{s} \left(\frac{\bar{\partial}}{\partial x_i} A_i(\nabla_x v + \nabla_\xi N^1, x, t) \ , \ \varphi\right) dx \ =$$

$$- \int_{B_{\varepsilon,\mu}\cap G} \left(\left(\frac{1}{M}\hat{L}_x v - R\frac{\partial^2 v}{\partial t^2} - S\frac{\partial v}{\partial t}\right), \varphi \right) dx \ +$$

$$+ \ O(\sqrt{\varepsilon}+\sqrt{\mu})\|\varphi\|_{H^1(B_{\varepsilon,\mu}\cap G)}\sqrt{mes(B_{\varepsilon,\mu}\cap G)}.$$

This implies finally

$$I_1 \ = \ \int_{B_{\varepsilon,\mu}\cap G} f(x,t)\varphi dx \ + \ O(\sqrt{\varepsilon}+\sqrt{\mu})\|\varphi\|_{H^1(B_{\varepsilon,\mu}\cap G)}\sqrt{mes(B_{\varepsilon,\mu}\cap G)} \ =$$

$$= \ \int_{B_{\varepsilon,\mu}\cap G} \left(\left(R\frac{\partial^2 u_{\varepsilon\mu}}{\partial t^2} + S\frac{\partial u_{\varepsilon\mu}}{\partial t}\right), \varphi \right) \ + \ \sum_{i=1}^{s} \left(A_i(\nabla_x u_{\varepsilon\mu}, x, t) \ , \ \frac{\partial\varphi}{\partial x_i}\right) dx \ +$$

$$+ \ O(\sqrt{\varepsilon}+\sqrt{\mu})\|\varphi\|_{H^1(B_{\varepsilon,\mu}\cap G)}\sqrt{mes(B_{\varepsilon,\mu}\cap G)}.$$

Theorem is proved.

Thus the sequence $u^{(1)}_{\varepsilon\mu} - u_{\varepsilon\mu}$ FLt-converges to zero .

5.4 L- and FL-Convergence in elasticity

Consider the elasticity system of equations

$$\sum_{i,j=1}^{s} \frac{\partial}{\partial x_i}(A_{ij}\frac{\partial u_{\varepsilon\mu}}{\partial x_j}) = f(x), \quad x \in B_{\varepsilon,\mu} \subset I\!R^s,$$

with the boundary condition

$$\sum_{i,j=1}^{s} n_i(A_{ij}\frac{\partial u_{\varepsilon\mu}}{\partial x_j}) = 0, \quad x \in \partial B_{\varepsilon,\mu},$$

where A_{ij} are constant $s \times s$–matrix-valued functions with the elements a_{ij}^{kl}, satisfying the conditions

1) $a_{ij}^{kl} = a_{il}^{kj} = a_{ji}^{lk}\ \forall\ k,i,l,j \in \{1,...,s\}$ and

2) there exists a constant $C_0 > 0$ such that

$$\sum_{i,j,k,l=1}^{s} a_{ij}^{kl}\eta_{lj}\eta_{ki} \geq C_0 \sum_{l,j=1}^{s} \eta_{lj}^2$$

for each symmetric constant matrix $(\eta_{lj})_{1\leq l,j\leq s}$.

In this case the estimation of Lemma 5.2.2 is not valid because of the dependence of the constant of Korn's inequality of μ. Indeed the following lemma takes place:

Lemma 5.4.1 (proved in [145]).*Let*

$$E_{B_{\varepsilon,\mu}\cap G}(\varphi) = \int_{B_{\varepsilon,\mu}\cap G} \sum_{i,j=1}^{s} (e_i^j(\varphi))^2\, dx, e_i^j(\varphi) = \frac{1}{2}(\frac{\partial \varphi_i}{\partial x_j} + \frac{\partial \varphi_j}{\partial x_i}).$$

Then for any lattice $B_{\varepsilon,\mu}$ and domain G, independent of ε and μ with piecewise-smooth boundary there exists such vector-valued function $v_{\varepsilon,\mu} \in H^1(B_{\varepsilon,\mu}\cap G)$ that

$$v_{\varepsilon,\mu}|_{B_{\varepsilon,\mu}\cap\partial G} = 0, \quad \|\nabla v_{\varepsilon,\mu}\|^2_{L^2(B_{\varepsilon,\mu}\cap G)} \geq c_1\mu^{s-1}\varepsilon^{s-2},$$

$$E_{B_{\varepsilon,\mu}\cap G}(v_{\varepsilon,\mu}) \leq c_2\mu^s\varepsilon^{s-2},$$

i.e.

$$E_{B_{\varepsilon,\mu}\cap G}(v_{\varepsilon,\mu}) \leq c_3\mu\|\nabla v_{\varepsilon,\mu}\|^2_{L^2(B_{\varepsilon,\mu}\cap G)},$$

where constants c_1, c_2, c_3 do not depend on ε and μ.

The proof is based on the analogous proposition for a bar $\pi = (-1,1) \times \beta_\mu$, where $\beta_\mu = \{(x'/\mu \in \beta\}$, β is $(s-1)$–dimensional bounded domain with a piecewise-smooth boundary (satisfying the cone condition; in the two-dimensional case β is an interval), i.e. there exist such constants c_1, c_2, $c_3 > 0$, independent of μ, and the s-dimensional vector-valued function $w_\mu(x)$, such that $w_\mu \in H^1(\pi)$ and

$$w_\mu|_{x_1=\pm 1} = 0, \quad \|w_\mu\|^2_{L^2(\pi)} \geq c_1 \mu^{s-1}, \quad \|\nabla w_\mu\|^2_{L^2(\pi)} \geq c_2 \mu^{s-1},$$

$$E_\pi(w_\mu) \leq c_3 \mu^s.$$

This example was constructed in subsection 4.4.1.

By the linear application we could build such function w_μ, which satisfies to all conditions of the proposition and vanishes in the vicinity of the planes $\{x_1 = \pm 1\}$. If now for each bar of the lattice structure $B_{\varepsilon,\mu}$ except of those that cross the boundary of the domain G we shall construct the corresponding function (with the replacement of μ by the product $\varepsilon\mu$ and of the length 2 of the bar by ε) we obtain the proof of this lemma.

Thus L-convergence for the elasticity equations is not the consequence of the $FL-$convergence.

Nevertheless the Korn inequality for the lattice structures with the constant estimated by a power μ^{-K} with some K was proved in [145] and then this constant was calculated more precisely in [129], [43] and then in [68]. The Korn inequality for lattices was used with boundary layer and homogenization methods for the construction of the asymptotic expansion of the solution of the elasticity system of equations, set in a lattice-like domain ([148]).

Here above we have introduced the L-convergence and the FL-convergence for the lattice-like domains and we have applied it to the homogenization of linear and non-linear problems in these domains. The techniques of proof of L-convergence is quite different of the techniques of extension of the solution of the problems for perforated domain inside the perforations [45]. More detailed information about the solution could be obtained from the asymptotic expansion of the solution, built in [136], [148].

Using the terminology of Ph.Ciarlet [38], one can say that lattice-like domains are multi-structures with a great number of junctions. Particularly the case of junction of two elastic rods was considered in [94], [95].

The extension techniques was developed for lattice-like structures in [46],[44], where the convergence results were obtained for the elasticity system of equations, set in these structures. The influence of the order of the passage to limit by small parameters was studied.

The boundary value problems considered above for lattice-like domains were of Neumann's type on $\partial B_{\varepsilon,\mu}$. The Dirichlet's problems for lattice-like domains are developed in [90] (periodic case) and [91] (non-periodic case).

5.5 Conductivity of a net

Consider the limit problem of section 4.4 on a periodic net. Define net $\mathcal{B}$ as a connected periodic in all $x_i (i = 1, \ldots, s)$ with period 1 union of segments in $\mathbb{R}^s$

:

$$\mathcal{B} = \bigcup_{n=-\infty}^{+\infty} e_n$$

such that
(i) any two segments $e_i, e_j (i \neq j)$ may have only one common point, which is an end-point for both segments. The end-points of the segments are called nodes.

(ii) the periodicity cube $Q = \{x \in \mathbb{R}^s | -\frac{1}{2} \leq x_i \leq \frac{1}{2}\}$ intersects finitely many segments $e_1, \ldots, e_M$ from $\mathcal{B}$; we assume that $\mathcal{B} \cap Q$ is connected.

Let $\mathcal{B}_\varepsilon = \{x \in \mathbb{R}^s | \frac{x}{\varepsilon} \in \mathcal{B}\}$. For any segment $e_i \subset \mathcal{B}$, denote $e_i^\varepsilon = \{x \in \mathbb{R}^s | \frac{x}{\varepsilon} \in e_i\}$. Introduce local longitudinal variable for any e_i^ε in a following way.

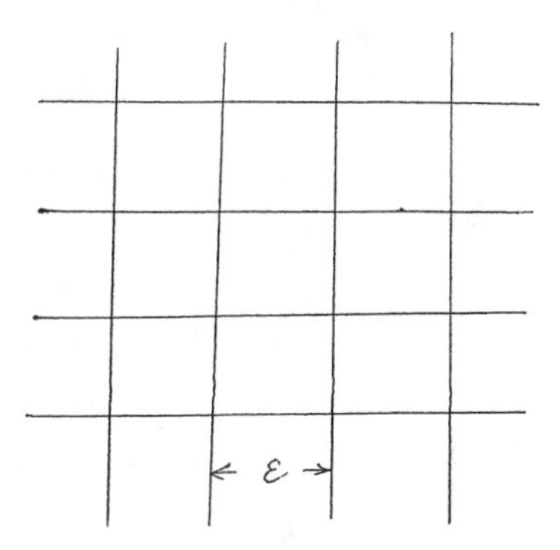

Figure 5.5.1. A net.

Fix an end point x_0 and define for any $x \in e_i^\varepsilon$ the variable $\widetilde{x}_1(x) = \sum_{i=1}^{s} \gamma_i (x_i - x_{0i})$ where $(\gamma_1, \ldots, \gamma_s)$ are the directing cosines of the vector e_i^ε with the origin at x_0; so $\widetilde{x}_1(x)$ is the first component of the vector $\alpha(x - x_0)$; here α is an orthogonal matrix ; α^* is the matrix of orthogonal space transformation $\widetilde{\Pi}$ such that the image of the vector $\overrightarrow{i}$ parallel to the x_1 axis of length $|e_i^\varepsilon|$ is the vector e_i^ε.

Consider the problem :

$$\frac{d^2 u_\varepsilon}{d\widetilde{x}_1^2} = f(x), \qquad x \in e_i^\varepsilon, \tag{5.5.1}$$

$$\sum_{e(x_0)} mes\ \beta_e \frac{du_{e_\varepsilon}}{d\widetilde{x}_1} = 0, \tag{5.5.2}$$

in the nodes x_0,

$$u_\varepsilon^{e_1} = u_\varepsilon^{e_2} \qquad \forall e_1, e_2 \tag{5.5.3}$$

with the common node x_0.
Here $\widetilde{x}_1$ is defined once for any segment e_i^ε (i.e. we fix direction once) ; $mes\beta_e$ are some positive numbers, associated with any segment e, 1 periodic on $\mathcal{B}_\varepsilon$.
Suppose that f is a $T-$periodic function of x such that $f \in C^\infty(\mathbb{R}^s)$, and that

$$\int_{Q_T} f(x)dx = 0, \int_{Q_T \cap \mathcal{B}_\varepsilon} f(x)d\widetilde{x}_1 = 0,$$

where $Q_T = (0,T)^s$, T is a multiple of ε.
We seek a $T-$periodic solution $u_\varepsilon(\widetilde{x}_1(x))$ on $\mathcal{B}_\varepsilon$; T is a multiple of ε.
The asymptotic expansion is sought in a form of series

$u_\varepsilon^{(\infty)} \sim \sum_{l=0}^{\infty} \varepsilon^l \sum_{|i|=l} \mathcal{N}_{i_1 \dots i_l}\left(\frac{\widetilde{x}_1(x)}{\varepsilon}\right) \frac{\partial^l v}{\partial x_{i_1} \dots \partial x_{i_s}}$. Applying the Bakhvalov's procedure described above in Chapter 2 we obtain equations on $\mathcal{N}_{i_1 \dots i_l}$:

$$\frac{d^2}{d\widetilde{\xi}_1^2}\mathcal{N}_{i_1 \dots i_l} + 2\gamma_{i_1}\frac{d\mathcal{N}_{i_2 \dots i_l}}{d\widetilde{\xi}_1} + \gamma_{i_1}\gamma_{i_2}\mathcal{N}_{i_3 \dots i_l} = h_{i_1 \dots i_l}, \ \xi \in e \subset \mathcal{B}, \tag{5.5.4}$$

$$\sum_{e(\xi_0)} mes\ \beta_e \left(\frac{d}{d\widetilde{\xi}_1}\mathcal{N}_{i_1 \dots i_l} + \gamma_{i_1}\mathcal{N}_{i_2 \dots i_l}\right) = 0, \tag{5.5.5}$$

in the nodes ξ_0 of $\mathcal{B}$ (the direction of $\widetilde{\xi}_1$ is chosen and fixed <u>once</u>) ; $\mathcal{N}_{i_1 \dots i_l}$ is continuous and 1 periodic on $\mathcal{B}$

$$h_{i_1 \dots i_l} = \sum_{ecQ_1 \cap \mathcal{B}_1} \int_e mes\ \beta_e \gamma_{i_1}^e \left(\frac{d}{d\widetilde{\xi}_1}\mathcal{N}_{i_2 \dots i_l} + \gamma_{i_2}^e \mathcal{N}_{i_3 \dots i_l}\right) d\widetilde{\xi}_1 / \sum_{ecQ \cap \mathcal{B}} mes\ \beta_e |e|; \tag{5.5.6}$$

for $l = 1$ we get

$$\frac{d}{d\xi_1}\left(\frac{d}{d\xi_1}\mathcal{N}_{i_1} + \gamma_{i_1}\right) = 0 \ , \ \xi \in e \subset \mathcal{B}, \tag{5.5.7}$$

$$\sum_{e(\xi_0)} mes\ \beta_e \left(\frac{d}{d\xi_1}\mathcal{N}_{i_1} + \gamma_{i_1}\right) = 0 \ , \ \xi = \xi_0 \ , \tag{5.5.8}$$

ξ_0 is a node of $\mathcal{B}$.

It can be proved that the matrix $(h_{i_1 i_2})_{1 \le i_1, i_2 \le s}$ is symmetric positive definite matrix (see below Lemma 5.7.3).
Thus we can organize a recurrent process of determining of coefficients of series

$$V = \sum_{j=0}^{\infty} \varepsilon^j V_j(x). \tag{5.5.9}$$

We obtain a recurrent chain of $T-$periodic problems

$$\widehat{L}V_j = f_j(x)\ , \qquad \widehat{L} = \sum_{i_1,i_2=1}^{s} h_{i_1 i_2} \frac{\partial^2}{\partial x_{i_1} \partial x_{i_2}} \tag{5.5.10}$$

where

$$f_j(x) = \sum_{k=0}^{j-1} \sum_{|i|=j-k} h_{i_1 \dots i_{j-k}} D^i V_k. \tag{5.5.11}$$

We prove by induction that $\displaystyle\int_{Q_T} f_j(x)dx = 0$; so the chain of problems is solvable.

Justification in a standard way gives an estimate for the difference of a $T-$periodic solution of problem (5.5.1)-(5.5.3) with vanishing average and of a partial sum

$$u_\varepsilon^{(K)} = \sum_{l=0}^{K+1} \varepsilon^l \sum_{i_1 \dots i_l \in \{1,\dots,s\}} \mathcal{N}_{i_1 \dots i_l} \left(\frac{\widetilde{x}_1(x)}{\varepsilon} \right) D^{i_1 \dots i_l} \sum_{j=0}^{k} \varepsilon^j V_j(x) \tag{5.5.12}$$

with vanishing average :

$$\|u_\varepsilon^{(k)} - u_\varepsilon\|_{H^1(Q_T \cap \mathcal{B}_\varepsilon)} = O(\varepsilon^k). \tag{5.5.13}$$

The main term is described by the equation

$$\widehat{L}V_0 = f(x) \tag{5.5.14}$$

with T periodicity condition ;
here operator $\widehat{L}$ is determined by relations (5.5.7),(5.5.8),(5.5.10).

5.6 Elasticity of a net

Consider the limit problem of section 4.5 on a periodic net $\mathcal{B}_\varepsilon$ in two-dimensional case. It consists of

(i) the equations

$$\alpha_e^* \begin{pmatrix} E \frac{d^2 \tilde{u}_\varepsilon^{1\ e}}{d\tilde{x}_1^2} \\ -E < \xi_2^2 >_\beta \mu^2 \frac{d^4 \tilde{u}_\varepsilon^{2\ e}}{d\tilde{x}_1^4} \end{pmatrix} = f(x), x \in e_\varepsilon, \tag{5.6.1}$$

where

$$\alpha_e = \begin{pmatrix} \gamma_1^e & \gamma_2^e \\ -\gamma_2^e & \gamma_1^e \end{pmatrix},$$

is an orthogonal matrix associated with any segment e. Here $\beta = (-1/2, 1/2)$, and $< \xi_2^2 >_\beta = 1/12$ and $mes\ \beta = 1$ with directing cosines (γ_1^e, γ_2^e).

The interface conditions are: for any node x_0,

$$\sum_{e(x_0)} mes\ \beta_e \gamma^e \frac{d\tilde{u}_\varepsilon^{1\ e}}{d\tilde{x}_1} = 0 \tag{5.6.2}$$

and

$$\sum_{e(x_0)} mes\ \beta_e \gamma^e < \xi_2^2 > \frac{d^2\tilde{u}_\varepsilon^{2\ e}}{d\tilde{x}_1^2} = 0; \tag{5.6.3}$$

moreover, $\gamma^e \tilde{u}_\varepsilon^1$, $\frac{d\tilde{u}_\varepsilon^{2\ e}}{d\tilde{x}_1}$ are continuous functions on $\mathcal{B}_\varepsilon$ and $\tilde{u}_\varepsilon^2 = 0$ in all nodes, $\gamma^e = (\gamma_1^e, \gamma_2^e)^*$.

Here E is the Young modulus, $f \in C^\infty$ is a $T-$periodic $s-$dimensional vector-valued function, T is a multiple of ε.

We assume as well that the following integrals of the right hand-side vanish:

$$\int_{(0,T)^2 \cap \mathcal{B}_\varepsilon} f(x)dx = 0, \quad \int_{(0,T)^2} f(x)dx = 0.$$

We seek the solution $u(x)$ in the class of $T-$periodic two-dimensional vector-valued functions.

The asymptotic expansion is sought in a form of series

$$u_\varepsilon^{(\infty)} \sim \sum_{l=0}^{\infty} \varepsilon^l \sum_{|i|=l} \mathcal{N}_{i_1,\dots,i_l}\left(\frac{\tilde{x}_1(x)}{\varepsilon}\right) \frac{\partial^l V(x)}{\partial x_{i_1}\dots\partial x_{i_l}}, \tag{5.6.4}$$

where $\mathcal{N}_i(\xi)$ are 2×2 $1-$periodic in ξ matrices, V is a two-dimensional vector,

$$\mathcal{N}_i(\xi) = \alpha_e^* \tilde{\mathcal{N}}_i^e(\xi)\alpha_e,$$

$$\mathcal{N}_\emptyset(\xi) = I,$$

$$\tilde{\mathcal{N}}_i^e(\xi) = \tilde{\mathcal{N}}_i^{e\ (1)}(\xi) + \mu^{-2}\tilde{\mathcal{N}}_i^{e\ (2)}(\xi), \tag{5.6.5}$$

where I stands for the identity matrix, $\tilde{\mathcal{N}}_i^{e\ (1)}(\xi)$ has a vanishing second line; $\tilde{\mathcal{N}}_i^{e\ (2)}(\xi)$ has a vanishing first line.

Applying the Bakhvalov procedure (section 2.2) we obtain the equations on the first component $(\tilde{\mathcal{N}}_{i_1,\dots,i_l}^{e\ (1)})_1$ of $\tilde{\mathcal{N}}_{i_1,\dots,i_l}^{e\ (1)}$:

$$E\left\{\frac{d^2(\tilde{\mathcal{N}}_{i_1,\dots,i_l}^{e\ (1)})_1}{d\tilde{\xi}_1^2} + 2\gamma_{i_1}\frac{d(\tilde{\mathcal{N}}_{i_2,\dots,i_l}^{e\ (1)})_1}{d\tilde{\xi}_1} + \gamma_{i_1}\gamma_{i_2}(\tilde{\mathcal{N}}_{i_3,\dots,i_l}^{e\ (1)})_1\right\} = (\alpha_e h_{i_1\dots i_l}\alpha_e^*)_1 \tag{5.6.6}$$

for any e,

$$\sum_{e(\xi_0)} E\ mes\ \beta_e\gamma_e\left\{\frac{d(\tilde{\mathcal{N}}^{e\ (1)}_{i_1,\ldots,i_l})_1}{d\tilde{\xi}_1}+\gamma_{i_1}(\tilde{\mathcal{N}}^{e\ (1)}_{i_2,\ldots,i_l})_1\right\}=0 \tag{5.6.7}$$

in the nodes ξ_0;

$\mathcal{N}^{e\ (1)}_i=\alpha^*_e\tilde{\mathcal{N}}^{e\ (1)}_i(\xi)\alpha_e$, are 1-periodic and continuous functions;

we obtain as well the following chain of problems for $\mathcal{N}^{e\ (2)}_i=\alpha^*_e\tilde{\mathcal{N}}^{e\ (2)}_i(\xi)\alpha_e$:

$$-E<\xi_2^2>_{\beta_e}\left\{\frac{d^4(\tilde{\mathcal{N}}^{e\ (2)}_{i_1,\ldots,i_l})_2}{d\tilde{\xi}_1^4}+4\gamma_{i_1}\frac{d^3(\tilde{\mathcal{N}}^{e\ (2)}_{i_2,\ldots,i_l})_2}{d\tilde{\xi}_1^3}+6\gamma_{i_1}\gamma_{i_2}\frac{d^2(\tilde{\mathcal{N}}^{e\ (2)}_{i_3,\ldots,i_l})_2}{d\tilde{\xi}_1^2}\right.$$

$$\left.+4\gamma_{i_1}\gamma_{i_2}\gamma_{i_3}\frac{d(\tilde{\mathcal{N}}^{e\ (2)}_{i_4,\ldots,i_l})_2}{d\tilde{\xi}_1}+\gamma_{i_1}\gamma_{i_2}\gamma_{i_3}\gamma_{i_4}(\tilde{\mathcal{N}}^{e\ (2)}_{i_5,\ldots,i_l})_2\right\}=(\alpha_e h_{i_1\ldots i_l}\alpha^*_e)_2 \tag{5.6.8}$$

for any e,

$$\sum_{e(\xi_0)} E\ <\xi_2^2>_{\beta_e}\ mes\ \beta_e\left\{\frac{d^2(\tilde{\mathcal{N}}^{e\ (2)}_{i_1,\ldots,i_l})_2}{d\tilde{\xi}_1^2}+2\gamma_{i_1}\frac{d(\tilde{\mathcal{N}}^{e\ (2)}_{i_2,\ldots,i_l})_2}{d\tilde{\xi}_1}+\gamma_{i_1}\gamma_{i_2}(\tilde{\mathcal{N}}^{e\ (2)}_{i_3,\ldots,i_l})_2\right\}=0 \tag{5.6.9}$$

in the nodes ξ_0;

$\frac{d(\tilde{\mathcal{N}}^{e\ (2)}_{i_1,\ldots,i_l})_2}{d\tilde{\xi}_1}+\gamma_{i_1}(\tilde{\mathcal{N}}^{e\ (2)}_{i_2,\ldots,i_l})_2$ is continuous on $\mathcal{B}$;

$\tilde{\mathcal{N}}^{e\ (2)}_i$ is a 1-periodic function vanishing in all nodes ξ_0;

$$h_{i_1,\ldots,i_l}=\left(\sum_{e\subset Q_1\cap\mathcal{B}}\int_e E\ mes\ \beta_e\alpha^*_e\left\{\gamma_{i_1}\frac{d(\tilde{\mathcal{N}}^{e\ (1)}_{i_1,\ldots,i_l})_1}{d\tilde{\xi}_1}+\right.\right.$$

$$\left.\left.+\gamma_{i_1}\gamma_{i_2}(\tilde{\mathcal{N}}^{e\ (1)}_{i_2,\ldots,i_l})_1\right\}\alpha_e d\tilde{\xi}_1\right)/\sum_{e\subset Q_1\cap\mathcal{B}} mes\ \beta_e|e|; \tag{5.6.10}$$

here $\mathcal{N}^e_\emptyset=I$ (identity matrix) and for any square matrix M, $(M)_k$ is the $k-$th line of M.

The solvability of problems for $\tilde{\mathcal{N}}^{e\ (1)}_i$ and $\tilde{\mathcal{N}}^{e\ (2)}_i$ are proved in section 5.8.

Assume that the operator $\hat{L}=\sum_{i_1,i_2=1}^2 h_{i_1i_2}D^{i_1i_2}$ is an operator of elasticity. Then we organize a recurrent procedure of determining of coefficients of series

$$V=\sum_{j=0}^{\infty}\varepsilon^j V_j(x) \tag{5.6.11}$$

as in the previous section.

The main term is

$$V_0 + (\frac{\varepsilon}{\mu})^2 \sum_{i_1,i_2=1}^{2} \mathcal{N}^{e\ (2)}_{i_1,i_2}\left(\frac{\tilde{x}_1(x)}{\varepsilon}\right)\frac{\partial^2 V_0}{\partial x_{i_1}\partial x_{i_2}}; \qquad (5.6.12)$$

here V_0 is a $T-$periodic solution of homogenized equation

$$\hat{L}V_0 = f(x), \qquad (5.6.13)$$

where the coefficients $h_{i_1i_2}$ of the operator $\hat{L}$ are determined as in [16] :

$$h_{i_1i_2} = \left(\sum_{e\subset Q_1\cap\mathcal{B}}\int_e E\ mes\ \beta_e\alpha_e^*\left\{\gamma_{i_1}\frac{d(\tilde{\mathcal{N}}^{e\ (1)}_{i_2})}{d\tilde{\xi}_1}+\right.\right.$$

$$\left.\left.+\gamma_{i_1}\gamma_{i_2}(\tilde{\mathcal{N}}^{e\ (1)}_{\emptyset})\right\}\alpha_e d\tilde{\xi}_1\right)/\sum_{e\subset Q_1\cap\mathcal{B}} mes\ \beta_e|e|,$$

here

$$\tilde{\mathcal{N}}^{e\ (1)}_{\emptyset} = (\alpha_e\alpha_e^*)_1 = \begin{pmatrix} 1 & 0 \\ 0 & 0 \end{pmatrix},$$

i.e.

$$h_{i_1i_2} =$$

$$\left(\sum_{e\subset Q_1\cap\mathcal{B}}\int_e E\ mes\ \beta_e\begin{pmatrix}\gamma_1^e\\ \gamma_2^e\end{pmatrix}\left\{\gamma_{i_1}\frac{d(\tilde{\mathcal{N}}^{e\ (1)}_{i_2})_1}{d\tilde{\xi}_1}\alpha_e+\right.\right.$$

$$\left.\left.\gamma_{i_1}\gamma_{i_2}\begin{pmatrix}(\gamma_1^e)^2 & \gamma_1^e\gamma_2^e\\ -\gamma_1^e\gamma_2^e & (\gamma_2^e)^2\end{pmatrix}\right\}d\tilde{\xi}_1\right)/\sum_{e\subset Q_1\cap\mathcal{B}} mes\ \beta_e|e|, \qquad (5.6.14)$$

where $\tilde{\mathcal{N}}^{e\ (1)}_{i_1}\alpha_e$ is a 1-periodic linear for any e solution to the problem

$$\sum_{e\subset Q_1\cap\mathcal{B}}\int_e E\ mes\ \beta_e\gamma_e\left\{\frac{d(\tilde{\mathcal{N}}^{e\ (1)}_{i_1})}{d\tilde{\xi}_1}\alpha_e+\gamma_{i_1}\begin{pmatrix}1 & 0\\ 0 & 0\end{pmatrix}\alpha_e\right\}=0 \qquad (5.6.15)$$

in the nodes ξ_0 such, that $\alpha_e^*\tilde{\mathcal{N}}^{e\ (1)}_{i_1}\alpha_e$ is a continuous function;
$\tilde{\mathcal{N}}^{e\ (2)}_{i_1i_2}\alpha_e$ is a 1-periodic solution to the problem

$$E\ <\xi_2^2>_{\beta_e}\frac{d^4(\tilde{\mathcal{N}}^{e\ (2)}_{i_1i_2})_2}{d\tilde{\xi}_1^4} = (\alpha_e h_{i_1i_2}\alpha_e^*)_2,\quad \xi\in e, \qquad (5.6.16)$$

$$\sum_{e(\xi_0)} E\ mes\ \beta_e\ <\xi_2^2>_{\beta_e}\left(\frac{d^2(\tilde{\mathcal{N}}^{e\ (2)}_{i_1i_2})}{d\tilde{\xi}_1^2}\alpha_e\right)=0,\quad \xi=\xi_0, \qquad (5.6.17)$$

and $\frac{d(\tilde{\mathcal{N}}^{e\ (2)}_{i_1i_2})_2}{d\tilde{\xi}_1}$ is a continuous on $\mathcal{B}$ function; $\alpha_e^*\tilde{\mathcal{N}}^{e\ (2)}_{i_1i_2}\alpha_e$ is a 1-periodic on $\mathcal{B}$ vanishing at all nodes matrix-valued function.

In particular for the net $\mathcal{B} = \mathcal{B}_{(1)} \cup \mathcal{B}_{(2)}$ with $\mathcal{B}_{(i)} = \cup_{k=-\infty}^{+\infty}\{\xi = (\xi_1, \xi_2) | \xi_i \in \mathbb{R},\ \xi_{3-i} = k\}$, here $mes\ \beta = 1$. We get (cf [16])

$$\hat{L}u = \begin{pmatrix} \frac{E}{2}\frac{\partial^2 u_1}{\partial x_1^2} \\ \frac{E}{2}\frac{\partial^2 u_2}{\partial x_2^2} \end{pmatrix} \tag{5.6.18}$$

and $\mathcal{N}_{i_1}^{e\ (1)} = 0$; $h_{i_1 i_2} = \delta_{i_1 i_2} \frac{E}{2} h_{(\frac{i_1+i_2}{2})}$ and $h_{(1)} = \begin{pmatrix} 1 & 0 \\ 0 & 0 \end{pmatrix}$, $h_{(2)} = \begin{pmatrix} 0 & 0 \\ 0 & 1 \end{pmatrix}$, and $\mathcal{N}_{i_1 i_2}^{e\ (2)} = 0$ for $i_1 \neq i_2$;

$$\tilde{\mathcal{N}}_{11}^{e\ (2)} = \begin{cases} (0,0) \text{ on } \mathcal{B}_{(1)} \\ (0, \phi(\tilde{\xi}_1)) \text{ on } \mathcal{B}_{(2)} \cap \{\xi_2 \in (0,1)\} \end{cases} \tag{5.6.19}$$

and

$$\tilde{\mathcal{N}}_{22}^{e\ (2)} = \begin{cases} (0,0) \text{ on } \mathcal{B}_{(2)} \\ (0, \phi(\tilde{\xi}_1)) \text{ on } \mathcal{B}_{(1)} \cap \{\xi_1 \in (0,1)\} \end{cases} \tag{5.6.20}$$

Here $\phi(\tilde{\xi}_1) = -\frac{1}{4}\tilde{\xi}_1^2(1-\tilde{\xi}_1)^2$.
So,

$$\mathcal{N}_{11}^{(2)} = \begin{cases} (0,0) \text{ on } \mathcal{B}_{(1)} \\ (\phi(\xi_2), 0) \text{ on } \mathcal{B}_{(2)} \cap \{\xi_2 \in (0,1)\} \end{cases} \tag{5.6.21}$$

and

$$\mathcal{N}_{22}^{(2)} = \begin{cases} (0, \phi(\xi_1)) \text{ on } \mathcal{B}_{(1)} \cap \{\xi_2 \in (0,1)\} \\ (0,0) \text{ on } \mathcal{B}_{(2)} \end{cases} \tag{5.6.22}$$

So, we get $< \mathcal{N}_{22}^{(2)} >_\beta = -(0, 1/240)$, $< \mathcal{N}_{11}^{(2)} >_\beta = (1/240, 0)$.
Thus,

$$\left\langle V_0 + (\frac{\varepsilon}{\mu})^2 \left(\mathcal{N}_{i_1,i_2}^{e\ (2)}) \frac{\tilde{x}_1(x)}{\varepsilon} \right) \frac{\partial^2 V_0}{\partial x_{i_1} \partial x_{i_2}} \right\rangle_\beta =$$

$$V_0 + (\frac{\varepsilon}{\mu})^2 (1/240) \left(\begin{pmatrix} 1 & 0 \\ 0 & 0 \end{pmatrix} \frac{\partial^2 V_0}{\partial x_1^2} - \begin{pmatrix} 0 & 0 \\ 0 & 1 \end{pmatrix} \frac{\partial^2 V_0}{\partial x_2^2} \right) =$$

$$V_0 + (1/240)(\frac{\varepsilon}{\mu})^2 \left(\frac{\partial^2 V_{01}}{\partial x_1^2}, \frac{\partial^2 V_{02}}{\partial x_2^2} \right)^* =$$

$$V_0 + (1/240)(\frac{\varepsilon}{\mu})^2 (2/E) f(x) =$$

$$V_0 + \frac{1}{120E}(\frac{\varepsilon}{\mu})^2 f(x). \tag{5.6.23}$$

We obtained thus the average of the leading term of the asymptotic in two-dimensional case. It satisfies the following homogenized equation:

$$\hat{L}(V_0 + \frac{1}{120E}(\frac{\varepsilon}{\mu})^2 f(x)) = f(x) - (1/240)(\frac{\varepsilon}{\mu})^2 \left(\frac{\partial^2 f_1}{\partial x_1^2}, \frac{\partial^2 f_2}{\partial x_2^2} \right)^* . \qquad (5.6.24)$$

So, the average of the main term is a solution to the homogenized equation with the operator obtained by principle of splitting ([134],[16]) with an additional term in the right hand side. The similar effect was obtained recently in [67] for a "regularized" elasticity equation, containing the term proportional to the displacement. The important difference of the elasticity equation in comparison to the conductivity equation is that there is a corrector in (5.6.12) of order $(\frac{\varepsilon}{\mu})^2$. It can be great or of order of 1. This corrector can be easily explained "mechanically": its presence is due to the orthogonal (with respect to every segment e) components of the force $f(x)$.

5.7 Conductivity of a lattice: an expansion

We consider equation

$$L_x\, u \;=\; \Delta u \;= f(x), \quad x \in B_{\mu,\varepsilon}, \qquad (5.7.1)$$

with boundary condition

$$\frac{\partial u}{\partial n} = 0, \quad x \in \partial B_{\mu,\varepsilon}, \qquad (5.7.2)$$

where $f \in C^\infty$ is a $T-$periodic function, T is a multiple of ε and as above $s = 2$ or $s = 3$. We seek the solution $u(x)$ in the class of $T-$periodic functions of $H^1_{T-per}(B_{\mu,\varepsilon})$. Here $H^1_{T-per}(B_{\mu,\varepsilon})$ stands for the completion of the set of $T-$periodic differentiable functions defined at $B_{\mu,\varepsilon}$ with respect to the $H^1(B_{\mu,\varepsilon} \cap (0,T)^s)-$norm.

Solution to this problem is defined as a function $u_{\mu,\varepsilon} \in H^1_{T-per}(B_{\mu,\varepsilon})$, such that

$$\forall v \in H^1_{T-per}(B_{\mu,\varepsilon}), \quad -\int_{B_{\mu,\varepsilon} \cap (0,T)^s} (\nabla u_{\mu,\varepsilon}, \nabla v)\; dx \;=\; \int_{B_{\mu,\varepsilon} \cap (0,T)^s} f(x)\; v\; dx.$$

Below we construct the asymptotic expansion of the solution to problem (5.7.1)-(5.7.2) as $\mu, \varepsilon \to 0$ under the assumption that the following integrals of the right hand-side vanish:

$$\int_{(0,T)^s} f(x)dx = 0, \quad \int_{B_{\mu,\varepsilon} \cap (0,T)^s} f(x)dx = 0.$$

Problem (5.7.1)-(5.7.2) simulates the conductivity of a frame structure (or lattice structure). A geometric frame model depending on two small parameters was first considered in [56] (it was called there "skeletal structure".)

Existence of solution is a well known corollary of the Riesz theorem on presentation of a linear bounded functional in a Hilbert space (or the Lax-Milgramm lemma) and the following Poincaré inequality:

Theorem 5.7.1 *There exist a constant $C_P > 0$, such that for all sufficiently small ε and μ, for all $u \in H^1_{T-per}(B_{\mu,\varepsilon})$,*

$$\|u\|^2_{L^2(B_{\mu,\varepsilon}\cap(0,T)^s)} \leq \frac{1}{mes\ B_{\mu,\varepsilon}\cap(0,T)^s}\left(\int_{B_{\mu,\varepsilon}\cap(0,T)^s} u\ dx\right)^2 +$$

$$+\frac{C_P}{\varepsilon^{s-1}}\int_{B_{\mu,\varepsilon}\cap(0,T)^s} (\nabla\ u)^2\ dx.$$

Proof is given in the Appendix to this Chapter.

Theorem 5.7.2*For all sufficiently small ε and μ, for all strictly positive α, for all $f_0, f_1, ..., f_s \in L^2(B_{\mu\varepsilon}\cap(-\alpha,\alpha)^s)$, $T-$periodic in $x_1, ..., x_s$, the problem*

$$\Delta u\ = f_0(x) + \sum_{i=1}^{s}\frac{\partial}{\partial x_i}f_i(x),\quad x\in B_{\mu,\varepsilon},$$

with boundary condition

$$\frac{\partial u}{\partial n} = \sum_{i=1}^{s} n_i f_i(x),\quad x\in\partial B_{\mu,\varepsilon},$$

has a $T-$periodic solution in $H^1_{T-per}(B_{\mu,\varepsilon})$ if and only if

$$\int_{B_{\mu,\varepsilon}\cap(0,T)^s} f_0(x)dx = 0.$$

There exist a constant independent of μ,ε, such that

$$\|\nabla u\|_{L^2(B_{\mu,\varepsilon}\cap(0,T)^s)} \leq \frac{C}{\varepsilon^{s-1}}\sum_{i=1}^{s}\|f_i\|_{L^2(B_{\mu,\varepsilon}\cap(0,T)^s)}.$$

Proof.

As it was mentioned above, the existence of solution is a well known corollary of the Riesz theorem on presentation of a linear bounded functional in a Hilbert space (or the Lax-Milgramm lemma). To this end we consider the subspace of $H^1_{T-per}(B_{\mu,\varepsilon})$ that is $\{u \in H^1_{T-per}(B_{\mu,\varepsilon}),\ \int_{B_{\mu,\varepsilon}\cap(0,T)^s} u(x)dx = 0\}$. The variational formulation with a test function equal to a solution to the problem gives the estimate

$$\|\nabla u\|^2_{L^2(B_{\mu,\varepsilon}\cap(0,T)^s)} \leq \left(\sum_{i=1}^{s}\|f_i\|_{L^2(B_{\mu,\varepsilon}\cap(0,T)^s)}\right)\|\nabla u\|_{L^2(B_{\mu,\varepsilon}\cap(0,T)^s)}+$$

$$+\|f_0\|_{L^2(B_{\mu,\varepsilon}\cap(0,T)^s)}\|u\|_{L^2(B_{\mu,\varepsilon}\cap(0,T)^s)}.$$

Applying now the Poincaré inequality of Theorem 5.7.1, we obtain the estimate of Theorem 5.7.2.

Let us describe the asymptotic procedure of construction of solution to problem (5.7.1)-(5.7.2).

At the first step for every segment e an asymptotic solution is sought in the form similar to that of section 4.3 and at the second step the boundary layer correctors are constructed in the same way as in section 4.3. Functions v_e set on the net satisfy the equations and the junction conditions analogous to that of section 5.5, therefore its asymptotic solution is sought in the form of series similar to $u_\varepsilon^{(\infty)}$ of section 5.5 (i.e. in the form of (5.7.7)). Of course this form is taken into account in the Taylor expansion procedure for the ansatz in the neighborhoods of the nodes (5.7.15). For the macroscopic unknown function V of the expansion (5.7.7) we obtain the homogenized equation similar to that of section 5.5. (according to the principle of splitting of the homogenized operator [134], [16]).

As usual, function V is expanded in the regular series (see (5.7.36)) and we obtain finally the recurrent chain of problems for the terms of this series. The justification is a standard procedure of the truncation of the expansion, its substitution into the equation and the boundary conditions and the estimation of the discrepancy. Then we apply the a priori estimate and get the estimate for the difference of the exact solution and the truncated expansion.

1. So, first , we construct the asymptotic corresponding to a segment $e = e_h^n$ in $\mathcal{B}$. We change to new variables $\tilde{x} = \alpha(x - \varepsilon h)$. In these new variables, the cylinder $\{x/\varepsilon \in \tilde{B}_{hj}^\alpha\}$ becomes the rod.

$$U_{\varepsilon\mu} = \{\tilde{x}_1 \in I\!R^s \mid 0 < \tilde{x} < l\varepsilon, \ \ \tilde{x}'/\varepsilon\mu \in \beta_j\};$$

we set

$$\Gamma_{\varepsilon\mu} = \{\tilde{x}_1 \in I\!R^s \mid 0 < \tilde{x} < l\varepsilon, \ \ \tilde{x}'/\varepsilon\mu \in \partial\beta_j\}.$$

In what follows $\beta = \beta_j$. We write $\tilde{f}(\tilde{x}) = f(\alpha^*\tilde{x}+\varepsilon h)$ and $\tilde{u}(\tilde{x}) = u(\alpha^*\tilde{x}+\varepsilon h)$

.

Then problem (5.7.1), (5.7.2) on $\tilde{B}_{hj}^\alpha$ becomes

$$\Delta\, \tilde{u} = \tilde{f}(\tilde{x}), \quad \tilde{x} \in U_{\varepsilon\mu}, \tag{5.7.3}$$

$$\frac{\partial \tilde{u}}{\partial n} = 0, \quad \tilde{x} \in \Gamma_{\varepsilon\mu}. \tag{5.7.4}$$

We make the substitution $\xi' = (\tilde{\xi}_2, \ldots, \tilde{\xi}_s), \tilde{\xi}_i = \tilde{x}_i/(\varepsilon\mu)$ and expand $\tilde{f}$ in an asymptotic series with respect to $\varepsilon\mu\tilde{\xi}_i$, of the form

$$\tilde{f} \sim \sum_{j=0}^{\infty} (\varepsilon\mu)^j \sum_{r=1}^{r_j} F_{jr}(\tilde{\xi}_2, \ldots, \tilde{\xi}_s)\tilde{\psi}_{jr}(\tilde{x}_1),$$

In this formula the functions $\tilde{\psi}_{jr}(\tilde{x}_1)$ are the values of the derivatives

$$\partial^j f/\partial x_1^{q_1}...\partial x_{s-1}^{q_{s-1}}\partial x_s{}^{j-q_1-...q_{s-1}}$$

at the point $x = \alpha^*(\tilde{x}, 0, \ldots, 0) + \varepsilon h$, $F_{jr}(\xi')$ are some polynomials. In particular, $\tilde{\psi}_{00}(\tilde{x}_1) = \tilde{f}(\tilde{x}_1, 0, ..., 0)$, and $F_{00} = 1$ and $r_0 = 1$. We represent F_{jr} in the form $\bar{F}_{jr} + \tilde{F}_{jr}$, where $\langle \tilde{F}_{jr} \rangle_\beta = 0$ and $\bar{F}_{jr} = \langle F_{jr} \rangle_\beta$.

Using superposition , we seek the solution to problem (5.7.3) , (5.7.4) in the form :

$$\tilde{u} \sim \sum_{j=0}^{(\infty)} (\varepsilon\mu)^j \sum_{r=1}^{r_j} \tilde{u}_{jr}, \tag{5.7.5}$$

where $\tilde{u}_{jr}$ is the solution of the problem for a right hand side of the form $F_{jr}\tilde{\psi}_{jr}$. From now on in this section we omit the $\sim$ that denotes the change of variables, fix j and r and do not mention them explicitly when this does not create ambiguity .

We seek a formal asymptotic solution to problem (5.7.3)-(5.7.4) with right hand side $F_{jr}\psi_{jr}$ in the form of a series

$$\tilde{u}_{jr}^{\infty} \sim \tilde{\omega}(\tilde{x}_1) + (\varepsilon\mu)^2 \sum_{l=0}^{\infty} (\varepsilon\mu)^l M_l\Big(\frac{\tilde{x}'}{\varepsilon\mu}\Big) \frac{d^l \tilde{\psi}_{jr}(\tilde{x}_1)}{d\tilde{x}_1^l} \tag{5.7.6}$$

where M_l are the functions of Remark of section 2.3;

$\tilde{\omega}(\tilde{x}_1(x)) = v_{\varepsilon\mu}(x) + \varepsilon^2 w_{\varepsilon\mu}(x)$, $v_{\varepsilon\mu}(x)$ and $w_{\varepsilon\mu}(x)$ are defined on $I\!R^s$, and

$$v_{\varepsilon\mu}(x) \sim V(x) + \sum_{m=1}^{\infty}\sum_{q=0}^{\infty} \varepsilon^m \mu^q \sum_{(i_1...i_m):\ i_n \in \{1...s\}} \tilde{\mathcal{N}}^q_{i_1...i_m}\Big(\frac{\tilde{x}_1(x)}{\varepsilon}\Big) \frac{\partial^m V(x)}{\partial x_{i_1} \ldots \partial x_{i_m}};$$

$$w_{\varepsilon\mu}(x) \sim \sum_{m=0}^{\infty}\sum_{q=0}^{\infty} \varepsilon^m \mu^q \sum_{(j_1...j_s):\ j_1+...+j_s=j}\ \sum_{(i_1...i_m):\ i_n \in \{1...s\}}$$

$$\times \tilde{\mathcal{M}}^q_{i_1...i_m, j_1...j_s}\Big(\frac{\tilde{x}_1(x)}{\varepsilon}\Big) \frac{\partial^m}{\partial x_{i_1} \ldots \partial x_{i_m}} \frac{\partial^j f}{\partial x_1^{j_1} \ldots \partial x_s^{j_s}}. \tag{5.7.7}$$

Here $V(x)$ is a smooth $T-$periodic function, $\tilde{\mathcal{N}}^q_{i_1...i_m}(\tilde{\xi}_1)$ and $\tilde{\mathcal{M}}^q_{i_1...i_m, j_1...j_s}(\tilde{\xi}_1)$ are functions defined on every segment e.

We note that $\tilde{\psi}_{jr}$ can be represented in the form

$$\tilde{\psi}_{jr}(\tilde{x}_1(x)) = \sum_{(j_1...j_s):\ j_1+...+j_s=j} \tilde{\mathcal{K}}_{j_1...j_s}\Big(\frac{x}{\varepsilon}\Big) \frac{\partial^j f(x)}{\partial x_1{}^{j_1} \ldots \partial x_s{}^{j_s}},$$

where $\tilde{\mathcal{K}}_{j_1 \ldots j_s}$ are constant on e .

The substitution of (5.7.6) in (5.7.3) and (5.7.4) is described in section 2.3 (with μ replaced by $\varepsilon\mu$). As a result, we find that (5.4.4) is satisfied asymptotically exactly and that (5.7.3) yields "the homogenized over the rod" equation, of the form

$$\frac{d^2\tilde{\omega}}{d\tilde{x}_1^2} - \overline{F}_{jr}\tilde{\psi}_{jr} + \sum_{l=0}^{\infty}(\varepsilon\mu)^l h_l^M \frac{d^l\tilde{\psi}_{jr}}{d\tilde{x}_1^l} \sim 0. \tag{5.7.9}$$

Here we have not yet used the fact that ω has the form (5.7.7). We substitute now (5.7.7) in (5.7.9) and gather together the coefficients of the same powers of $\varepsilon\mu$ and the derivatives

$$D^i V = \frac{\partial^m V}{\partial x_{i_1} \ldots \partial x_{i_m}}, D^j f = \frac{\partial^j f}{\partial x_1^{j_1} \ldots \partial x_s^{j_s}}.$$

We obtain an asymptotic equality of the form

$$\sum_{m=1}^{\infty}\sum_{q=0}^{\infty} \varepsilon^{m-2}\mu^q \sum_{(i_1 \ldots i_m)} \left(\mathcal{H}^{N\ q}_{i_1 \ldots i_m}\left(\frac{\tilde{x}_1(x)}{\varepsilon}\right) \frac{\partial^m V}{\partial x_{i_1} \ldots \partial x_{i_m}} \right.$$

$$\left. + \sum_{(j_1 \ldots j_s):j_1+\ldots+j_s=j} \mathcal{H}^{M\ q}_{i_1 \ldots i_m,\ j_1 \ldots j_s}\left(\frac{\tilde{x}_1(x)}{\varepsilon}\right) \frac{\partial^m}{\partial x_{i_1} \ldots \partial x_{i_m}} \frac{\partial^j f}{\partial x_1^{j_1} \ldots \partial x_s^{j_s}} \right) \sim 0, \tag{5.7.10}$$

$$\mathcal{H}^{N\ q}_{i_1 \ldots i_m} = \mathcal{R}\tilde{\mathcal{N}}^q_{i_1 \ldots i_m}(\tilde{\xi}_1) + \mathcal{T}^q_{i_1 \ldots i_m}(\mathcal{N}),$$

$$\mathcal{H}^{M\ q}_{i_1 \ldots i_m,\ j_1 \ldots j_s} = \mathcal{R}\tilde{\mathcal{M}}^q_{i_1 \ldots i_m,\ j_1 \ldots j_s}(\tilde{\xi}_1) + \mathcal{T}^q_{i_1 \ldots i_m,\ j_1 \ldots j_s}(\mathcal{M}), \tag{5.7.11}$$

$\mathcal{T}^q_{i_1 \ldots i_m}(\mathcal{N})(\mathcal{T}^q_{i_1 \ldots i_m,\ j_1 \ldots j_s}(\mathcal{M}))$ are linear combinations of the

$$\mathcal{N}^q_{i_1 \ldots i_m}(\tilde{\mathcal{M}}^q_{i_1 \ldots i_m,\ j_1 \ldots j_s})$$

and their derivatives with multi-index $(i_1 \ldots i_m)$ of length m_1 less than m or $m_1 = m$ with superscript $q_1 < q$, and $\mathcal{R}$ is a differential operator along the variable $\tilde{\xi}_1 = \tilde{x}_1(x)/\varepsilon$, of the form

$$\mathcal{R} = \frac{\partial^2}{\partial \tilde{\xi}_1^2}.$$

We equate $\tilde{\mathcal{H}}^N_{j_1 \ldots j_s}(\xi_1)$ and $\tilde{\mathcal{H}}^M_{i_1 \ldots i_s,\ j_1 \ldots j_s}(\xi_1)$ with the constants $h^q_{i_1 \ldots i_m}$ and $h^q_{i_1 \ldots i_m,\ j_1 \ldots j_s}$. We obtain a homogenized equation for V of the form:

$$\sum_{m=1}^{\infty}\sum_{q=0}^{\infty}\varepsilon^{m-2}\mu^{q}\left(\sum_{(i_1\ldots i_m)} h^{q}_{i_1\ldots i_m}\frac{\partial^m V}{\partial x_{i_1}\ldots\partial x_{i_m}}\right.$$

$$\left.+\sum_{(i_1\ldots i_m)}\ \sum_{(j_1\ldots j_s):\ j_1+\ldots+j_s=j} h^{q}_{i_1\ldots i_m,\ j_1\ldots j_s}\frac{\partial^m}{\partial x_{i_1}\ldots\partial x_{i_m}}\frac{\partial^j f}{\partial x_1^{j_1}\ldots\partial x_s^{j_s}}\right)\sim 0. \tag{5.7.12}$$

The coefficients $h^{q}_{i_1\ldots i_m}$, and $h^{q}_{i_1\ldots i_m,\ j_1\ldots j_s}$ of this equation will be determined later.

Thus the $\tilde{\mathcal{N}}^{q}_{i_1\ldots i_m}$ and $\tilde{\mathcal{M}}^{q}_{i_1\ldots i_m,\ j_1\ldots j_s}$ are determined from the equations

$$\mathcal{R}\,\tilde{\mathcal{N}}^{q}{}_{i_1\ldots i_m}=\tilde{\mathcal{T}}^{q}_{i_1\ldots i_m}(N)+h^{q}_{i_1\ldots i_m}; \tag{5.7.13}$$

$$\mathcal{R}\,\tilde{\mathcal{M}}^{q}{}_{i_1\ldots i_m,\ j_s\ldots j_s}=\tilde{\mathcal{T}}^{q}_{i_1\ldots i_m,\ j_s\ldots j_s}(M)+h^{q}_{i_1\ldots i_m} \tag{5.7.14}$$

and some interface conditions formulated below.

Let $\tilde{u}^{(\infty)}_{\varepsilon}=\tilde{u}^{(\infty)}_{jr}(x)$ the asymptotic series (5.7.6), (5.7.7), which is the solution of (5.7.1), (5.7.2) in the cylinder $\{x:\frac{x}{\varepsilon}\in\tilde{B}^{\alpha}_{hj}\}$.

We expand the functions

$$\frac{\partial^m V}{\partial x_{i_1}\ldots\partial x_{i_m}},\quad \frac{\partial^m}{\partial x_{i_1}\ldots\partial x_{i_m}}\frac{\partial^j f}{\partial x_1^{j_1}\ldots\partial x_s^{j_s}},\quad \frac{\partial^l\psi_{jr}}{\partial\tilde{x}_i^l},$$

$\tilde{\mathcal{N}}_{i_1\ldots i_m}(\frac{\tilde{x}_1(x)}{\varepsilon})$ and $\tilde{\mathcal{M}}_{i_1\ldots i_m,j_1\ldots j_s}(\frac{\tilde{x}_1(x)}{\varepsilon})$ in Taylor series in the neighborhood of the point x_0 corresponding to the node $\frac{x_0}{\varepsilon}$, and substitute these expansions in (5.7.6) . We obtain

$$\tilde{u}^{(\infty)}_{\varepsilon}\sim\sum_{l=0}^{\infty}\sum_{r=0}^{\infty}\varepsilon^{l}\mu^{r}\sum_{(i_1\ldots i_l)}(N^{e\ \ r}_{i_1\ldots i_l}(\frac{x-x_0}{\varepsilon\mu})\frac{\partial^l V}{\partial x_{i_1}\ldots\partial x_{i_l}}(x_0)$$

$$+\varepsilon^2\sum_{(j_1\ldots j_s),\ j_1+\ldots+j_l=j} M^{e\ \ r}_{i_1\ldots i_l,\ j_1\ldots j_s}(\frac{x-x_0}{\varepsilon\mu})$$

$$\times\frac{\partial^{l+j}f}{\partial x_{i_1}\ldots\partial x_{i_l}\partial x_1^{j_1}\ldots\partial x_s^{j_s}}(x_0))$$

$$+(\varepsilon\mu)^2\sum_{l=0}^{\infty}(\varepsilon\mu)^l\sum_{(i_1\ldots i_l)}\ \sum_{(j_1\ldots j_s),\ j_1+\ldots+j_l=j} K^{e\ \ r}_{i_1\ldots i_l,\ j_1\ldots j_s}(\frac{x-x_0}{\varepsilon\mu})$$

$$\times\frac{\partial^{l+j}f}{\partial x_{i_1}\ldots\partial x_{i_l}\partial x_1^{j_1}\ldots\partial x_s^{j_s}}(x_0). \tag{5.7.15}$$

Here

$$N^{e\quad r}_{i_1\dots i_m}(z)=\tilde{\mathcal{N}}^{r}_{i_1\dots i_l}(\frac{\tilde{x}_1(x_0)}{\varepsilon})+\Delta_N(\tilde{\mathcal{N}}). \qquad (5.7.16)$$

where $z=\frac{x-x_0}{\varepsilon\mu}$, $\Delta_N(\tilde{\mathcal{N}})$ is defined in terms of the functions $\tilde{\mathcal{N}}$ with multi-indices $(i_1\dots i_l$ of length less than l or with multi index ($(i_1\dots i_l$ but with superscripts not exceeding $r-1$.

An analogue of the representations (5.7.16) can also be found for the matrices $\mathcal{M}_{i_1\dots i_l,\ j_1\dots j_s}$.

A formal asymptotic solution of the problem near a pre-nodal component Π_{x_0} corresponding to the node x_0/ε is sought in the form

$$\tilde{u}^{(\infty)}_{\Pi_{x_0}}\sim(\sum_{l=0}^{\infty}\sum_{r=0}^{\infty}\varepsilon^l\mu^r\sum_{(i_1\dots i_l)}(N^{0\quad r}_{i_1\dots i_l}(\frac{x-x_0}{\varepsilon\mu})\frac{\partial^l V}{\partial x_{i_1}\dots\partial x_{i_l}}(x_0)$$

$$+\varepsilon^2\sum_{(j_1\dots j_s),\ j_1+\dots+j_l=j}M^{0\quad r}_{i_1\dots i_l,\ j_1\dots j_s}(\frac{x-x_0}{\varepsilon\mu})$$

$$\times\frac{\partial^{l+j}f}{\partial x_{i_1}\dots\partial x_{i_l}\partial x_1^{j_1}\dots\partial x_s^{j_s}}(x_0))$$

$$+\sum_{e(x_0)}\chi_e(x)\hat{\rho}(\alpha_e\frac{x-x_0}{\varepsilon\mu})u_e^{(\infty)}; \qquad (5.7.17)$$

here the summation in the last sum in (5.7.17) extends over all segments e having x_0/ε as an end-point, $\chi_e(x)$ is the characteristic function of the section of e, α_e the matrix of the transformation α which corresponds to e, and $\hat{\rho}(t)=\rho(t_1)$ a differentiable function that vanishes for $|t_1|\le c_0$, is equal to unity for $|t_1|\ge c_0+1$, and such that $0\le\hat{\rho}(t_1)\le 1$, where $t=(t_1,\dots,t_s)$.

We set $z=(x-x_0)/(\varepsilon\mu)$ and denote by $\Pi_{x_0,z,0}$ the image of Π_{x_0}. Let $\tilde{\tilde{B}}^{\alpha}_{hj}$ be a cylinder whose intersection with Π_{x_0} is nonempty. We extend it to a semi-infinite cylinder whose basis has no common points with Π_{x_0}. We denote by $B^{\alpha}_{hj\infty}$ this extended cylinder, and by $\Pi_{x_0\infty}$ the union of all such cylinders $B^{\alpha}_{hj\infty}$ and Π_{x_0}. Moreover, we denote by $\Pi_{x_0,z,\infty}$ the image of $\Pi_{x_0,\infty}$ under the transformation $z=(x-x_0)/(\varepsilon\mu)$.

Let $\tilde{\chi}_e(z)$ be the characteristic function of the dilated under transformation $z=(x-x_0)/(\varepsilon\mu)$ domain $B^{\alpha}_{hj\infty}$ corresponding to the section e.

Substituting (5.7.17) into (5.7.1), (5.7.2), we find that $N^{0\quad r}_{i_1\dots i_l}$ is the solution to the problem (similar to (4.3.7) of Chapter 4):

$$L_zN^{0\quad r}_{i_1\dots i_l}=\sum_{e(x_0)}\bar{\chi}_e(z)(L_z((1-\hat{\rho}(\alpha_e z))N^{e\quad r}_{i_1\dots i_l}(z))$$

$$-(1-\hat{\rho}(\alpha_e z))L_zN^{e\quad r}_{i_1\dots i_l}(z)),\ z\in\Pi_{x_0,z,\infty},$$

$$\frac{\partial}{\partial n_z} N^{0\ \ r}_{i_1\ldots i_l} = \sum_{e(x_0)} \bar{\chi}_e(z)\Big(\frac{\partial}{\partial \nu_z}((1-\hat{\rho}(\alpha_e z))N^{e\ \ r}_{i_1\ldots i_l}(z))$$

$$-(1-\hat{\rho}(\alpha_e z))\frac{\partial}{\partial n_z} N^{e\ \ r}_{i_1\ldots i_l}(z)\Big),\ z \in \partial\Pi_{x_0,z,\infty}. \qquad (5.7.18)$$

Similar problems are obtained for $M^{0\ \ r}_{i_1\ldots i_l,\ j_1\ldots j_s}$. These problems as well as problem (5.7.18) are of the form

$$L_z N^0 = \sum_{e(x_0)} \bar{\chi}_e(z) L_z \phi_e(z) + \phi_0(z), \quad z \in \Pi_{x_0,z,\infty}.$$

$$\frac{\partial}{\partial n_z} N^0 = \sum_{e(x_0)} \bar{\chi}_e(z)\left(\frac{\partial}{\partial n_z}\phi_e(z) + \sum_{i=1}^{s} n_i \phi_{i\ e}(z)\right), \quad z \in \partial\Pi_{x_0,z,\infty}, \qquad (5.7.19)$$

where ϕ_e are smooth functions and ϕ_0 and $\phi_{i\ e}$ piecewise smooth functions with compact support.

Lemma 5.7.1. *Problem (5.7.19) is solvable in the class of functions that stabilize exponentially to a constant (in the sense of [88]-[90]) if and only if*

$$\int_{\Pi_{x_0,z,\infty}} \phi_0(z)dz =$$

$$= \sum_{e(x_0)} \left(\int_{\partial\Pi_{x_0,z,\infty}} \sum_{k=1}^{s} n_k \phi_{k\ e}(z)\, ds + \int_{\partial\Pi_{x_0,z,0}\setminus\partial\Pi_{x_0,z,\infty}} \frac{\partial}{\partial \nu_z}\phi_e(z)\, ds \right)$$

where $(n_1,\ldots,n_s)$ is the outward normal to $\Pi_{x_0,z,0}$.

The proof of the lemma is fully analogous to that of the main theorem in [89]. Lemma 5.7.1 and (5.7.17) yield the conditions

$$\sum_{e(x_0)} \varepsilon\mu \int_{\Pi_{x_0,z,\infty}} \bar{\chi}_e(z)(1-\hat{\rho}(\alpha_e z))\frac{\partial}{\partial \nu_z}\Delta_x u_e^{N\ (\infty)} dz =$$

$$= -\sum_{e(x_0)} \int_{\partial\Pi_{x_0,z,0}\setminus\partial\Pi_{x_0,z,\infty}} \bar{\chi}_e(z)\frac{\partial}{\partial n_z} u_e^{N\ (\infty)}\, ds =\sim 0, \qquad (5.7.20)$$

and similar conditions for $u_e^{M\ (\infty)}$; here $u_e^{N\ (\infty)}$ and $u_e^{M\ (\infty)}$ are the first and second sums in (5.7.16).

Taking into account the fact

$$\Delta_x u_e^{N\ (\infty)} \sim \frac{d^2\tilde{v}}{d\tilde{x}_1^2},$$

$$\frac{\partial}{\partial n_x} u_e^{N\ (\infty)} \sim \frac{d\tilde{v}}{d\tilde{x}_1},$$

and considering the analogous expansions for $\Delta_x u_e^{M\,(\infty)}$ and $\frac{\partial}{\partial n_x} u_e^{M\,(\infty)}$ and Remark 2.2.3, we find that $\tilde{\mathcal{N}}^{e\ r}_{i_1\ldots i_l}$ and $\tilde{\mathcal{M}}^{e\ r}_{i_1\ldots i_l,\ j_1\ldots j_s}$ satisfy at the nodes the conditions (following from (5.7.20))

$$\hat{\Lambda}\tilde{\mathcal{N}}^{e\ r}_{i_1\ldots i_l}|_{\tilde{x}_1(x_0)/\varepsilon} = \Psi^r_{N,\ i_1\ldots i_l}(x_0/\varepsilon),$$

$$\hat{\Lambda}\tilde{\mathcal{M}}^{e\ r}_{i_1\ldots i_l, j_1\ldots j_s}|_{\tilde{x}_1(x_0)/\varepsilon} = \Psi^r_{M,\ i_1\ldots i_l, j_1\ldots j_s}(x_0/\varepsilon), \tag{5.7.21}$$

where the right-hand sides of (5.7.21) are defined in terms of $\tilde{\mathcal{N}}^{e\ q}_{i_1\ldots i_l}$ and $\tilde{\mathcal{M}}^{e\ q}_{i_1\ldots i_l,\ j_1\ldots j_s}$ with indices $(l_1, q) < (l, r)$ (that is, $l_1 \leq l,\ q \leq r$ and $l_1 \neq l$ or $q \neq r$),

$$\hat{\Lambda}_e = \sum_{e(x_0)} \Lambda_e mes \beta_e, \tag{5.7.22}$$

$\Lambda_e = \frac{\partial}{\partial \tilde{\xi}_1}$ and $\tilde{\xi} = \alpha(\xi - \xi_0),\ \xi_0 = x_0/\varepsilon$.

If (5.7.21) holds then, by Lemma 5.7.1, problems (5.7.18) have solutions that stabilize exponentially, as $|z| \to \infty$, to some "limit" constants. For every segment $e(x_0)$ and its corresponding half-bounded cylinder $B^{\alpha}_{hj\infty}$, denote the limit constant C_e.

From the representation (5.7.16) it follows that if we change the value of $\tilde{\mathcal{N}}^{e\ q}_{i_1\ldots i_l}|_{\tilde{x}_1(x_0)/\varepsilon}$ by $\delta\tilde{\mathcal{N}}^{e\ q}_{i_1\ldots i_l}|_{\tilde{x}_1(x_0)/\varepsilon}$ then the value of C_e, also changes by

$$\delta\tilde{\mathcal{N}}^{e\ q}_{i_1\ldots i_l}|_{\tilde{x}_1(x_0)/\varepsilon}$$

(as in section 4.3)

The value of $\tilde{\mathcal{M}}^{e\ q}_{i_1\ldots i_l,\ j_1\ldots j_s}|_{\tilde{x}_1(x_0)/\varepsilon}$ changes according the same rule.

Suppose that for any $e_1(x_0)$ and $e_2(x_0)$, at the point $\tilde{x}_1(x_0)/\varepsilon$ we have

$$\tilde{\mathcal{N}}^{e_1\ q}_{i_1\ldots i_l}|_{\tilde{x}_1(x_0)/\varepsilon} = \tilde{\mathcal{N}}^{e_2\ q}_{i_1\ldots i_l}|_{\tilde{x}_1(x_0)/\varepsilon},$$

$$\tilde{\mathcal{M}}^{e_1\ q}_{i_1\ldots i_l,\ j_1\ldots j_s}|_{\tilde{x}_1(x_0)/\varepsilon} = \tilde{\mathcal{M}}^{e_2\ q}_{i_1\ldots i_l,\ j_1\ldots j_s}|_{\tilde{x}_1(x_0)/\varepsilon}, \tag{5.7.23}$$

and that under these conditions, for every $e = e(x_0)$ the solution of (5.7.18) stabilizes to C_e^N, and the solution of the corresponding problem for $M^0_{i_1\ldots i_l,\ j_1\ldots j_s}$ stabilizes to C_e^M on the cylinder $B^{\alpha}_{hj\infty}$ corresponding to e. Then we redefine the values $\tilde{\mathcal{N}}^{e\ q}_{i_1\ldots i_l}|_{\tilde{x}_1(x_0)/\varepsilon}$ and $\tilde{\mathcal{M}}^{e\ q}_{i_1\ldots i_l,\ j_1\ldots j_s}|_{\tilde{x}_1(x_0)/\varepsilon}$, so that their corresponding solutions of the problems in $\Pi_{x_0,z,\infty}$ stabilize to one and the same constant for all segments $e(x_0)$ with the same end-point x_0. To this end , it suffices to set the new (improved) values of $\tilde{\mathcal{N}}^{e\ q}_{i_1\ldots i_l}|_{\tilde{x}_1(x_0)/\varepsilon}$ and $\tilde{\mathcal{M}}^{e\ q}_{i_1\ldots i_l,\ j_1\ldots j_s}|_{\tilde{x}_1(x_0)/\varepsilon}$, equal to the old ones minus C_e^N and C_e^M, respectively. In other words, (5.7.23) needs to be replaced by conditions of the form

$$\tilde{\mathcal{N}}^{e_1\ q}_{i_1\ldots i_l}|_{\tilde{x}_1(x_0)/\varepsilon} + \delta\tilde{\mathcal{N}}^{e_1\ q}_{i_1\ldots i_l} = \tilde{\mathcal{N}}^{e_2\ q}_{i_1\ldots i_l}|_{\tilde{x}_1(x_0)/\varepsilon} + \delta\tilde{\mathcal{N}}^{e_2\ q}_{i_1\ldots i_l}, \tag{5.7.24}$$

$$\tilde{\mathcal{M}}^{e_1\ q}_{i_1\ldots i_l,\ j_1\ldots j_s}|_{\tilde{x}_1(x_0)/\varepsilon} + \delta\tilde{\mathcal{M}}^{e_1\ q}_{i_1\ldots i_l,\ j_1\ldots j_s} = \tilde{\mathcal{M}}^{e_2\ q}_{i_1\ldots i_l,\ j_1\ldots j_s}|_{\tilde{x}_1(x_0)/\varepsilon} + \delta\tilde{\mathcal{M}}^{e_2\ q}_{i_1\ldots i_l,\ j_1\ldots j_s},$$

for any segments e_1 and e_2 having the common end-point x_0. Here the constants $\delta\tilde{\mathcal{N}}^{e\ q}_{i_1\ldots i_l}$ and $\delta\tilde{\mathcal{M}}^{e\ q}_{i_1\ldots i_l,\ j_1\ldots j_s}$ are chosen so that (5.7.18) has solutions that stabilize to zero; in what follows we consider precisely these solutions of (5.7.18) (here $\delta\tilde{\mathcal{N}}^{e\ q}_{i_1\ldots i_l} = C^N_e$ and $\delta\tilde{\mathcal{M}}^{e\ q}_{i_1\ldots i_l,\ j_1\ldots j_s} = C^M_e$).

From the above arguments we derive the following algorithm for the recursive computation of the matrices $\tilde{\mathcal{N}}^{e\ q}_{i_1\ldots i_l}$, $\tilde{\mathcal{M}}^{e\ q}_{i_1\ldots i_l,\ j_1\ldots j_s}$, $N^{0\ q}_{i_1\ldots i_l}$ and $M^{0\ q}_{i_1\ldots i_l,\ j_1\ldots j_s}$.

The basis of the recursive process is $\tilde{\mathcal{N}}^{e\ 0}_{\emptyset} = 1, \tilde{\mathcal{M}}^{e\ 0}_{\emptyset} = 0$; for $q < 0$, we formally set $\tilde{\mathcal{N}}^{e\ q}_{i_1\ldots i_l} = \tilde{\mathcal{M}}^{e\ q}_{i_1\ldots i_l,\ j_1\ldots j_s} = 0$; $N^{0\ 0}_{i_1\ldots i_l}$ are the solutions of problems (5.7.18) from which

$$N^{0\ 0}_{i_1\ldots i_l} = -\sum_{e(x_0)} \bar{\chi}_e(z)\hat{\rho}(\alpha_e z)\tilde{\mathcal{N}}^{e\ 0}_{i_1\ldots i_l}\left(\frac{\tilde{x}_1(x_0)}{\varepsilon}\right) + const,$$

$$M^{0\ 0}_{i_1\ldots i_l,\ j_1\ldots j_s} = -\sum_{e(x_0)} \bar{\chi}_e(z)\hat{\rho}(\alpha_e z)\tilde{\mathcal{M}}^{e\ 0}_{i_1\ldots i_l,\ j_1\ldots j_s}\left(\frac{\tilde{x}_1(x_0)}{\varepsilon}\right) + const.$$

Afterwards, the functions $\tilde{\mathcal{N}}^{e\ q}_{i_1\ldots i_l}$, $\tilde{\mathcal{M}}^{e\ q}_{i_1\ldots i_l,\ j_1\ldots j_s}$, $N^{0\ q}_{i_1\ldots i_l}$ and $M^{0\ q}_{i_1\ldots i_l,\ j_1\ldots j_s}$. are determined successively, in alphabetical order, by induction on $(m+q, q)$ from (5.7.13),(5.7.14),(5.7.21),(5.7.24) and (5.7.18). Problems (5.7.13), (5.7.21) (5.7.24) for $\tilde{\mathcal{N}}^{e\ q}_{i_1\ldots i_l}$ are written in the form:

$$\frac{d^2(\tilde{\mathcal{N}}^{e\ q}_{i_1,\ldots,i_l})}{d\tilde{\xi}_1^2} = -2\gamma^e_{i_1}\frac{d(\tilde{\mathcal{N}}^{e\ q}_{i_2,\ldots,i_l})}{d\tilde{\xi}_1} - \gamma^e_{i_1}\gamma^e_{i_2}\tilde{\mathcal{N}}^{e\ q}_{i_3,\ldots,i_l} + $$
$$+h^q_{i_1\ldots i_l} + \Psi^{e\ q}_{i_1,\ldots,i_l}(\tilde{\xi}_1) \qquad (5.7.25)$$

for any e, and

$$\sum_{e(\xi_0)} mes\ \beta_e\left\{\frac{d\tilde{\mathcal{N}}^{e\ q}_{i_1,\ldots,i_l}}{d\tilde{\xi}_1} + \gamma^e_{i_1}\tilde{\mathcal{N}}^{e\ q}_{i_2,\ldots,i_l}\right\} = \Psi^{\xi_0\ q}_{i_1,\ldots,i_l} \qquad (5.7.26)$$

and

$$\tilde{\mathcal{N}}^{e_1\ q}_{i_1,\ldots,i_l} = \tilde{\mathcal{N}}^{e_2\ q}_{i_1,\ldots,i_l} + \delta^{e_1,e_2\ q}_{i_1,\ldots,i_l} \qquad (5.7.27)$$

in the nodes ξ_0 for any e_1 and e_2 with the end-point in ξ_0.

Here $\Psi^{e\ q}_{i_1,\ldots,i_l}(\tilde{\xi}_1)$ are determined by $N^{q_1}_j$ with $q_1 < q$ and $\Psi^{\xi_0\ q}_{i_1,\ldots,i_l}$ and $\delta^{e_1,e_2\ q}_{i_1,\ldots,i_l}$ are also determined by $N^{q_1}_j$ with $q_1 < q$, $(\gamma^e_1, ..., \gamma^e_s)$ are the directing cosines of the vector e with the initial point in ξ_0. Constants $h^q_{i_1\ldots i_l}$ are chosen in such a way that every problem (5.4.25)-(5.4.27) has a $1-$periodic solution. The possibility of such choice is a consequence of the following lemma.

Lemma 5.7.2 *Let $\Psi(\tilde{\xi}_1)$ be any 1-periodic function defined on $\mathcal{B}$*
- which has generalized derivative along e for every $e \subset \mathcal{B}$,
- which is continuous on $\mathcal{B}$ and such that

$$\sum_e \int_{e\cap Q} \Psi^e(\tilde{\xi}_1)d\tilde{\xi}_1 = 0, \qquad (5.7.28)$$

where the summation extends over all segments e of the periodicity cell $Q = (0,1)^s$ (here the parts of the segments e lying on the boundary of Q and differing by a period are taken only once). Then the following inequality holds:

$$\sum_e \int_{e\cap Q} (\Psi^e(\tilde{\xi}_1))^2 d\tilde{\xi}_1 \le c \sum_e \int_{e\cap Q} (d\Psi^e/d\tilde{\xi}_1(\tilde{\xi}_1))^2 d\tilde{\xi}_1 \qquad (5.7.29)$$

with a constant c depending only on $\mathcal{B}$.

Proof.

Condition (5.7.28) and continuity of Ψ yield that there exist a point $\xi^0 \in \mathcal{B} \cap Q$, (remind that $Q = (0,1)^s$) such that $\Psi(\xi^0) = 0$. Then applying the Newton-Leibnitz formula for any connected path $P_{\xi^0,\bar{\xi}}$ of segments in $\mathcal{B} \cap Q$, with the initial point in ξ^0 and the final point in $\bar{\xi}$,

$$\int_{P_{\xi^0,\bar{\xi}}} \frac{d\Psi}{dl} dl = \Psi(\bar{\xi})$$

we obtain the estimate

$$\int_{P_{\xi^0,\bar{\xi}}} \Psi^2 dl \le |P_{\xi^0,\bar{\xi}}| \int_{P_{\xi^0,\bar{\xi}}} \left(\frac{d\Psi}{dl}\right)^2 dl,$$

where $|P_{\xi^0,\bar{\xi}}|$ is the length of the path $P_{\xi^0,\bar{\xi}}$.

Adding these inequalities for some pathes covering $\mathcal{B}\cap Q$, we obtain estimate (5.7.29).

Applying Lemma 5.7.2 and the Riesz theorem on representation of linear functionals in Hilbert space, we obtain the solvability of problems(5.7.25)-(5.7.27) if and only if

$$\sum_e \int_{e\cap Q} \left(- \gamma^e_{i_1} \frac{d(\tilde{\mathcal{N}}^{e\ q}_{i_2,\dots,i_l})}{d\tilde{\xi}_1} - \gamma^e_{i_1}\gamma^e_{i_2}\tilde{\mathcal{N}}^{e\ q}_{i_3,\dots,i_l}) + h^q_{i_1\dots i_l} + \right.$$

$$\left. + \Psi^{e\ q}_{i_1,\dots,i_l}(\tilde{\xi}_1) \right) d\tilde{\xi}_1 mes\ \beta_e + \sum_{\xi_0:\xi_0\in\bar{Q}} \Psi^{\xi_0\ q}_{i_1,\dots,i_l} = 0. \qquad (5.7.30)$$

Here in integrals $\int_{e\cap Q}$ we choose any of the two ends of the delated segment e as an end-point ξ_0 to define $\tilde{\xi}$ as $\alpha(\xi - \xi_0)$). In the sum $\sum_{\xi_0:\xi_0\in\bar{Q}}$ periodic points of the boundary of Q are taken only once.

Because of equations (5.7.25), we obtain that (5.7.10) (and respectively (5.7.1)) is transformed into the homogenized equation

$$\sum_{m=2}^{\infty}\sum_{q=0}^{\infty} \varepsilon^{m-2}\mu^q \sum_{(i_1\dots i_m)} h^{N\ q}_{i_1\dots i_m} \frac{\partial^m V}{\partial x_{i_1}\dots\partial x_{i_m}}$$

$$+ \sum_{m=1}^{\infty}\sum_{q=0}^{\infty} \varepsilon^m \mu^q \sum_{(i_1\dots i_m)} \sum_{(j_1\dots j_s):j_1+\dots+j_s=j} h^{M\ q}_{i_1\dots i_m,\ j_1\dots j_s} \frac{\partial^m}{\partial x_{i_1}\dots\partial x_{i_m}} \frac{\partial^j f}{\partial x_1^{j_1}\dots\partial x_s^{j_s}} \sim$$

$$\sim 0, \tag{5.7.31}$$

where the main part of the first sum is

$$\sum_{i_1,i_2=1}^{s} h^0_{i_1 i_2} \frac{\partial^2}{\partial x_{i_1} \partial x_{i_2}} V,$$

where $h^0_{i_1 i_2}$ are defined by relations

$$h^0_{i_1 i_2} =$$

$$\frac{\sum_{e \subset Q_1 \cap \mathcal{B}} \int_e mes\ \beta_e \gamma^e_{i_1} \left(\frac{d(\tilde{\mathcal{N}}^{e\ 0}_{i_2})}{d\tilde{\xi}_1} + \gamma^e_{i_2} \right) d\tilde{\xi}_1}{\sum_{e \subset Q_1 \cap \mathcal{B}} mes\ \beta_e |e|},$$

where $\mathcal{N}^0_{i_2}$ is a 1-periodic continuous solution to the cell problem

$$\frac{d^2(\tilde{\mathcal{N}}^{e\ 0}_{i_2} + \xi_{i_2})}{d\tilde{\xi}^2_1} = 0; \tag{5.7.32}$$

for any e, and

$$\sum_{e(\xi_0)} mes\ \beta_e \frac{d(\tilde{\mathcal{N}}^{e\ 0}_{i_2} + \xi_{i_2})}{d\tilde{\xi}_1} = 0 \tag{5.7.33}$$

and

$$\tilde{\mathcal{N}}^{e_1\ 0}_{i_2} = \tilde{\mathcal{N}}^{e_2\ 0}_{i_2} \tag{5.7.34}$$

in the nodes ξ_0 for any e_1 and e_2 with the end-point in ξ_0. Thus

$$h^0_{i_1 i_2} =$$

$$\frac{\sum_{e \subset Q \cap \mathcal{B}} \int_e mes\ \beta_e \frac{d(\tilde{\mathcal{N}}^{e\ 0}_{i_1} + \xi_{i_1})}{d\tilde{\xi}_1} d\tilde{\xi}_1}{\sum_{e \subset Q \cap \mathcal{B}} mes\ \beta_e |e|} \tag{5.7.35}$$

and $\tilde{\mathcal{N}}^{e\ 0}_{i_1}$ are linear on every e. This formula coincides with the algorithm for calculation the homogenized operator according to the principle of splitting in section 8.3 of [16].

Denote

$$\hat{L} = \sum_{i_1,i_2=1}^{s} h^0_{i_1 i_2} \frac{\partial^2}{\partial x_{i_1} \partial x_{i_2}}.$$

Lemma 5.7.3 *The homogenized operator $\hat{L}$ has a symmetric positive definite matrix $h^0 = (h^0_{i_1 i_2})_{i_1,i_2=1,\ldots,s}$.*

Proof.

Applying the variational formulation to problem (5.7.32)-(5.7.34), we obtain

$$\sum_{e\subset Q\cap\mathcal{B}}\int_{e\cap Q} mes\ \beta_e \frac{d(\tilde{\mathcal{N}}_{i_1}^{e\ 0}+\xi_{i_1})}{d\tilde{\xi}_1}\frac{d\tilde{\mathcal{N}}_{i_2}^{e\ 0}}{d\tilde{\xi}_1}d\tilde{\xi}_1 \ = \ 0,$$

so

$$h^0_{i_1 i_2} =$$

$$= \frac{\sum_{e\subset Q\cap\mathcal{B}}\int_{e\cap Q} mes\ \beta_e \frac{d(\tilde{\mathcal{N}}_{i_1}^{e\ 0}+\xi_{i_1})}{d\tilde{\xi}_1}\frac{d(\tilde{\mathcal{N}}_{i_2}^{e\ 0}+\xi_{i_2})}{d\tilde{\xi}_1}d\tilde{\xi}_1}{\sum_{e\subset Q\cap\mathcal{B}} mes\ \beta_e|e|}, \tag{5.7.36}$$

and so

$$h^0_{i_1 i_2} = h^0_{i_2 i_2}, \quad i_1, i_2 \in \{1, ..., s\}.$$

Moreover, for any $\eta = (\eta_1, ..., \eta_s)^T \in I\!R^s$,

$$\sum_{e\subset Q\cap\mathcal{B}}\int_{e\cap Q} mes\ \beta_e d\tilde{\xi}_1 \sum_{i_1,i_2=1}^{s} h^0_{i_1 i_2}\eta_{i_2}\eta_{i_1} \ =$$

$$= \sum_{e\subset Q\cap\mathcal{B}}\int_{e\cap Q} mes\ \beta_e \left(\frac{d\sum_{i_1=1}^{s}(\tilde{\mathcal{N}}_{i_1}^{e\ 0}+\xi_{i_1})\eta_{i_1}}{d\tilde{\xi}_1}\right)^2 d\tilde{\xi}_1 \ \geq \ 0.$$

Let it be zero. Then for any e,

$$\frac{d\sum_{i_1=1}^{s}(\tilde{\mathcal{N}}_{i_1}^{e\ 0}+\xi_{i_1})\eta_{i_1}}{d\tilde{\xi}_1} \ = \ 0;$$

since $\mathcal{N}_{i_1}^{e\ 0}$ is a continuous function, for any e,

$$\sum_{i_1=1}^{s}(\tilde{\mathcal{N}}_{i_1}^{e\ 0}+\xi_{i_1})\eta_{i_1} \ = \ const$$

and since $\mathcal{B}$ is a connected 1-periodic set, it is the same constant for all e.

Moreover, since $\sum_{i_1=1}^{s}\mathcal{N}_{i_1}^{e\ 0}\eta_{i_1}$ is a linear 1-periodic function, it is a constant. So, $\sum_{i_1=1}^{s}\xi_{i_1}\eta_{i_1}$ is a constant on $\mathcal{B}$. It is equal to zero for $\xi = 0$, therefore $\sum_{i_1=1}^{s}\xi_{i_1}\eta_{i_1} = 0$. Taking now for any $k \in \{1, ..., s\}$, $\xi = (\delta_{1k}, ..., \delta_{sk})$, we prove that $\eta_k = 0$. Thus, $\eta = 0$, and h^0 is a positive definite matrix. Lemma is proved.

Thus, after all the functions $\tilde{\mathcal{N}}^{e\ q}_{i_1...i_l}$, $\tilde{\mathcal{M}}^{e\ q}_{i_1...i_l,\ j_1...j_s}$, $N^{0\ q}_{i_1...i_l}$ and $M^{0\ q}_{i_1...i_l,\ j_1...j_s}$. are constructed, we can proceed with the construction of V, the formal asymptotic solution of the system of equations (5.4.12), which we seek in the form of a regular series in ε and μ :

$$V \ = \ \sum_{p_1,p_2=0}^{\infty} \varepsilon^{p_1}\mu^{p_2}V_{p_1p_2}(x). \tag{5.7.36}$$

where the coefficients of this series are solutions of the recursive sequence of problems

$$\hat{L}V_{p_1p_2} = f_{p_1p_2}(x). \tag{5.7.37}$$

where $f_{p_1p_2}(x)$ are defined in terms of the $V_{q_1q_2}$ with indices $(q_1, q_2) < (p_1, p_2)$ that is, $q_i \leq p_i$, $i = 1, 2$ and $q_1 \neq p_1$ or $q_2 \neq p_2$, and $f_{00} = f(x)$; thus, the above equation coincides with the homogenized one obtained in section 5.1 by means of principle of splitting of the homogenized operator. It can be proved by induction that the $f_{p_1p_2}(x)$ are $T-$periodic in x and that

$$\int_{(0,T)^s} f_{p_1p_2}(x)dx = 0;$$

consequently, the following assertion holds

Lemma 5.7.3. *The sequence of problems (5.7.37) is recursively solvable in class of $T-$periodic functions $V_{q_1q_2}$ with zero averages, that is, such that* $\int_{(0,T)^s} V_{p_1p_2}(x)dx = 0$.

We denote by $u^{(J)}$ the truncated (partial) sum of the asymptotic series (5.7.5)-(5.7.7), (5.7.17), (5.7.36), obtained by neglecting the terms of order $O(\varepsilon^J + \mu^J)$ (in the $H^1(B_T)-$norm, $B_T = B_{\mu\varepsilon} \cap (0, T)^s$). Choosing J sufficiently large, we can show for any K that equation (5.7.1) and condition (5.7.2) are satisfied with remainders of order $O(\varepsilon^K + \varepsilon^{-1}\mu^K)$ (in the $L^2(B_T)-$norm).

Hence , using the a priori estimate for problem (5.7.1),(5.7.2) given by Theorem 5.7.2, we obtain the following assertion.

Theorem 5.7.3. *If there are numbers α_1 and α_2 such that $\varepsilon = O(\mu^{\alpha_1})$ and $\mu = O(\varepsilon^{\alpha_2})$ as $\varepsilon, \mu \to 0$, then*

$$\|u - \tilde{u}^{(J)}\|_{H^1(B_T)} = O(\varepsilon^J + \mu^J),$$

where $\tilde{u}^{(J)} = u^{(J)} - < u^{(J)} >_T$, u is the solution of problem (5.7.1),(5.7.2), and

$$< u^{(J)} >_T = 0, \qquad < . >_T = \int_{B_T} .dx.$$

Thus,(5.7.5)-(5.7.7), (5.7.17), (5.7.36) is an asymptotic approximation of the solution of problem (5.7.1),(5.7.2).

Remark 5.7.1 From the structure of the formal asymptotic solution (5.7.5)-(5.7.7), (5.7.17), (5.7.36) it follows that, as $\varepsilon, \mu \to 0$, the solution converges in the norm

$$\|.\|_{H^1(B_T)}/\sqrt{mes B_T}$$

to a function V_{00} independent of ε and μ. This function is a $T-$periodic solution to the homogenized problem

$$\hat{L}V_{00} = f(x).$$

5.8 High order homogenization of elastic lattices

We consider the system of equations of elasticity theory in $B_{\mu,\varepsilon}$

$$L_x\, u \;=\; \sum_{i,j=1}^{s} \frac{\partial}{\partial x_i}\left(A_{ij}\frac{\partial u}{\partial x_j}\right) = f(x), \quad x \in B_{\mu,\varepsilon}, \tag{5.8.1}$$

$$\frac{\partial u}{\partial \nu} = 0, \quad x \in \partial B_{\mu,\varepsilon}, \tag{5.8.2}$$

where

$$\frac{\partial u}{\partial \nu} = \sum_{i,j=1}^{s} A_{ij}\frac{\partial u}{\partial x_j} n_i,$$

and A_{ij} are constant $s \times s$ matrices, the elements a_{ij}^{kl} of the A_{ij} satisfy the relations

$$a_{ij}^{kl} = (\delta_{ij}\delta_{kl} + \delta_{il}\delta_{jk})\mathcal{M} + \lambda\delta_{ik}\delta_{jl},$$

Here $\mathcal{M}$ and λ are the Lamé coefficients, $(n_1, \ldots, n_s)$ is the normal to $\partial B_{\mu,\varepsilon}$, $f \in C^\infty$ is a $T-$periodic $s-$dimensional vector-valued function, T is a multiple of ε. As above $s = 2$ or $s = 3$. We seek the solution $u(x)$ in the class of $T-$periodic $s-$dimensional vector-valued functions of $H^1_{T-per}(B_{\mu,\varepsilon})$. Here $H^1_{T-per}(B_{\mu,\varepsilon})$ stands for the completion of the set of $T-$periodic differentiable functions defined at $B_{\mu,\varepsilon}$ with respect to the $H^1(B_{\mu,\varepsilon} \cap (0,T)^s)-$norm.

Below we construct the asymptotic expansion of the solution to problem (5.8.1)-(5.8.2) as $\mu, \varepsilon \to 0$ under some assumptions PF_1 and PF_2 of geometrical rigidity of the lattice structure. These assumptions will be formulated below for the limit problem. We assume as well that the following integrals of the right hand-side vanish:

$$\int_{(0,T)^s} f(x)dx = 0, \quad \int_{B_{\mu,\varepsilon}\cap(0,T)^s} f(x)dx = 0.$$

Theorem 5.8.1. *There exist a unique solution to problem (5.8.1),(5.8.2) such that*

$$\int_{B_{\mu,\varepsilon}\cap(0,T)^s} u(x)dx = 0.$$

and the estimate holds: there exist constants C and q independent of μ and ε such that

$$\|u\|^2_{H^1_{T-per}(B_{\mu,\varepsilon}\cap(0,T)^s)} \le C\mu^{-q}\|f\|^2_{L^2(B_{\mu,\varepsilon}\cap(0,T)^s)}.$$

This means that any solution of problem (5.8.1),(5.8.2) satisfies the estimate

$$\|\nabla u\|^2_{L^2(B_{\mu,\varepsilon}\cap(0,T)^s)} \le C\mu^{-q}\|f\|^2_{L^2(B_{\mu,\varepsilon}\cap(0,T)^s)}.$$

The result on existence and uniqueness of solution to a mixed boundary value problem of elasticity is well known (see for example [55]).

Let us prove the a priori estimate. As it was shown in section 5.4, the Korn inequality constant for structures depends on the small parameter. Composing the example of Lemma 5.4.1 for every rod, we obtain an $\varepsilon-$periodic function that has the Korn inequality constant of order μ^{-1}. On the other hand the theorem of Appendix shows that this Korn inequality constant is not worse than μ^{-q} $(q \in I\!R)$. Moreover, the Poincaré inequality constant is also not worse than some power of μ. So, finally we obtain the estimate of Theorem 5.4.1.

The asymptotic of the solution to problem (5.8.1), (5.8.2) are constructed in several steps. These steps are similar to the procedure of the previous section. The new elements are: the presence in the asymptotic expansion of the term of order $(\frac{\varepsilon}{\mu})^2$ (see section 5.6) and the dependency of the Korn inequality constant on μ^{-q} (this dicrepancy is not dangerous for the justification of the truncated asymptotic expansion because we apply the same improving of the accuracy procedure as in section 4.4).

1. First , we construct the asymptotic corresponding to a segment $e = e_h^n$ in $\mathcal{B}$. We change to new variables $\tilde{x} = \alpha(x - \varepsilon h)$. In these new variables, the cylinder $\{x/\varepsilon \in \tilde{B}_{hj}^{\alpha}\}$ becomes the rod.

$$U_{\varepsilon\mu} = \{\tilde{x}_1 \in I\!R^s \mid 0 < \tilde{x} < l\varepsilon, \ \ \tilde{x}'/\varepsilon\mu \in \beta_j\};$$

we set

$$\Gamma_{\varepsilon\mu} = \{\tilde{x}_1 \in I\!R^s \mid 0 < \tilde{x} < l\varepsilon, \ \ \tilde{x}'/\varepsilon\mu \in \partial\beta_j\}.$$

In what follows $\beta = \beta_j$. We write $\tilde{f}(\tilde{x}) = \alpha f(\alpha^* \tilde{x} + \varepsilon h)$ and $\tilde{u}(\tilde{x}) = \alpha u(\alpha^* \tilde{x} + \varepsilon h)$.

Then problem (5.8.1), (5.8.2) on $\tilde{B}_{hj}^{\alpha}$ becomes

$$P\tilde{u} = -\sum_{i,j=1}^{s} \frac{\partial}{\partial x_i}(A_{ij}\frac{\partial \tilde{u}}{\partial x_j}) = \tilde{f}(\tilde{x}), \quad \tilde{x} \in U_{\varepsilon\mu}, \tag{5.8.3}$$

$$\frac{\partial \tilde{u}}{\partial \nu} = \sum_{i,j=1}^{s} A_{ij}\frac{\partial \tilde{u}}{\partial x_j} n_i = 0, \quad \tilde{x} \in \Gamma_{\varepsilon\mu}. \tag{5.8.4}$$

We make the substitution $\xi' = (\tilde{\xi}_2, \ldots, \tilde{\xi}_s), \tilde{\xi}_i = \tilde{x}_i/(\varepsilon\mu)$ and expand $\tilde{f}$ in an asymptotic series with respect to $\varepsilon\mu\tilde{\xi}_i$, of the form

$$\tilde{f} \sim \sum_{j=0}^{\infty} (\varepsilon\mu)^j \sum_{r=1}^{r_j} \Phi F_{jr}(\tilde{\xi}_2, \ldots, \tilde{\xi}_s)\tilde{\psi}_{jr}(\tilde{x}_1),$$

In this formula the d-dimensional vectors $\tilde{\psi}_{jr}(\tilde{x}_1)$ are the values of the derivatives $\partial^j f/\partial x_1^{q_1} ... \partial x_{s-1}^{q_{s-1}} \partial x_s{}^{j-q_1-\ldots q_{s-1}}$ at the point $x = \alpha^*(\tilde{x}, 0, \ldots, 0) + \varepsilon h$, $F_{jr}(\xi')$ are matrices of polynomials and Φ is a matrix of rigid displacements. In particular , $\tilde{\psi}_{00}(\tilde{x}_1) = (\tilde{f}(\tilde{x}_1, 0, 0), 0)^*$ for $s = 3$, and $\tilde{\psi}_{00}(\tilde{x}_1, 0) = \tilde{f}(\tilde{x}_1, 0)$ for $s = 2$, , $F_{00} = I$ (the identity matrix) and $r_0 = 1$. We represent F_{jr} in the form $\bar{F}_{jr} + \tilde{F}_{jr}$, where $\langle \Phi^* \Phi \tilde{F}_{jr} \rangle_\beta = 0$ and $\bar{F}_{jr} = \langle \Phi^* \Phi F_{jr} \rangle_\beta$.

Using superposition , we seek the solution to problem (5.8.3) , (5.8.4) in the form:

$$\tilde{u} \sim \sum_{j=0}^{(\infty)} (\varepsilon\mu)^j \sum_{r=1}^{r_j} \tilde{u}_{jr}, \tag{5.8.5}$$

where $\tilde{u}_{jr}$ is the solution of the problem for a right hand side of the form $\Phi F_{jr}\tilde{\psi}_{jr}$. From now on in this section we omit the $\sim$ that denotes the change of variables, fix j and r and do not mention them explicitly when this does not create ambiguity .

We seek a formal asymptotic solution to problem (5.8.3) , (5.8.4) with right hand side $\Phi F_{jr}\psi_{jr}$ in the form of a series

$$\tilde{u}_{jr}^{\infty} \sim \sum_{l=0}^{\infty} (\varepsilon\mu)^l N_l\Big(\frac{\tilde{x}'}{\varepsilon\mu}\Big)\frac{d^l\tilde{\omega}(\tilde{x}_1)}{dx_1^l} + (\varepsilon\mu)^2 \sum_{l=0}^{\infty} (\varepsilon\mu)^l M_l\Big(\frac{\tilde{x}'}{\varepsilon\mu}\Big)\frac{d^l\tilde{\psi}_{jr}(\tilde{x}_1)}{d\tilde{x}_1^l} \tag{5.8.8}$$

where $N_l$and M_l are the matrix valued functions of section 2.2; $\tilde{\omega}(\tilde{x}_1) = \tilde{v}_{\varepsilon\mu}(x) + \varepsilon^2\tilde{w}_{\varepsilon\mu}(x)$, $v_{\varepsilon\mu}(x) = \alpha^*\tilde{v}_{\varepsilon\mu}(x)$, and $w_{\varepsilon\mu}(x) = \alpha^*\tilde{w}_{\varepsilon\mu}(x)$ are defined on $I\!R^s$, and

$$\tilde{v}_{\varepsilon\mu}(x) \sim E_0\alpha V(x) + \sum_{m=1}^{\infty}\sum_{q=0}^{\infty} \varepsilon^m\mu^q \sum_{(i_1\ldots i_m):\ i_n\in\{1\ldots s\}} \tilde{\mathcal{N}}^q_{i_1\ldots i_m}\Big(\frac{\tilde{x}_1(x)}{\varepsilon}\Big)\frac{\partial^m V(x)}{\partial x_{i_1}\ldots\partial x_{i_m}}$$

$$\tilde{w}_{\varepsilon\mu}(x) \sim \sum_{m=0}^{\infty}\sum_{q=0}^{\infty} \varepsilon^m\mu^q \sum_{(j_1\ldots j_s):\ j_1+\ldots+j_s=j}\ \sum_{(i_1\ldots i_m):\ i_n\in\{1\ldots s\}}$$

$$\times\tilde{\mathcal{M}}^q_{i_1\ldots i_m,j_1\ldots j_s}\Big(\frac{\tilde{x}_1(x)}{\varepsilon}\Big)\frac{\partial^m}{\partial x_{i_1}\ldots\partial x_{i_m}}\frac{\partial^j f}{\partial x_1^{j_1}\ldots\partial x_s^{j_s}} \tag{5.8.9}$$

Here $V(x)$ is an s-dimensional vector $\tilde{\mathcal{N}}^q_{i_1\ldots i_m}(\tilde{\xi}_1)$ and $\tilde{\mathcal{M}}^q_{i_1\ldots i_m,j_1\ldots j_s}(\tilde{\xi}_1)$ are $d\times s$ matrices of the form

$$\tilde{\mathcal{N}}^q_{i_1\ldots i_m} = \hat{\alpha}\mathcal{N}^q_{i_1\ldots i_m}\alpha^* = \tilde{\mathcal{N}}^{q(1)}_{i_1\ldots i_m} + \mu^{-1}\tilde{\mathcal{N}}^{q(4)}_{i_1\ldots i_m} + \mu^{-2}\tilde{\mathcal{N}}^{q(2)}_{i_1\ldots i_m}$$

$$\tilde{\mathcal{M}}^q_{i_1\ldots i_m,j_1\ldots j_s} = \alpha\mathcal{M}^q_{i_1\ldots i_m,j_1\ldots j_s}\alpha^*$$

$$= \tilde{\mathcal{M}}^{q(1)}_{i_1\ldots i_m,j_1\ldots j_s} + \mu^{-1}\tilde{\mathcal{M}}^{q(4)}_{i_1\ldots i_m,j_1\ldots j_s} + \mu^{-2}\tilde{\mathcal{M}}^{q(2)}_{i_1\ldots i_m,j_1\ldots j_s}, \tag{5.8.10}$$

$E_0 = I$ for $s = 2$,

$$E_0 = \begin{pmatrix} & I & \\ 0 & 0 & 0 \end{pmatrix}$$

for $s = 3$ (here I is the unit $s \times s$ matrix); $\hat{\alpha} = \alpha$ for $s = 2$, and

$$\hat{\alpha} = \begin{pmatrix} & & & 0 \\ & \alpha & & 0 \\ & & & 0 \\ 0 & 0 & 0 & 1 \end{pmatrix}$$

for $s = 3$, and $\tilde{x}_1(x)/\varepsilon$ is the projection of x/ε on e ; all the rows of the matrices $\tilde{\mathcal{N}}^{q(1)}_{i_1\ldots i_m}$ and $\tilde{\mathcal{M}}^{q(1)}_{i_1\ldots i_m, j_1\ldots j_s}$ except the first one, all the rows of $\tilde{\mathcal{N}}^{q(4)}_{i_1\ldots i_m}$ and $\tilde{\mathcal{M}}^{q(4)}_{i_1\ldots i_m, j_1\ldots j_s}$ except the fourth one for $s = 3$, and all the rows of $\tilde{\mathcal{N}}^{q(2)}_{i_1\ldots i_m}$ and $\tilde{\mathcal{M}}^{q(2)}_{i_1\ldots i_m, j_1\ldots j_s}$ except the second one (and the third one for $s = 3$) , are zero. We remark that $\tilde{\psi}_{jr}$ can be represented in the form

$$\tilde{\psi}_{jr}(\tilde{x_1}(x)) = \sum_{(j_1\ldots j_s):\ j_1+\ldots+j_s=j} \tilde{\mathcal{K}}_{j_1\ldots j_s}(\frac{x}{\varepsilon})\frac{\partial^j f(x)}{\partial {x_1}^{j_1}\ldots \partial {x_s}^{j_s}}, \tag{5.8.11}$$

where $\tilde{\mathcal{K}}_{j_1\ldots j_s}$ are constant for every e (this constant changes if one passes from one segment e to another).

The substitution of (5.8.8) in (5.8.3) and (5.8.4) is described in section 2.2 (with μ replaced by $\varepsilon\mu$). As a result, we find that (5.8.4) is satisfied asymptotically exactly and that (5.5.3) yields a system of d equations "homogenized over the rod" , of the form

$$h_2^N \frac{d^2\tilde{\omega}}{d\tilde{x}_1^2} + (\varepsilon\mu)^2 h_4^N \frac{d^4\tilde{\omega}}{d\tilde{x}_1^4} - \overline{F}_{jr}\tilde{\psi}_{jr}$$

$$+\sum_{l=6}^{\infty}(\varepsilon\mu)^{l-2}h_l^N\frac{d^l\tilde{\omega}}{d\tilde{x}_1^l} + \sum_{l=0}^{\infty}(\varepsilon\mu)^l h_l^M \frac{d^l\tilde{\psi}_{jr}}{d\tilde{x}_1^l} \sim 0. \tag{5.8.12}$$

Here we have not yet used the fact that ω has the form (5.8.9). We substitute (5.8.9) in (5.8.12) multiplied on the left by $\hat{\alpha}^*$ and gather together the coefficients of the same powers of $\varepsilon\mu$ and the derivatives

$$D^i V = \frac{\partial^m V}{\partial x_{i_1}\ldots\partial x_{i_m}}, D^j f = \frac{\partial^j f}{\partial x_1^{j_1}\ldots\partial x_s^{j_s}}.$$

We obtain an asymptotic equality of the form

$$\sum_{m=1}^{\infty}\sum_{q=0}^{\infty}\varepsilon^{m-2}\mu^q \sum_{(i_1\ldots i_m)} \Bigg(\mathcal{H}^{N\ q}_{i_1\ldots i_m}\Big(\frac{\tilde{x}_1(x)}{\varepsilon}\Big)\frac{\partial^m V}{\partial x_{i_1}\ldots\partial x_{i_m}}$$

$$+ \sum_{(j_1 \dots j_s): j_1 + \dots + j_s = j} \mathcal{H}^{M\ q}_{i_1 \dots i_m,\ j_1 \dots j_s} \left(\frac{\tilde{x}_1(x)}{\varepsilon} \right) \frac{\partial^m}{\partial x_{i_1} \dots \partial x_{i_m}} \frac{\partial^j f}{\partial x_1^{j_1} \dots \partial x_s^{j_s}} \Bigg) \sim 0 \tag{5.8.13}$$

$$\mathcal{H}^{N\ q}_{i_1 \dots i_m} = \hat{\alpha}_e^* \mathcal{R} \tilde{\mathcal{N}}^q_{i_1 \dots i_m}(\tilde{\xi}_1) \alpha_e + \hat{\alpha}_e^* \mathcal{T}^q_{i_1 \dots i_m}(\mathcal{N}) \alpha_e,$$

$$\mathcal{H}^{M\ q}_{i_1 \dots i_m,\ j_1 \dots j_s} = \hat{\alpha}_e^* \mathcal{R} \tilde{\mathcal{M}}^q_{i_1 \dots i_m,\ j_1 \dots j_s}(\tilde{\xi}_1) \alpha_e + \hat{\alpha}_e^* \mathcal{T}^q_{i_1 \dots i_m,\ j_1 \dots j_s}(\mathcal{M}) \alpha_e, \tag{5.8.14}$$

$\mathcal{T}^q_{i_1 \dots i_m}(\mathcal{N})(\mathcal{T}^q_{i_1 \dots i_m,\ j_1 \dots j_s}(\mathcal{M}))$ are linear combinations of the $\mathcal{N}^q_{i_1 \dots i_m}$ $(\tilde{\mathcal{M}}^q_{i_1 \dots i_m,\ j_1 \dots j_s})$ and their derivatives with multi-index $(i_1 \dots i_m)$ of length m_1 less than m or with the multi-index $(i_1, ..., i_m)$ with $m = m_1$ and with superscript $q_1 < q$, and $\mathcal{R}$ is a differential operator along the variable $\tilde{\xi_1} = \tilde{x}_1(x)/\varepsilon$, of the form

$$\mathcal{R} = \begin{pmatrix} \bar{E} \frac{\partial^2}{\partial \tilde{\xi}_1^2} & 0 \\ 0 & -\bar{E} \langle \tilde{\xi}_2^2 \rangle_\beta \frac{\partial^4}{\partial \tilde{\xi}_1^4} \end{pmatrix}$$

for $s = 2$,

$$\mathcal{R} = \begin{pmatrix} \bar{E} \frac{\partial^2}{\partial \tilde{\xi}_1^2} & 0 & 0 & 0 \\ 0 & -\bar{E} \langle \tilde{\xi}_2^2 \rangle_\beta \frac{\partial^4}{\partial \tilde{\xi}_1^4} & 0 & 0 \\ 0 & 0 & -\bar{E} \langle \tilde{\xi}_3^2 \rangle_\beta \frac{\partial^4}{\partial \tilde{\xi}_1^4} & 0 \\ 0 & 0 & 0 & \bar{\mathcal{M}} \frac{\partial^2}{\partial \tilde{\xi}_1^2} \end{pmatrix}$$

for $s = 3$.

We equate $\tilde{\mathcal{H}}^N_{i_1 \dots i_s}(\xi_1)$ and $\tilde{\mathcal{H}}^M_{i_1 \dots i_s,\ j_1 \dots j_m}(\xi_1)$ with the constant matrices $h^q_{i_1 \dots i_m}$ and $h^q_{i_1 \dots i_m,\ j_1 \dots j_s}$ so that for $s = 3$ the fourth row and the fourth column are zero. We denote by $\overline{h^q_{i_1 \dots i_m}}$, and $\overline{h^q_{i_1 \dots i_m,\ j_1 \dots j_s}}$ the minors of these matrices obtained by removing their fourth rows and columns. We obtain a homogenized equation for V of the form:

$$\sum_{m=1}^{\infty} \sum_{q=0}^{\infty} \varepsilon^{m-2} \mu^q \sum_{(i1 \dots im)} \overline{h^q_{i_1 \dots i_m}} \frac{\partial^m V}{\partial x_{i_1} \dots \partial x_{i_m}}$$

$$+ \sum_{m=1}^{\infty} \sum_{q=0}^{\infty} \varepsilon^{m-2} \mu^q \sum_{(i_1 \dots i_m)} \sum_{(j_1 \dots j_s):\ j_1 + \dots + j_s = j} \overline{h^q_{i_1 \dots i_m,\ j_1 \dots j_s}} \frac{\partial^m}{\partial x_{i_1} \dots \partial x_{i_m}} \frac{\partial^j f}{\partial x_1^{j_1} \dots \partial x_s^{j_s}}$$

$$\sim 0. \tag{5.8.15}$$

The coefficients $\overline{h^q_{i_1\dots i_m}}$, and $\overline{h^q_{i_1\dots i_m,\ j_1\dots j_s}}$ of this equation will be determined later.

Thus the $\tilde{\mathcal{N}}^q_{i_1\dots i_m}$ and $\tilde{\mathcal{M}}^q_{i_1\dots i_m,\ j_1\dots j_s}$ satisfy the equations

$$\mathcal{R}\,\tilde{\mathcal{N}}^q{}_{i_1\dots i_m} = \tilde{\mathcal{T}}^q_{i_1\dots i_m}(N) + \hat{\alpha}_\varepsilon h^q_{i_1\dots i_m}\hat{\alpha}^*_\varepsilon \tag{5.8.16}$$

$$\mathcal{R}\,\tilde{\mathcal{M}}^q{}_{i_1\dots i_m,\ j_s\dots j_s} = \tilde{\mathcal{T}}^q_{i_1\dots i_m,\ j_s\dots j_s}(M) + \hat{\alpha}_\varepsilon h^q_{i_1\dots i_m}\hat{\alpha}^*_\varepsilon. \tag{5.8.17}$$

Let $\tilde{u}^{(\infty)}_\varepsilon = \tilde{u}^{(\infty)}_{jr}(x)$ be the asymptotic series (5.8.8), (5.8.9), which is the solution of (5.8.1) (5.8.2) in the cylinder $\{x : \frac{x}{\varepsilon} \in \tilde{B}^\alpha_{hj}\}$.

We expand the functions

$$\frac{\partial^m V}{\partial x_{i_1}\dots\partial x_{i_m}},\quad \frac{\partial^m}{\partial x_{i_1}\dots\partial x_{i_m}}\frac{\partial^j f}{\partial x_1^{j_1}\dots\partial x_s^{j_s}},\quad \frac{\partial^l\psi_{jr}}{\partial\tilde{x}^l_i},$$

$\tilde{\mathcal{N}}_{i_1\dots i_m}(\frac{\tilde{x}_1(x)}{\varepsilon})$ and $\tilde{\mathcal{M}}_{i_1\dots i_m,j_1\dots j_s}(\frac{\tilde{x}_1(x)}{\varepsilon})$ in Taylor series in the neighborhood of the point x_0 corresponding to the node $\frac{x_0}{\varepsilon}$, and substitute these expansions in (5.8.8) . We obtain

$$\tilde{u}^{(\infty)}_\varepsilon \sim \alpha(\sum_{l=0}^{\infty}\sum_{r=-2}^{\infty}\varepsilon^l\mu^r\sum_{(i_1\dots i_l)}(N^{e\ \ r}_{i_1\dots i_l}(\frac{x-x_0}{\varepsilon\mu})\frac{\partial^l V}{\partial x_{i_1}\dots\partial x_{i_l}}(x_0)$$

$$+\varepsilon^2\sum_{(j_1\dots j_s),\ j_1+\dots+j_l=j} M^{e\ \ r}_{i_1\dots i_l,\ j_1\dots j_s}(\frac{x-x_0}{\varepsilon\mu})$$

$$\times\frac{\partial^{l+j}f}{\partial x_{i_1}\dots\partial x_{i_l}\partial x_1^{j_1}\dots\partial x_s^{j_s}}(x_0))$$

$$+(\varepsilon\mu)^2\sum_{l=0}^{\infty}(\varepsilon\mu)^l\sum_{(i_1\dots i_l)}\ \sum_{(j_1\dots j_s),\ j_1+\dots+j_l=j} K^{e\ \ r}_{i_1\dots i_l,\ j_1\dots j_s}(\frac{x-x_0}{\varepsilon\mu})$$

$$\times\frac{\partial^{l+j}f}{\partial x_{i_1}\dots\partial x_{i_l}\partial x_1^{j_1}\dots\partial x_s^{j_s}}(x_0)). \tag{5.8.18}$$

Let $\tilde{z}_1 = \tilde{x}_1/(\varepsilon\mu)$. Then

$$N^{e\ \ r}_{i_1\dots i_m}(z) = \alpha^*\Bigg(N_0(\tilde{z}')\Big(\tilde{\mathcal{N}}^{r(1)}_{i_1\dots i_l}(\frac{\tilde{x}_1(x_0)}{\varepsilon}) + \tilde{\mathcal{N}}^{r+1\ (4)}_{i_1\dots i_l}(\frac{\tilde{x}_1(x_0)}{\varepsilon})$$

$$+\tilde{\mathcal{N}}^{r+2\ (2)}_{i_1\dots i_l}(\frac{\tilde{x}_1(x_0)}{\varepsilon}) + \frac{\partial\tilde{\mathcal{N}}^{r+1\ (2)}_{i_1\dots i_l}}{\partial\tilde{\xi}_1}(\frac{\tilde{x}_1(x_0)}{\varepsilon})\tilde{z}_1\Big)\alpha$$

$$+N_1(\tilde{z}')\frac{\partial\tilde{\mathcal{N}}^{r+1\ (2)}_{i_1\dots i_l}}{\partial\tilde{\xi}_1}(\frac{\tilde{x}_1(x_0)}{\varepsilon})\alpha\Bigg) + \Delta_N(\tilde{\mathcal{N}}). \tag{5.8.19}$$

where $\Delta_N(\tilde{\mathcal{N}})$ is defined in terms of the functions $\tilde{\mathcal{N}}$ with multi-indices $(i_1 \dots i_l)$ of length less than l or with multi index $(i_1 \dots i_l)$ but with superscripts not exceeding $r-1$, or in terms of $\mathcal{N}^{r(2)}_{i_1\dots i_l}$, $\mathcal{N}^{r(4)}_{i_1\dots i_l}$.

An analogue of the representations (5.8.19) can also be found for the matrices $M^{e\ r}_{i_1\dots i_l,\ j_1\dots j_s}$.

A formal asymptotic solution of the problem near a pre-nodal component Π_{x_0} corresponding to the node x_0/ε is sought in the form

$$\tilde{u}^{(\infty)}_{\Pi_{x_0}} \sim (\sum_{l=0}^{\infty}\sum_{r=-2}^{\infty} \varepsilon^l \mu^r \sum_{(i_1\dots i_l)} (N^{0\ \ r}_{i_1\dots i_l}(\frac{x-x_0}{\varepsilon\mu})\frac{\partial^l V}{\partial x_{i_1}\dots\partial x_{i_l}}(x_0)$$

$$+\varepsilon^2 \sum_{(j_1\dots j_s),\ j_1+\dots+j_l=j} M^{0\ \ r}_{i_1\dots i_l,\ j_1\dots j_s}(\frac{x-x_0}{\varepsilon\mu})$$

$$\times \frac{\partial^{l+j} f}{\partial x_{i_1}\dots\partial x_{i_l}\partial x_1^{j_1}\dots\partial x_s^{j_s}}(x_0))$$

$$+\sum_{e(x_0)} \chi_e(x)\hat{\rho}(\alpha_e\frac{x-x_0}{\varepsilon\mu})u^{(\infty)}_e; \tag{5.8.20}$$

here $u^{(\infty)}_e = \alpha^* \tilde{u}^{(\infty)}_e$, the summation in the last sum in (5.8.20) extends over all segments e having x_0/ε as an end-point, $\chi_e(x)$ is the characteristic function of the section of e, α_e the matrix of the transformation α which corresponds to e, and $\hat{\rho}(t) = \rho(t_1)$ a differentiable function that vanishes for $|t_1| \le c_0$, is equal to unity for $|t_1| \ge c_0 + 1$, and such that $0 \le \hat{\rho}(t_1) \le 1$, where $t = (t_1, \dots, t_s)$.

As in the previous section we set $z = (x-x_0)/(\varepsilon\mu)$ and denote by $\Pi_{x_0,z,0}$ the image of Π_{x_0}. Let $\tilde{\tilde{B}}^{\alpha}_{hj}$ be a cylinder whose intersection with Π_{x_0} is nonempty. We extend it to a semi-infinite cylinder whose basis has no common points with Π_{x_0}. We denote by $B^{\alpha}_{hj\infty}$ this extended cylinder, and by $\Pi_{x_0\infty}$ the union of all such cylinders $B^{\alpha}_{hj\infty}$ and Π_{x_0}. Moreover, we denote by $\Pi_{x_0,z,\infty}$ the image of $\Pi_{x_0,\infty}$ under the transformation $z = (x-x_0)/(\varepsilon\mu)$.

Let $\tilde{\chi}_e(z)$ be the characteristic function of $B^{\alpha}_{hj\infty}$ corresponding to the section e.

Substituting (5.8.40) in (5.8.1), (5.8.2), we find that $N^{0\ \ r}_{i_1\dots i_l}$ is the solution to the problem (similar to (4.3.7) of Chapter 4).

$$L_z N^{0\ \ r}_{i_1\dots i_l} = \sum_{e(x_0)} \tilde{\chi}_e(z)(L_z((1-\hat{\rho}(\alpha_e z))N^{e\ \ r}_{i_1\dots i_l}(z))$$

$$-(1-\hat{\rho}(\alpha_e z))L_z N^{e\ \ r}_{i_1\dots i_l}(z)),\ z \in \Pi_{x_0,z,\infty},$$

$$\frac{\partial}{\partial \nu_z} N^{0\ \ r}_{i_1\dots i_l} = \sum_{e(x_0)} \tilde{\chi}_e(z)(\frac{\partial}{\partial \nu_z}((1-\hat{\rho}(\alpha_e z))N^{e\ \ r}_{i_1\dots i_l}(z))$$

$$-(1-\hat{\rho}(\alpha_e z))\frac{\partial}{\partial \nu_z} N^{e\ \ r}_{i_1\dots i_l}(z)),\ z \in \partial\Pi_{x_0,z,\infty}. \tag{5.8.21}$$

Similar problems are obtained for $M^{0\ r}_{i_1 \dots i_l,\ j_1 \dots j_s}$. These problems as well as problem (4.8.21) are of the form

$$L_z N^0 = \sum_{e(x_0)} \bar{\chi}_e(z) L_z \phi_e(z) + \phi_0(z), \quad z \in \Pi_{x_0,z,\infty}.$$

$$\frac{\partial}{\partial \nu_z} N^0 = \sum_{e(x_0)} \bar{\chi}_e(z) \left(\frac{\partial}{\partial \nu_z} \phi_e(z) + \sum_{i=1}^{s} n_i \phi_{i\ e}(z) \right), \quad z \in \partial \Pi_{x_0,z,\infty}, \quad (5.8.22)$$

where ϕ_e are smooth functions and ϕ_0 and $\phi_{i\ e}$ piecewise smooth functions with compact support.

Let $\tilde{\mathcal{J}}(z)$ be the matrix in section 2.2 and let $\mathcal{J}$ be the linear span of columns of $\tilde{\mathcal{J}}(z)$.

Lemma 5.8.1 *Problem (5.8.22) is solvable in the class of vector-valued functions that stabilize exponentially to a vector-valued function in $\mathcal{J}$ (in the sense of [130]) if and only if*

$$\int_{\Pi_{x_0,z,\infty}} \tilde{\mathcal{J}}^*(z) \phi_0(z) dz$$

$$= \sum_{e(x_0)} \left(\int_{\partial \Pi_{x_0,z,\infty}} \tilde{\mathcal{J}}^*(z) \sum_{k=1}^{s} n_k \phi_{k\ e}(z)\, ds + \int_{\partial \Pi_{x_0,z,0} \setminus \partial \Pi_{x_0,z,\infty}} \tilde{\mathcal{J}}^*(z) \frac{\partial}{\partial \nu_z} \phi_e(z)\, ds \right)$$

where $(n_1, \dots, n_s)$ is the outward normal to $\Pi_{x_0,z,0}$.

The proof of the lemma is fully analogous to that of the main theorem in [130]. Lemma 5.8.1 and (5.8.20) yield the conditions

$$\sum_{e(x_0)} \varepsilon\mu \int_{\Pi_{x_0,z,\infty}} \bar{\chi}_e(z)(1 - \hat{\rho}(\alpha_e z)) \frac{\partial}{\partial \nu_z} \tilde{\mathcal{J}}^*(z) L_x u_e^{N\ (\infty)} dz$$

$$= - \sum_{e(x_0)} \int_{\partial \Pi_{x_0,z,0} \setminus \partial \Pi_{x_0,z,\infty}} \bar{\chi}_e(z) \tilde{\mathcal{J}}^*(z) \frac{\partial}{\partial \nu_z} u_e^{N\ (\infty)}\, ds\,, \quad (5.8.23)$$

and similar conditions for $u_e^{M\ (\infty)}$; here $\alpha^* u_e^{N\ (\infty)}$ and $\alpha^* u_e^{M\ (\infty)}$ are the first and second sums in (5.8.8), and

$$L_x = \sum_{i,j=1}^{s} \frac{\partial}{\partial x_i} \left(A_{ij} \frac{\partial}{\partial x_j} \right), \quad \frac{\partial}{\partial \nu_z} = \sum_{i,j=1}^{s} n_i \left(A_{ij} \frac{\partial}{\partial x_j} \right).$$

Taking into account the fact

$$L_x u_e^{N\ (\infty)} \sim \sum_{l=2}^{\infty} (\varepsilon\mu)^{l-2} h_l^N \frac{d^l \tilde{v}}{d\tilde{x}_1^l},$$

$$\frac{\partial}{\partial \nu_x} u_e^{N\ (\infty)} \sim \alpha_e^* \sum_{l=1}^{\infty} (\varepsilon\mu)^{l-1} \left(\sum_{j=1}^{s} A_{1j} \frac{\partial N_l}{\partial \xi_j} + A_{11} N_{l-1} \right) \frac{d^l \tilde{v}}{d\tilde{x}_1^l},$$

and considering the analogous expansions for $L_x u_e^{M\ (\infty)}$ and $\frac{\partial}{\partial \nu_x} u_e^{M\ (\infty)}$ and relation of Remark 2.3.2, we find that $\tilde{\mathcal{N}}^{e\ r}_{i_1 \dots i_l}$ and $\tilde{\mathcal{M}}^{e\ r}_{i_1 \dots i_l,\ j_1 \dots j_s}$ satisfy at the nodes the conditions (following from (5.8.23))

$$\hat{\Lambda} \tilde{\mathcal{N}}^{e\ r}_{i_1 \dots i_l} |_{\tilde{x}_1(x_0)/\varepsilon} = \Psi^r_{N,\ i_1 \dots i_l}(x_0/\varepsilon),$$

$$\hat{\Lambda} \tilde{\mathcal{M}}^{e\ r}_{i_1 \dots i_l, j_1 \dots j_s} |_{\tilde{x}_1(x_0)/\varepsilon} = \Psi^r_{M,\ i_1 \dots i_l, j_1 \dots j_s}(x_0/\varepsilon), \tag{5.8.24}$$

where the right-hand sides of (5.8.24) are defined in terms of $\tilde{\mathcal{N}}^{e\ q}_{i_1 \dots i_l}$ and $\tilde{\mathcal{M}}^{e\ q}_{i_1 \dots i_l,\ j_1 \dots j_s}$ with indices $(l_1, q) < (l, r)$ (that is, $l_1 \leq l$, $q \leq r$ and $l_1 \neq l$ or $q \neq r$),

$$\hat{\Lambda}_e = \sum_{e(x_0)} \Gamma_e \Lambda_e mes \beta_e, \tag{5.8.25}$$

Λ_e is the operator in (4.4.17); Γ_e the $d \times d$ matrix of the transformation

$$\tilde{\mathcal{J}}^*(z) \alpha_e^* = \Gamma_e \tilde{\mathcal{J}}^*(\alpha_e z),$$

where $\tilde{\mathcal{J}}(z)$ is the matrix of rigid displacements of section 4.4, and β_e the cross-section of the cylinder corresponding to e. As it was proved in Chapter 4

$$\Gamma_e = \begin{pmatrix} & & 0 \\ & \alpha_e^* & 0 \\ 0 & 0 & 1 \end{pmatrix}$$

for $s = 2$, and

$$\Gamma_e = \begin{pmatrix} \alpha_e^* & O \\ O & \bar{\Gamma}_e \end{pmatrix} \tag{5.8.26}$$

for $s = 3$, where $\bar{\Gamma}_e$ is an orthogonal 3×3 matrix and O the 3×3 zero matrix.

If (5.8.24) holds then, by Lemma 5.8.1, problems (5.8.21) have solutions that stabilize exponentially, as $|z| \to \infty$, to some "limit" functions in $\mathcal{J}$. For every segment $e(x_0)$ and its corresponding half-bounded cylinder $B^{\alpha}_{hj\infty}$, the limit function can be represented in the form $\tilde{\mathcal{J}}(z) C_e$, where C_e is a constant $\bar{d} \times s$ matrix with $\bar{d} = 3$ for $s = 2$ and $\bar{d} = 6$ for $s = 3$; here $\tilde{\mathcal{J}}(z)$ is a representative of a linear span $\mathcal{J}$. On the other hand, we consider the $\bar{d} \times s$ matrices $\tilde{S}^{e\ q}_{\mathcal{N},\ i_1 \dots i_l}$ and $\tilde{S}^{e\ q}_{\mathcal{M},\ i_1 \dots i_l,\ j_1 \dots j_s}$ constructed from the matrices $\tilde{\mathcal{N}}^{e\ r}_{i_1 \dots i_l}$. and $\tilde{\mathcal{M}}^{e\ r}_{i_1 \dots i_l,\ j_1 \dots j_s}$ as follows: for $s = 2$, the first two rows of the matrix $\tilde{S}^{e\ q}_{\mathcal{N},\ i_1 \dots i_l}$ coincide with the matrix $\tilde{\mathcal{N}}^{e\ q\ (1)}_{i_1 \dots i_l} + \tilde{\mathcal{N}}^{e\ q+2\ (2)}_{i_1 \dots i_l}$, and the third row with the second row of $\frac{\partial}{\partial \xi_1} \tilde{\mathcal{N}}^{e\ q+1}_{i_1 \dots i_l}$; for $s = 3$, the first four rows of $\tilde{S}^{e\ q}_{\mathcal{N},\ i_1 \dots i_l}$ coincide with the matrix $\tilde{\mathcal{N}}^{e\ q\ (1)}_{i_1 \dots i_l} + \tilde{\mathcal{N}}^{e\ q+2\ (2)}_{i_1 \dots i_l} + \tilde{\mathcal{N}}^{e\ q+1\ (4)}_{i_1 \dots i_l} a_e$, and the last two rows with the second and third rows of $\frac{\partial}{\partial \tilde{\xi}_1} \tilde{\mathcal{N}}^{e\ q+1}_{i_1 \dots i_l}$. Here a_e is normalization factor (2.3.3). The matrices $\tilde{S}^{e\ q}_{\mathcal{M},\ i_1 \dots i_l,\ j_1 \dots j_s}$ are constructed analogously from $\tilde{\mathcal{M}}^{e\ r}_{i_1 \dots i_l,\ j_1 \dots j_s}$. If we change the value of $\tilde{S}^{e\ q}_{\mathcal{N},\ i_1 \dots i_l} |_{\tilde{x}_1(x_0)/\varepsilon}$ by $\delta \tilde{S}^{e\ q}_{\mathcal{N},\ i_1 \dots i_l}$, then the value of $\tilde{\mathcal{J}}(z) C_e$, changes by $\alpha_e^* \tilde{\mathcal{J}}(\alpha_e z) \delta \tilde{S}^{e\ q}_{\mathcal{N},\ i_1 \dots i_l} \alpha_e$, that is, the value of C_e changes by $\Gamma_e \delta \tilde{S}^{e\ q}_{\mathcal{N},\ i_1 \dots i_l} \alpha_e$. The value of $\tilde{S}^{e\ q}_{\mathcal{M},\ i_1 \dots i_l,\ j_1 \dots j_s} |_{\tilde{x}_1(x_0)/\varepsilon}$ changes according the same rule.

Suppose that for any $e_1(x_0)$ and $e_2(x_0)$, at the point $\tilde{x}_1(x_0)/\varepsilon$ we have

$$\Gamma_{e_1}\tilde{S}^{e_1\ q}_{\mathcal{N},\ i_1...i_l}\alpha_{e_1} = \Gamma_{e_2}\tilde{S}^{e_2\ q}_{\mathcal{N},\ i_1...i_l}\alpha_{e_2},$$

$$\Gamma_{e_1}\tilde{S}^{e_1\ q}_{\mathcal{M},\ i_1...i_l\ j_1...j_s}\alpha_{e_1} = \Gamma_{e_2}\tilde{S}^{e_2\ q}_{\mathcal{M},\ i_1...i_l\ j_1...j_s}\alpha_{e_2}, \tag{5.8.27}$$

and that under these conditions, for every $e = e(x_0)$ the solution of (5.8.21) stabilizes to $\tilde{\mathcal{J}}(z)C^N_e$, and the solution of the corresponding problem for $M^0_{i_1...i_l,\ j_1...j_s}$ stabilizes to $\tilde{\mathcal{J}}(z)C^M_e$ on the cylinder $B^\alpha_{hj\infty}$ corresponding to e. Then we redefine $\tilde{S}^{e\ q}_{\mathcal{N},\ i_1...i_l}|_{\tilde{x}_1(x_0)/\varepsilon}$ and $\tilde{S}^{e\ q}_{\mathcal{M},\ i_1...i_l,\ j_1...j_s}|_{\tilde{x}_1(x_0)/\varepsilon}$ so that their corresponding solutions of the problems in $\Pi_{x_0,z,\infty}$ stabilize to one and the same rigid displacement in $\mathcal{J}$ for all segments $e(x_0)$ with the same end-point x_0. To this end , it suffices to set the new (improved) values of $\tilde{S}^{e\ q}_{\mathcal{N},\ i_1...i_l}|_{\tilde{x}_1(x_0)/\varepsilon}$ and $\tilde{S}^{e\ q}_{\mathcal{M},\ i_1...i_l,\ j_1...j_s}|_{\tilde{x}_1(x_0)/\varepsilon}$ equal to the old ones minus $\Gamma^*_e C^N_e\alpha^*_e$ and $\Gamma^*_e C^M_e\alpha^*_e$, respectively. In other words, (5.8.27) needs to be replaced by conditions of the form

$$\Gamma_{e_1}\left(\tilde{S}^{e_1\ q}_{\mathcal{N},\ i_1...i_l} + \delta\tilde{S}^{e_1\ q}_{\mathcal{N},\ i_1...i_l}\right)\alpha_{e_1} = \Gamma_{e_2}\left(\tilde{S}^{e_2\ q}_{\mathcal{N},\ i_1...i_l} + \delta\tilde{S}^{e_2\ q}_{\mathcal{N},\ i_1...i_l}\right)\alpha_{e_2},$$

$$\Gamma_{e_1}\left(\tilde{S}^{e_1\ q}_{\mathcal{M},\ i_1...i_l\ j_1...j_s} + \delta\tilde{S}^{e_1\ q}_{\mathcal{M},\ i_1...i_l\ j_1...j_s}\right)\alpha_{e_1}$$

$$= \Gamma_{e_2}\left(\tilde{S}^{e_2\ q}_{\mathcal{M},\ i_1...i_l\ j_1...j_s} + \delta\tilde{S}^{e_2\ q}_{\mathcal{M},\ i_1...i_l\ j_1...j_s}\right)\alpha_{e_2}, \tag{5.8.28}$$

for any segments e_1 and e_2 having the common end-point x_0. Here the constant $\bar{d}\times s$ matrices $\delta\tilde{S}^{e\ q}_{\mathcal{N},\ i_1...i_l}$ and $\delta\tilde{S}^{e_2\ q}_{\mathcal{M},\ i_1...i_l\ j_1...j_s}$ are chosen so that (5.8.21) has solutions that stabilize to zero; in what follows we consider precisely these solutions of (5.8.21) (here $\delta\tilde{S}^{e\ q}_{\mathcal{N},\ i_1...i_l} = \Gamma^*_e C^N_e\alpha^*_e$, and $\delta\tilde{S}^{e_2\ q}_{\mathcal{M},\ i_1...i_l\ j_1...j_s} = \Gamma^*_e C^M_e\alpha^*_e$, and (5.8.28) contains now the "new" values of $\tilde{S}^{e\ q}_{\mathcal{N},\ i_1...i_l}$ and $\tilde{S}^{e\ q}_{\mathcal{M},\ i_1...i_l,\ j_1...j_s}$).

From the above arguments we derive the following algorithm for the recursive computation of the matrices $\tilde{\mathcal{N}}^{e\ q}_{i_1...i_l}$, $\tilde{\mathcal{M}}^{e\ q}_{i_1...i_l,\ j_1...j_s}$, $N^{0\ q}_{i_1...i_l}$ and $M^{0\ q}_{i_1...i_l,\ j_1...j_s}$.

The basis of the recursive process is $\tilde{\mathcal{N}}^{e\ 0}_{\emptyset} = I, \tilde{\mathcal{M}}^{e\ 0}_{\emptyset} = 0$; for $q < 0$, we formally set $\tilde{\mathcal{N}}^{e\ q}_{i_1...i_l} = \tilde{\mathcal{M}}^{e\ q}_{i_1...i_l,\ j_1...j_s} = 0$; $N^{0\ -2}_{i_1...i_l}$ are the solutions of problems (5.8.18), (5.8.19),(5.8.21) from which

$$N^{0\ -2}_{i_1...i_l} = -\sum_{e(x_0)}\bar{\chi}_e(z)\hat{\rho}(\alpha_e z)\alpha^*_e N_0(\alpha_e z)\tilde{\mathcal{N}}^{e\ 0\ (2)}_{i_1...i_l}\left(\frac{\tilde{x}_1(x_0)}{\varepsilon}\right)\alpha_e + const,$$

$$M^{0\ -2}_{i_1...i_l,\ j_1...j_s} = -\sum_{e(x_0)}\bar{\chi}_e(z)\hat{\rho}(\alpha_e z)\alpha^*_e N_0(\alpha_e z)\tilde{\mathcal{M}}^{e\ 0\ (2)}_{i_1...i_l,\ j_1...j_s}\left(\frac{\tilde{x}_1(x_0)}{\varepsilon}\right)\alpha_e + const.$$

Afterwards, the matrix-valued functions $\tilde{\mathcal{N}}^{e\ q}_{i_1...i_l}$, $\tilde{\mathcal{M}}^{e\ q}_{i_1...i_l,\ j_1...j_s}$, $N^{0\ q}_{i_1...i_l}$ and $M^{0\ q}_{i_1...i_l,\ j_1...j_s}$. are determined successively, in alphabetical order, by induction on

$(m+q,q)$ from (5.8.16),(5.8.17),(5.8.24),(5.8.28). Problems (5.8.16),(5.8.24),(5.8.28) for $\tilde{\mathcal{N}}^{e\ q}_{i_1\dots i_l}$ split into recursively solvable pairs of problems: for $\tilde{\mathcal{N}}^{e\ q\ (1)}_{i_1\dots i_l}\alpha_e$, consisting of the first equation (5.8.16), the s conditions

$$\sum_{e(x_0)} \alpha_e^* \Lambda_e \tilde{\mathcal{N}}^{e\ r}_{i_1\dots i_l} \alpha_e mes \beta_e|_{\tilde{x}_1(x_0)/\varepsilon} = \overline{\Psi^r_{\mathcal{N},\ i_1\dots i_l}}\left(\frac{x_0}{\varepsilon}\right)\alpha_e \tag{5.8.29}$$

(we denote by $\bar{A}$the first s rows of the matrix A) and

$$\alpha^*_{e_1}\left(\tilde{\mathcal{N}}^{e_1\ r\ (1)}_{i_1\dots i_l} + \lambda^{(2)}_{e_1}\right)\alpha_{e_1}|_{\tilde{x}_1(x_0)/\varepsilon} = \alpha^*_{e_2}\left(\tilde{\mathcal{N}}^{e_2\ r\ (1)}_{i_1\dots i_l} + \lambda^{(2)}_{e_2}\right)\alpha_{e_2}|_{\tilde{x}_1(x_0)/\varepsilon}, \tag{5.8.30}$$

where $\lambda^{(2)}_e$ is a linear $d \times s$ matrix-valued function on e whose first (and fourth, for $s = 3$) row is zero, and for $\tilde{\mathcal{N}}^{e\ q\ (2)}_{i_1\dots i_l} + \tilde{\mathcal{N}}^{e\ q\ (4)}_{i_1\dots i_l}$ (for $s = 2$ the last term is missing), consisting of the remaining equations (5.8.16) and conditions (5.8.24), as well as conditions (5.8.28). A similar split also occurs for the problem for $\tilde{\mathcal{M}}^{e\ q}_{i_1\dots i_l,\ j_1\dots j_s}\alpha_e$.

Suppose that the following Condition PF_1, analogous to the Poincaré- Friedrichs inequality, holds.

Condition PF_1. Let $\tilde{\Psi}(\tilde{\xi}_1)$ be any 1-periodic s-dimensional vector-valued function defined on $\mathcal{B}$

- which has generalized derivative along e for every $e \subset \mathcal{B}$,
- which is such that all its components, except the first, are linear on every segment e,
- which satisfies at all nodes x_0 the matching conditions $\alpha^*_{e_1}\tilde{\Psi}^{e_1} = \alpha^*_{e_2}\tilde{\Psi}^{e_2}$ for any two segments $e_1(x_0)$ and $e_2(x_0)$, having the common end-point x_0, and
- which satisfies the condition

$$\sum_e \int_{e\cap Q} \alpha_e^* \tilde{\Psi}^e(\tilde{\xi}_1) d\tilde{\xi}_1 = 0, \tag{5.8.31}$$

where the summation extends over all segments e of the periodicity cell $Q = (0,1)^s$ (here the parts of the segments e lying on the boundary of Q and differing by a period are taken only once). Then *it is assumed* that the following inequality holds:

$$\sum_e \int_{e\cap Q} (\tilde{\Psi}^e_1(\tilde{\xi}_1))^2 d\tilde{\xi}_1 \le c \sum_e \int_{e\cap Q} (d\tilde{\Psi}^e_1/d\tilde{\xi}_1(\tilde{\xi}_1))^2 d\tilde{\xi}_1$$

with a constant c depending only on $\mathcal{B}$. Here $\tilde{\Psi}^e_1$ is the first component of the vector-valued function $\tilde{\Psi}^e$.

Thus, Condition PF_1 is satisfied if for example only one node lies on the periodicity cell $\mathcal{B} \cap \mathcal{Q}$.

Lemma 5.8.2 *Problem (5.8.16) (the first equation), (5.8.29), (5.8.30) is solvable if and only if*

$$\sum_e \left(\int_{e\cap Q} mes\beta_e(-\alpha_e^* \overline{\tilde{T}^q_{i_1\dots i_l}(\mathcal{N})}\alpha_e + \overline{h^q_{i_1\dots i_l}(\mathcal{N})})d\tilde{\xi}_1 \right.$$

$$\left. + \sum_{x_0\cap Q} \overline{\Psi^q_{\mathcal{N},\ i_1\dots i_l}}(\frac{x_0}{\varepsilon})\alpha_e \right) = 0. \tag{5.8.32}$$

Here the summation in the second sum extends over all the nodes of the periodicity cell. The nodes lying on the boundary of Q and differing by a period are taken only once. The solution is determined up to an arbitrary additive constant $s \times s$ matrix.

Proof follows from the Lax-Milgram theorem.

Suppose that for $s = 3$ the following Condition PF_2, analogous to the Poincaré- Friedrichs inequality, holds.

Condition PF_2. Let $\tilde{\Psi}(\tilde{\xi}_1)$ be any three-dimensional 1-periodic vector-valued function defined on $\mathcal{B}$

- whose first component $\tilde{\Psi}_1$ has generalized derivative along e for every segment $e \subset \mathcal{B}$,

- whose second and third components $\tilde{\Psi}_k$, $k = 2, 3$, have two generalized derivatives along e for every segment $e \subset \mathcal{B}$,

- whose second and third components vanish at all nodes, and

- which satisfies at all nodes x_0 the matching conditions of the form

$$\bar{\Gamma}^*_{e_1} \begin{pmatrix} \tilde{\Psi}_1^{e_1} \\ d\tilde{\Psi}_2^{e_1}/d\tilde{\xi}_1 \\ d\tilde{\Psi}_3^{e_1}/d\tilde{\xi}_1 \end{pmatrix} = \bar{\Gamma}^*_{e_2} \begin{pmatrix} \tilde{\Psi}_1^{e_2} \\ d\tilde{\Psi}_2^{e_2}/d\tilde{\xi}_1 \\ d\tilde{\Psi}_3^{e_2}/d\tilde{\xi}_1 \end{pmatrix} \tag{5.8.33}$$

for any two segments $e_1(x_0)$ and $e_2(x_0)$, having the common end-point x_0.

Then *it is assumed* that the following inequality holds:

$$\sum_e \int_{e\cap Q} \left(\tilde{\Psi}_1^e(\tilde{\xi}_1))^2 + (\frac{d\tilde{\Psi}_2^e}{d\tilde{\xi}_1}(\tilde{\xi}_1))^2 + (\frac{d\tilde{\Psi}_3^e}{d\tilde{\xi}_1}(\tilde{\xi}_1))^2 \right) d\tilde{\xi}_1$$

$$\leq c \sum_e \int_{e\cap Q} \left(\frac{d\tilde{\Psi}_1^e}{d\tilde{\xi}_1}(\tilde{\xi}_1))^2 + (\frac{d^2\tilde{\Psi}_2^e}{d\tilde{\xi}_1^2}(\tilde{\xi}_1))^2 + (\frac{d^2\tilde{\Psi}_3^e}{d\tilde{\xi}_1{}^2}(\tilde{\xi}_1))^2 \right) d\tilde{\xi}_1$$

with a constant c depending only on $\mathcal{B}$.

We suppose also for simplicity of presentation that for any node there are three non-coplanar (if $s = 3$), respectively two non-collinear (if $s = 2$), segments e having this node as an end-point. This condition will be called the non-coplanarity condition.

Lemma 5.8.3 *If $s = 3$ and Condition PF_2 holds, then the problem for $\tilde{\mathcal{N}}^{e\ q\ (2)}_{i_1\dots i_l} + \tilde{\mathcal{N}}^{e\ q\ (4)}_{i_1\dots i_l}$ (consisting of the last three equations (5.8.16), the last three matching conditions (5.8.24), and matching conditions (5.8.28)) is also solvable.*

Analogues of Lemmas 5.8.2 and 5.8.3 also hold for problems (5.8.17),(5.8.24), (5.8.28) for $\tilde{\mathcal{M}}^{e\ q}_{i_1\ldots i_l,\ j_1\ldots j_s}$.

The solvability of the problems in Lemmas 5.8.2 and 5.8.3 is proved by means of the Riesz representation theorem for a bounded linear functional on a Hilbert space.

An analysis of the right-hand sides of problems (5.8.17),(5.8.24),(5.8.28) shows that the problem for $\tilde{\mathcal{N}}^{e\ 0\ (2)}_{i_1}$ ($\tilde{\mathcal{N}}^{e\ 0\ (4)}_{i_1}$) is homogeneous, and that $\tilde{\mathcal{N}}^{e\ 0\ (2)}_{i_1} = 0$ ($\tilde{\mathcal{N}}^{e\ 0\ (4)}_{i_1} = 0$); let us denote $\tilde{\mathcal{N}}^{e\ 0\ \alpha}_{i_1} = \tilde{\mathcal{N}}^{e\ 0\ (1)}_{i_1}\alpha_e$; then the first row $\tilde{n}^{e\ 0\ \alpha\ (1)}_{i_1}$ of this matrix-valued function satisfies the equations

$$\frac{d^2\tilde{n}^{e\ 0\ \alpha\ (1)}_{i_1}}{d\tilde{\xi}_1^2} = 0, \quad \tilde{\xi}_1 \in e, \tag{5.8.34}$$

and the matching conditions at all nodes: (5.8.30), and

$$\sum_{e(x_0)} \bar{E}\left(\gamma_e \frac{d\tilde{n}^{e\ 0\ \alpha\ (1)}_{i_1}}{d\tilde{\xi}_1} + \alpha_e^* \gamma_{ei_1} I_{1,1}\alpha_e\right) mes\beta_e = 0, \tag{5.8.34'}$$

where γ_e is a column vector with components γ_{ei_1} that are the cosines of the angles between e and the axis Ox_{i_1}, and $I_{1,1}$ is the $s \times s$ matrix whose $(1,1)-$th element is one while the others are zero.

This last condition (5.8.34') can be represented in the form

$$\sum_{e(x_0)} \bar{E}\left(\gamma_e \frac{d\tilde{n}^{e\ 0\ \alpha\ (1)}_{i_1}}{d\tilde{\xi}_1} + \gamma_{ei_1}\mathcal{G}_e\right) mes\beta_e = 0, \tag{5.8.35}$$

where $\mathcal{G}_e$ is the $s \times s$ matrix whose $(i,j)-$th element is $\gamma_{e\ i}\gamma_{e\ j}$. For the frames considered in section 8.2 in [16] we have $\tilde{n}^{e\ 0\ \alpha\ (1)}_{i_1} = 0$. The boundary layer problems for $N^{e\ 0}_{i_1}$ are solvable. We also have

$$\overline{h_{i_1i_2}} = \sum_e \gamma_{e\ i_1} \int_{e\cap Q} mes\beta_e \bar{E}\left(\gamma_e \frac{d\tilde{n}^{e\ 0\ \alpha\ (1)}_{i_2}}{d\tilde{\xi}_1} + \gamma_{ei_2}\mathcal{G}_e\right) d\tilde{\xi}_1 / \sum_e \int_{e\cap Q} mes\beta_e d\tilde{\xi}_1,$$

which coincides with the algorithm for computing the homogenized operator according to the principle of splitting of the averaged operator in section 8.2 in [16] (proposed in [134]).

Sufficient condition are also indicated there for

$$\hat{L} = \sum_{i_1,i_2=1}^{s} \overline{h_{i_1i_2}} \frac{\partial^2}{\partial x_{i_1}\partial x_{i_2}}$$

to be the operator of elasticity theory. Suppose that it is the case. Then, after all the matrix-valued functions $\tilde{\mathcal{N}}^{e\ q}_{i_1\ldots i_l}$, $\tilde{\mathcal{M}}^{e\ q}_{i_1\ldots i_l,\ j_1\ldots j_s}$, $N^{0\ q}_{i_1\ldots i_l}$ and $M^{0\ q}_{i_1\ldots i_l,\ j_1\ldots j_s}$. are constructed, we can proceed with the construction of V, the formal asymptotic solution of the system of equations (5.8.15), which we seek in the form of a regular series in ε and μ :

$$V = \sum_{p_1, p_2=0}^{\infty} \varepsilon^{p_1} \mu^{p_2} V_{p_1 p_2}(x). \tag{5.8.36}$$

where the coefficients of this series are solutions of the recursive sequence of problems

$$\hat{L} V_{p_1 p_2} = f_{p_1 p_2}(x). \tag{5.8.37}$$

where $f_{p_1 p_2}(x)$ are defined in terms of the $V_{q_1 q_2}$ with indices $(q_1, q_2) < (p_1, p_2)$ that is, $q_i \leq p_i$, $i = 1, 2$ and $q_1 \neq p_1$ or $q_2 \neq p_2$, and $f_{00} = f(x)$; thus, the above equation coincides with the homogenized one obtained in section 5.3 by means of principle of decomposition of the homogenized operator. It can be proved by induction that the $f_{p_1 p_2}(x)$ are $T-$periodic in x and that $\int_{(0,T)^s} f_{p_1 p_2}(x) dx = 0$; consequently, the following assertion holds

Lemma 5.8.4 *The sequence of problems (5.8.37) is recursively solvable in class of $T-$periodic functions $V_{q_1 q_2}$ with zero averages, that is, such that $\int_{(0,T)^s} V_{p_1 p_2}(x) dx = 0$.*

We denote by $u^{(J)}$ the truncated (partial) sum of the asymptotic series (5.8.7)-(5.8.9), (5.8.20), (5.8.36), obtained by neglecting the terms of order $O(\mu^{-2}\varepsilon^J + \mu^J)$ (in the $H^1(B_T)-$norm, $B_T = B_{\mu\varepsilon} \cap (0,T)^s$). Choosing J sufficiently large, we can show for any K that equation (5.8.1) and condition (5.8.2) are satisfied with remainders of order $O(\mu^{-2}\varepsilon^K + \varepsilon^{-1}\mu^K)$ (in the $L^2(B_T)-$norm. Hence , using the a priori estimate for problem (5.8.1),(5.8.2) given by Theorem 5.8.1, we obtain the following assertion.

Theorem 5.8.2 *If there are numbers α_1 and α_2 such that $\varepsilon = O(\mu^{\alpha_1})$ and $\mu = O(\varepsilon^{\alpha_2})$ as $\varepsilon, \mu \to 0$, then*

$$\|u - \tilde{u}^{(J)}\|_{H^1(B_T)} = O(\mu^{-2}\varepsilon^J + \mu^J),$$

where $\tilde{u}^{(J)} = u^{(J)} - < u^{(J)} >_T$, u is the solution of problem (5.8.1),(5.8.2), and

$$< u^{(J)} >_T = 0, \qquad < . >_T = \int_{B_T} .dx.$$

Thus, (5.8.7)-(5.8.9), (5.8.20), (5.8.36) is a formal asymptotic solution of problem (5.8.1),(5.8.2).

Remark 5.8.1 From the structure of the formal asymptotic solution (5.8.7)-(5.8.9), (5.8.20), (5.8.36) it follows that, as $\varepsilon, \mu \to 0$, the solution does not always converge in the norm

$$\|.\|_{H^1(B_T)} / \sqrt{mes B_T}$$

to a function V_{00} independent of ε and μ; this convergence happens only if $\varepsilon/\mu \to 0$. For $\varepsilon/\mu = const$ or $\varepsilon/\mu \to +\infty$ this kind of convergence, generally speaking, does not occur. However, for any relationship between ε and μ the function V_{00} satisfies a homogenized equation of the form (5.8.37). The main term of the asymptotic expansion coincides with that of section 5.6 where the functions $\mathcal{N}_{i_1 \ldots i_l}$ have to be extended to $B_{\varepsilon\mu}$ as constants in every hyperplane orthogonal to the segments e inside of sections and as constants equal to the value $\mathcal{N}_{i_1\ldots i_l}(\xi_0)$ the value in the nodal domain of a node ξ_0.

The asymptotic analysis of the elasticity equation in lattice structures by means of the two-scale convergence technique was developed in [67] and it gives the analogous main term.

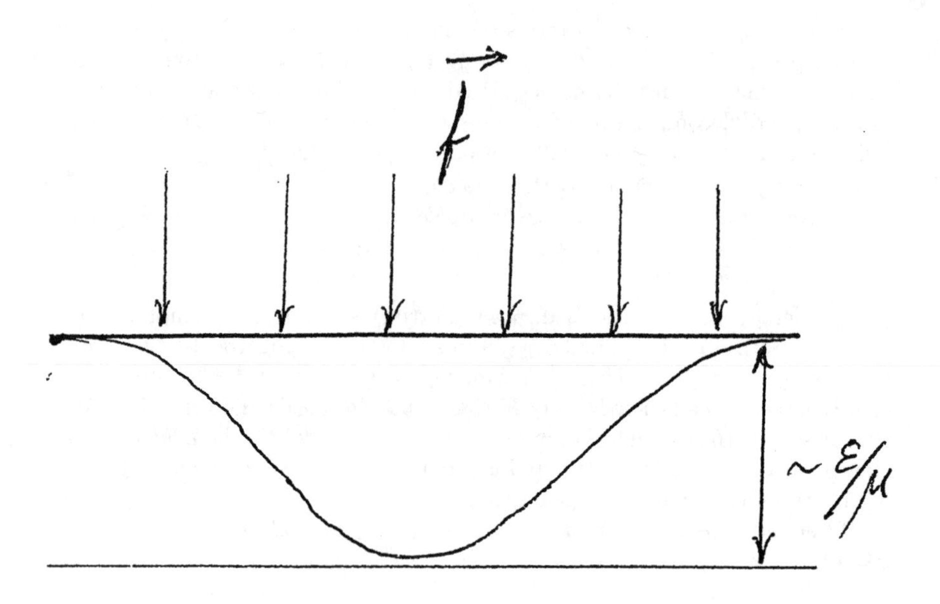

Figure 5.8.1. Magnitude of the micro-fluctuations of the displacement field.

5.9 Random coefficients on a lattice

The Poisson's equation with random coefficients set in a lattice type domain of a small measure is considered. Such domains simulate the system of fissures and depend on two small parameters: the ratio of the period of the system to the characteristic size of the domain and the ratio of the width of a fissure to the period of the system. The asymptotic analysis is developed.

5.9.1 The Simplest Lattice. The main result.

Consider the simplest model of lattice-structures: two-dimensional rectangular lattice as in section 5.2.

Let G be a domain with the boundary $\partial G \in C^\infty$ which is independent of ε and μ.

Consider a lattice $B_{\varepsilon,\mu}$ and for each strip $B_k^j = \{(x_1, x_2) \in I\!R^2 \mid \mid x_{3-j} - k\varepsilon \mid < \varepsilon\mu/2\}, j = 1, 2$ we associate a random-valued constant $(2 \times 2)-$matrix $A_{B_k^j}$, independent of ε and μ. All $A_{B_k^j}$ are independent in aggregate and have the same discrete distribution:

$$P\{A_{B_k^j} = A^{(s)}\} = p_s, \quad s = 1, ..., r,$$

where

$$\sum_{s=1}^{r} p_s = 1, \quad A^{(s)} = (A^{(s)})^T > 0.$$

Let $a_{ij}^{(s)}$ be the elements of the constant matrices $A^{(s)}$.

Let Π be an intersection of the strips:

$$\Pi = \cup_{k,j=-\infty}^{+\infty} (B_k^1 \cap B_j^2),$$

and $A^{(0)}$ be a fixed constant matrix such that $A^{(0)} = (A^{(0)})^T > 0$. We pose $A = A^{(0)}$ on the set Π and $A = A_{B_k^j}$ in each strip $B_k^j \backslash \Pi$ without Π.

Let G be a domain with the boundary $\partial G \in C^\infty$ which is independent of ε and μ, $f \in C^1(\bar{G})$. Consider the problem with random coefficients

$$-div\ (\ A\ grad\ u_{\varepsilon,\mu}) = f(x)\ ,\ for\ x \in B_{\varepsilon,\mu} \cap G \tag{5.9.1}$$

$$(\ A\ grad\ u_{\varepsilon,\mu}\ ,\ n\ ) = 0,\ for\ x \in \partial B_{\varepsilon,\mu} \cap G, \tag{5.9.2}$$

$$u_{\varepsilon,\mu} = 0,\ for\ x \in \bar{B}_{\varepsilon,\mu} \cap \partial G. \tag{5.9.3}$$

Problems (5.9.1)-(5.9.3) simulate a problem of permeability of a fissured rock filled with porous substance , with A being the permeability tensor of the substance in the fissures , $u_{\varepsilon,\mu}$ is a microscopic pressure, and u_0 is a macroscopic pressure. As it was just noticed in section 5.2, numerical solution of problems

(5.9.1)-(5.9.3), with $\varepsilon << 1$, $\mu << 1$, is very difficult since the step size of the grid must have an order much less than ε. The application of the standard homogenization procedure is also impeded, since the problem on a cell depends on the small parameter μ, and in order to solve it numerically, we must select the step size of the grid to be much less than μ. Hence, an asymptotic investigation of the problem is needed.

Introduce the notation $\bar{A} = (\bar{a}_{ij})$,

$$\bar{a}_{11} = \sum_{s=1}^{r} p_s 0.5(a_{11}^{(s)} - a_{12}^{(s)}(a_{22}^{(s)})^{-1}a_{21}^{(s)}),\ \bar{a}_{12} = 0,$$

$$\bar{a}_{22} = \sum_{s=1}^{r} p_s 0.5(a_{22}^{(s)} - a_{21}^{(s)}(a_{11}^{(s)})^{-1}a_{12}^{(s)}),\ \bar{a}_{21} = 0.$$

Let u_0 be a solution of the homogenized problem

$$-div\ (\ \bar{A}\ grad\ u_0)\ =\ f(x)\ ,\ x \in G,\quad u_0|_{\partial G}\ =\ 0. \tag{5.9.4}$$

This solution exists, it is unique and it belongs to the space $C^3(\bar{G})$. Then the following theorem holds true.

Theorem 5.9.1 *For each $\delta \in (0,1)$ the estimate takes place*

$$P\{\frac{\|u_{\varepsilon,\mu} - u_0\|_{L^2(B_{\varepsilon,\mu}\cap G)}}{\sqrt{mes(B_{\varepsilon,\mu}\cap G)}}\ >\ (\sqrt{\varepsilon}+\sqrt{\mu})^{1-\delta}\} \le c\varepsilon^{\delta}.$$

The L-convergence result for constant coefficients was formulated and proved in above . Theorem 5.9.1 is a probabilistic version of L-convergence result (see [134],[135],[153]).

5.9.2 Proof of theorem 5.9.1

Let us introduce the function $u^{(1)}_{\varepsilon,\mu}$ as in section 5.2:

$$u^{(1)}_{\varepsilon,\mu}(x) = \begin{cases} u_0 + (-a_{22}^{-1}a_{21}\frac{\partial u_0}{\partial x_1} - \frac{\partial u_0}{\partial x_2})(x_2 - k\varepsilon)\rho(\frac{x_1}{\varepsilon\mu})\psi(\frac{x}{\varepsilon}) & \text{for } x \in B_k^1 \cap G, \\ u_0 + (-a_{11}^{-1}a_{12}\frac{\partial u_0}{\partial x_2} - \frac{\partial u_0}{\partial x_1})(x_1 - k\varepsilon)\rho(\frac{x_2}{\varepsilon\mu})\psi(\frac{x}{\varepsilon}) & \text{for } x \in B_k^2 \cap G, \end{cases}$$

where $B_k^j = \{|x_{3-j} - k\varepsilon| < \frac{\varepsilon\mu}{2}\}, \rho(\frac{x_j}{\varepsilon\mu})$ is an $\varepsilon-$ periodic function coinciding on the period $[-\varepsilon/2, \varepsilon/2]$ with the function $\bar{\rho}(\frac{x_j}{\varepsilon\mu})$, and

$$\bar{\rho}(t) = \begin{cases} 0 & \text{for } |t| \le 3/4, \\ 1 & \text{for } |t| \ge 1, \end{cases}$$

$\bar{\rho}(t) \in C^1$,

$$\psi(\frac{x}{\varepsilon}) = \begin{cases} 0 & \text{in } \varepsilon- \text{ vicinity of } \partial G, \\ 1 & \text{out of } 2\varepsilon- \text{ vicinity of } \partial G, \end{cases}$$

$|\nabla\psi(\xi)| < c_0, \quad 0 \le \psi \le 1, \quad \psi \in C^1, \quad c_0$ is a positive constant.

Our goal is to prove the estimate of the FL-norm of the difference $u_{\varepsilon,\mu} - u^{(1)}_{\varepsilon,\mu}$:

$$P\{\|u_{\varepsilon,\mu} - u^{(1)}_{\varepsilon,\mu}\|_{FL(B_{\varepsilon,\mu}\cap G)} > (\sqrt{\varepsilon} + \sqrt{\mu})^{1-\delta}\} \le C\varepsilon^\delta.$$

Denote for every $B_k^j = \{ \mid x_{3-j} - k\varepsilon \mid < \varepsilon\mu/2 , x_j \in R\}, \ j = 1, 2$

$$N_1(x, \xi_{3-j}) = (-a_{3-j,3-j}^{-1}a_{3-j,j}\frac{\partial u_0}{\partial x_j} - \frac{\partial u_0}{\partial x_{3-j}})(\xi_{3-j} - k),$$

then

$$\nabla u^{(1)}_{\varepsilon,\mu}(x) =$$

$$\nabla_x u_0 + \nabla_\xi(N_1(x, \xi_{3-j})\rho(\xi_j)\psi(\xi)) + \varepsilon\nabla_x(N_1(x, \xi_{3-j})\rho(\xi_j)\psi(\xi)) =$$

$$\nabla_x u_0 + \nabla_\xi N_1(x, \xi_{3-j}) + R_1,$$

where

$$R_1 = \nabla_\xi(N_1(x, \xi_{3-j}) \ (\rho(\xi_j)\psi(\xi) - 1)) + \varepsilon\nabla_x(N_1(x, \xi_{3-j})\rho(\xi_j)\psi(\xi)),$$

$$\|R_1\|^2_{L^2(B_{\varepsilon,\mu}\cap G)} = O(\varepsilon + \mu)mes(B_{\varepsilon,\mu} \cap G)$$

because the measure of the support of the first term is $O(\varepsilon + \mu)mes(B_{\varepsilon,\mu} \cap G)$, and it is bounded and the second term has a value $O(\varepsilon)$.

Denote $\tilde{A}_i(\nabla u) = \sum_{r=1}^2 a_{i,r}\frac{\partial u}{\partial x_r}$ and estimate the integral

$$I = \int_{B_{\varepsilon,\mu}\cap G} \tilde{A}_i(\nabla_x u^{(1)})\frac{\partial\varphi}{\partial x_i}dx =$$

$$\sum_{j=1}^{2}\int_{B^j_{\varepsilon,\mu}\cap G} A_i(\nabla_x u_0 + \nabla_\xi N_1)\frac{\partial \varphi}{\partial x_i}dx \ + \ \delta_1, \tag{5.9.5}$$

where $A_i(\nabla u) \ = \ \sum_{r=1}^{2} a^{j,k}_{i,r}\frac{\partial u}{\partial x_r}$, and $a^{j,k}_{i,r}$ are the elements of the random matrices $A_{B^j_k}$ in each strip B^j_k, and

$$|\delta_1| \ = \ O(\sqrt{\varepsilon}+\sqrt{\mu})\sqrt{mes(B_{\varepsilon,\mu}\cap G)}\|\varphi\|_{H^1(B_{\varepsilon,\mu}\cap G)}.$$

The integral

$$I_j \ = \ \int_{B^j_{\varepsilon,\mu}\cap G}\sum_{i=1}^{2} A_i(\nabla_x u_0 + \nabla_\xi N_1)\frac{\partial \varphi}{\partial x_i}dx \ =$$

$$- \int_{B^j_{\varepsilon,\mu}\cap G}\sum_{i=1}^{2}\frac{\partial}{\partial x_i}A_i(\nabla_x u_0 + \nabla_\xi N_1)|_{\xi=x/\varepsilon}\varphi dx,$$

because $A_{3-j}(\nabla_x u_0 + \nabla_\xi N_1) \ = \ 0$ in $B^j_{\varepsilon,\mu}$.

Therefore

$$I_j \ = \ -\int_{B^j_{\varepsilon,\mu}\cap G}\frac{\partial}{\partial x_j}\hat{A}_j(\nabla_x v)\varphi dx,$$

where

$$\hat{A}_j(\nabla_x u_0) \ = \ A_j(\nabla_x u_0 + \nabla_\xi N_1) \ = \ (a^{kj}_{jj} - a^{kj}_{j,3-j}(a^{kj}_{3-j,3-j})^{-1}a^{kj}_{3-j,j})\frac{\partial u_0}{\partial x_j}.$$

Let $\tilde{\xi}$ be a random variable with the mathematical expectation $E\xi$ and the variance $D\tilde{\xi}$ then for each real $\eta \in (0,1)$ the Tchebyshev's inequality takes place

$$P\{|\tilde{\xi} - E\tilde{\xi}| \geq \sqrt{D\tilde{\xi}/\eta}\} \leq \eta.$$

Then for $\tilde{\xi} = \sum_{j=1}^{2} I_j$ we obtain

$$P\{ \ |\sum_{j=1}^{2} I_j - \sum_{j=1}^{2} EI_j| \geq \sqrt{\sum_{j=1}^{2} DI_j/\eta} \ \ \} \leq \eta,$$

where

$$DI_j \ = \ E\Big(\int_{B^j_{\varepsilon,\mu}\cap G}(\frac{\partial}{\partial x_j}\hat{A}_j(\nabla_x v) \ - \ \frac{\partial}{\partial x_j}E\hat{A}_j(\nabla_x v))\varphi dx\Big)^2 \ =$$

$$\sum_{k=-\infty}^{+\infty} E\Big(\int_{B^j_k\cap G}(\frac{\partial}{\partial x_j}\hat{A}_j(\nabla_x v) \ - \ \frac{\partial}{\partial x_j}E\hat{A}_j(\nabla_x v))\varphi dx\Big)^2 .$$

Applying Cauchy-Bunyakowskii-Schwartz inequality we obtain that the last sum is not more than

$$\sum_{k=-\infty}^{+\infty} E\Big(\int_{B_k^j\cap G}(\frac{\partial}{\partial x_j}\hat{A}_j(\nabla_x v) \ - \ \frac{\partial}{\partial x_j}E\hat{A}_j(\nabla_x v))^2 dx \int_{B_k^j\cap G}\varphi^2 dx\Big) \leq$$

$$sup_{x\in G}\Big(\frac{\partial}{\partial x_j}\hat{A}_j(\nabla_x v) \ - \ \frac{\partial}{\partial x_j}E\hat{A}_j(\nabla_x v))\Big)^2 sup_{j,k} mes(B_k^j\cap G)\|\varphi\|^2_{L^2(B^j_{\varepsilon,\mu}\cap G)} \ \leq$$

$$C(\mu\varepsilon)\|\varphi\|^2_{L^2(B_{\varepsilon,\mu}\cap G)}.$$

Here C is a positive constant independent of ε, μ. Thus

$$P\{|\sum_{j=1}^2 I_j - \sum_{j=1}^2 EI_j| \geq \sqrt{2C(\mu\varepsilon)\|\varphi\|^2_{L^2(B_{\varepsilon,\mu}\cap G)}/\eta}\} \leq \eta.$$

Taking

$$\eta \ = \ \frac{2C(\mu\varepsilon)}{mes(B_{\varepsilon,\mu}\cap G)(\sqrt{\varepsilon}+\sqrt{\mu})^{2(1-\delta)}}$$

we obtain the inequality

$$P\{|\sum_{j=1}^2 I_j - \sum_{j=1}^2 EI_j| \geq \sqrt{mes(B_{\varepsilon,\mu}\cap G)}(\sqrt{\varepsilon}+\sqrt{\mu})^{1-\delta}\|\varphi\|_{H^1(B_{\varepsilon,\mu}\cap G)}\} \leq$$

$$\frac{C_1\varepsilon}{(\sqrt{\varepsilon}+\sqrt{\mu})^{2(1-\delta)}} \ \leq \ C_1\varepsilon^{\delta}, \tag{5.9.6}$$

where $C_1 \geq 2C\mu \ / \ mes(B_{\varepsilon,\mu}\cap G)$.

Consider the intersections of squares $Q_{k_1k_2} \ = \ \{ \ | \ x_i - k_i\varepsilon \ | \ < \ \varepsilon/2 \ , \ i = 1,2\}$, with the domain $B^j_{\varepsilon,\mu}\cap G$. We obtain the equality

$$\sum_{j=1}^2 EI_j \ = \ -\sum_{j=1}^2\int_{B^j_{\varepsilon,\mu}\cap G}\frac{\partial}{\partial x_j}E\hat{A}_j(\nabla_x u_0)\varphi dx \ =$$

$$- \sum_{k_1,k_2:Q_{k_1k_2}\subset G}\sum_{j=1}^2\int_{B^j_{\varepsilon,\mu}\cap Q_{k_1k_2}}\frac{\partial}{\partial x_j}E\hat{A}_j(\nabla_x u_0)\varphi dx \ + \ \delta_2,$$

where

$$|\delta_2| \ = \ O(\sqrt{\varepsilon}+\sqrt{\mu})\sqrt{mes(B_{\varepsilon,\mu}\cap G)}\|\varphi\|_{H^1(B_{\varepsilon,\mu}\cap G)}.$$

Introduce notation

$$\langle\varphi\rangle_{k_1k_2} \ = \ \frac{\int_{B_{\varepsilon,\mu}\cap Q_{k_1k_2}}\varphi dx}{mes(B_{\varepsilon,\mu}\cap Q_{k_1k_2})},$$

then the following equality is valid:

$$\varphi = \langle\varphi\rangle_{k_1k_2} + (\varphi - \langle\varphi\rangle_{k_1k_2}), \tag{5.9.7}$$

and from Lemma 5.2.4 we obtain the estimate

$$\|\varphi - \langle\varphi\rangle_{k_1k_2}\|_{L^2(B_{\varepsilon,\mu}\cap Q_{k_1k_2})} \le C\varepsilon\|\nabla\varphi\|_{L^2(B_{\varepsilon,\mu}\cap Q_{k_1k_2})},$$

where C does not depend on ε, μ. Therefore

$$\sum_{j=1}^{2} EI_j = - \sum_{k_1,k_2:Q_{k_1k_2}\subset G} \sum_{j=1}^{2} \int_{B^j_{\varepsilon,\mu}\cap Q_{k_1k_2}} \frac{\partial}{\partial x_j} E\hat{A}_j(\nabla_x u_0)dx \ \langle\varphi\rangle_{k_1k_2} +$$

$$\delta_2 + \delta_3,$$

where

$$|\delta_3| = O(\sum_{k_1,k_2:Q_{k_1k_2}\subset G} \sqrt{mes(B_{\varepsilon,\mu}\cap Q_{k_1k_2})}\|\varphi - \langle\varphi\rangle_{k_1k_2}\|_{L^2(B_{\varepsilon,\mu}\cap Q_{k_1k_2})}) =$$

$$O(\varepsilon^2\sqrt{\mu} \sum_{k_1,k_2:Q_{k_1k_2}\subset G} \|\nabla\varphi\|_{L^2(B_{\varepsilon,\mu}\cap Q_{k_1k_2})}.$$

From Cauchy-Bunyakovskii-Schwartz inequality for sums we obtain then, that

$$|\delta_3| = O(\varepsilon^2\sqrt{\mu}\sqrt{\varepsilon^{-2} \sum_{k_1,k_2:Q_{k_1k_2}\subset G} \|\nabla\varphi\|^2_{L^2(B_{\varepsilon,\mu}\cap Q_{k_1k_2})}}),$$

and therefore

$$|\delta_3| = O(\varepsilon\sqrt{\mu}\|\varphi\|_{H^1(B_{\varepsilon,\mu}\cap G)}) = O(\varepsilon)\sqrt{mes(B_{\varepsilon,\mu}\cap G)}\|\varphi\|_{H^1(B_{\varepsilon,\mu}\cap G)}.$$

Thus

$$\sum_{j=1}^{2} EI_j =$$

$$- \sum_{k_1,k_2:Q_{k_1k_2}\subset G} \int_{B_{\varepsilon,\mu}\cap Q_{k_1k_2}} (1/2)\sum_{j=1}^{2} \frac{\partial}{\partial x_j}\hat{A}_j(\nabla_x u_0)|_{x_1=k_1\varepsilon,x_2=k_2\varepsilon}dx \ \langle\varphi\rangle_{k_1k_2} +$$

$$\delta_2 + \delta_3 + \delta_4, \tag{5.9.8}$$

where

$$|\delta_4| = O((\varepsilon+\mu) \sum_{k_1,k_2:Q_{k_1k_2}\subset G} mes(B_{\varepsilon,\mu}\cap Q_{k_1k_2})|\langle\varphi\rangle_{k_1k_2}|) =$$

$$O((\varepsilon+\mu)\int_{B_{\varepsilon,\mu}\cap G}|\varphi|dx) \;=\; O((\varepsilon+\mu)\sqrt{mes(B_{\varepsilon,\mu}\cap G)}\|\varphi\|_{H^1(B_{\varepsilon,\mu}\cap G)}).$$

Now we use once more the representation (5.9.7) and Taylor's expansion for

$$\frac{\partial}{\partial x_j}E\hat{A}_j(\nabla_x u_0)$$

and obtain that

$$\sum_{j=1}^{2} EI_j \;=\; -\int_{B_{\varepsilon,\mu}\cap G}(1/2)\sum_{j=1}^{2}\frac{\partial}{\partial x_j}E\hat{A}_j(\nabla_x u_0)\varphi dx \;+\; \delta_5 \;=,$$

$$-\int_{B_{\varepsilon,\mu}\cap G} div(\hat{A}\nabla_x u_0)\varphi dx \;+\; \delta_5,$$

where

$$|\delta_5| \;=\; O(\sqrt{\varepsilon}+\sqrt{\mu})\sqrt{mes(B_{\varepsilon,\mu}\cap G)}\|\varphi\|_{H^1(B_{\varepsilon,\mu}\cap G)}.$$

From the homogenized equation (5.9.4) we can change this integral by

$$\sum_{j=1}^{2} EI_j \;=\; \int_{B_{\varepsilon,\mu}\cap G} f\varphi dx \;+\; \delta_5,$$

and therefore

$$\sum_{j=1}^{2} EI_j \;-\; \int_{B_{\varepsilon,\mu}\cap G} f\varphi dx \;=$$

$$\sum_{j=1}^{2} EI_j \;-\; \int_{B_{\varepsilon,\mu}\cap G}(A\nabla(u_{\varepsilon\mu})\;,\;\nabla\varphi)dx \;=$$

$$O(\sqrt{\varepsilon}+\sqrt{\mu})\sqrt{mes(B_{\varepsilon,\mu}\cap G)}\|\varphi\|_{H^1(B_{\varepsilon,\mu}\cap G)}.$$

Applying now the estimates (5.9.5), (5.9.6) we obtain:

$$P\{\frac{|I \;-\; \int_{B_{\varepsilon,\mu}\cap G}(A\nabla(u_{\varepsilon\mu})\;,\;\nabla\varphi)dx|}{\sqrt{mes(B_{\varepsilon,\mu}\cap G)}\|\varphi\|_{H^1(B_{\varepsilon,\mu}\cap G)}} \;>\; (\sqrt{\varepsilon}+\sqrt{\mu})^{1-\delta}\} \le C\varepsilon^{\delta},$$

where C does not depend on ε,μ,φ i.e.

$$P\{\|u_{\varepsilon,\mu}-u^{(1)}_{\varepsilon,\mu}\|_{FL(B_{\varepsilon,\mu}\cap G)} \;>\; (\sqrt{\varepsilon}+\sqrt{\mu})^{1-\delta}\} \le C\varepsilon^{\delta}.$$

Applying now Lemma 5.2.2 and the estimate of section 5.2

$$\frac{\|u^{(1)}_{\varepsilon,\mu} \;-\; u_0\|_{L^2(B_{\varepsilon,\mu}\cap G)}}{\sqrt{mes(B_{\varepsilon,\mu}\cap G)}} \;=\; O(\sqrt{\varepsilon}+\sqrt{\mu})$$

we get the final estimate

$$P\{\frac{\|u_{\varepsilon,\mu} - u_0\|_{L^2(B_{\varepsilon,\mu}\cap G)}}{\sqrt{mes(B_{\varepsilon,\mu} \cap G)}} > (\sqrt{\varepsilon} + \sqrt{\mu})^{1-\delta}\} \leq c\varepsilon^{\delta}.$$

So the theorem is proved.

The same consideration is developed for more general types of lattice structures, modelling the systems of fissures oriented in a lot of directions, capillary systems with varying width etc. [135]. The Dirichlet's problem for Poisson's equation set in non-periodic lattice structure is considered in [92].

5.10 Bibliographical Remark

The lattice-like structures (or skeletal structures) were introduced in [134], where the asymptotic analysis was developed and the principle of splitting the homogenized operator was formulated and justified. The complete asymptotic expansion for a conductivity problem was obtained in [136]. Random structures were studied in [135],[153]. Later lattice-like structures were studied by D.Cioranescu and J. Saint Jean Paulin [46-48]. They considered the result of the passage to limit as the small parameters asymptotic analysis of the lattice-like structures tend to zero in different orders. The L-convergence was introduced in [16] and [155]. The complete asymptotic expansion of a solution to the elasticity equations set in a lattice-like structure was constructed in [145],[148] where the essentially different behavior of the solution was detected in case when the ratio μ/ε is small, finite or great. This result confirms the non-commutativity of the consecutive passages to limit in μ and ε for the elasticity equation described in [44] and [62]; D.Cioranescu and J. Saint Jean Paulin study there the different orders of passage to the limit in so called gridworks , tall structures and honeycomb structures depending on three small parameters μ, ε and e (see Figures 5.9.1 and 5.9.2).

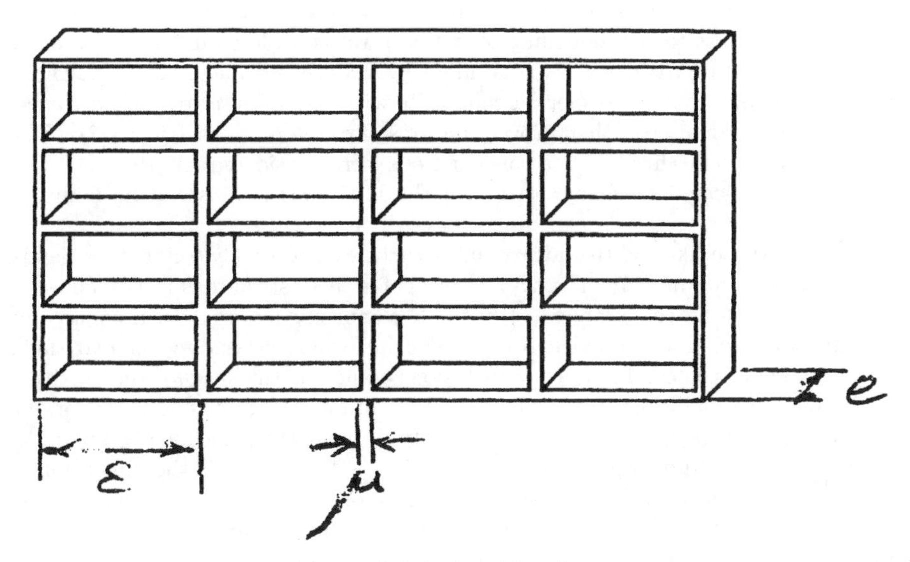

Figure 5.9.1. A gridwork

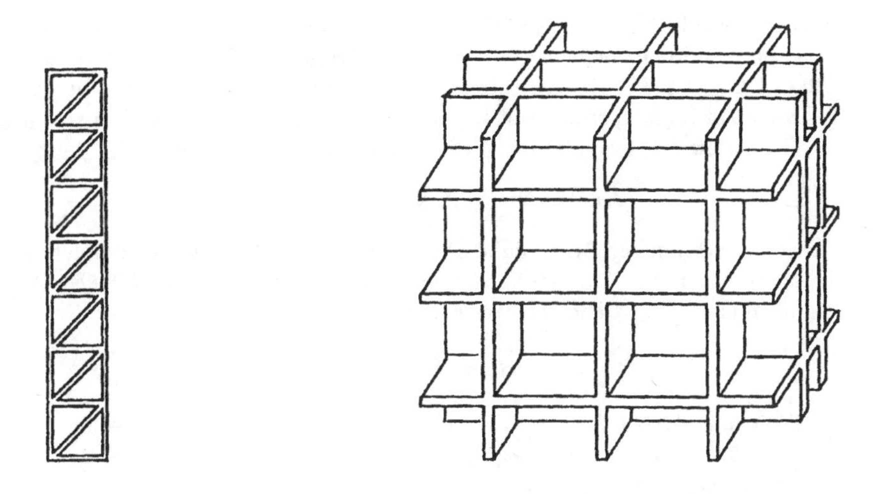

Figure 5.9.2. A tall structure and a honeycomb structure.

Nowadays the researches still keep their interest to the lattice-like structures (see, for example, [169] where the non-linear elasticity is considered and justified) and also [66], [67] where the "regularized" (by some artificial potential) elasticity equation in the rectangular lattice is studied by means of the two-scale convergence.

Let us discuss now some models close to the lattice-like structures. The equations on a net were considered in [104]; the asymptotic passage in the difference operators (close to the modelling of lattices) was considered in [126]. Recently the network approximation was studied by L.Berlyand and A.G.Kolpakov in [26] for a high contrast dispersed composite with a small inter-particle distance. The error estimate was then proved in terms of the Voronoy metric in [27]. The fractal-like hierarchical lattice constructions were considered in [79].

The structural mechanics study of the lattice structures can be found in [171].

The lattice-like and the honeycomb structures in a weakly conductive matrix were considered in [17]. The typical result for these structures is as follows.

Consider the cell problem [16] for a composite material with square (or cubic in three-dimensional case) periodic inclusions depending on two small parameters : the conductivity of inclusions δ and the thickness of the partition between the inclusions μ. In this case we obtain the explicit formula for the main term of the coefficient of conductivity $\widehat{K}$ using the principle of splitting of the homogenized operator [16] : $\widehat{K} = \delta + \mu(s-1)$, where s is the dimension of the space (2 or 3). Let us prove the estimate of the error of this formula. The cases of a finite and of a large conductivity of inclusions are also considered.

Consider the following cell problem

$$div(K(x)grad(N_1 + x_1)) = 0, \; x \in I\!R^s$$

for the 1-periodic function N_1 of the variables $x = (x_1, \ldots, x_s)$ with the piecewise-constant 1-periodic coefficient K, taking the values : δ inside the cube $\{|x_i - 1/2| < 1/2 - \mu/2, i = 1, \ldots, s\}$, and 1 on the other part of the unit cube $[0,1]^s$. The values of $K(x)$ in each point $x \in I\!R^s$ are obtained by the periodic extension.

The variational formulation of the cell problem has a form :

$$\langle (K(x)\nabla(N_1 + x_1), \nabla\varphi) \rangle = 0, \; \forall \varphi \in H^1_{per},$$

where $\langle . \rangle$ is the integral over cube $Q = [-1/2, 1/2]^s$, H^1_{per} is the completion of the space of smooth 1-periodic functions with respect to norm $H^1(Q)$, N_1 is sought in class H^1_{per}. The effective conductivity is defined as follows:

$$\widehat{K} = \langle K(x)\partial(N_1 + x_1)/\partial x_1 \rangle.$$

Theorem 4.10.1 $\widehat{K} = \delta + (s-1)\mu + O(\mu(\delta + \mu))$ *if* $\mu \to 0, \; \delta \to 0,$
$\widehat{K} = \delta + O(\mu)$ *if* $\mu \to 0, \; \delta = const,$
$\widehat{K} = \delta + O(\delta^2\mu)$ *if* $\mu \to 0, \delta \to +\infty, \; \delta\mu \to 0,$
$\widehat{K} = \delta/(1 + \delta\mu) + O(1)$ *if* $\mu \to 0, \; \delta \to +\infty, \; \delta\mu = const,$
$\widehat{K} = 1/\mu + O((\delta\mu^2)^{-1} + 1)$ *if* $\mu \to 0, \; \delta \to +\infty, \; \delta\mu \to +\infty.$

Proof.

Denote $\langle \rangle_i$ the integral $\int_{-1/2}^{1/2} dx_i$. In [16] Chapter 4, section 4.2 (and earlier in the non-published paper by N. S. Bakhvalov) the estimate has been obtained :

$$\langle\langle\langle K^{-1}\rangle_1^{-1}\rangle_2\rangle_3 \le \widehat{K} \le \langle\langle\langle K\rangle_2\rangle_3^{-1}\rangle_1^{-1}.$$

For $s = 3$ we calculate

$$\langle\langle\langle K^{-1}\rangle_1^{-1}\rangle_2\rangle_3 = ((\delta^{-1}(1-\mu)+\mu)^{-1}(1-\mu)+\mu)(1-\mu)+\mu,$$

and

$$\langle\langle\langle K\rangle_2\rangle_3^{-1}\rangle_1^{-1} = (((\delta(1-\mu)+\mu)(1-\mu)+\mu)^{-1}(1-\mu)+\mu)^{-1}.$$

We see that the right side and left side of this bilateral estimate have the same asymptotic leading term which is declared in the theorem. Therefore $\widehat{K}$ has also the same asymptotic leading term. The estimate in the two-dimensional case is proved in a same way.

The two-scale convergence approach on some domains of degenerating measure was applied to the "networks", i.e. the composite materials with some degenerating concentration of a very rigid component [33].

5.11 Appendices

5.A1. Appendix 1: the Poincaré and the Friedrichs inequalities for lattices

Let $\widetilde{\beta}^{\mu}_{1(0,d_1)}$, β^{μ}_3, $\widetilde{\beta}^{\mu}_{2(0,d_2)}$ be sets in lemma 4.A2.4 and let β be its union.

Lemma 5.A1.1 *In assumptions of lemma 4.A2.4* $\mu^{s-1}|\{u\}^{d_1}_{1,\mu}-\{u\}^{d_2}_{2,\mu}|^2 \leq C_{10}\int_\beta (\nabla u)^2 dx$, *where* $\{u\}^{d_i}_{i,\mu} = \dfrac{1}{mes\,\widetilde{\beta}^{\mu}_{i(0,d_i)}}\displaystyle\int_{\widetilde{\beta}^{\mu}_{i(0,d_i)}} u\,dx$ *and* C_{10} *does not depend on* μ.

Proof.

$|\{u\}^{d_1}_{1,\mu}-\{u\}^{d_2}_{2,\mu}| \leq |\{u\}^{d_1}_{1,\mu}-\{u\}^{h_1\mu}_{1,\mu}| + |\{u\}^{h_1\mu}_{1,\mu}-\{u\}_{3,\mu}| + |\{u\}_{3,\mu}-\{u\}^{h_2\mu}_{2,\mu}| + |\{u\}^{h_2\mu}_{2,\mu}-\{u\}^{d_2}_{2,\mu} \leq 4\sqrt{\mu^{1-s}C'_{10}\int_\beta(\nabla u)^2 dx}$, C'_{10} does not depend on μ (Lemmas 4.A2.2, 4.A2.3).
Lemma is proved.

Lemma 5.A1.2 *Let* $B_{\varepsilon\mu}$ *be a lattice and* $B_\mu = \{\xi|\epsilon\xi \in B_{\varepsilon\mu}\}$, *let in notations of Theorem 4.A2.1* $\widetilde{\mathcal{B}} = B_\mu \cap Q_N, Q_N = (0,N)^s, \beta_i, \beta_{3j}$ *do not depend on* N, *and let* $\widetilde{\beta}^{\mu}_{1(0,d_1)}$, $\widetilde{\beta}^{\mu(\bar K)}_{1(0,d_1)} \subset \widetilde{\mathcal{B}}$, *where* $\widetilde{\beta}^{\mu(\bar K)}_{1(0,d_1)} = \{x \in I\!R^s | x-\bar K \in \widetilde{\beta}^{\mu}_{1(0,d_1)}\}$, $\bar K = (K_1,\ldots,K_s), 0 \leq K_i \leq N$.
Then

$$\mu^{s-1}|\{u\}^{d_1}_{1,\mu}-\{u\}^{d_1(\bar K)}_{1,\mu}|^2 \leq C_{11}N\int_{\widetilde{\mathcal{B}}}(\nabla u)^2 dx,$$

where $\{u\}^{d_1(\bar K)}_{1,\mu} = \dfrac{1}{mes\,\widetilde{\beta}^{\mu}_{1(0,d_1)}}\displaystyle\int_{\widetilde{\beta}^{\mu(\bar K)}_{1(0,d_1)}} u\,dx$, C_{11} *is independent of* μ, N.

Proof.

Let $\widetilde{\beta}^{\mu}_{3,j_1}, \widetilde{\beta}^{\mu}_{i_1(0,d_{i1})}, \ldots, \widetilde{\beta}^{\mu}_{3,j_q}$ be a path (without repeating) connecting $\widetilde{\beta}^{\mu}_{1(0,d_1)}$ with $\widetilde{\beta}^{\mu(\bar K)}_{1(0,d_1)}$. Then applying lemma 5.A1.1 to every two consecutive sets $\widetilde{\beta}^{\mu}_{i_l(0,d_{il})}$ of the path separated by some set $\widetilde{\beta}^{\mu}_{3j_l}$, we have $\mu^{s-1}|\{u\}^{d_l}_{i_l,\mu}-\{u\}^{d_{l+1}}_{i_{l+1},\mu}|^2 \leq C_{10}\displaystyle\int_{\widetilde{\beta}^{\mu}_{i_l(0,d_{i_l})}\cup\widetilde{\beta}^{\mu}_{3j_l}\cup\widetilde{\beta}^{\mu}_{i_{l+1}(0,d_{i_{l+1}})}}(\nabla u)^2 dx$ with constant C_{10} from lemma 5.A1.1 which does not depend of μ, N.

Here $i_0 = 1$, $\widetilde{\beta}^{\mu}_{i_{q+1}(0,d_{i_{q+1}})} = \widetilde{\beta}^{\mu(N)}_{1(0,d_1)}$. Adding these inequalities and applying

Cauchy-Bunyakowskii-Schwartz inequality for sums we have

$$\mu^{s-1}|\{u\}_{1,\mu}^{d_1} - \{u\}_{1,\mu}^{d_1(N)}|^2 \le \mu^{s-1} q \sum_{l=1}^{q} |\{u\}_{i_l,\mu}^{d_l} - \{u\}_{i_{l+1},\mu}^{d_{l+1}}|^2 \le$$

$$\le 3C_{10} q \int_{\widetilde{\mathcal{B}}} (\nabla u)^2 dx \le C_{11} N \int_{\widetilde{\mathcal{B}}} (\nabla u)^2 dx$$

. Lemma is proved. Here the length of the path is bounded by const N.

Lemma 5.A1.3 *Let assumptions of Lemma 5.A1.2 be true and let for any $K \in \{0, 1, \dots, N-1\}^s$, set $B_K = \widetilde{\mathcal{B}} \cap \{\xi \in I\!R^s,\ \xi - K \in (0,1)^s\}$ be connected and contain $\widetilde{\beta}^{\mu(K)}_{1(0,d_1)}$; let the assumptions of Theorem 4.A2.1 with respect to set B_K (instead of set $\widetilde{\mathcal{B}}$ of Theorem 4.A2.1) be satisfied. Then*

$$\textit{1)}\quad \|u\|^2_{L^2(\widetilde{\mathcal{B}})} \le C_{12} \left(\frac{N^s}{mes \widetilde{\beta}^{\mu}_{1(0,d_1)}} \left(\int_{\widetilde{\beta}^{\mu}_{1(0,d_1)}} u\, dx \right)^2 + N^{s+1} \int_{\widetilde{\mathcal{B}}} (\nabla u)^2 dx \right);$$

2) and if $\widetilde{\mathcal{B}}_1 = \widetilde{\mathcal{B}} \cap \{\xi : (\xi_2, \dots, \xi_s) \in (0,1)^{s-1}\}$,be a connected set then

$$\|u\|^2_{L^2(\widetilde{\mathcal{B}}_1)} \le C_{12} \left(\frac{N}{mes \widetilde{\beta}^{\mu}_{1(0,d_1)}} \left(\int_{\widetilde{\beta}^{\mu}_{1(0,d_1)}} u\, dx \right)^2 + N^2 \int_{\widetilde{\mathcal{B}}_1} (\nabla U)^2 dx \right),$$

where C_{12} does not depend on μ, N.

Proof

According to Lemmas 4.A2.2 - 4.A2.4 and Lemma 5.A1.2, if $\widetilde{\beta} = \widetilde{\beta}^{\mu}_{i(0,d_i)}$ or $\widetilde{\beta} = \widetilde{\beta}^{\mu}_{3,j}$ and $\widetilde{\beta} \subset B_K$, then

$$\|u\|^2_{L^2(\widetilde{\beta})} \le C_1 \left(\frac{1}{mes \widetilde{\beta}} \left(\int_{\widetilde{\beta}} u\, dx \right)^2 + \|\nabla u\|^2_{L^2(\widetilde{\beta})} \right) \le$$

$$\le C_2 \left(\frac{1}{mes \beta^{\mu(K)}_{1,(0,d_1)}} \left(\int_{\beta^{\mu(K)}_{1(0,d_1)}} u\, dx \right)^2 + \|\nabla u\|^2_{L^2(B_K)} \right) =$$

$$= C_3 (\mu^{s-1} d_1 mes \beta_1 (\{u\}^{d_1(K)}_{1,\mu})^2 + \|\nabla u\|^2_{L^2(B_K)}) \le$$

$$\le 2C_3 (\mu^{s-1} d_1 mes \beta_1 (\{u\}^{d_1}_{1,\mu})^2 + \mu^{s-1} d_1 mes\ \beta_1 (\{u\}^{d_1}_{1,\mu} - \{u\}^{d_1(K)}_{1,\mu})^2 + \|\nabla u\|^2_{L^2(B_K)}) \le$$

$$2C_3 \left(\frac{1}{mes \widetilde{\beta}^{\mu}_{1(0,d_1)}} \times \left(\int_{\beta^{\mu}_1(0,d_1)} u\, dx \right)^2 + C_{11} N \int_{\widetilde{\mathcal{B}}} (\nabla u)^2 dx\ d_1 mes \beta_1 + \|\nabla u\|^2_{L^2(B_K)} \right),$$

where C_3, C_{12} do not depend on K, μ.

So $\|u\|^2_{L^2(B_K)} \le C_{12}\left(\frac{1}{mes\widetilde{\beta}^{\mu}_{1(0,d_1)}}\left(\int_{\widetilde{\beta}^{\mu}_{1(0,d_1)}} u\,dx\right)^2 + N\int_{\widetilde{\mathcal{B}}}(\nabla u)^2 dx\right)$;

$\|u\|^2_{L^2(\widetilde{\mathcal{B}})} \le C_{12}\left(\frac{N^s}{mes\beta^{\mu}_{1(0,d_1)}}\left(\int_{\widetilde{\beta}^{\mu}_{1(0,d_1)}} u\,dx\right)^2 + N^{s+1}\int_{\widetilde{\mathcal{B}}}(\nabla u)^2 dx\right)$

and

$$\|u\|^2_{L^2(\widetilde{\mathcal{B}}_1)} \le C_{12}\left(\frac{N}{mes\widetilde{\beta}^{\mu}_{1(0,d_1)}}\left(\int_{\widetilde{\beta}^{\mu}_{1(0,d_1)}} u\,dx\right)^2 + N^2\int_{\widetilde{\mathcal{B}}_1}(\nabla u)^2 dx\right).$$

Proof of Theorem 5.7.1

Let us take in the last lemma $\mu_r = \int_{\widetilde{\beta}^{\mu}_{1(0,d_1)}} u\,dx$.

Then,

$\|u-\mu_r\|^2_{L^2(\widetilde{\mathcal{B}})} \le C_4 N^{s+1}\int_{\widetilde{\mathcal{B}}}(\nabla u)^2 dx.$

Let us find now $\min_Z \int_{\beta}(u-Z)^2 dx$. Denote $F(Z) = \int_{\widetilde{\mathcal{B}}}(u-Z)^2 dx$. Then $F'(Z) = 2(mes\widetilde{\mathcal{B}})Z - 2\int_{\widetilde{\mathcal{B}}} u\,dx$ and $\min_Z F(Z) = F(\int_{\widetilde{\mathcal{B}}} u\,dx/mes\widetilde{\mathcal{B}}) = \int_{\widetilde{\mathcal{B}}} u^2 dx - \frac{1}{mes\widetilde{\mathcal{B}}}\left(\int_{\widetilde{\mathcal{B}}} u\,dx\right)^2$.

So it is less than $\|u-\mu_r\|^2_{L^2(\mathcal{B})} \le C_4 N^{s+1}\int_{\mathcal{B}}(\nabla u)^2 dx$. Now contracting $\widetilde{\mathcal{B}}$ $1/\epsilon$ times we obtain the estimate of Theorem 5.7.1.

Proposition 5.A1.1 *Let $u \in H^1(G\cap B_{\varepsilon,\mu}), G = (-1,1)^s$, $u = 0$ if $x \in \partial G \cap \bar{B}_{\varepsilon,\mu}$. Then $\|u\|_{L^2(G\cap B_{\varepsilon,\mu})} \le C\|\nabla u\|_{L^2(G\cap B_{\varepsilon,\mu})}$ where C is independent of ε, μ.*

Proof.

Extend u by zero on $\widetilde{G}\cap B_{\varepsilon,\mu}$, where $\widetilde{G} = (-2,2)^s$, then for all cylinders $B^{\alpha+}_{hj} \subset \widetilde{G}\backslash G$ we have $\int_{B^{\alpha+}_{hj}} u\,dx = 0$.

Applying the second assertion of Lemma 5.A1.3 to the "bar" $(-2,2)\times(0,\varepsilon)^{s-1}\cap B_{\varepsilon,\mu}$ dilated $1/\varepsilon$ times we obtain that

$$\|u\|^2_{L^2(\widetilde{G}\cap B_{\varepsilon,\mu})} \le C_{12}\left(\frac{1}{\varepsilon\, mesB^{\alpha+}_{hj}}\left(\int_{B^{\alpha+}_{hj}} u\,dx\right)^2 + \int_{\widetilde{G}\cap B_{\varepsilon,\mu}}(\nabla_x u)^2 dx\right) =$$

$$= C_{12}\int_{\widetilde{G}\cap B_{\varepsilon,\mu}}(\nabla_x u)^2 dx,$$

, where C_{12} is a constant of lemma 5.A1.3 part 2. This yields the Proposition. This assertion is valid when $u = 0$ only on one of the faces of cube G.

5.A2. Appendix 2: the Korn inequality for lattices

Consider $B_\mu = \{\varepsilon x \in B_{\varepsilon,\mu}\}$, a 1-periodic set.

The objective of this appendix is to prove the following theorem.

Theorem 5.A2.1 *For sufficiently small $\mu > 0$ there exist two positive constants q and C such that, for any 1-periodic vector valued function u defined on B_μ, $u \in H^1_{per\,1}(B_\mu)$, the inequality holds*

$$\|\nabla u\|^2_{L^2(B_\mu \cap (0,1)^s)} \le C_K E_{B_\mu \cap (0,1)^s}(u)$$

where $C_K = \mu^q C$.

After the $\frac{1}{\varepsilon}$ times contraction, we will obtain the same inequality for $B_{\varepsilon,\mu}$:

Theorem 5.A2.1' *For sufficiently small $\mu > 0$ there exist two positive constants q and C such that, for any T-periodic function u defined on $B_{\varepsilon,\mu}$, $u \in H^1_{per\,T}(B_{\varepsilon,\mu})$, the inequality holds*

$$\|\nabla u\|^2_{L^2(B_{\varepsilon,\mu} \cap (0,T)^s)} \le C_K E_{B_{\varepsilon,\mu} \cap (0,T)^s}(u)$$

where $C_K = \mu^q C$.

We assume here that for a finite sufficiently small μ this assertion is proved. (The sketch of this prove will be given further in Remark 5.A2.7).

Let $\mathcal{R}$ be the space of rigid displacements

Lemma 5.A2.1 *Let G be a domain such that there exists a constant C_K (depending on G) such that for any $u \in (H^1(G))^s$, there exists a rigid displacement $\eta \in \mathcal{R}$ such that $\|\nabla(u - \eta)\|^2_{L^2(G)} \le C_K E_G(u)$*
(i.e. the Korn inequality holds with a constant C_K). Let $G_{(a,b)}$ be a domain $G_{(a,b)} = \{x \in \mathbb{R}^s, (\frac{x_1}{a}, \frac{x_2}{b}, \ldots, \frac{x_s}{b}) \in G\}, a, b > 0$. Then for any (a_1, a_2, b_1, b_2) such that $0 < a_1 \le a_2, 0 < b_1 \le b_2$, there exist a constant $C_z > 0$, depending only on a_1, a_2, b_1, b_2 such that for any $a \in [a_1, a_2]$,
$b \in [b_1, b_2]$, for any $u \in (H^1(G_{(a,b)}))^s$, there exists a rigid displacement $\eta \in \mathcal{R}$ such that $\|\nabla(u - \eta)\|^2_{L^2(G_{(a,b)})} \le C_z C_K E_G(u)$.

Proof. Let us make the change $\hat{x} = (\frac{x_1}{a}, \frac{x_2}{b}, \ldots, \frac{x_s}{b}), \hat{u} = (au_1, bu_2, \ldots, bu_s)$

We have

$$\frac{\partial \hat{u}_i}{\partial \hat{x}_j} = \begin{cases} b^2 \frac{\partial u_i}{\partial x_j} & if\ i, j \ne 1, \\ ab \frac{\partial u_i}{\partial x_j} & if\ i = 1, j \ne 1 \text{ or } i \ne 1, j = 1, \\ a^2 \frac{\partial u_1}{\partial x_1} & if\ i = j = 1\ ; \end{cases}$$

let now $\hat{\eta} \in \mathcal{R}$, $\hat{\eta} = \sum_{\lambda=1}^{d-s} \alpha_\lambda \hat{\eta}_\lambda$, where $\hat{\eta}_1(\hat{x}) = (-\hat{x}_2, \hat{x}_1)^T$, $(s = 2)$ and

$$\hat{\eta}_1(\hat{x}) = (-\hat{x}_2, \hat{x}_1, 0)^T,\ \ \hat{\eta}_2(\hat{x}) = (-\hat{x}_3, 0, \hat{x}_1)^T,\ \ \hat{\eta}_3(\hat{x}) = (0, -\hat{x}_3, \hat{x}_2)^T,\ \ (s = 3)$$

then

$$\eta(x) =$$

$$\left(a^{-1}\hat{\eta}_1\left(\frac{x_1}{a},\frac{x_2}{b},\dots,\frac{x_s}{b}\right), b^{-2}\hat{\eta}_2\left(\frac{x_1}{a},\frac{x_2}{b},\dots,\frac{x_s}{b}\right),\dots, b^{-1}\eta_s\left(\frac{x_1}{a},\frac{x_2}{b},\dots,\frac{x_s}{b}\right)\right) =$$

$$= \begin{cases} (-\alpha_1\frac{x_2}{ab},\alpha_1\frac{x_1}{ab}) \in \mathcal{R} \quad (s=2) \\ (-\alpha_1\frac{x_2}{ab},\alpha_1\frac{x_1}{ab},0) + (-\alpha_2\frac{x_3}{ab},0,\alpha_2\frac{x_1}{ab}) + (0,\alpha_3\frac{-x_3}{b^2},\alpha_3\frac{x_2}{b^2}) \in \mathcal{R} \quad (s=3) \end{cases}$$

So we have

$$\|\nabla(u-\eta)\|^2_{L^2(G_{(a,b)})} \le ab^{s-1}(max(a,b))^4\|\nabla_{\widehat{x}}(\widehat{u}-\widehat{\eta}\|^2_{L^2(G)} \le ab^{s-1}(\max(a,b))^4$$

$$C_K E_G(\widehat{u}) \le (\max(a,b))^4 C_K \quad (\max\left(\frac{1}{a},\frac{1}{b}\right))^4 E_{G_{(a,b)}}(u)$$

with $(\max(a,b)\max\left(\frac{1}{a},\frac{1}{b}\right))^4 = (\max\left(\frac{a}{b},\frac{b}{a}\right))^4 \le \max^4\left(\frac{a_2}{b_1},\frac{b_2}{a_1}\right)$

Remark 5.A2.1 For homothetic mapping $(a=b)C_z = 1$.

Lemma 5.A2.2 *Let G_1, G_2 be two domains such that $G_1 \subset G_2$ and such that $\exists C_E > 0$ such that $\forall u \in (H^1(G_1))^s$, there exists an extension $P_E u$ over G_2 such that $P_E u \in (H^1(G_2))^s$, $P_E u = u$ in G_1, and $E_{G_2}(P_E u) \le C_E\, E_{G_1}(u)$.*
Let $G_{1(a,b)}$ and $G_{2(a,b)}$ be the same mappings of G_1 and G_2 as in Lemma 5.A2.1. Then for any (a_1,a_2,b_1,b_2) such that $0 < a_1 \le a_2, 0 < b_1 \le b_2$, there exists a constant $C_z > 0$ depending only on a_1,a_2,b_1,b_2 (the same that in Lemma 5.A2.1) such that $\forall a \in [a,a_2], b \in [b_1,b_2], \forall u \in (H^1(G_{1(a,b)}))^s$, there exists an extension $P_{E(a,b)}u$ over $G_{2(a,b)}$ such that

$$E_{G_{2(a,b)}}(P_{E(a,b)}u) \le C_z C_E E_{G_{1(a,b)}}(u).$$

Proof. Indeed making the same change as in Lemma 5.A2.1, we consider $P_E\hat{u}$, an extension of $\widehat{u}$ from G_1 over G_2, and then we make the inverse change and obtain the necessary extension $P_{E(a,b)}$ and

$$E_{G_{2(a,b)}}(P_{E(a,b)}u) \le ab^{s-1}(\max(a,b))^4 E_{G_2}(P_E\widehat{u}) \le$$
$$ab^{s-1}(\max(a,b))^4 C_E E_{G_1}(\widehat{u}) \le (\max(a,b))^4 C_E(\max\left(\tfrac{1}{a},\tfrac{1}{b}\right))^4 E_{G_1(a,b)}$$

Thus Lemma 5.A2.2 is proved with the same constant C_z as in Lemma 5.A2.1.

Lemma 5.A2.3 *Let G_1, G_2 be two domains such that $G_1 \subset G_2$, and such that there exists a positive constant C_{H^1} such that $\forall u \in H^1(G_1)$ there exists an extension $\widetilde{u} \in H^1(G_2), \widetilde{u}(x) = u(x)$ in G_1, such that $\|\nabla\widetilde{u}\|_{L^2(G_2)} \le C_{H^1}\|\nabla u\|_{L^2(G_1)}$. Assume that $\forall u \in H^1(G_1)$ there exist a rigid displacement $\eta \in \mathcal{R}$ such that*

$$\|\nabla(u-\eta)\|^2_{H^1(G_1)} \le C_K E_{G_1}(u).$$

Then there exist the extension of u from G_1 over G_2, $P_E u = \widetilde{u-\eta}+\eta$ such that $E_{G_2}(\widetilde{u-\eta}+\eta) \le C_E E_{G_1}(u)$.

Proof.

$$E_{G_2}(\widetilde{u-\eta}+\eta) = E_{G_2}(\widetilde{u-\eta}) \le \|\nabla(\widetilde{u-\eta})\|^2_{L^2(G_2)} \le (C_{H^1})^2\|\nabla(u-\eta)\|^2_{L^2(G_1)} \le$$

$$\le (C_{H^1})^2 C_K E_{G_1}(u-\eta) = (C_{H^1})^2 C_K E_{G_1}(u),$$

so, $C_E = (C_{H^1})^2 C_K$.

Remark 5.A2.2 The assertion of the Lemma 9.A2.3 is valid if $H^1(G_1)$ and $H^1(G_2)$ is replaced by subspaces $\widetilde{H}^1(G_1)$ and $\widetilde{H}^1(G_2)$ such that $\forall u \in \widetilde{H}^1(G_2)$ its restriction on G_1 belongs to $\widetilde{H}^1(G_1)$.

Corollary 5.A2.1 *Let β be a bounded star-shaped domain in $\mathbb{R}^{s-1}$ with a piecewise smooth boundary. Consider the cylinders $C_{r,\mu\beta} = (0,r) \times \mu\beta$, where $\mu\beta = \{x \in \mathbb{R}^{s-1}; \frac{x}{\mu} \in \beta\}$. Let a_1, a_2, b_1, b_2, positive numbers such that $a_1 \le a_2$, $b_1 \le b_2$.*
There exists a unform constant C_z of Lemmas 5.A2.1, 5.A2.2, such that $\forall r \in [a_1, a_2]$, $\forall \mu \in [b_1, b_2]$,
$\forall u \in (H^1(C_{r,\mu\beta}))^s$, $\exists \eta \in \mathcal{R}$ such that

$$\|\nabla(u-\eta)\|^2_{L^2(C_{r,\mu\beta})} \le C_z C_K E_{C_{r,\mu\beta}}(u), \tag{5.A2.1}$$

C_K depends only on β.

Proof. Indeed, the Korn inequality for the cylinder $C_{1,\beta}$ follows from [55]. Then applying Lemma 5.A2.1, we obtain this result.

Remark 5.A2.3. Let $(H^1_{00}(C_{r,\mu\beta}))^s$ be the subspace of $(H^1((C_{r,\mu\beta}))^s$ such that its elements vanish at the base $x_1 = 0$. Then the inequality (1) transforms into

$$\|\nabla u\|^2_{L^2(C_{r,\mu\beta})} \le C_z C_K E_{C_{r,\mu\beta}}(u).$$

Corollary 5.A2.2 *Let a_1, a_2, b_1, b_2 be positive numbers $a_1 \le a_2$, $b_1 \le b_2$. There exists a constant C_z of Lemmas 5.A2.1, 5.A2.2, such that $\forall r \in [a_1, a_2]$, $\forall \mu \in [b_1, b_2]$,*
1) there exists such an extension $P_{E_{(r,\mu)}}$ from $H^1(C_{r,\mu\beta})$ to $H^1(C_{r,(2\mu)\beta})$ such that

$$E_{C_{r,\mu\beta}}(P_{E_{(r,\mu)}} u) \le C_z C_E E_{C_{r,(2\mu)\ \beta}}(u), \tag{5.A2.2}$$

where C_E depends only on β.

2) there exists such an extension $P_{E(r,\mu)}$ from $H^1((\overline{C_{r,(2\mu)\beta}} \cup \overline{C_{(r,2r),\mu\beta}} \cup \overline{C_{(2r,3r),2\mu\beta}})')$ to $H^1(C_{3r,(2\mu)\beta})$ such that estimate (5.A2.2) holds true with C_E depending only on β; here $C_{r,\mu\beta} = (0,r) \times \mu\beta$ (a cylinder), a bar $\bar{\cdot}$ is a symbol of closure, A' is a set of interior points of A.

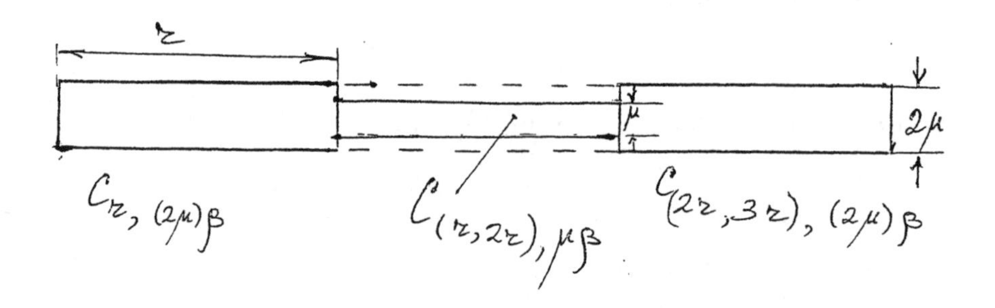

Figure 5.A2.1. The extension.

Let $2\mu_0$ be a positive number such that B_μ is presented as a union of sections S_0 and of maximal connected subsets $\prod_{x_0}$ (x_0 are the nodal points) in such a way that the number of sections and of maximal subsets $\prod_{x_0}$ in the unit cube $(0,1)^s$ does not change for all $\mu \in (0, 2\mu_0)$. (To this end we can chose μ_0 from the inequality

$$\frac{2\mu_0 \max\limits_j \, diam\beta_j}{\min_{x_0} \min_{e_1^\alpha(x_0), e_2(x_0)} \sin\left(\frac{\widehat{e_1 e_2}}{2}\right)} < \frac{1}{3} \min_{\alpha\in\Delta, h\in H_\alpha} |e_h^\alpha|,$$

where $\min\limits_{x_0}$ is a minimum over all nodes x_0 and $\min\limits_{e_1(x_0), e_2(x_0)}$ is a minimum over all segments with the end point x_0, $\widehat{e_1 e_2}$ is the angle between e_1 and e_2). Thus, μ_0 is small enough but finite!

Lemma 5.A2.4 *Let us cut every segment e_h^α in the middle point by a perpendicular plane and consider then the connected subdomain of B_μ containing the pre-nodal set Π_{x_0}. Denote this part of B_μ as $\widehat{\prod}_{x_0,\mu}$. Then for any $\mu \in [\frac{\mu_0}{2}, \mu_0]$ there exists an extension P from $H^1(\widehat{\prod}_{x_0,\mu})$ to $H^1(\widehat{\prod}_{x_0,2\mu})$ such that for any $u \in H^1(\widehat{\prod}_{x_0,\mu}), Pu = u$ if $x \in \widehat{\prod}_{x_0,\mu}$,*
$Pu \in H^1(\widehat{\prod}_{x_0,2\mu})$, and $E_{\widehat{\prod}_{x_0,2\mu}}(Pu) \le C_4 E_{\widehat{\prod}_{x_0,\mu}}(u)$
where constant C_4 does not depend on μ and u ; it depends on μ_0.

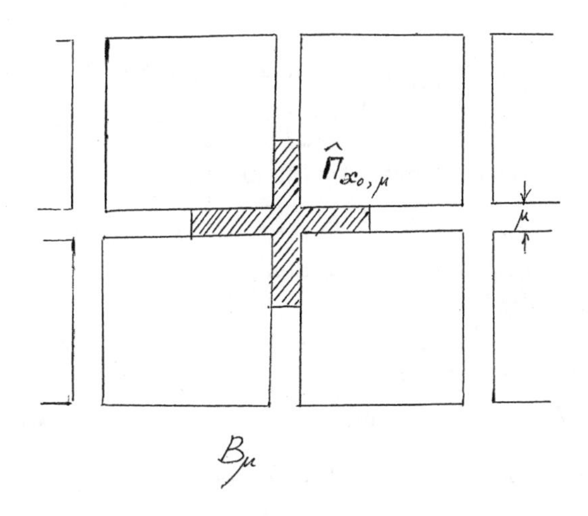

Figure 5.A2.2. Set B_μ.

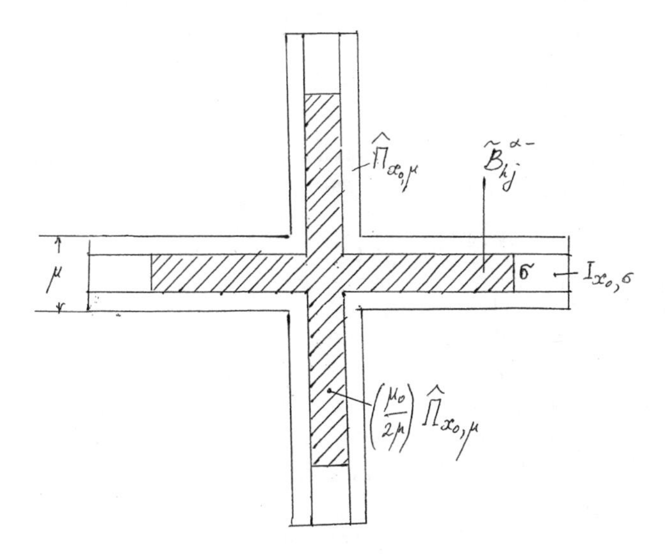

Figure 5.A2.3. Sets of the proof.

Proof If $\mu = \frac{\mu_0}{2}$ the assertion is evident : we construct an extension [85] $\sim$ from $H^1(\widehat{\prod}_{x_0,\frac{\mu_0}{2}})$ to $H^1(\widehat{\prod}_{x_0,\mu_0})$ such that $\|\nabla \widetilde{u}\|_{L^2(\widehat{\prod}_{x_0,\mu_0})} \leq \widehat{C}_{\mu_0} \|\nabla u\|_{L^2(\widehat{\prod}_{x_0,\frac{\mu_0}{2}})}$ and then take $Pu = \widetilde{u-\eta} + \eta$, where η is a rigid displacement such that Korn's inequality holds true for $u - \eta$ on $\widehat{\prod}_{x_0,\frac{\mu_0}{2}}$ with a constant which depends only on μ_0 (see Lemma 5.A2.3).

If $\mu > \frac{\mu_0}{2}$ then we make a change of variables first $\widehat{x} - x_0 = \frac{\mu_0}{2\mu}(x - x_0)$, $\widehat{u} = \frac{2\mu}{\mu_0}u$, and then we extend $\widehat{u}$ from the homothetic image $\left(\frac{\mu_0}{2\mu}\right)\widehat{\prod}_{x_0,\mu}$ of $\widehat{\prod}_{x_0,\mu}$ to $\prod_{x_0,\mu_0/2}$.
This extension from $\left(\frac{\mu_0}{2\mu}\right)\widehat{\prod}_{x_0,\mu}$ to $\widehat{\prod}_{x_0,\frac{\mu_0}{2}}$ is as follows. Let σ be truncated base of one of the cylinders of $\left(\frac{2\mu}{\mu_0}\right)^{-1}\widehat{\prod}_{x_0,\mu}$. It means that σ does not contain x_0 ; let σ belongs to the cylinder $\widetilde{B}^{\alpha}_{hj}$ and divides it on two parts $\widetilde{B}^{\alpha-}_{hj}$ (the part of $\left(\frac{2\mu}{\mu_0}\right)^{-1}\widehat{\prod}_{x_0,\mu}$) and $\widetilde{B}^{\alpha+}_{hj} = \widetilde{B}^{\alpha}_{hj}\backslash\overline{B^{\alpha-}_{hj}}$. First we construct the even extension $\mathcal{P}_\sigma$ of $\widehat{u}$ from $\widetilde{B}^{\alpha-}_{hj}$ to the intersection $I_{x_0,\sigma} = \widetilde{B}^{\alpha+}_{hj} \cap \prod_{x_0,\frac{\mu_0}{2}}$ (even with respect to σ) ; and then we take as usual the continuation $\mathcal{P}_\sigma(\widehat{u} - \eta) + \eta$ where η is a rigid displacement such that the Korn inequality holds true on $\widetilde{B}^{\alpha}_{hj}$ with a constant independent of μ and u (Lemma 5.A2.1).

When such an extension is constructed for every truncating base σ of $\left(\frac{2\mu}{\mu_0}\right)^{-1}$ $\widehat{\prod}_{x_0,\mu}$ we obtain an extension P_- of $\widehat{u}$ from $\left(\frac{2\mu}{\mu_0}\right)^{-1}\widehat{\prod}_{x_0,\mu}$ to $\widehat{\prod}_{x_0,\frac{\mu_0}{2}}$ such that $E_{\widehat{\prod}_{x_0,\frac{\mu_0}{2}}}(P_-\widehat{u}) \le C_{\mu_0}\ E(\widehat{u})$ with the constant C_{μ_0} independent of μ. At the second stage we construct an extension of $P_-\widehat{u}$ from $\widehat{\prod}_{x_0,\frac{\mu_0}{2}}$ to $\widehat{\prod}_{x_0,\mu_0}$ as in the beginning of the proof and then we make the inverse change of variables $\widehat{x} \to x$. Lemma is proved.

Our goal now is to prove that there exist a uniform constant C_{μ_0} such that for all $\mu \in [\mu_0, 2\mu_0]$, the Korn inequality holds

$$\|\nabla u\|^2_{L^2(B_\mu \cap (0,1)^s)} \le C_{\mu_0} E_{B_\mu \cap (0,1)^s}(u)$$

for all 1−periodic functions of $H^1_{per}(B_\mu)$.

Let us trace the cutting planes separating the sections S_0 on 3 parts of the same height h. For any node x_0 denote $\widetilde{\prod}_{x_0}$ the nodal domain x_0 reunited with the nearest pieces of sections (the common boundary of $\prod_{x_0}$ and these pieces also enter in $\widetilde{\prod}_{x_0}$). For any node x_0 make the change $\widehat{x} - x_0 = \frac{\mu_0}{2\mu}(x - x_0)$. For any segment e denote $\widetilde{e}$ the intersection of e with the middle piece of the section corresponding to e. Let $\overline{x}_0$ be the middle point of $\widetilde{e}$. Make the change $\widehat{x} - \overline{x}_0 = \frac{\mu_0}{2\mu}(x - x_0)$. After this change the neighbor pieces of sections : S_{x_0} (joint to $\widetilde{\prod}_{x_0}$) and $S_{\overline{x}_0}$ transform into the cylinders $\widetilde{S}_{x_0}$ and $\widetilde{S}_{\overline{x}_0}$ respectively. Consider now the minimal cylinder $\widetilde{S}_{\overline{x}_0}$ having the same cross section and including $\widetilde{S}_{x_0}$ and $\widetilde{S}_{\overline{x}_0}$. Its bases are the farthest bases of $\widetilde{S}_{x_0}$ and $\widetilde{S}_{\overline{x}_0}$.

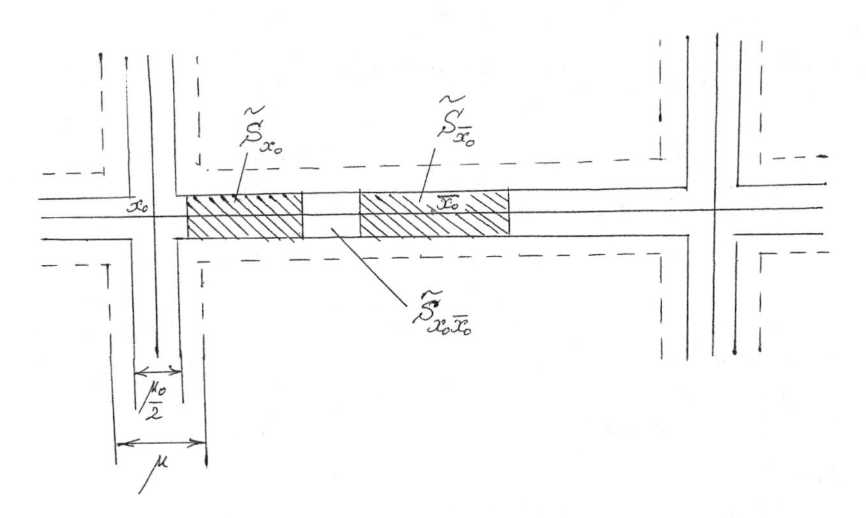

Figure 5.A2.4. Extension from $\widetilde{S}_{x_0} \cup \widetilde{S}_{\overline{x}_0}$ to $\widetilde{S}_{x_0\overline{x}_0}$.

Denote $\widetilde{S}_{x_0\overline{x}_0}$ the open part of $\widetilde{C}_{x_0\overline{x}_0}$ between the cylinders $\widetilde{S}_{x_0}$ and $\widetilde{S}_{\overline{x}_0}$. The heights of $\widetilde{S}_{x_0}$ and $\widetilde{S}_{\overline{x}_0}$ are $\frac{\mu_0}{2\mu}h$, while the height of $\widetilde{S}_{x_0\overline{x}_0}$ is $(1-\frac{\mu_0}{2\mu})(l+\frac{h}{2})$, where l is the initial distance from x_0 to the nearest cutting surface, i.e. $l = \text{dist}(x_0, \overline{x}_0) - \frac{h}{2}$: so the height of $\widetilde{S}_{x_0\overline{x}_0}$ is,

$$dist(x_0, \overline{x}_0)(1 - \frac{\mu_0}{2\mu}), \quad (1 \geq 1 - \frac{\mu_0}{2\mu} \geq \frac{1}{2}).$$

Let $G = (\overline{S}_{x_0} \cup \overline{S}_{\overline{x}_0})'$ and let u be a vector-valued function of $H^1(G)$.
Make a change $\widehat{u} = \frac{2\mu}{\mu_0}u$. Note that ∇u does not change after the passage from u, x to $\widehat{u}, \widehat{x}$.
Let us prove that there exist such extensions P, P_E of $\widehat{u}(\widehat{x})$ from $\widetilde{S}_{x_0} \cup \widetilde{S}_{\overline{x}_0}$ to $\widetilde{C}_{x_0,\overline{x}_0}$ that $P_E\eta = \eta \quad \forall \eta \in \mathcal{R}$ and $\|\nabla P\widehat{u}\|^2_{L^2(\widetilde{C}_{x_0,\overline{x}_0})} \leq C_1\|\nabla \widehat{u}\|^2_{L^2(\widetilde{S}_{x_0}\cup\widetilde{S}_{\overline{x}_0})}$, $E_{\widetilde{C}_{x_0,\overline{x}_0}}(P_E\widehat{u}) \leq C_2 E_{\widetilde{S}_{x_0}\cup\widetilde{S}_{\overline{x}_0}}(\widehat{u})$, where $C_1, C_2 > 0$ are uniform with respect to $\mu \in [\mu_0, 2\mu_0]$.

Indeed, let us extend $\widehat{u}$ from $\widetilde{S}_{x_0}$ to $\widetilde{S}_{x_0\,\overline{x}_0}$ in such a way that the extension be pair with respect to the common base of the cylinders $\widetilde{S}_{x_0}$ and $\widetilde{S}_{x_0\,\overline{x}_0}$ and multiply this extension by a function χ_{μ_0} such that depends only on the longitudinal variable x_1^e (its axis coincides with the direction of e). This function depends only on μ_0, it is equal to 1 in some neighborhood of the common base of $\widetilde{S}_{x_0}$ and $\widetilde{S}_{x_0\,\overline{x}_0}$ and it is equal to 0 at the distance $\frac{l}{2}$ from this base and farther. In the same way we make the extension from $\widetilde{S}_{x_0}$ to $\widetilde{S}_{x_0\,\overline{x}_0}$.

For this extension $\widetilde{\widehat{u}}$ we have the estimate

$$\|\nabla\widetilde{\widehat{u}}\|_{L^2(\widetilde{C}_{x_0\,\overline{x}_0})} \leq \overline{C}_1\|\nabla\widehat{u}\|_{L^2(\widetilde{S}_{x_0}\cup\widetilde{S}_{\overline{x}_0})}$$

where $\overline{C}_1$ do not depend on μ (it depends on μ_0).
Let now η be such a rigid displacement that

$$\|\nabla(u-\eta)\|^2_{L^2((\overline{S_{x_0}\cup S_{\overline{x}_0}})')} \leq \overline{C}_2 E_{\overline{S_{x_0}\cup S_{\overline{x}_0}}}(u-\eta)$$

$\overline{C}_2$ does not depend on μ (Lemma 5.A2.1). Then

$$\|\nabla(\widehat{u}-\widehat{\eta}(\widehat{x}))\|^2_{L(\widetilde{S}_{x_0}\cup\widetilde{S}_{\overline{x}_0})} \leq \overline{C}_2 E_{\widetilde{S}_{x_0}\cup\widetilde{S}_{\overline{x}_0}}(\widehat{u}-\widehat{\eta}).$$

Let us construct the above extension $\widetilde{\widehat{u}-\widehat{\eta}}$ for $\widehat{u}-\widehat{\eta}$ and add the rigid displacement $\widehat{\eta}$. We have then
$E_{\widetilde{C}_{x_0\,\overline{x}_0}}(\widetilde{\widehat{u}-\widehat{\eta}}+\widehat{\eta}) = E_{\widetilde{C}_{x_0\,\overline{x}_0}}(\widetilde{\widehat{u}-\widehat{\eta}}) \leq \|\nabla(\widetilde{\widehat{u}-\widehat{\eta}})\|^2_{L^2(\widetilde{C}_{x_0\,\overline{x}_0})} \leq \overline{C}_1\|\nabla(\widehat{u}-\widehat{\eta})$
$\|^2_{L^2(\widetilde{S}_{x_0}\cup\widehat{S}_{\overline{x}_0})} \leq \overline{C}_1\overline{C}_2 E_{\widetilde{S}_{x_0}\cup\widetilde{S}_{\overline{x}_0}}(\widehat{u}-\widehat{\eta}) = \overline{C}_1\overline{C}_2 E_{\widetilde{S}_{x_0}\cup\widetilde{S}_{\overline{x}_0}}(\widehat{u})$.
Thus $P_E\widehat{u} = \widehat{u}$ on $\widetilde{S}_{x_0}\cup\widetilde{S}_{\overline{x}_0}$
$P_E\widehat{u} = \widetilde{\widehat{u}-\widehat{\eta}}+\widehat{\eta}$ on $\widetilde{S}_{x_0\,\overline{x}_0}$.

Remark 5.A2.4 The extensions P and P_E can be constructed in such a manner that $P = P_E$. Indeed, let us construct first an extension $\widetilde{\widehat{u}}$ described above. For this extension $\widetilde{\widehat{u}}$ we have the estimates

$$\|\widetilde{\widehat{u}}\|_{H^1(\widetilde{C}_{x_0\,\overline{x}_0})} \leq \overline{\overline{C}}_1\|\widehat{u}\|_{H^1(\widetilde{S}_{x_0}\cup\widetilde{S}_{\overline{x}_0})}$$

and

$$\|\nabla\widetilde{\widehat{u}}\|_{L^2(\widetilde{C}_{x_0\,\overline{x}_0})} \leq \overline{\overline{C}}_2\|\nabla\widehat{u}\|_{L^2(\widetilde{S}_{x_0}\cup\widetilde{S}_{\overline{x}_0})}$$

where $\overline{\overline{C}}_1, \overline{\overline{C}}_2$ do not depend on μ (they depend only on μ_0).

Now construct the solution V of the problem
$\Delta V + 2graddivV = 0$ in $\widetilde{S}_{x_0\,\overline{x}_0}$
$V = \widetilde{\widehat{u}}$ on $\partial\widetilde{S}_{x_0}\cap\partial\widetilde{S}_{x_0\,\overline{x}_0}$ and on $\partial\widetilde{S}_{\overline{x}_0}\cap\partial S_{x_0\,\overline{x}_0}$
$\sum_{i,k=1}^{s} B_{ik}\dfrac{\partial V}{\partial x_k}n_i = 0$ on $\partial\widetilde{S}_{x_0\,\overline{x}_0}\cap\partial\widetilde{C}_{x_0\,\overline{x}_0}$, $B_{ik} = (b^{jl}_{ik})$, $b^{jl}_{ik} = \delta_{ij}\delta_{kl}+\delta_{ik}\delta_{jl}+$ $\delta_{il}\delta_{jk}$, $(n_1,\ldots,n_s)$ is a normal vector.

We obtain the estimate

$$\|\nabla V\|_{L^2(\widetilde{S}_{x_0\,\overline{x}_0})} \leq \overline{\overline{C}}_3\|\nabla\widetilde{\widehat{u}}\|_{L^2(\widetilde{S}_{x_0\,\overline{x}_0})}$$

where $\overline{\overline{C}}_3$ is uniform with respect to $\mu\in[\mu_0, 2\mu_0]$ (cf Lemma 5.A2.1 and its corollaries, $\overline{\overline{C}}_3$ depends only on β,μ_0,l and h). So

$$\|\nabla V\|^2_{L^2(\widetilde{S}_{x_0\,\overline{x}_0})} \leq \overline{\overline{C}}_4\|\nabla\widehat{u}\|^2_{L^2(\widetilde{S}_{x_0}\cup\widetilde{S}_{\overline{x}_0})}$$

Consider now

$$Pv = \begin{cases} v \text{ for } x \in \widetilde{S}_{x_0} \cup \widetilde{S}_{\overline{x}_0} \\ V \text{ for } x \in \widetilde{S}_{x_0\,\overline{x}_0} \end{cases}$$

Let now η be such a rigid displacement that

$$\|\nabla(u-\eta)\|^2_{L^2(S_{x_0}\cup S_{\overline{x}_0})} \leq \overline{C}_5 E_{S_{x_0}\cup S_{\overline{x}_0}}(u-\eta)$$

$\overline{C}_5$ does not depend on μ. (Lemma 5.A2.1)

Then

$$\|\nabla(\widehat{u}-\widehat{\eta}(\widehat{x}))\|^2_{L^2(\widetilde{S}_{x_0}\cup\widetilde{S}_{\overline{x}_0})} \leq \overline{C}_5 E_{\widetilde{S}_{x_0}\cup\widetilde{S}_{\overline{x}_0}}(\widehat{u}-\widehat{\eta}).$$

So $\|\nabla(V-\widehat{\eta}(\widehat{x}))\|^2_{L^2(\widetilde{S}_{x_0\,\overline{x}_0})} \leq \overline{C}_4\overline{C}_5 E_{\widetilde{S}_{x_0}\cup\widetilde{S}_{\overline{x}_0}}(\widehat{u}-\widehat{\eta})$
so $E_{\widetilde{S}_{x_0\,\overline{x}_0}}(V-\widehat{\eta}) \leq \overline{C}_4\overline{C}_5 E_{\widetilde{S}_{x_0}\cup\widetilde{S}_{\overline{x}_0}}(\widehat{u}-\widehat{\eta})$
and so, $E_{\widetilde{S}_{x_0\,\overline{x}_0}}(V) \leq \overline{C}_4\overline{C}_5 E_{\widetilde{S}_{x_0}\cup\widetilde{S}_{\overline{x}_0}}(\widehat{u})$.
So the lemma is proved.
Here the constant $\overline{C}_4\overline{C}_5$ is uniform with respect to $\mu \in [\mu_0, 2\mu_0]$.

Theorem 5.A2.2 *Let $\mu \in [\mu_0, 2\mu_0]$. Then for any 1-periodic vector-valued function u defined on B_μ, $u \in H^1_{per\,1}(B_\mu)$, the inequality holds*

$$\|\nabla u\|^2_{L^2(B_\mu\cap(0,1)^s)} \leq C_{\mu_0} E_{B_\mu\cap(0,1)^s}(u)$$

where constant C_{μ_0} is uniform with respect to μ.

Proof. Make a change as above and extend the function $\widehat{u}(\widehat{x})$ to the cylinders $\widetilde{S}_{x_0\,\overline{x}_0}$ as above. Apply now the Korn inequality for $B_{\mu_0/2}$ (for finite μ_0 it is proved as in [55], [129]). Its constant $\overline{C}_{\mu_0/2}$ depends on μ_0

$$\|\nabla\widehat{u}\|^2_{L^2(B_{\mu_0/2}\cap(0,1)^s)} \leq \overline{C}_{\mu_0/2} E_{B_{\mu_0/2}\cap(0,1)^s}(\widehat{u}).$$

In particular, we have the same estimate for

$$\|\nabla\widehat{u}\|^2_{L^2((B_{\mu_0/2}\cap(0,1)^s)\setminus\bigcup_{x_0,\overline{x}_0\in(0,1)^s}\widetilde{S}_{x_0\,\overline{x}_0})}.$$

On the other hand, the extension was made in such a way that

$$E_{B_{\mu_0/2}\cap(0,1)^s}(\widehat{u}) \leq (2\overline{C}_1\overline{C}_2+1)E_{B_{\mu_0/2}\cap(0,1)^s\setminus\bigcup_{x_0,\overline{x}_0\in(0,1)^s}\widetilde{S}_{\widetilde{x}_0\,\overline{x}}}(\widehat{u})$$

where $\overline{C}_1, \overline{C}_2$ are uniform for $\mu \in [\mu_0, 2\mu_0]$.
Making the inverse change and passing from $\widehat{u}, \widehat{x}$ to u, x we obtain the estimate of the theorem with

$$C_{\mu_0} = \overline{C}_{\mu_0/2}(2\overline{C}_1\overline{C}_2+1).$$

Remark 5.A2.5 In the same way the same estimate can be proved for a finite rod structure when $u = 0$ on the base of one of cylinders.

Now let us prove the main Theorem 5.A2.1'.

Let k be chosen in such a way that $\frac{\min\limits_{e\in\mathcal{B}}|e|}{2^k} \geq 2(c_0+1)\mu \geq \frac{\min\limits_{e\in\mathcal{B}}|e|}{2^{k+1}}$ and let $\mu_0 = \frac{\min\limits_{e\in\mathcal{B}}|e|}{32(c_0+1)}$, where c_0 is the constant of Definition 5.1.1.

Subdivide every cylinder $\widetilde{B}^{\alpha}_{hj}$ on 2^k parts $\widetilde{B}^{\alpha,k}_{hj,q}(q = 1,\ldots,2^k)$ of the same heights by the planes parallel to the bases of the cylinder $\widetilde{B}^{\alpha}_{hj}$ in such a way that $\overline{\widetilde{B}^{\alpha,k}_{hj,1}}$ contains x_{00}, $\overline{\widetilde{B}^{\alpha,k}_{hj,2^k}}$ contains x_{01}, where x_{00} and x_{01} are the ends of the segment e^{α}_h, and $\widetilde{B}^{\alpha k}_{hj,q}$ has a common base with $\widetilde{B}^{\alpha k}_{hj,q+1}$, $q = 1,\ldots,2^k-1$.

Let $\Pi^k_{x_0}$ be the connected part of B_μ cut from B_μ by the nearest cutting planes to the nodal point x_0. We will make now $k-3$ extensions of $u \in H^1_{per}(B_\mu)$ from B_μ to $B_{2\mu}$ from $B_{2\mu}$ to $B_{2^2\mu},\ldots$, and finally from $B_{2^{k-4}\mu}$ to $B_{2^{k-3}\mu}$ in such a way that the extensions $\widetilde{u}_l$ from $B_{2^{l-1}\mu}$ to $B_{2^l\mu}$ belong to $H^1_{per}(B_{2^l\mu})$ and $E_{B_{2^l\mu}}(\widetilde{u}_l) \leq C\, E_{B_{2^{l-1}\mu}}(\widetilde{u}_{l-1})$ with a constant C independent of μ and of l. Let us describe the construction for a passage from $B_{2^{l-1}\mu}$ to $B_{2^l\mu}$ $(l \leq k-4)$. At this stage the cylinders $\widetilde{B}^{\alpha}_{hj}$ (corresponding to the value of small parameters that is $2^{l-1}\mu$) are divided by the planes parallel to the bases on $2^{k-(l-1)}$ parts. We will keep the notation $\widetilde{B}^{\alpha,k-(l-1)}_{hj,q}$ for this parts as well as the notation $\Pi^{k-(l-1)}_{x_0}$ for the truncated part corresponding to the nodal point x_0 that is a homothetic extension of $\Pi^k_{x_0}$ in 2^{l-1} times (x_0 is a homothety center). First we extend $\widetilde{u}_{l-1}$ to $\Pi^{k-l}_{x_0}$ that is a homothetic image of $\Pi^{k-(l-1)}_{x_0}$ with the coefficient of homothety 2 and the center x_0. This extension is made as follows. Denote $\widehat{\Pi}^{k-(l-1)}_{x_0}$ the part of $B_{2^{l-1}\mu}$ containing $\Pi^{k-(l-1)}_{x_0}$ and all the parts $\widetilde{B}^{\alpha,k-(l-1)}_{hj,g}$ having the common boundary with $\Pi^{k-(l-1)}_{x_0}$.

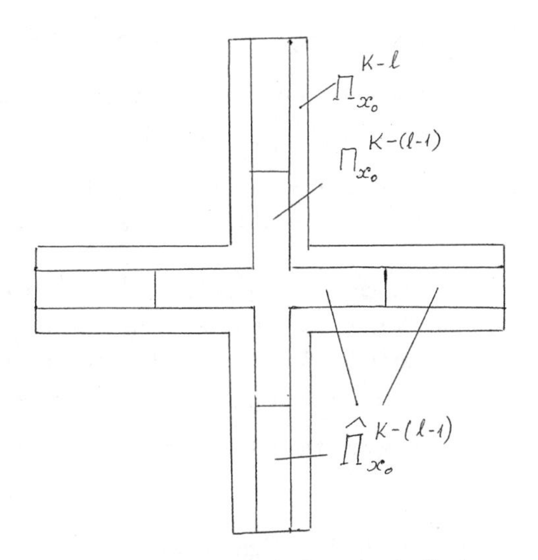

Figure 5.A2.5. Sets $\Pi_{x_0}^{k-l)}$ *and* $\Pi_{x_0}^{k-(l-1)}$.

$$\widetilde{u}_l(x) = \begin{cases} \widetilde{u}_{l-1}(x)\ ,\ x \in \widehat{\Pi}_{x_0}^{k-(l-1)} \\ \widetilde{w}_l(Y(x))\ ,\ x \in \Pi_{x_0}^{k-l} \backslash \overline{\widehat{\Pi}_{x_0}^{k-(l-1)}} \end{cases}$$

where $\widetilde{w}_l$ is an extension of the function $y \longmapsto \widetilde{u}_{l-1}(X(y))$ from $\widehat{\Pi}_{x_0,\mu_0}$ to $\Pi_{x_0,2\mu_0}$; here $\widehat{\Pi}_{x_0,\mu_0}$ and $\Pi_{x_0,2\mu_0}$ are the homothetic images of $\widehat{\Pi}_{x_0}^k$ and $\Pi_{x_0}^k$ for a homothety with the center x_0 and the coefficients respectively $\dfrac{\mu_0}{\mu}$ and $\dfrac{2\mu_0}{\mu}$;
$y = Y(x)$ and $x = X(y)$ are changes of variables such that

$$Y - x_0 = \frac{2\mu_0}{2^{(k-l)}\mu} \quad (x - x_0)$$

and respectively

$$y - x_0 = \frac{2\mu_0}{2^{(k-l)}\mu} \quad (X - x_0).$$

Lemma 5.A2.4 yields that there exist a constant C_{μ_0} of the extension of any function of $H^1(\widehat{\Pi}_{x_0,\mu_0})$ from $\widehat{\Pi}_{x_0,\mu_0}$ to $\Pi_{x_0,2\mu_0}$ such that the integral $E_{\Pi_{x_0,2\mu_0}}$ is estimated by the integral $E_{\widehat{\Pi}_{x_0,\mu_0}}$ multiplied by this constant C_{μ_0} which does not depend on the function. It depends on μ_0 only.

We apply this extension procedure to the function $y \longmapsto \widetilde{u}_{l-1}(X(y))$, obtain the function $y \longmapsto \widetilde{w}_l(y)$ and then make a back change and obtain $x \longmapsto \widetilde{w}_l(Y(x))$. Evidently (as always for a homothetic change of variables) we have the same inequality for $\widetilde{u}_{l-1}$ and $\widetilde{w}_l \circ Y$, i.e. $E_{\Pi_{x_0}^{k-l} \backslash \Pi_{x_0}^{k-(l-1)}}(\widetilde{w}_l \circ Y) \le C_{\mu_0} E_{\widehat{\Pi}_{x_0}^{k-(l-1)}}$ $(\widetilde{u}_{l-1})$. This constant C_{μ_0} does not depend on μ or l.

Now consider the extension procedure for the $2^{k-(l-1)}-4$ parts of B^{α}_{hj} that were not included into $\bigcup\limits_{x_0}\Pi^{k-l}_{x_0}$, i.e. $B^{\alpha,l-1}_{hj,q}$, $q=3,\ldots,2^{k-(l-1)}-2$. Let us denote C_{l-1} the cylinder composed of two parts $B^{\alpha,l-1}_{hj,q}$, $q=4,5$.
For C_{l-1} as well as for every $B^{\alpha,l-1}_{hj,q}$, q is odd, $7\leq q\leq 2^{k-(l-1)}-3$, make a change of variables $(\widetilde{\widetilde{X}}^{e^{\alpha}_h})'=2(\widetilde{X}^{e^{\alpha}_h})'$ where $\widetilde{X}^{e^{\alpha}_h}$ is a local coordinate system corresponding to the segment e^{α}_h that is as above $\widetilde{X}^{e^{\alpha}_h}=\alpha(X-\epsilon h)$, and $(X^{e^{\alpha}_h})'$ are the components of $\widetilde{X}^{l^{\alpha}_h}$ number $2,\ldots,s$.

Denote $\widetilde{C}_l$ and $\widetilde{B}^{\alpha,l-1}_{hj,q}$ the maps of C_l and $B^{\alpha,l-1}_{hj,q}$ with respect to this change. Make an extension from C_l to $\widetilde{C}_l$ as well as from $B^{\alpha,l-1}_{hj,q}$ to $\widetilde{B}^{\alpha,l-1}_{hj,q}$ with the constant C_{μ_0} such that the extended function has the estimate of $E_{\widetilde{C}_l}$ and $E_{\widetilde{B}^{\alpha,l-1}_{hj,q}}$ by E_{C_l} and $E_{B^{\alpha,l-1}_{hj,q}}$ respectively multiplied by this constant C_{μ_0}. Such an estimate can be obtained in a way that is similar to the extension from $\widehat{\Pi}^{k-(l-1)}_{x_0}$ to $\Pi^{k-l}_{x_0}$.

Indeed, we make the homothetic change of variables of a coefficient $\dfrac{2\mu_0}{2^{(k-l)}\mu}$ with the center in the middle point of the height of the cylinder (the height that is a part of the segment l^{α}_h).
Applying corollary 5.A2.2 (part 1) we obtain such an extension in dilated variables. The back change leaves the same constant for the passage from $\Pi^{k-(l-1)}_{x_0}$ to $\Pi^{k-l}_{x_0}$ that is uniform with respect to μ and l. Then for every triplet $B^{\alpha,l-1}_{hj,q}$, $q=2r+1,2r+2,2r+3,\ 2\leq r\leq 2^{k-(l-1)-1}-2$ as well as for $B^{\alpha,l-1}_{hj,q}$ $q=2,3,4$ we make as above with help of corollary 2 (part 2) an extension from

$$H^1(\overline{(\widetilde{B}^{\alpha,l-1}_{hj,2r+1}\cup B^{\alpha,l-1}_{hj,2r+2}\cup\widetilde{B}^{\alpha,l-1}_{hj,2r+3})'})$$

to

$$H^1(\overline{(\widetilde{B}^{\alpha,l-1}_{hj,2r+1}\cup\widetilde{B}^{\alpha,l-1}_{hj,2r+2}\cup\widetilde{B}^{\alpha,l-1}_{hj,2r+3})'})$$

as well as from

$$H^1(\overline{(\widetilde{B}^{\alpha,l-1}_{hj,2}\cup B^{\alpha,l-1}_{hj,3}\cup\widetilde{B}^{\alpha,l-1}_{hj,4})'})$$

to

$$H^1(\overline{(\widetilde{B}^{\alpha,l-1}_{hj,2}\cup\widetilde{B}^{\alpha,l-1}_{hj,3}\cup\widetilde{B}^{\alpha,l-1}_{hj,4})'})$$

"conserving" the E.

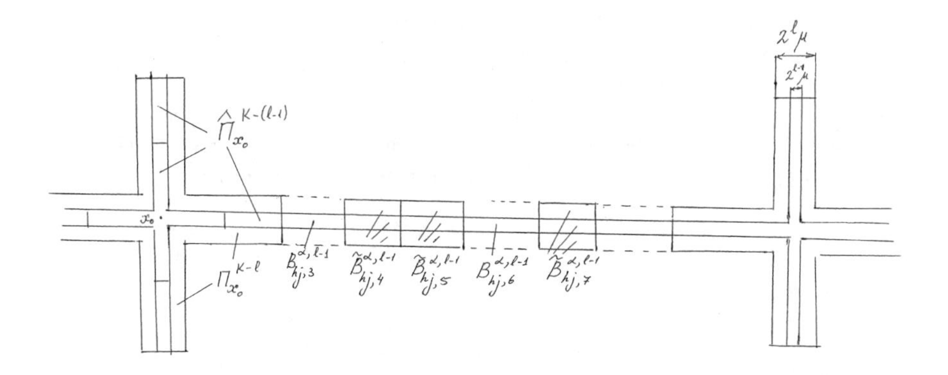

Figure 5.A2.6. Extension from $\Pi_{x_0}^{k-(l-1)}$ *to* $\Pi_{x_0}^{k-l)}$.

At the step $k-4$ we extend u from $\widehat{\Pi}_{x_0}^4$ to $\Pi_{x_0}^3$ and then to $B_{2^{k-3}\mu}$ and then finally we obtain the extension to $B_{2^{k-3}\mu}$, where $\mu_0 \le 2^{k-3}\mu \le 2\mu_0$.
Thus we obtain an extension P_k from B_μ to $B_{2^{k-3}\mu}$ such that $E_{B_{2^{k-3}\mu}\cap(0,1)^s}(u) \le C_0 C^k E_{B_\mu \cap(0,1)^s}(u)$, where C does not depend on μ.
According to Theorem 5.A2.1 we have $\|\nabla u\|^2_{L^2(B_\mu\cap(0,1)^s)} \le \|\nabla P_k u\|^2_{L^2(B_{2^{k-3}\mu}\cap(0,1)^s)} \le C_{\mu_0} E_{(B_{2^{k-3}\mu}\cap(0,1)^s)}(P_k u) \le C_0 C^k C_{\mu_0}\ E_{B_\mu\cap(0,1)^s}(u)$ and the Korn inequality is obtained with a constant of order $C^k = e^{k\ lnC}$, where $k = O(|ln\ \mu|)$ $(\mu \to 0)$, i.e. $k \le \widehat{C}|ln\ \mu|$,
so
$C^k \le \mu^{-q}$,
where q does not depend on μ.

Thus the theorem on the Korn inequality with a polynomial dependency of the constant on μ is proved.

Remark 5.A2.6 In the same way we can prove the result for $H^1(B_\mu)$, where B_μ is a finite rod structure and a function vanishes on some part of $\partial B\mu$ that is a base of cylinders constituting B_μ. We apply the same idea of the extension of a vector-valued function vanishing on $\partial_1 B_\mu$ from B_μ to $B_{2^{k-3}\mu}$ (vanishing on $\partial_1 B_\mu$ that is the 2^{k-3} times dilated base $\partial_1 B_\mu$).

Remark 5.A2.7 Let us prove the Korn inequality for $H^1_{perN}(B_{\overline{\mu}})$ for a finite sufficiently small $\overline{\mu} > 0$, for any $N \in \mathbb{N}$ with a constant independent of N. To

this end we construct an extension from $B_{\overline{\mu}}$ to $\mathbb{R}^s$ in $s+1$ steps ("conserving" the energy with a constant independent of N)
Denote $Q_{i_1\dots i_s}$ the unit cube with the center in the point $(i_1,\dots,i_s)\in\mathbb{N}^s$, $Q_{i_1\dots i_s}=\{x\in\mathbb{R}^s, \mid x_j-i_j\mid<\frac{1}{2}, j=1,\dots,s\}$; denote $B^a_{i_1\dots i_s}$ the ball of a radius equal to a with the center $(i_1,\dots,i_s)$ where $(2i_1,\dots,2i_s)\in\mathbb{N}^s$, $B^a_{i_1\dots i_s}=\{x\in\mathbb{R}^s,\sum_{j=1}^s(x_j-i_j)^2<(a)^2\}$. In case $s=3$ consider the cylinders

$$C^{k,a}_{i_1\dots i_{k-1}i_{k+1}\dots i_s}=\{x\in\mathbb{R}^s,\sum_{j\neq k}(x_j-i_j)^2<a^2\},$$

$$(2i_1,\dots,2i_{k-1},2i_{k+1},\dots,2i_s)\in\mathbb{N}^{s-1};$$

Denote

$$B^a=\bigcup_{i_1,\dots,i_n\in\mathbb{N}}B^a_{(i_1+1/2,\dots,i_s+1/2)},$$

$$C^a=\bigcup_{k=1}^{s}\bigcup_{i_1,\dots,i_{k-1},i_{k+1},\dots i_s\in\mathbb{N}}C^{k,a}_{(i_1+1/2,\dots,i_s+1/2)}$$

In case $s=2$, N even, at the first step we extend the function from $B_{\overline{\mu}}$ to

$$\bigcup_{(i_1,i_2):i_1+i_2\ \ odd}Q_{i_1i_2}\backslash B^a$$

for some finite $a<\frac{1}{4}$; at the second step we extend the function further to $\mathbb{R}^2\backslash B^a$; and finally further to $\mathbb{R}^2$. Mention that these first step extensions are made independently for every $Q_{i_1i_2}\backslash B^a$ by the same extension operator P_1 ; the second step extensions are made independently to every $Q_{i_1i_2}\backslash B^a$ with even i_1+i_2 from the neighboring $Q_{j_1j_2}\backslash B^a$ with odd j_1+j_2 (and with the same extension operator P_2). The third step extensions are made to every $B^a_{i_1+1/2,i_2+1/2}$ from the neighboring $Q_{j_1j_2}\backslash B^a$ independently by the same extension operator P_3. The operators P_1,P_2,P_3 are those of lemma 5.A2.3 or as in [129].

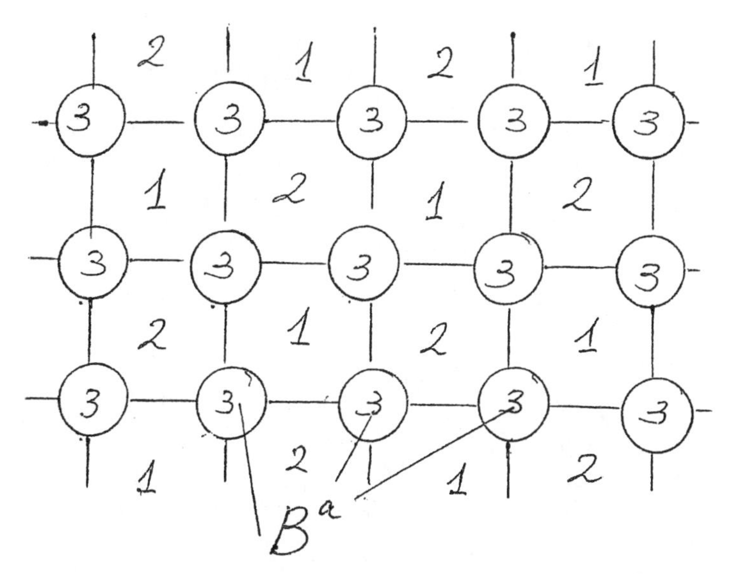

Figure 5.A2.7. The three steps of extension from B_μ to $\mathbb{R}^2$.

If N is odd then the first step set should be

$$\bigcup_{(i_1,i_2):i_1+i_2+[\frac{i_1}{N}]+[\frac{i_2}{N}] \quad odd} Q_{i_1 i_2}\backslash B^a$$

with a special extension operators for the squares $Q_{i_1 i_2}$ with i_1 or i_2 multiples of N.

In case $s = 3$ (N even) at the first step we extend the function from $B_{\overline{\mu}}$ to

$$\bigcup_{(i_1,i_2,i_3)\in\mathbb{N}^3:i_1+i_2+i_3 \quad odd} Q_{i_1 i_2 i_3}\backslash(C^b \bigcup B^a)$$

for some $0 < b < a < \frac{1}{4}$; further to $\mathbb{R}^3\backslash(C^b \bigcup B^a)$; then to $\mathbb{R}^3\backslash B^a$ and finally to $\mathbb{R}^3$.

Every step of extension (from G_1 to G_2) has been made in such a way that the extension operator P satisfies the estimates $E_{G_2\cap(0,N)^s}(u) \le CE_{G_1\cap(0,N)^s}(u)$, ($C$ is independent of N), and therefore finally we have constructed an N−periodic extension from $B_{\overline{\mu}}$ to $\mathbb{R}^s$ such that, $E_{(0,N)^s}(Pu) \le CE_{B_{\overline{\mu}}\cap(0,N)^s}(u)$; so $\|\nabla u\|^2_{L^2(B_{\overline{\mu}}\cap(0,N)^s)} \le \|\nabla(Pu)\|^2_{L^2(B_{\overline{\mu}}\cap(0,N)^s)} \le C_K E_{((0,N)^s)}(u) \le C_K C E_{B_{\overline{\mu}}\cap(0,N)^s}(u)$,where C_K is the Korn constant for a periodic cube and it does not depend on N.

Chapter 6

The Multi-Scale Domain Decomposition

The method of asymptotic partial decomposition of domain (MAPDD) introduced in [154] is applied here below to partial differential equations, set in rod structures, i.e. in some connected unions of thin cylinders, described in Chapter 4. This method is based on the information about the structure of the asymptotic solution in different parts of such complicated domain. The principal idea of the method is to extract the subdomain of singular behavior of the solution and to reduce dimension of the problem in the subdomain of regular behavior of the solution. The special interface conditions are set on the common boundary of these partially decomposed subdomains. This approach can be easily implemented in the frame of one of modifications of finite element method.

Thus this method makes a dimensional (or scale) zoom of some parts of the domain, where the solution has a singular behavior.

Rod structures (as it was defined in Chapter 4) are finite connected unions of thin cylinders. The direct numerical solution of partial derivative equations in such domains is very expensive because a complicated geometry demands a large number of nodes in the grid.

Another decision is to reduce the dimension and pass to locally one-dimensional model: i.e. each rod is one-dimensional with some junction conditions in the ends. This approach was represented in by sections 4.2, 4.3 but it ignores the boundary layers in neighborhoods of the ends of the rods, while it is well known that these boundary layers are sometimes very important from the point of view of calculation of gradients of solutions: the failure of the rod structure often begins from these neighborhoods because of great concentration of stresses.

These stresses can be calculated by constructing of the asymptotic expansion of the solution as in sections 4.7, 4.8, but it is not too easy.

Therefore in the present chapter we propose a hybrid method which uses a combined 3D-1D models: it is three-dimensional in the boundary layer domain and it is one-dimensional outside of the boundary layer domain. We cut the rods at some distance from the ends of the rods, we keep the dimension three in the neighborhood of the ends and we reduce dimension on the truncated (main) part of rods. Of course the most important question is: what are the interface conditions between 3D and 1D parts? Here below we formulate two approaches of construction of such hybrid models and justify the closeness of the partially decomposed model and initial model.

The main principles of construction of such conditions are as follows:

i) the asymptotic expansion of the exact solution of the initial problem should satisfy these interface conditions with great accuracy;

ii) the hybrid 3D-1D problem with the interface conditions (i.e. partially decomposed problem) should be well posed, i.e. it should have the unique solution and it should be stable with respect to small perturbations in the right hand side.

For example, consider the structure at Figure 6.0.1. It is a "thin domain", where the partial derivative equations are set with some boundary conditions. The structural mechanics approach reduces it to a completely one-dimensional object (Figure 6.0.2) with some ordinary differential equations along the segments and some junction conditions in the nodes. Although this passage can be justified asymptotically as the small parameter μ tends to zero, the two-dimensional information in the neighborhoods of junctions will be completely lost. This information is nevertheless important in the fracture analysis because fractures often start from these domains where the stresses are locally concentrated. From this point of view the method of asymptotic partial decomposition of domain preferable because it passes to a one-dimensional model in the main part of the domain and it keeps the information in some small neighborhoods of the diameter of order $\varepsilon ln(\varepsilon)$. Some special interface conditions are set on the contact surfaces between one-dimensional and two-dimensional parts. So, this method can be interpreted as a multi-scale model with the zoom in the neighborhoods of the junctions (Figure 6.0.3).

In section 6.1 the differential version of the method of partial asymptotic domain decomposition (MAPDD) is described on some model examples; the general description is given. The variational version is given in section 6.2. The general scheme is discussed; the main theorem about the estimate of the difference between the exact solution and the solution to the partially decomposed problem is proved. The main theorem is then applied to the modelling of thin structures (the finite rod structures). These structures simulate, in particular, the mechanical behavior of the human or animal blood circulatory system [172]. The dimension reduction in such a modelling is a natural approach, although the full-dimensional models have to be kept in the neighborhoods of the bifurcations or junctions. So the MAPDD gives the asymptotically exact answer what should be the correct interface conditions. The flows in such structures are discussed in section 6.3. Other applications (such as an extrusion process) are as well discussed in section 6.3.

In section 6.4 the MAPDD is applied to the homogenization problems. The classical homogenization problem [12],[22],[49],[177] of the Dirichlet problem

$$div(A(\frac{x}{\varepsilon})grad u_\varepsilon) = f(x), \quad x \in G, u_\varepsilon|_{\partial G} = 0,$$

is considered; here $A(\xi)$ is an $s \times s$ positive definite symmetric matrix 1−periodic in ξ, and ε is a small parameter. Although the asymptotic behavior of the asymptotic solution "inside" (i.e. at some distance from the boundary) is well studied, the boundary layers are still remain an open problem in the homogenization. And these boundary layers are of great importance for the flux description, for the fracture analysis etc. So, it would be natural to homogenize the problem inside of G at some distance from the boundary keeping the initial formulation in some thin boundary strip. The implementation of this idea arises the question about the asymptotically correct interface conditions between the homogenized and non-homogenized subdomains. We study this question in a model situation when G is a thick layer in $I\!R^s$, and the asymptotic expansion of the solution is known (it was constructed in [133],[16]). The MAPDD is implemented here below as the variational formulation of the initial problem on the subspace of functions having the form of the asymptotic approximation of order K at some distance from the boundary. Some small corrections transform the MAPDD into the partial homogenization method, prescribing some special interface conditions coupling the homogenized and the non-homogenized parts. Theorem 6.4.1 justifies the partial homogenization by estimating the difference between the asymptotic solution and the solution of the partially homogenized problem.

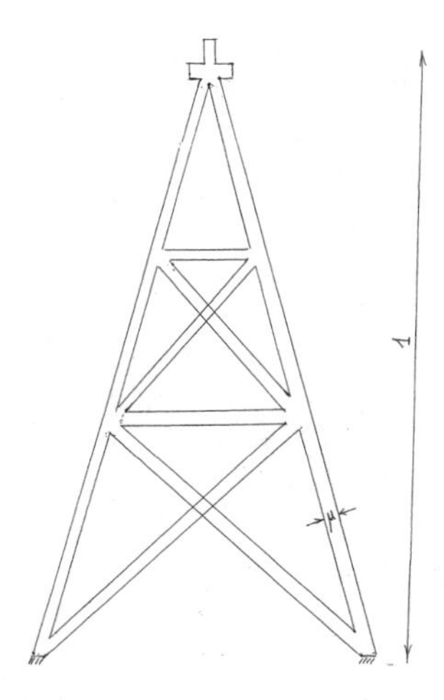

Figure 6.0.1. A finite rod structure.

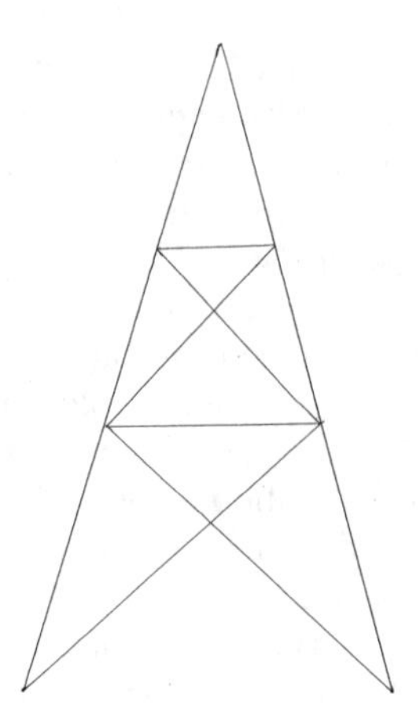

Figure 6.0.2. The structural mechanics approach: passage to the limit as μ tends to zero (the dimensional reduction).

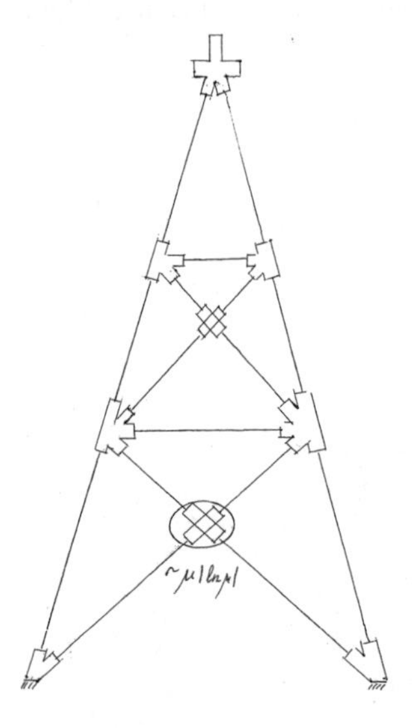

Figure 6.0.3. Asymptotic partial decomposition of the domain: a hybrid 2D - 3D multi-scale model with the two- dimensional zoom.

Some similar hybrid models appeared earlier in mechanics and computations. They are based on some heuristic approaches (see for example, a numerical simulation of shallow water equation in the system "lake-river" [30]). The open questions were: what is the relation between the original completely 3D model and the hybrid model? what is the error? what are the mathematical principles of construction of precise hybrid models, especially the precise interface conditions?

Below we shall analyze these questions for hybrid models of conductivity and elasticity equations stated in rod structures.

We consider below two versions of the method of asymptotic partial decomposition of domain. The first version is "differential", i.e. we work with the differential formulation of the initial problem , we obtain the 1D differential equation in the reduced part of the rod structure and we add the differential interface conditions on the boundary between 3D and 1D parts (as in [26]). Of course we can pass to a variational formulation of the partially decomposed problem but it is generated by a differential one.

The second version is a direct variational approach, when the 3D integral identity for the original problem is restated for a special subspace of functions having a form of the ansatz of the asymptotic solution in the regular thin part of the rod structure. This approach is similar to I.Babuska's idea of the projection with the dimensional reduction, but the partial asymptotic decomposition keeps the initial functional set in the boundary layer domain

In this chapter we always use the notation ε for a small parameter, so sometimes it stands for the parameter denoted μ in chapters 1-4.

6.1 Differential version

6.1.1 General description of the differential version

Consider the abstract problem

$$\Lambda_\varepsilon u_\varepsilon = F_\varepsilon,$$

which depends on a small parameter ε. Here Λ_ε is a linear operator, F_ε is a known right hand side and u_ε is the unknown solution.

Let $u_\varepsilon, F_\varepsilon$ be the functions of the vector valued variable $x = (x_1, ..., x_s)$ varying in the domain G^ε. Suppose that the asymptotic behavior of the solution is essentially different in some parts of the domain G^ε, i.e. it is a "regular asymptotic expansion" in a subdomain G_1^ε of the domain G^ε and it is "asymptotically singular function" in the subdomain $G^\varepsilon \backslash G_1^\varepsilon$. For example it depends only on the variable x_1 in G^ε, and in $G^\varepsilon \backslash G_1^\varepsilon$ it has the form of the boundary layer (as it will be in subsection 1.1.2), fig.6.1.1. Usually the part $G^\varepsilon \backslash G_1^\varepsilon$ is much smaller than the part G_1^ε. Then it is natural to decompose the initial problem to a simplified equation in the part G_1^ε of G^ε (for example, by dimensional reduction as it will be shown in section 1.2) and to the original equation stated in the "small" reduced part $G^\varepsilon \backslash G_1^\varepsilon$. Of course we should add the conditions on the interface of two parts G_1^ε and $G^\varepsilon \backslash G_1^\varepsilon$. The interface conditions should be "correct", i.e. the asymptotic solution has to satisfy these conditions, and the whole problem containing the simplified equation set in G_1^ε, the original equation posed in the "small" subdomain $G^\varepsilon \backslash G_1^\varepsilon$ and the interface conditions has to be well posed. Let us develop now this idea. Let u_ε be the solution of some problem depending on a small parameter ε stated in the domain $G^\varepsilon \subset I\!R^s$; let the structure of the asymptotic expansion of the solution is known in the "regular" subdomain G_1^ε of the domain G^ε , i.e.

$$u_\varepsilon \sim \sum_{l=l_0}^{\infty} \varepsilon^l (v_l(x) + u_l(x, \frac{x}{\varepsilon})), \tag{6.1.1}$$

where u_l are bounded and defined identically by the "previous" functions $v_{l_0}, \ldots, v_{l-1}, u_{l_0}, \ldots, u_{l-1}$.

Assume that we know the relation of u_l on these "previous" functions but we do not know explicitly $v_{l_0}, \ldots, v_{l-1}$.

Suppose that the function u_ε is expanded asymptotically in the subdomain $G^\varepsilon \backslash G_1^\varepsilon$ as follows :

$$u_\varepsilon \sim \sum_{l=l_0}^{\infty} \varepsilon^l U_{l\varepsilon}, \tag{6.1.2}$$

where $U_{l\varepsilon}$ are also bounded but not known.

Let each pair of partial sums

$$(v^{(K)}, U^{(K)}) = (\sum_{l=l_0}^{N(K)} \varepsilon^l v_l, \sum_{l=l_0}^{N(K)} \varepsilon^l U_{l\varepsilon})$$

be a solution of the problem

$$L_\varepsilon(v^{(K)}, U^{(K)}) = F_\varepsilon^{(K)} + O(\varepsilon^K), \tag{6.1.3}$$

where the linear operator $L_\varepsilon : H_1^\varepsilon \to H_2^\varepsilon$ is known and $H_1^\varepsilon, H_2^\varepsilon$ are the Hilbert spaces and $F_\varepsilon^{(K)}$ is an $O(\varepsilon^K)$ approximation of F_ε. Assume that for each $F \in H_2^\varepsilon$ there exists a unique solution of the problem of type (6.1.3)

$$L_\varepsilon V = F$$

and assume that the a priori estimate (stability condition) holds true

$$\|V\|_{H_1^\varepsilon} \leq C\varepsilon^{-r} \|F\|_{H_2^\varepsilon}$$

where the constants C and r do not depend on ε.

Then clearly the solution $(\bar{v}^{(K)}, \bar{U}^{(K)})$ of each problem

$$L_\varepsilon(\bar{v}^{(K)}, \bar{U}^{(K)}) = F_\varepsilon^{(K)} \tag{6.1.4}$$

is close to the pair $(v^{(K)}, U^{(K)})$, i.e.

$$\|(v^{(K)}, U^{(K)}) - (\bar{v}^{(K)}, \bar{U}^{(K)})\|_{H_1^\varepsilon} = O(\varepsilon^{K-r}).$$

Thus the calculation of the terms of the asymptotic expansions (6.1.1) and (6.1.2) up to $O(\varepsilon^{K-r})$ is reduced to the solution of the problem (6.1.4).

One can build some algorithms based on this idea if the structure of the asymptotic expansions is known . We should only state such a problem (6.1.4)

a) which is uniquely solvable and stable and

b) for which the partial sums of the asymptotic expansions $v^{(K)}$ and $U^{(K)}$ are the solutions.

6.1.2 Model example

Consider the Poisson equation

$$\Delta u_\varepsilon = f(x_1), \tag{6.1.5}$$

with the right hand side from $L^2(0,1)$, stated in the thin rectangle

$$G_\varepsilon = (0,1) \times (-\frac{\varepsilon}{2}, \frac{\varepsilon}{2}) \ (\varepsilon << 1)$$

with the boundary conditions

$$\begin{gathered} u_\varepsilon = 0 \text{ for } x \in \gamma_1 = \{x_1 = 1, x_2 \in (-\frac{\varepsilon}{2}, \frac{\varepsilon}{2})\}, \\ u_\varepsilon = 0 \text{ for } x \in \gamma_2 = \{x_1 = 0, x_2 \in (-\frac{\varepsilon}{4}, \frac{\varepsilon}{4})\}, \\ \frac{\partial u_\varepsilon}{\partial n} = 0 \text{ for } x \in \gamma_3 = \partial\Pi \backslash (\gamma_1 \cup \gamma_2). \end{gathered} \tag{6.1.6}$$

The asymptotic expansion of the solution has the following structure (up to $O(e^{-c/\varepsilon}), c > 0)$:

$$U_\varepsilon \sim \sum_{l=0}^{\infty} \varepsilon^l (v_l(x_1) + U_l(\frac{x}{\varepsilon})). \tag{6.1.7}$$

Here $v_0(x_1)$ is the solution of the boundary value problem

$$\left\{ \begin{array}{l} v_0'' = f, \ x_1 \in (0,1), \\ v_0(0) = 0, \ v_0(1) = 0; \end{array} \right. \tag{6.1.8}$$

$U_l(\xi)$ are the boundary layer terms, i.e. the exponentially decaying (as $\xi_1 \to +\infty$) solutions of the chain of problems $(l = 1, 2, \ldots)$

$$\left\{ \begin{array}{r} \Delta U_l = 0, \ \xi \in \Omega = (0, +\infty) \times (-\frac{1}{2}, \frac{1}{2}), \\ U_l = c_l \text{ for } \xi_1 = 0, \ \xi_2 \in (-\frac{1}{4}, \frac{1}{4}), \\ \frac{\partial U_l}{\partial \xi_2} = 0 \text{ for } \xi_2 = \pm\frac{1}{2}, \\ \frac{\partial U_l}{\partial \xi_1} = -v_{l-1}'(0) \text{ for } \xi_1 = 0, \ |\xi_2| \geq \frac{1}{4}, \end{array} \right. \tag{6.1.9}$$

where c_l are the constants defined by the condition $U_l \to 0$ as $\xi_1 \to +\infty$; $v_l(x_1)$ are the solutions of the chain of problems $(l = 1, 2, \ldots)$

$$\left\{ \begin{array}{l} v_l'' = 0, \ x_1 \in (0,1), \\ v_l(0) = -c_l, \ v_l(1) = 0 \end{array} \right. \tag{6.1.10}$$

i.e. $v_l(x_1) = c_l(x_1 - 1)$; and $U_0 = 0$.

Let us state the partially decomposed problem for function

$$v^K(x_1) = \sum_{l=0}^{K} \varepsilon^l v_l(x_1)$$

defined in the domain $(\delta, 1)$ $(\delta << 1)$ and for the function

$$U^{(K)}(x) = \sum_{l=0}^{K} \varepsilon^l (v_l(x_1) + U_l(\frac{x}{\varepsilon})) \tag{6.1.7'}$$

in the domain $(0, \delta) \times (-\frac{\varepsilon}{2}, \frac{\varepsilon}{2})$.

Let $\gamma_\delta = \{x \in \Pi \mid x_1 = \delta\}$. Since $U_l(\frac{x}{\varepsilon})$ are the boundary layer functions they vanish up to $O(\varepsilon^J)$ for any J on the interval γ_δ with $\delta = \varepsilon^{1-\alpha}(\alpha > 0)$. Then the following two relations hold true on γ_δ up to $O(\varepsilon^J)$

$$U^{(K)}(\delta, x_2) = v^{(K)}(\delta),$$

$$\frac{1}{\varepsilon}\int_{-\varepsilon/2}^{\varepsilon/2} \frac{\partial U^{(K)}}{\partial x_1}(\delta, x_2) dx_2 = v^{(K)'}(\delta). \tag{6.1.11}$$

Moreover, for any J there exist $J_1 \in I\!R$ such that if $\delta = J_1 \varepsilon\ |ln\varepsilon|$ then

$$U_l(\frac{\delta}{\varepsilon}, \frac{x_2}{\varepsilon}) \ = \ O(\varepsilon^J)$$

and therefore (6.1.11) holds true up to $O(\varepsilon^J)$. Below we take $J = K$.

The information about the structure of asymptotic solution (6.1.7),(6.1.8), (6.1.10) and the relation (6.1.11), true for this asymptotic solution, gives us the conjecture of construction of partially decomposed problem: we keep 3D problem in the neighborhood of end $x_1 = 0$

$$\Delta U = f(x_1),\ x \in (0, \delta) \times (-\frac{\varepsilon}{2}, \frac{\varepsilon}{2}),$$

$$\frac{\partial U}{\partial n} = 0,\ x \in \gamma_3 \cap \{x_1 \leq \delta\}$$

$$U = 0,\ x \in \gamma_2,$$

we reduce the dimension at the distance $\delta = O(\varepsilon|ln\varepsilon|)$ of the end $x_1 = 0$:

$$v'' = f(x_1),\ x \in (\delta, 1),$$

$$v(1) = 0,$$

and we state the interface conditions induced by (6.1.11)

$$U(\delta, x_2) = v(\delta),\ \ \frac{1}{\varepsilon}\int_{-\varepsilon/2}^{\varepsilon/2} \frac{\partial U}{\partial x_1}(\delta, x_2) dx_2 = v'(\delta), \tag{6.1.11'}$$

i.e. the partially decomposed problem is

$$v'' = f(x_1),\ x \in (\delta, 1),$$
$$v(1) = 0,$$
$$\Delta U = f(x_1),\ x \in (0,\delta) \times (-\frac{\varepsilon}{2}, \frac{\varepsilon}{2}),$$
$$\frac{\partial U}{\partial n} = 0,\ x \in \gamma_3 \cap \{x_1 \leq \delta\} \tag{6.1.12}$$
$$U = 0,\ x \in \gamma_2$$

$$U(\delta, x_2) = v(\delta),\ \ \frac{1}{\varepsilon}\int_{-\varepsilon/2}^{\varepsilon/2} \frac{\partial U}{\partial x_1}(\delta, x_2)dx_2 = v'(\delta).$$

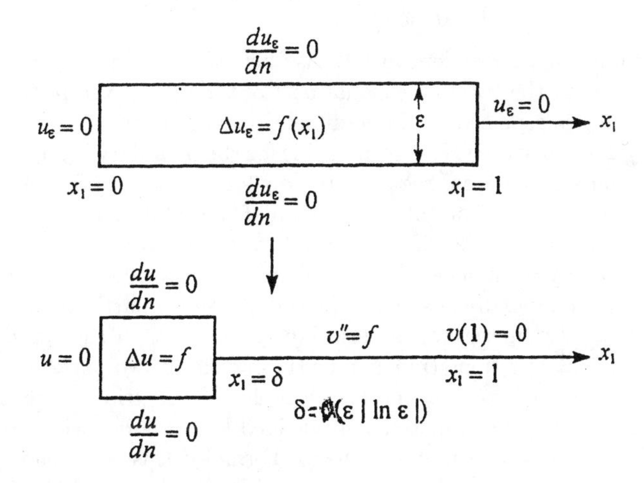

Figure 6.1.1. The "cutting" procedure in the method of partial asymptotic decomposition of domain.

The pair $(v^{(K)}, U^{(K)})$ satisfies (6.1.12) exactly with the exception of the condition

$$\frac{\partial U}{\partial n} = 0 \text{ for } x_1 = 0,\ |x_2| \geq \frac{\varepsilon}{4}, \tag{6.1.13}$$

which is satisfied up to the term $O(\varepsilon^K)$, i.e.

$$\frac{\partial U^{(K)}}{\partial x_1} = -\varepsilon^K v_K'(0),$$

and with exception of the conditions (6.1.11 ') satisfied up to $O(\varepsilon^K)$.

So the pair $(v^{(K)}, U^{(K)})$ satisfies the problem of the type (6.1.4).

The variational formulation of the problem (6.1.12) is as follows.

Let $H^1((0,\delta)\times(-\frac{\varepsilon}{2},\frac{\varepsilon}{2}),(\delta,1))$ be the space of pairs of functions $(R(x_1,x_2), \rho(x_1))$ such that

$$R \in H^1((0,\delta)\times(-\frac{\varepsilon}{2},\frac{\varepsilon}{2})), \quad \rho \in H^1((\delta,1)), \quad R(\delta,x_2)=\rho(\delta), \quad R|_{\gamma_2}=0, \rho(1)=0.$$

We seek such an element $(U,v) \in H^1((0,\delta)\times(-\frac{\varepsilon}{2},\frac{\varepsilon}{2}),(\delta,1))$, that for any

$$(R,\rho) \in H^1((0,\delta)\times(-\frac{\varepsilon}{2},\frac{\varepsilon}{2}),(\delta,1)),$$

$$-\int_0^\delta \int_{-\frac{\varepsilon}{2}}^{\frac{\varepsilon}{2}} (\nabla U, \nabla R) dx_1 dx_2 \; - \; \int_\delta^1 \varepsilon v' \rho' dx_1 \; =$$

$$= \int_0^\delta \int_{-\frac{\varepsilon}{2}}^{\frac{\varepsilon}{2}} f R dx_1 dx_2 \; + \; \int_\delta^1 \varepsilon f \rho dx_1. \tag{6.1.14}$$

Remark 6.1.1 Problem (6.1.12) is "better" than (6.1.5), (6.1.6) because of the reducing of the two dimensional part of the problem from G_ε to $G_{1\varepsilon}$. The part $G_\varepsilon \backslash G_{1\varepsilon}$ is replaced by a one-dimensional segment.

The problem set in $G_{1\varepsilon}$ is responsible for the boundary layer and it is coupled with the one-dimensional regular problem stated in $(\delta,1)$. It contains *complete* information on the asymptotic expansion of the solution of the problem (6.1.5), (6.1.6). Now this coupled 2D - 1D problem (6.1.12) could be solved numerically by the finite element method. To this end we consider the standard triangle partition of the two dimensional part and a uniform partition of the one dimensional part . Enumerate all nodes. We introduce then the standard piecewise linear base hat functions (independently in each of two parts; the hat function number n is equal to 1 in one of the nodes of the grid and it vanishes in all other nodes; these hat functions are defined for all nodes with exception of the nodes belonging to the interface line γ_δ. Formally, this set of base functions corresponds to two independent sets of standard hat functions defined in 2D part and in 1D part with the Dirichlet condition on γ_δ. Then we complete the union of these two sets of 2D and 1D hat functions by one special hybrid hat function ("super-element") which is defined on the union $G_\varepsilon \cup (\delta,1)$ and it is partially 2D and partially 1D hat function: it is piecewise linear, it is equal to 1 on γ_δ (i.e. for all nodes of γ_δ) and it is equal to zero in all other nodes. Finally, we extend all standard 2D hat functions by zero on the 1D part, we extend all standard 1D hat functions by zero on the 2D part, and then use all hat functions (the hybrid "super-element" included) to form a base of finite element subspace H^1_{fe} and we state the problem (6.1.14) on this subspace, i.e. $(R,\rho) \in H^1_{fe}$. We developed the direct numerical study of the model 2D problem (6.1.5),(6.1.6) by finite element method and the numerical study of the partially decomposed problem (6.1.14). The comparison of results of these numerical experiments shows the excellent precision of the partially decomposed problem with respect to the original 2D problem.

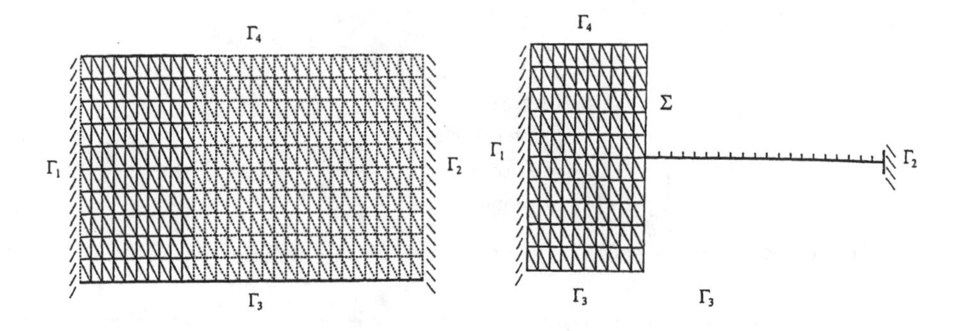

Figure 6.1.2. Two dimensional triangulation for direct finite element method and for a partially decomposed domain.

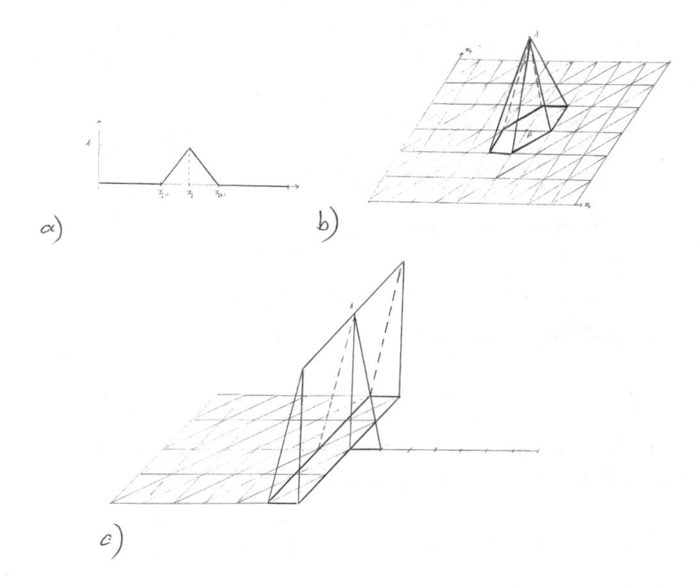

Figure 6.1.3. The finite element implementation: a) a 1D basis "hat"-function; b) a 2D basis "hat"-function; c) the hybrid 2D-1D basis "hat"-function.

The same numerical experiment for a one bundle rod structure (see definition below in section 6.1.3) confirms the closeness of the original and the hybrid models.

Now to justify the passage from the problem (6.1.5), (6.1.6) to the problem (6.1.12), we should prove the existence and uniqueness of the solution of the problem (6.1.12) (which is not standard) and prove its stability with respect to perturbations in the right hand sides of the conditions (6.1.13) and (6.1.11'). Note that the question of stability could be reduced to the stability of the condition (6.1.11') only. Indeed let U^c be the solution of the following boundary layer problem :

$$\begin{cases} \Delta U^c = 0 \text{ for } \xi \in \Omega, \\ U^c = 0 \text{ for } \xi_1 = 0, \xi_2 \in (-\frac{1}{4}, \frac{1}{4}), \\ \frac{\partial U^c}{\partial \xi_2} = 0 \text{ for } \xi_2 = \pm\frac{1}{2}, \\ \frac{\partial U^c}{\partial \xi_1} = 1 \text{ for } \xi_1 = 0, |\xi_2| \geq \frac{1}{4}, \\ U^c \to \text{ const as } \xi_1 \to +\infty. \end{cases}$$

Add the corrector $\varepsilon^K v(0) U^c(\frac{x}{\varepsilon})$ to $U^{(K)}$ in the domain $G_{1\varepsilon} = (0,\delta) \times (-\frac{\varepsilon}{2}, \frac{\varepsilon}{2})$.

Note that the norm $H^1(G_{1\varepsilon})$ of this corrector is $O(\varepsilon^{K-1}\sqrt{\varepsilon\delta})$ and that the corrected solution satisfies to all conditions (6.1.12) exactly and to conditions (6.1.11') up to the term $O(\varepsilon^K)$. Thus the justification of the passage from the problem (6.1.5), (6.1.6) to the problem (6.1.12) is reduced to the questions of existence and uniqueness of the solution (6.1.12) and its stability with respect to the perturbations in the conditions (6.1.11') only.

Let us prove now that the problem (6.1.12) is uniquely solvable.

Consider the variational formulation (6.1.14) (the other proof is given for classical formulation in [156]).

The Poincaré-Friedrichs inequality gives:

$$\|R\|_{L^2((0,\delta)\times(-\frac{\varepsilon}{2},\frac{\varepsilon}{2}))} \leq C\delta \|\nabla R\|_{L^2((0,\delta)\times(-\frac{\varepsilon}{2},\frac{\varepsilon}{2}))},$$

where C does not depend on δ, ε, and

$$\|\rho\|_{L^2((\delta,1))} \leq C\|\rho'\|_{L^2((\delta,1))}.$$

So the right hand side functional is bounded by

$$C\sqrt{\varepsilon}\|f\|_{L^2((0,1))}(\|\nabla R\|^2_{L^2((0,\delta)\times(-\frac{\varepsilon}{2},\frac{\varepsilon}{2}))} + \varepsilon\|\rho'\|^2_{L^2((\delta,1))})^{1/2}.$$

Applying the Lax-Milgram lemma we obtain the existence and uniqueness of the solution. We obtain also the a priori estimate

$$\|U, v\|_1 = \sqrt{\|U\|^2_{H^1((0,\delta)\times(-\frac{\varepsilon}{2},\frac{\varepsilon}{2}))} + \varepsilon\|v\|^2_{H^1((\delta,1))}} \leq$$

$$C_1\sqrt{\varepsilon}\|f\|_{L^2((0,1))}.$$

The perturbations in the conditions (6.1.11') could be implemented to the right hand side by simple subtraction of a pair of functions $\Psi \in H^1((0,\delta) \times (-\frac{\varepsilon}{2}, \frac{\varepsilon}{2}))$ and $\psi \in H^1((\delta, 1))$ such that

1) this pair satisfies the boundary conditions $\partial\Psi/\partial n = 0, \ x \in \gamma_3 \cap \{x_1 \leq \delta\}$, $\Psi = 0$ on γ_2 and $\psi(\delta) = \psi(1) = 0$;

2) the trace of the function $\Psi(\delta, x_2)$ is equal to the given perturbation in the first condition (11') and the value $\psi'(\delta) - \frac{1}{\varepsilon}\int_{-\varepsilon/2}^{\varepsilon/2} \frac{\partial\Psi}{\partial x_1}(\delta, x_2)dx_2$ is equal to the given perturbation in the second condition (6.1.11')

The structure of the asymptotic solution U_ε shows that for any K, there exists $K_1 \in I\!R$ such that if $\delta = K_1\varepsilon\ |ln\varepsilon|$ then the perturbations in (6.1.11') are of order $O(\varepsilon^K)$.

Therefore again for any K, there exist $K_1 \in I\!R$ such that if $\delta = K_1\varepsilon\ |ln\varepsilon|$ then

$$\|U_\varepsilon^{(K)} - U\ ,\ v^{(K)} - v\|_1 \ = \ O(\varepsilon^K).$$

On the other hand,

$$\|U_\varepsilon^{(K)} - u_\varepsilon\|_{H^1(B^\varepsilon)} \ = \ O(\varepsilon^K).$$

So for any K, there exist $K_1 \in I\!R$ such that if $\delta = K_1\varepsilon\ |ln\varepsilon|$ then

$$\|u_\varepsilon - U\|_{H^1((0,\delta)\times(-\frac{\varepsilon}{2},\frac{\varepsilon}{2}))} \ = \ O(\varepsilon^K),$$

$$\|u_\varepsilon - v\|_{H^1((\delta,1)\times(-\frac{\varepsilon}{2},\frac{\varepsilon}{2}))} \ = \ O(\varepsilon^K).$$

6.1.3 Poisson equation in a rod structure

In the same way the partially decomposed problem can be set for the Poisson equation in a rod structure (the asymptotic expansion of the solution was built in Chapter 4 and [148]). It could be proved that the same estimate holds true. Describe the method in a simple case of a rod structure containing one bundle.

Let $e_1,\ \dots\ , e_n$ be n closed segments in $I\!R^s$ $(s = 2, 3)$, which have a single common point O (i.e. the origin of the coordinate system), and let it be the common end point of all these segments. Let $\beta_1, \dots, \beta_n$ be n bounded (s-1)-dimensional domains in $I\!R^s$, which belong to n hyper-planes containing the point O. Let β_j be orthogonal to e_j. Let β_j^ε be the image of β_j obtained by a homothetic contraction in $1/\varepsilon$ times with the center O. Denote B_j^ε the open cylinders with the bases β_j^ε and with the heights e_j, denote also $\hat{\beta}_j^\varepsilon$ the second base of each cylinder B_j^ε and let O_j be the end of the segment e_j which belongs to the base $\hat{\beta}_j^\varepsilon$. Define the bundle of segments e_j centered in O as $B \ = \ \cup_{j=1}^n\ e_j$.

Denote below $O_0 = O$. Define the one bundle rod structure associated with the bundle B as the set B^ε of inner points of the union $\cup_{j=1}^n\ \bar{B}_j^\varepsilon$. Here the bar means the operation of the closure.

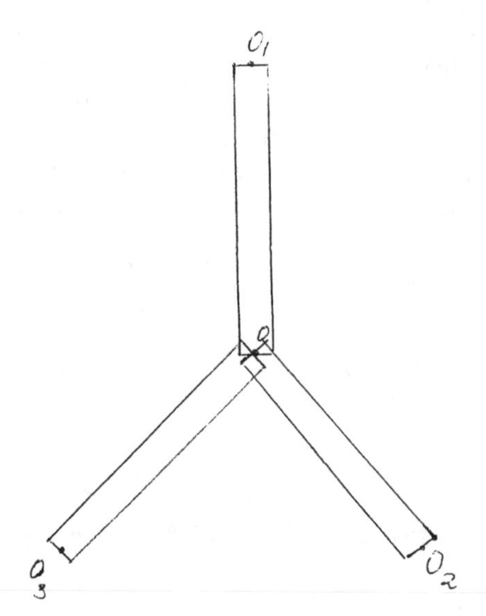

Figure 6.1.4. A one bundle rod structure.

Introduce the local coordinate system $Ox_1^{e_{0j}}...x_s^{e_{0j}}$ associated with a segment e_j such that the direction of the axis $Ox_1^{e_{0j}}$ coincides with the direction of the segment OO_j, i.e. $x_1^{e_{0j}}$ is a longitudinal coordinate. The axes $Ox_1^{e_{0j}}, ..., Ox_s^{e_{0j}}$ form a cartesian coordinate system.

Define some cutting functions χ_j on $I\!R_+$. Let $d_j\varepsilon$ be the minimal distance from the base of the cylinders B_j^ε to a parallel cross-section σ_j^ε of B_j^ε such that this cross-section does not contain points of other cylinders (i.e. $\sigma_j^\varepsilon \cap B_i^\varepsilon = \emptyset, \ i \neq j$). Then let $\chi_j = 0$ for $x_1^{e_{0j}} \leq d_j$ and $\chi_j = 1$ for $x_1^{e_{0j}} \geq d_j + 1$; we suppose that the function $\chi_j \in C^2$ and it does not depend on ε.

In the same way we can introduce local coordinate systems $O_j x_1^{e_{j0}}...x_s^{e_{j0}}$ associated with e_j such that the direction of the axis $O_j x_1^{e_{j0}}$ coincides with the direction of the segment O_jO, and so $x_1^{e_{0j}} = |e_j| - x_1^{e_{j0}}$.

Consider the Poisson equation

$$\Delta u_\varepsilon = f(x), \quad x \in B^\varepsilon \tag{6.1.15}$$

with the boundary conditions

$$u_\varepsilon = 0 \text{ for } x \in \hat{\beta}_j^\varepsilon, \ j = 1, ..., n, \tag{6.1.16}$$

$$\frac{\partial u_\varepsilon}{\partial n} = 0 \text{ for } x \in \partial B^\varepsilon \setminus \cup_{j=1}^n \hat{\beta}_j^\varepsilon, \tag{6.1.17}$$

where f is equal to $f_j(x_1^{e_{0j}})\chi_j(x_1^{e_{0j}}/\varepsilon)$ for each $e_j, \ \ f_j \in C^\infty((0, |e_j|))$.

The asymptotic analysis of this problem developed in Chapter 4 shows that the boundary layers decay exponentially.

Cut off the cylinders B_j^ε at the distance $\delta = const\ \varepsilon\ ln\ (\varepsilon)$ from the nodes by the planes perpendicular to the segments and replace the inner parts of the cylinders by the corresponding parts of the segments e_j. We obtain the set $B^{\varepsilon,\delta}$. Denote $B_i^{\varepsilon,\delta}$ the connected truncated part of B^ε, containing the node O_i, and denote e_{ij} the part of one segment $e = O_iO_j$ among $e_1, ..., e_n$ which connects $B_i^{\varepsilon,\delta}$ and $B_j^{\varepsilon,\delta}$; let S_{ij} be a cross-section of the truncated cylinder corresponding to e_{ij}, such that it belongs to $\partial B_i^{\varepsilon,\delta}$.

Consider the equations for each $B_i^{\varepsilon,\delta}$:

$$\Delta U = f, \tag{6.1.18}$$

$$U = 0, \quad x \in \hat{\beta}_j^\varepsilon, \tag{6.1.19}$$

$$\frac{\partial U_\varepsilon}{\partial n} = 0, \quad x \in (\partial B_i^{\varepsilon,\delta} \cap \partial B^\varepsilon) \setminus \cup_{j=1}^n \hat{\beta}_j^\varepsilon, \tag{6.1.20}$$

the equation for each segment e_{j_1,j_2} :

$$v''_{j_1,j_2} = f(x_1^{e_{j_1,j_2}}), \tag{6.1.21}$$

and the interface conditions on each cross-section S_{j_1,j_2}

$$U = v_{j_1,j_2}, \quad x_1^{e_{j_1,j_2}} = \delta; \tag{6.1.22}$$

$$\frac{1}{mesS_{j_1,j_2}} \int_{S_{j_1,j_2}} \frac{\partial U}{\partial n} ds = v'_{j_1,j_2}(\delta), \tag{6.1.23}$$

where v_{j_1,j_2} are unknown functions of the variable $x_1^{e_{j_1,j_2}}$ defined on each segment e_{j_1,j_2}. Of course

$$v_{j_1,j_2}(x_1^{e_{j_1,j_2}}) = v_{j_2,j_1}(|e_{j_1,j_2}| + 2\delta - x_1^{e_{j_1,j_2}}). \tag{6.1.24}$$

Let us give the variational formulation for this problem. Let

$$H^1(B_0^{\varepsilon,\delta}, ..., B_n^{\varepsilon,\delta}, e_1, ..., e_n)$$

be the space of ordered collections $(U_0, ..., U_n, w_1, ..., w_n)$, where U_i are the functions from $H^1(B_i^{\varepsilon,\delta})$, equal to zero on $\partial B_i^{\varepsilon,\delta} \cap \cup_{j=1}^n \hat{\beta}_j^\varepsilon$, and $w_1, ..., w_n$ are the functions of the variable $x_1^{e_j}$ associated with the segments $e_1, ..., e_n$ such that for each S_{j_1,j_2} we have

$$U_{j_1} = w_j, \tag{6.1.25}$$

where $e_j = e_{j_1,j_2}$.

The scalar product is defined as

$$((U_0, ..., U_n, w_1, ..., w_n), (V_0, ..., V_n, v_1, ..., v_n))_P =$$

$$\sum_{i=0}^n \int_{B_i^{\varepsilon,\delta}} \sum_{r=1}^s \frac{\partial U_i}{\partial x_r} \frac{\partial \Phi_i}{\partial x_r} dx + \varepsilon^{s-1} mes\beta_j \sum_{j=1}^n \int_{e_j \setminus \cup_{i=0}^n B_i^{\varepsilon,\delta}} v'_j w'_j dx_1^{e_{0j}}.$$

The norm is defined as

$$\|(U_0, ..., U_n, w_1, ..., w_n)\|_P = ((U_0, ..., U_n, w_1, ..., w_n), (U_0, ..., U_n, w_1, ..., w_n))_P^{1/2}.$$

Then the variational formulation is as follows:
find $(U_0, ..., U_n, v_1, ..., v_n) \in H^1(B_0^{\varepsilon,\delta}, ..., B_n^{\varepsilon,\delta}, e_1, ..., e_n)$ such that for any

$$(\Phi_0, ..., \Phi_n, w_1, ..., w_n) \in H^1(B_0^{\varepsilon,\delta}, ..., B_n^{\varepsilon,\delta}, e_1, ..., e_n)$$

the integral identity holds true:

$$-\sum_{i=0}^{n} \int_{B_i^{\varepsilon,\delta}} \sum_{r=1}^{s} \frac{\partial U_i}{\partial x_r} \frac{\partial \Phi_i}{\partial x_r} dx \; - \; \varepsilon^{s-1} mes\beta_j \sum_{j=1}^{n} \int_{e_j \setminus \cup_{i=0}^n B_i^{\varepsilon,\delta}} v_j' w_j' dx_1^{e_{0j}}$$

$$= \sum_{i=0}^{n} \int_{B_i^{\varepsilon,\delta}} (f, \Phi) dx$$

$$+ \varepsilon^{s-1} mes\beta_j \sum_{j=1}^{n} \int_{e_j \setminus \cup_{i=1}^n B_i^{\varepsilon,\delta}} f w_j \; dx_1^{e_{0j}}. \tag{6.1.26}$$

The asymptotic analysis of the problem (6.1.15)- (6.1.17) shows that for any K we can find such $K_1 \geq K$ that if $\delta = K_1 \varepsilon \; |ln\varepsilon|$ then the truncated sum $u^{(K)}$ of order K of the asymptotic expansion (see section 4.7) adopted to a scalar case and defined on $B_i^{\varepsilon,\delta}$ and the truncated sum $v_j^{(K)}$ of order K of the regular asymptotic expansion (depending on $x_1^{e_{0j}}$ only) $j = 1, ..., n$ satisfy to the same integral identity with the discrepancy

$$\sum_{i=0}^{n} \int_{\partial B_i^{\varepsilon,\delta} \setminus \partial B^{\varepsilon}} \varepsilon^K \rho_1 \Phi_i ds \; + \; \sum_{i=0}^{n} \int_{B_i^{\varepsilon,\delta}} \varepsilon^K \rho_2 \Phi_i dx$$

$$+ \varepsilon^{K+(s-1)} \sum_{j=1}^{n} \int_{e_j \setminus \cup_{i=1}^n B_i^{\varepsilon,\delta}} r_{3,j} w_j' \; dx_1^{e_{0j}}. \tag{6.1.27}$$

There is also a discrepancy in the condition (6.1.22) of order $O(exp(-c\delta/\varepsilon))$, i.e.

$$u^{(K_1)} = v_j^{(K_1)} + exp(-c\delta/\varepsilon) r_3, \quad x \in S_{j_1, j_2}, \tag{6.1.28}$$

where $c > 0, \rho_1, \rho_2, r_{3,j}$ are bounded vector valued functions. Moreover the discrepancy $exp(-c\delta/\varepsilon)r_3$ can be continued from S_{j_1,j_2} to $B_i^{\varepsilon,\delta}$ in such a way that this extension has an order $O(exp(-c_1\delta/\varepsilon))$, $c_1 > 0$ in H^1.

Subtracting this extension from the truncated series $u^{(K)}$ we obtain the integral identity with the final discrepancy of a form (6.1.27).

The functional (6.1.27) is bounded in the norm P due to the estimate

$$\|\Phi_i\|_{L^2(B^{\varepsilon,\delta})}, \quad \varepsilon^{s-1} mes\beta_j \|w_j\|_{L^2(e_j \setminus \cup_{i=1}^n B_i^{\varepsilon,\delta})}, \quad \|\Phi_i\|_{L^2(\partial B^{\varepsilon,\delta} \setminus \partial B^{\varepsilon})}$$

$$\leq \quad C\|\Phi_0, ..., \Phi_n, w_1, ..., w_n\|_P \tag{6.1.29}$$

with the constant C independent of ε, δ.

This estimate is a corollary of the Poincaré - Friedrichs inequality for finite rod structures (see Appendix 4.A2).

$$\|\Phi\|_{L^2(B^\varepsilon)} \leq C\|\nabla\Phi\|_{L^2(B^\varepsilon)} \tag{6.1.30}$$

if we apply it to the function Φ defined as

$$\Phi(x) = \left\{ \begin{array}{c} \Phi_i(x), \quad x \in B_i^{\varepsilon,\delta}, \ i = 0, ..., n, \\ w_j(x_1^{e_{0j}}), \quad x \in B_j^\varepsilon \backslash \cup_{i=1}^n B_i^{\varepsilon,\delta}, \ j = 1, ..., n. \end{array} \right.$$

For $\|\Phi_i\|_{L^2(S_{j_1,j_2})}$ we should also use the standard estimate

$$\|\Phi_i\|_{L^2(S_{j_1,j_2})} \leq C\|\Phi\|_{H^1(B_{j_1,j_2}^{\varepsilon,\delta} \backslash \cup_{i=1}^n B_i^{\varepsilon,\delta})},$$

where $B_{j_1,j_2}^{\varepsilon,\delta}$ is the cylinder $B_j^{\varepsilon,\delta}$ which corresponds to the surface S_{j_1,j_2} (i.e. contains this surface).

The integral identity (6.1.26) has the unique solution due to the Lax-Milgram lemma.

Taking into consideration the discrepancies (6.1.27) we obtain for the difference a following estimate

$$\|(u^{(K)}|_{B_0^{\varepsilon,\delta}} - U_0, ..., u^{(K)}|_{B_n^{\varepsilon,\delta}} - U_n, v_1^{(K)} - v_1, ..., v_n^{(K)} - v_n)\|_P = O(\varepsilon^K), \tag{6.1.31}$$

Combining the estimate (6.1.31) with the estimate of the difference between the exact and asymptotic solutions we obtain that for any K there exist such $\hat{K}$ independent of ε that if $\delta = \hat{K}\varepsilon|ln(\varepsilon)|$ then the estimate holds true

$$\|U - u\|_{H^1(B^\varepsilon)} = O(\varepsilon^K), \tag{6.1.32}$$

where

$$U(x) = \left\{ \begin{array}{c} U_i(x), \ if \ x \in B_i^{\varepsilon,\delta}, \ i = 0, ..., n, \\ v_j(x_1^{e_{0j}}), \ if \ x \in B_j^\varepsilon \backslash \cup_{i=0}^n B_i^{\varepsilon,\delta}, \ j = 1, ..., n, \end{array} \right.$$

where the cylinder B_j^ε corresponds to e_j. The estimate (6.1.32) justifies the method of asymptotic partial decomposition of domain (MAPDD).

Remark 6.1.2. This construction can be easily generalized to the case of a finite rod structure with m bundles of bars:

$$B_1 = \cup_{j=1}^{n_1} e_{j,1}, ..., B_m = \cup_{j=1}^{n_m} e_{j,m}.$$

We suppose that all common points of these bundles are end points of some segments of these bundles. Let the union of all bundles be connected. Consider the rod structure B_α^ε associated with the bundle B_α and let

$$B^\varepsilon \;=\; \cup_{\alpha=1}^m \; B_\alpha^\varepsilon$$

be connected.

Then the same asymptotic decomposition of the domain gives the estimate (6.1.31). We can also add some small domains in the neighborhoods of the nodes, as it was done in section 4.1.

6.2 Variational version

6.2.1 General description of the variational version

Let H_ε be a family of Hilbert spaces (depending on a small parameter ε). Consider a variational problem:

find $u_\varepsilon \in H_\varepsilon$ such that

$$\forall w \in H_\varepsilon, \;\; B(u_\varepsilon, w) \;=\; (f, w). \tag{6.2.1}$$

Here $B(.,.)$ is a bilinear symmetric coercive form and $(f,.)$ is a linear bounded functional with the norm $\|f\|$. We suppose that

$$\forall w \in H_\varepsilon, \;\; B(w,w) \;\geq\; c_1 \varepsilon^r \|w\|^2, \tag{6.2.2}$$

where $c_1 > 0$ and $r \geq 0$ do not depend on ε,

$$\forall u, w \in H_\varepsilon, \;\; B(u,w) \;=\; B(w,u). \tag{6.2.3}$$

On the other hand, by definition, we get

$$\forall w \in H_\varepsilon, \;\; |(f,w)| \;\leq\; \|f\| \|w\|. \tag{6.2.4}$$

Then it is well known that there exists a unique solution u_ε of problem (6.2.1).

Let $H_{\varepsilon,dec}$ be a linear subspace of H_ε. Let u_ε^a be an asymptotic solution such that

(i) $u_\varepsilon^a \in H_{\varepsilon,dec}$, and such that:

(ii) there exists a functional $(\psi_\varepsilon, .) \in H_\varepsilon^*$ such that its norm $\|\psi_\varepsilon\| \leq c_2$, where c_2 does not depend on ε, and

$$\forall w \in H_\varepsilon, \;\; B(u_\varepsilon^a, w) \;=\; (f,w) + \varepsilon^K (\psi_\varepsilon, w), \tag{6.2.5}$$

where $K > r$.

Subtracting (6.2.1) from (6.2.5) we obtain

$$\forall w \in H_\varepsilon, \;\; B(u_\varepsilon^a - u_\varepsilon, w) \;=\; \varepsilon^K (\psi_\varepsilon, w),$$

i.e. for $w = u^a_\varepsilon - u_\varepsilon$ we obtain

$$c_1\varepsilon^r \|u^a_\varepsilon - u_\varepsilon\| \le \varepsilon^K \|\psi_\varepsilon\| \le \varepsilon^K c_2,$$

i.e.

$$\|u^a_\varepsilon - u_\varepsilon\| \le (c_2/c_1)\varepsilon^{K-r},$$

i.e.

$$\|u^a_\varepsilon - u_\varepsilon\| = O(\varepsilon^{K-r}). \tag{6.2.6}$$

Let u^d_ε be a solution (its existence is assumed)of the partially decomposed problem, i.e. of the identity (6.2.1) restricted onto the subspace $H_{\varepsilon,dec}$:

$$\forall w \in H_{\varepsilon,dec}, \quad B(u^d_\varepsilon, w) = (f, w). \tag{6.2.7}$$

We assume the existence and uniqueness of solution u^d_ε of problem (6.2.7). We assume that subspace $H_{\varepsilon,dec}$ has a more simple structure than H_ε, for example the functions of $H_{\varepsilon,dec}$ are polynomial on the regular part G^ε_1 of the domain. Therefore the problem (6.2.7) is in some sense easier than problem (6.2.1). So the variational version is related to a special choice of a simple subspace $H_{\varepsilon,dec}$ (i.e. special restriction of the original problem), satisfying the conditions (i), (ii).

To justify the variational version let us subtract the identity (6.2.7) from (6.2.5) for any $w \in H_{\varepsilon,dec}$. Then we obtain

$$\forall w \in H_{\varepsilon,dec}, \quad B(u^a_\varepsilon - u^d_\varepsilon, w) = \varepsilon^K(\psi_\varepsilon, w),$$

i.e. for $w = u^a_\varepsilon - u^d_\varepsilon$ we obtain

$$c_1\varepsilon^r \|u^a_\varepsilon - u^d_\varepsilon\| \le \varepsilon^K \|\psi_\varepsilon\| \le \varepsilon^K c_2,$$

i.e.

$$\|u^a_\varepsilon - u^d_\varepsilon\| \le (c_2/c_1)\varepsilon^{K-r},$$

i.e.

$$\|u^a_\varepsilon - u^d_\varepsilon\| = O(\varepsilon^{K-r}). \tag{6.2.8}$$

Comparing the estimates (6.2.6) and (6.2.8) we obtain

$$\|u_\varepsilon - u^d_\varepsilon\| = O(\varepsilon^{K-r}). \tag{6.2.9}$$

This estimate justifies the method.

We can also consider a more general problem, when $B_{\varepsilon,t}$ is an arbitrary (possibly non-linear) mapping from $\tilde{H}_{\varepsilon,T} \times H_{\varepsilon,T}$ to $I\!R$ satisfying instead of the above inequality the following relation:

$$\forall w_1, w_2 \in \tilde{H}_{\varepsilon,T},$$

$$\sup_{t\in(0,T)} \left(B_{\varepsilon,t}(w_1, w_1 - w_2) - B_{\varepsilon,t}(w_2, w_1 - w_2)\right) \geq c_1\varepsilon^r \|w_1 - w_2\|_T^{1+\alpha}, \quad (6.2.2')$$

here α is a positive constant.

Now the above proof of estimate (6.2.9) is modified as follows.

Let $H_{\varepsilon,dec}$ be a subspace of $\tilde{H}_{\varepsilon,T}$. Let u_ε^a be an asymptotic solution such that

(i) $u_\varepsilon^a \in H_{\varepsilon,dec}$, and such that

(ii) there exists $\psi_\varepsilon \in H^*_{\varepsilon,T}$, such that $\|\psi_\varepsilon\| \leq c_2$, where c_2 does not depend on ε, and such that for almost all $t \in (0,T)$,

$$\forall w \in H_{\varepsilon,T}, \quad B_{\varepsilon,t}(u_\varepsilon^a, w) = (f_t, w) + \varepsilon^K(\psi_\varepsilon, w), \quad (6.2.5')$$

where $K > r$. Subtracting (6.2.1) from (6.2.5') we get

$$\forall t \in (0,T), \ \forall w \in H_{\varepsilon,T}, \quad B_{\varepsilon,t}(u_\varepsilon^a, w) - B_{\varepsilon,t}(u_\varepsilon, w) = \varepsilon^K(\psi_\varepsilon, w),$$

i.e. for $w = u_\varepsilon^a - u_\varepsilon$ passing to $\sup_{t\in(0,T)}$ we have

$$c_1\varepsilon^r \|u_\varepsilon^a - u_\varepsilon\|_T^\alpha \leq \varepsilon^K \|\psi_\varepsilon\|_T \leq \varepsilon^K c_2,$$

i.e.

$$\|u_\varepsilon^a - u_\varepsilon\|_T = O(\varepsilon^{\frac{K-r}{\alpha}}). \quad (6.2.6')$$

Let u_ε^d be the solution of the partially decomposed problem, i.e. of the identity restricted onto the subspace $H_{\varepsilon,dec}$:

$$\forall w \in H_{\varepsilon,dec}, \quad B_{\varepsilon,t}(u_\varepsilon^d, w) = (f_t, w), \quad (6.2.7')$$

As above, we assume that this subspace has a more simple structure than $H_{\varepsilon,T}$. Let us subtract this identity from (6.2.1) written for any $w \in H_{\varepsilon,dec}$. Then we obtain

$$\forall t \in (0,T), \ \forall w \in H_{\varepsilon,dec}, \quad B_{\varepsilon,t}(u_\varepsilon^a, w) - B_{\varepsilon,t}(u_\varepsilon^d, w) = \varepsilon^K(\psi_\varepsilon, w),$$

i.e. for $w = u_\varepsilon^a - u_\varepsilon^d$ passing to $\sup_{t\in(0,T)}$ we obtain

$$c_1\varepsilon^r \|u_\varepsilon^a - u_\varepsilon^d\|_T^\alpha \leq \varepsilon^K \|\psi_\varepsilon\|_T \leq \varepsilon^K c_2,$$

i.e.

$$\|u_\varepsilon^a - u_\varepsilon^d\|_T = O(\varepsilon^{\frac{K-r}{\alpha}}). \quad (6.2.8')$$

Comparing estimates (6.2.8') to (6.2.6') we get the following assertion.

Theorem 6.2.1 *The estimate holds*

$$\|u_\varepsilon - u_\varepsilon^d\|_T = O(\varepsilon^{\frac{K-r}{\alpha}}).$$

This theorem generalizes estimate (6.2.9) and justifies the MAPDD for the non-linear and non-steady-state case.

6.2.2 Model example

Returning to the example (6.1.5),(6.1.6), we can define H_ε as a subspace of functions of $H^1((0,1)\times(-\frac{\varepsilon}{2},\frac{\varepsilon}{2}))$ vanishing on $\gamma_1\cup\gamma_2$. Then the identity (6.2.1) is

$$\int_0^1\int_{-\frac{\varepsilon}{2}}^{\frac{\varepsilon}{2}}(\nabla u_\varepsilon,\nabla w)dx_1dx_2 \;=\; -\int_0^1\int_{-\frac{\varepsilon}{2}}^{\frac{\varepsilon}{2}} fwdx_1dx_2. \tag{6.2.10}$$

Consider now the subspace

$$H_{\varepsilon,dec} \;=\; \{w\in H_\varepsilon,\; w(x_1,x_2)=\rho(x_1)\;\; in\;\; (\delta,1)\times(-\frac{\varepsilon}{2},\frac{\varepsilon}{2}),\rho\in H^1((\delta,1))\}.$$

Evidently, the partially decomposed problem in this second version coincides with the problem (6.1.14) up to the notations:

$$u_\varepsilon(x_1,x_2) \;=\; \begin{cases} U(x_1,x_2) & if\ x_1<\delta,\\ v(x_1) & if\ x_1\ge\delta,\end{cases}$$

$$w(x_1,x_2) \;=\; \begin{cases} R(x_1,x_2) & if\ x_1<\delta,\\ \rho(x_1) & if\ x_1\ge\delta.\end{cases}$$

The asymptotic solution u_ε^a should be slightly corrected because in a form (6.1.7') it does not belong to the subspace $H_{\varepsilon,dec}$ (it is not a constant on the rectangle $(\delta,1)\times(-\frac{\varepsilon}{2},\frac{\varepsilon}{2})$ although the error is exponentially small). Therefore we multiply the boundary layer functions U_l at the domain where they are exponentially small by a cutting function χ; i.e. let χ be a smooth function defined on $I\!R$ such that

$$\chi(t) \;=\; \begin{cases} 1 & if\ t<-1,\\ 0 & if\ t\ge 0,\end{cases}$$

$|\chi(t)|\le 1$, then

$$u_\varepsilon^a(x)=\sum_{l=0}^{K}\varepsilon^l(v_l(x_1)+U_l(\frac{x}{\varepsilon})\chi(\frac{x_1-\delta}{\varepsilon})).$$

Clearly

$$\|u_\varepsilon^a-U^{(K)}\|_{H^1(G_\varepsilon)} \;=\; O(e^{-c_3\frac{\delta}{\varepsilon}}),\;\; c_3>0$$

due to exponential decaying of $U_l(\xi)$ as $\xi_1\to+\infty$, and this correction gives a discrepancy of the same order in the variational formulation (6.2.10). So for any K we can find K_1 such that if $\delta=K_1\varepsilon\;|ln\;\varepsilon|$ then

$$e^{-c_3\frac{\delta}{\varepsilon}} \;=\; O(\varepsilon^K),$$

and the conditions (i) and (ii) of the subsection 6.2.1 are satisfied and we again obtain the estimate

$$\|u_\varepsilon^a - u_\varepsilon^d\|_{H^1(G_\varepsilon)} = O(\varepsilon^K), \tag{6.2.11}$$

confirming the result of the previous section 6.1. (Here $r = 0$ due to the Poincaré-Friedrichs inequality for rod structures). Of course the same approach is good for the general rod structures.

6.2.3 Elasticity equations

Consider the elasticity system of equations set in

$$G_\varepsilon = (0,1) \times (-\frac{\varepsilon}{2}, \frac{\varepsilon}{2})$$

(below the convention of the summation from 1 to 2 in repeating indices is accepted):

$$\frac{\partial}{\partial x_r}(A_{rm}\frac{\partial u_\varepsilon}{\partial x_m}) = f(\frac{x}{\varepsilon}), \quad x \in G_\varepsilon, \tag{6.2.12}$$

where A_{rm} are constant 2×2 matrices with the components a_{rm}^{kl}:

$$a_{rm}^{kl} = \lambda\delta_{rk}\delta_{lm} + \mu(\delta_{rm}\delta_{kl} + \delta_{rl}\delta_{km}),$$

λ and μ are the positive constants ; here u_ε, f are two- dimensional vector-valued functions, *supp* $f(\xi)$ belongs to the square $(0,1) \times (-\frac{1}{2}, \frac{1}{2})$, and $f \in L^2$.

Let us set the following boundary conditions: free lateral boundary and fixed ends:

$$A_{2m}\frac{\partial u_\varepsilon}{\partial x_m} = 0, \quad x_2 = \pm\varepsilon/2 \tag{6.2.13}$$

$$u_\varepsilon = 0, \quad x_1 = 0, 1. \tag{6.2.14}$$

The variational formulation is the integral identity (6.2.1) where

$$B(u_\varepsilon, w) = \int_{G_\varepsilon} (\frac{\partial w}{\partial \xi_i})^T A_{ij} \frac{\partial u_\varepsilon}{\partial \xi_j} dx,$$

$$(f, w) = -\int_{G_\varepsilon} f w dx, \tag{6.2.15}$$

H_ε is the subspace of vector valued functions of $[H^1(G_\varepsilon)]^2$ vanishing on the segments $\{x_1 = 0\}$ and $\{x_1 = 1\}$.

The complete asymptotic expansion of the solution U_ε is constructed in Chapter 2. We apply it here.

It consists of two exponentially decaying boundary layers u_P^0 and u_P^1 and regular polynomial expansion u_B, i.e it has a form :

$$u_\varepsilon^{(\infty)} = u_B + u_P^0 + u_P^1 \tag{6.2.16}$$

where

$$u_B = \sum_{l=0}^{\infty} \varepsilon^l N_l(\frac{x_2}{\varepsilon}) \frac{d^l v}{dx_1^l},$$

$$u_P^0 = \sum_{l=0}^{\infty} \varepsilon^l N_l^0(\frac{x}{\varepsilon}) \frac{d^l v}{dx_1^l},$$

$$u_P^1 = \sum_{l=0}^{\infty} \varepsilon^l N_l^1(\frac{x_1 - 1}{\varepsilon}, \frac{x_2}{\varepsilon}) \frac{d^l v}{dx_1^l},$$

$N_l(\xi_2)$ are 2×2 matrix valued solutions of cell problems

$$\left\{ \begin{array}{c} \frac{\partial}{\partial \xi_2}(A_{22}\frac{\partial N_l}{\partial \xi_2}) + \frac{\partial}{\partial \xi_2}(A_{21}N_{l-1}) \\ +A_{12}\frac{\partial}{\partial \xi_2}N_{l-1} + A_{11}N_{l-2} = h_l, \ \xi_2 \in (-\frac{1}{2}, \frac{1}{2}), \\ A_{22}\frac{\partial N_l}{\partial \xi_2} + A_{21}N_{l-1} = 0, \ \xi_2 = \pm\frac{1}{2}, \\ h_l = \int_{-1/2}^{1/2}(A_{12}\frac{\partial N_{l-l}}{\partial \xi_2} + A_{11}N_{l-2})d\xi_2, \\ l = 1, 2 \ldots, \end{array} \right. \tag{6.2.17}$$

N_l^0, N_l^1 are some exponentially decaying in the first variable 2×2 matrix valued solutions of the chain of the boundary layer problems and $N_0 = I = \begin{pmatrix} 1 & 0 \\ 0 & 1 \end{pmatrix}$, h_l are diagonal matrices,

$$h_{2k+1} = 0, \ \ h_2 = \begin{pmatrix} E & 0 \\ 0 & 1 \end{pmatrix},$$

$$h_4 = \begin{pmatrix} 0 & 0 \\ 0 & -E/12 \end{pmatrix}, \quad E = \frac{(\lambda + 2\mu)^2 - \lambda^2}{\lambda + 2\mu},$$

$v = (v^1, v^2)^T$ is the solution of the homogenized equation

$$\left\{ \begin{array}{c} E(v^1)'' = 0, \\ -\frac{E}{12}(v^2)'''' = 0. \end{array} \right. \tag{6.2.18}$$

Here h_l^{11} and h_l^{22} are the corresponding elements of the matrix h_l. So v^j and N_l are polynomials and therefore u_B has a form

$$\begin{pmatrix} v_1 - x_2 v_2' + \{\frac{1}{6}(\frac{E}{\mu} - \frac{\lambda}{\lambda+2\mu})x_2^3 + \varepsilon^2 \frac{1}{8}(\frac{\lambda}{3(\lambda+2\mu)} - \frac{E}{\mu})x_2\}v_2''' \\ v_2 - \frac{\lambda}{\lambda+2\mu}x_2 v_1' + \frac{\lambda}{2(\lambda+2\mu)}(x_2^2 - \frac{1}{12}\varepsilon^2)v_2'' \end{pmatrix}, \tag{6.2.19}$$

where

$$v_1(x_1) \ = \ ax_1 + b, \tag{6.2.20}$$

$$v_2(x_1) \;=\; cx_1^3 + dx_1^2 + ex_1 + g \tag{6.2.21}$$

are polynomials with some undetermined coefficients a, b, c, d, e, g . The constants a, b, c, d, e, g could be defined after calculation of the boundary layers. Introduce the subspace $H_{\varepsilon,dec}$ of partially decomposed problem as the subspace of H_ε, such that its elements have the form (6.2.19) for all x of the rectangle $(\delta, 1-\delta) \times (-\frac{\varepsilon}{2}, \frac{\varepsilon}{2})$.

The partially decomposed problem has a form:

find $u_\varepsilon \in H_{\varepsilon,dec}$, such that for all $w \in H_{\varepsilon,dec}$,

$$\int_{((0,\delta)\cup(1-\delta,1))\times(-\frac{\varepsilon}{2},\frac{\varepsilon}{2})} (\frac{\partial w}{\partial \xi_i})^T A_{ij} \frac{\partial u_\varepsilon}{\partial \xi_j} dx +$$

$$+ \int_\delta^{1-\delta} (\varepsilon E v_1' s_1' \;+\; \varepsilon^3 \frac{E}{12} v_2'' s_2'' \;+\; \varepsilon^5 \frac{E^2}{120\mu} v_2''' s_2''') dx_1 \;=$$

$$= \;-\int_{((0,\delta)\cup(1-\delta,1))\times(-\frac{\varepsilon}{2},\frac{\varepsilon}{2})} fw dx,$$

where v and s are polynomials from (6.2.20)-(6.2.21) corresponding to u_ε and w respectively.

Representing the vector valued function u_B for $x_1 \in [\delta, 1-\delta]$ as a linear combination $\sum_{i=1}^{6} \alpha_i \Phi_i$, where

$$\Phi_1(x_1,x_2) \;=\; \begin{pmatrix} 1 \\ 0 \end{pmatrix}, \quad , \; \Phi_2(x_1,x_2) \;=\begin{pmatrix} x_1 \\ -(\frac{\lambda}{\lambda+2\mu})x_2 \end{pmatrix}, \quad , \; \Phi_3(x_1,x_2) \;=\; \begin{pmatrix} 0 \\ 1 \end{pmatrix}, \quad ,$$

$$\Phi_4(x_1,x_2) \;=\; \begin{pmatrix} -x_2 \\ x_1 \end{pmatrix}, \quad , \; \Phi_5(x_1,x_2) \;=\; \begin{pmatrix} -2x_1x_2 \\ x_1^2 + (\frac{\lambda}{\lambda+2\mu})(x_2^2 - \frac{\varepsilon^2}{12}) \end{pmatrix},$$

$$\Phi_6(x_1,x_2) \;=\; \begin{pmatrix} -3x_2x_1^2 + (\frac{E}{\mu} - \frac{\lambda}{\lambda+2\mu})x_2^3 + \varepsilon^2 \frac{3}{4}(\frac{\lambda}{3(\lambda+2\mu)} - \frac{E}{\mu})x_2 \\ x_1^3 + 3\frac{\lambda}{2(\lambda+2\mu)}(x_2^2 - \frac{1}{12})x_1 \end{pmatrix}, \quad ,$$

we can deduce the junction conditions on the truncations $\{x_1 = \delta\}$ and $\{x_1 = 1-\delta\}$:

$$\sum_{x_1=\delta,1-\delta} \sum_{j=1}^{2} \int_{-\varepsilon/2}^{\varepsilon/2} [A_{1j} \frac{\partial u_\varepsilon^d}{\partial x_j}]|_{x_1} \Phi_i(x_1,x_2) dx_2 \;=\; 0$$

for $i = 1, 2, 3, 4, 5, 6$.

Multiplying the boundary layer functions u_P^0 and u_P^1 by cutting functions $\chi(\frac{x_1-\delta}{\varepsilon})$ and $\chi(\frac{1-\delta-x_1}{\varepsilon})$ respectively, we obtain a corrected asymptotic solution

$$u_\varepsilon^a = u_B + u_P^0 \chi(\frac{x_1-\delta}{\varepsilon}) + u_P^1 \chi(\frac{1-\delta-x_1}{\varepsilon}),$$

which satisfies the conditions (i) and (ii) of the subsection 6.2.1 and therefore as in the subsection 6.2.1 we can obtain the estimate (6.2.19), i.e. for any K there exist such $\hat{K} \in I\!R$ independent of ε that if $\delta = \hat{K}\varepsilon|ln(\varepsilon)|$ then the estimate (6.2.19) holds true. (In order to prove the coercivity (6.2.2) we should use the Korn inequality for thin domains; it gives the estimate (6.2.2) with some $r \leq 3$.

The main result of this section can be generalized for the rod structures B^ε described in the subsection 6.1.3.

Consider the equation

$$\frac{\partial}{\partial x_r}\left(A_{rm}\frac{\partial u_\varepsilon}{\partial x_m}\right) = \sum_{i=0}^{n} f_i\left(\frac{x - y_i}{\varepsilon}\right), \quad x \in G_\varepsilon, \tag{6.2.22}$$

where $y_0 = 0$ and $y_i \neq 0$, $i = 1, ..., n$, are the ends of the segments e_i. We suppose that *supp* f_i belong to the unitary disc $\{\xi_1^2 + \xi_2^2 \leq 1\}$.

We state the boundary conditions:

$$u_\varepsilon = 0 \text{ for } x \in \hat{\beta}_j^\varepsilon, \; j = 1, ..., n, \tag{6.2.23}$$

$$n_r A_{rm}\frac{\partial u_\varepsilon}{\partial x_m} = 0, \text{ for } x \in \partial B^\varepsilon \setminus \cup_{j=1}^n \hat{\beta}_j^\varepsilon. \tag{6.2.24}$$

Here (n_1, n_2) is the normal vector.

Let H_ε be the subspace of vector valued functions of $[H^1(G_\varepsilon)]^2$ vanishing on the segments $\hat{\beta}_j^\varepsilon$, $j = 1, ..., n$. The variational formulation of the problem (6.2.22)-(6.2.24) is the same as (6.2.15) with B_ε instead of G_ε.

Consider the subspace $H_{\varepsilon,dec}$ of H_ε, such that its elements have the form

$$\begin{pmatrix} v_1 - x_2^{e_{0j}} v_2' + \{\frac{1}{6}(\frac{E}{\mu} - \frac{\lambda}{\lambda+2\mu})(x_2^{e_{0j}})^3 + \varepsilon^2\frac{1}{8}(\frac{\lambda}{3(\lambda+2\mu)} - \frac{E}{\mu})x_2^{e_{0j}}\}v_2''' \\ v_2 - \frac{\lambda}{\lambda+2\mu}x_2^{e_{0j}}v_1' + \frac{\lambda}{2(\lambda+2\mu)}((x_2^{e_{0j}})^2 - \frac{1}{12}\varepsilon^2)v_2'' \end{pmatrix} \tag{6.2.25}$$

on $B_{0,j}^{\varepsilon,\delta}$, where

$$v_1(x_1^{e_{0j}}) = a_{e_{0j}}x_1^{e_{0j}} + b_{e_{0j}}, \tag{6.2.26}$$

$$v_2(x_1^{e_{0j}}) = c_{e_{0j}}(x_1^{e_{0j}})^3 + d_{e_{0j}}(x_1^{e_{0j}})^2 + e_{e_{0j}}x_1^{e_{0j}} + g_{e_{0j}} \tag{6.2.27}$$

are polynomials with coefficients $a_{e_{0j}}, b_{e_{0j}}, c_{e_{0j}}, d_{e_{0j}}, e_{e_{0j}}, g_{e_{0j}}$. Here $B_{0,j}^{\varepsilon,\delta}$ is the part of the cylinders $B_j^{\varepsilon,\delta}$ concluded between $S_{0,j}$ and $S_{j,0}$.

In this case we can easily check the hypothesis (i) and (ii) and prove the same result as in the case of the rectangle G_ε if the hypothesis of section 4.4 are satisfied. This result can be generalized for a multi bundle rod structure (in particular, three dimensional rod structure satisfying to the conditions of Chapter 4). The Korn inequality for rod structures is proved in Appendices 4.A3 and 5.A2.

Let us mention that it would be reasonable to apply to the partially decomposed problem the classical domain decomposition method for all 1-dimensional and multi-dimensional parts. This idea allows to parallelize iteratively computations in all these parts. Although in the linear case such parallelization can be obtained by a direct method.

The comparison of the direct implementation of the finite elements to problem (6.2.12)-(6.2.14) to the MAPDD finite element implementation is shown in the Figure 6.2.1 (computed by R.Paz and V.Ruas); here the data are taken as follows: $\varepsilon = 0.05$, $\mu = \lambda = 1$, and $f = (0,1)^T$. The mesh contains twelve thousand elements.

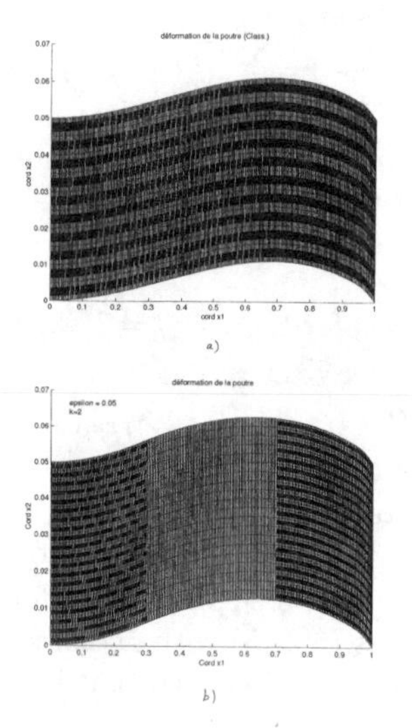

Figure 6.2.1. The direct and the MAPDD finite element implementation for the elasticity problem in a thin rectangle.

Remark 6.2.1 The case when the right hand side f is different from zero and depends on the longitudinal variable in a regular part G_1^ε, can be reduced to the previous case. Indeed, consider again the elasticity equation stated in

$$G_\varepsilon = (0,1) \times (-\frac{\varepsilon}{2}, \frac{\varepsilon}{2}) :$$

$$\frac{\partial}{\partial x_r}(A_{rm}\frac{\partial u_\varepsilon}{\partial x_m}) = f_\varepsilon(x_1), \quad x \in G_\varepsilon, \tag{6.2.28}$$

where

$$f_\varepsilon = (f^1(x_1), \varepsilon^2 f^2(x_1)), \quad f^j \in C^\infty([0,1]).$$

Let us set the boundary conditions (6.2.13),(6.2.14). The complete asymptotic expansion of the solution U_ε again has a form (6.2.16) with the same N_l, N_l^0, N_l^1 as above(see (6.2.17)), but v_1, v_2 now are the series

$$v_\alpha = \sum_{j=0}^{\infty} \varepsilon^j v_{\alpha,j}, \quad \alpha = 1, 2, \tag{6.2.29}$$

asymptotically satisfying the homogenized equations

$$\sum_{l=2}^{\infty} \varepsilon^{l-2} h_l^{11} v_1^{(l)} = f_1(x_1), \tag{6.2.30}$$

$$\sum_{l=4}^{\infty} \varepsilon^{l-2} h_l^{22} v_2^{(l)} = \varepsilon^2 f_2(x_1). \tag{6.2.31}$$

Here h_l^{11} and h_l^{22} are the corresponding elements of the matrix h_l from (6.2.27). Substituting (6.2.29) into (6.2.30),(6.2.31), we obtain the equations for $v_{\alpha,j}$:

$$E(v_{1,j})'' = -\sum_{k=0}^{j-1} h_{j-k+2}^{11} (v_{1,k})^{(j-k+2)} + \delta_{j0} f_1(x_1) \tag{6.2.32}$$

$$-\frac{E}{12}(v_{2,j})'''' = -\sum_{k=0}^{j-1} h_{j-k+4}^{22} (v_{2,k})^{(j-k+4)} + \delta_{j0} f_2(x_1) \tag{6.2.33}$$

It could be proved by induction that

$$(v_{1,j})'' = A_j f_1^{(j)}, \quad (v_{2,j})'''' = B_j f_2^{(j)},$$

where

$$A_0 = 1/E, \quad B_0 = -12/E,$$

$$E A_j = -\sum_{k=0}^{j-1} h_{j-k+2}^{11} A_k,$$

$$-\frac{E}{12} B_j = -\sum_{k=0}^{j-1} h_{j-k+4}^{22} B_k,$$

where $j \geq 1$.

So,

$$v_1'' = \sum_{j=0}^{\infty} \varepsilon^j A_j f_1^{(j)}$$

$$v_2'''' = \sum_{j=0}^{\infty} \varepsilon^j B_j f_2^{(j)} \tag{6.2.34}$$

and then

$$v_1^{(m)} = \sum_{j=0}^{\infty} \varepsilon^j A_j f_1^{(j+m-2)}, \quad m \geq 2$$

$$v_2^{(m)} = \sum_{j=0}^{\infty} \varepsilon^j B_j f_2^{(j+m-4)}, \quad m \geq 4 \tag{6.2.35}$$

where v_α are defined by (6.2.34) up to 6 free constants.

Substituting the truncated series (up to ε^K included) into the variational formulation (6.2.1) we make the changing of unknown functions : we introduce ν_α according to the relations

$$v_1(x_1) = \nu_1(x_1) + \int_\delta^{x_1} \int_\delta^{\theta} \sum_{j=0}^{K} \varepsilon^j A_j f_1^{(j)}(t) dt d\theta, \tag{6.2.36}$$

$$v_2(x_1) = \nu_2(x_1) + \int_\delta^{x_1} \int_\delta^{\theta_1} \int_\delta^{\theta_2} \int_\delta^{\theta_3} \sum_{j=0}^{K} \varepsilon^j B_j f_2^{(j)}(t) dt d\theta_1 d\theta_2 d\theta_3, \tag{6.2.37}$$

where the new unknown functions ν_α are polynomials:

$$\begin{aligned} \nu_1(x_1) &= a x_1 + b, \\ \nu_2(x_1) &= c x_1^3 + d x_1^2 + e x_1 + g. \end{aligned}$$

So the partially decomposed problem is reduced to the problem of the type (6.2.3) with replacement of v_α by the relations (6.2.36), (6.2.37).

The same modifications can be done in case of rod structures with the right hand side function depending on the longitudinal variable (as in Chapter 4).

The other method to solve the problem with the right hand side function depending on x_1 is to consider the standard variational formulation of the problem (6.2.25),(6.2.13),(6.2.14) restricted to the Hilbert subspace $H_{\varepsilon,dec}$ of H_ε, such that its elements have the form of the truncated series u_B, i.e.

$$\sum_{l=0}^{K+1} \varepsilon^l N_l(\frac{x_2}{\varepsilon}) \frac{d^l v}{dx_1^l},$$

for all x of the rectangle $(\delta, 1-\delta) \times (-\frac{\varepsilon}{2}, \frac{\varepsilon}{2})$. Here v is an arbitrary vector valued function of $H^{K+2}([\delta, 1-\delta])$, and the matrices N_l are the "known" (pre-calculated) solutions of the chain of problems (6.2.27).

This approach can be applied for asymptotic partial decomposition of some homogenization problems (see section 6.4).

Remark 6.2.2 This method of partial asymptotic decomposition can be applied to multi-structures, [38, 118, 119, 178] .

6.3 Decomposition of a flow in a tube structure

Here below we discuss the application of MAPDD for the Navier-Stokes problem (4.5.1)-(4.5.3) with ε standing for μ. For simplicity we consider the case of $g = 0$. We associate to problem (4.5.1)-(4.5.3) set in B^ε with a right hand side of form (4.5.36) the partially decomposed problem. To this end we cut the cylinders at the distance $\delta = const\ \varepsilon\ ln\ (\varepsilon)$ from the nodes by the planes perpendicular to the segments and replace the inner parts of the cylinders by the corresponding parts of the segments e_j. We obtain the set $B^{\varepsilon,\delta}$. Denote $B_i^{\varepsilon,\delta}$ the connected truncated part of B^{ε}, containing the node O_i, e_{ij} the part of the segment connecting $B_i^{\varepsilon,\delta}$ and $B_j^{\varepsilon,\delta}$; let S_{ij} be a cross-section of the truncated cylinder corresponding to e_{ij}, such that it belongs to $\partial B_i^{\varepsilon,\delta}$.

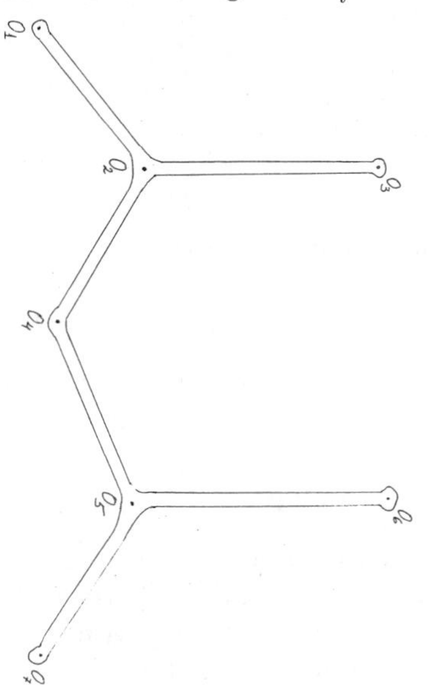

Figure 6.3.1. A tube structure.

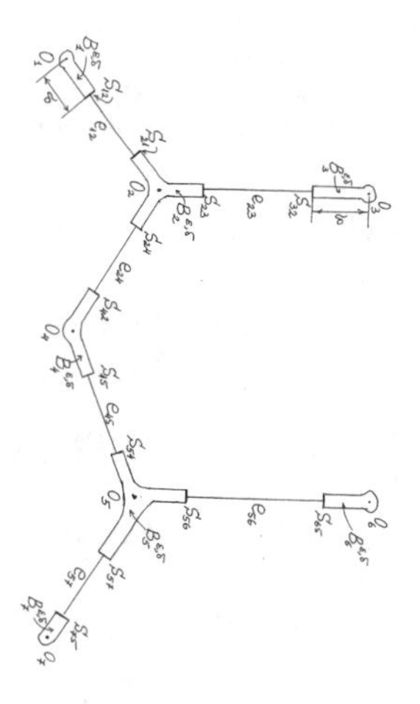

Figure 6.3.2. The asymptotically partially decomposed domain; $\delta = O(\varepsilon|ln(\varepsilon)|)$.

Consider the equations for each $B_i^{\varepsilon,\delta}$:

$$\nu\Delta U - (U,\nabla)U - \nabla P \ = \ f, \tag{6.3.1}$$

$$div\ U \ = \ 0, \ \ x \in B_i^{\varepsilon,\delta} \tag{6.3.2}$$

$$U \ = \ 0 \ \ x \in \partial B_i^{\varepsilon,\delta} \cap \partial B^{\varepsilon}, \ i = 1, ..., N, \tag{6.3.3}$$

the equation for each segment e_{j_1,j_2} :

$$-p'_{j_1,j_2} \ = (\Gamma^T_{j_1,j_2} f)^1 \ - w_{j_1,j_2} \tag{6.3.4}$$

and the interface conditions on each cross-section S_{j_1,j_2}

$$U \ = \ \varepsilon^2 w_{j_1,j_2} \Gamma_{j_1,j_2} (\tilde{u}^{e_{j_1,j_2}}, 0, ..., 0)^T, \tag{6.3.5}$$

$$\int_{S_{j_1,j_2}} ((\nu \frac{\partial U}{\partial n} - (U,n)U - Pn), \Gamma_{j_1,j_2}(\tilde{u}^{e_{j_1,j_2}}, 0, ..., 0)^T) ds$$

$$= \ - \int_{S_{j_1,j_2}} (n, \Gamma_{j_1,j_2}(\tilde{u}^{e_{j_1,j_2}}, 0, ..., 0)^T) \ ds \ p_{j_1,j_2} \tag{6.3.6}$$

where $\tilde{u}^{e_{j_1,j_2}}$ is the solution of the Dirichlet problem for Poisson equation (4.5.10) on the cross-section of the rod e_{j_1,j_2} and Γ_{j_1,j_2} is the matrix of passage to the local base corresponding to the segment e_{j_1,j_2} with the origin in O_{j_1}, w_{j_1,j_2} are unknown constants, p_{j_1,j_2} is a function of the variable $x_1^{e_{j_1,j_2}}$ defined on each segment e_{j_1,j_2}. And

$$p_{j_1,j_2}(x_1^{e_{j_1,j_2}}) \ = \ p_{j_2,j_1}(|e_{j_1,j_2}| + 2\delta - x_1^{e_{j_1,j_2}}), \tag{6.3.7}$$

$$w_{j_1,j_2} \ = \ - w_{j_2,j_1}. \tag{6.3.8}$$

Taking into account (6.3.5), relation (6.3.6) can be rewritten in a form

$$- \int_{S_{j_1,j_2}} ((\varepsilon^4 w^2_{j_1,j_2} \tilde{u}^{e_{j_1,j_2}} \Gamma_{j_1,j_2}(\tilde{u}^{e_{j_1,j_2}}, 0, ..., 0)^T -$$

$$+Pn), \Gamma_{j_1,j_2}(\tilde{u}^{e_{j_1,j_2}}, 0, ..., 0)^T) ds \ =$$

$$= - \int_{S_{j_1,j_2}} (n, \Gamma_{j_1,j_2}(\tilde{u}^{e_{j_1,j_2}}, 0, ..., 0)^T) \ ds \ p_{j_1,j_2}. \tag{6.3.6'}$$

Let us give the variational formulation for this problem. Let $H_{div=0}(B_1^{\varepsilon,\delta}, ..., B_N^{\varepsilon,\delta}, e_1, ..., e_M)$ be the space of ordered collections $(U_1, ..., U_N, w_1, ..., w_M)$, where U_i are the vector valued functions from $H_{div=0}(B_i^{\varepsilon,\delta})$, equal to zero on $\partial B_i^{\varepsilon,\delta} \cap \partial B^{\varepsilon}$, and $w_1, ..., w_M$ are the constants associated with the segments $e_1, ..., e_M$ such that for each e_j, containing e_{j_1,j_2}, and connecting $B_{j_1}^{\varepsilon,\delta}$ with $B_{j_2}^{\varepsilon,\delta}$, we have

$$U_{j_1} \ = \ sign(j_2 - j_1) \varepsilon^2 w_j \Gamma_{j_1,j_2}(\tilde{u}^{e_{j_1,j_2}}, 0, ..., 0)^T, \ on \ S_{j_1,j_2} \ and \ on \ S_{j_2,j_1}. \tag{6.3.9}$$

The scalar product is defined as

$$((U_1,...,U_N,w_1,...,w_M),(V_1,...,V_N,v_1,...,v_M))_P =$$

$$\sum_{i=1}^N \int_{B_i^{\varepsilon,\delta}} \nu \sum_{r=1}^s (\frac{\partial U_i}{\partial x_r}, \frac{\partial \Phi_i}{\partial x_r})\, dx + \varepsilon^{2+(s-1)} \sum_{j=1}^M (|e_j| - 2\delta) \int_{\beta_j} (-\tilde{u}^{e_j}(\xi^L)) d\xi w_j v_j.$$

Note that $\int_{\beta_j} \tilde{u}^{e_j}(\xi^L) d\xi < 0$ because $\tilde{u}^{e_j}(\xi^L) < 0$ by the principle of maximum for elliptic equation (4.5.10).

The norm is defined as

$$\|(U_1,...,U_N,w_1,...,w_M)\|_P = ((U_1,...,U_N,w_1,...,w_M),(U_1,...,U_N,w_1,...,w_M))_P^{1/2}.$$

Then the variational formulation is as follows: find $(U_1,...,U_N,w_1,...,w_M) \in H_{div=0}(B_1^{\varepsilon,\delta},...,B_N^{\varepsilon,\delta},e_1,...,e_M)$ and such that for any $(\Phi_1,...,\Phi_N,v_1,...,v_M) \in H_{div=0}(B_1^{\varepsilon,\delta},...,B_N^{\varepsilon,\delta},e_1,...,e_M)$ the integral identity holds true:

$$-\sum_{i=1}^N \int_{B_i^{\varepsilon,\delta}} \nu \sum_{r=1}^s (\frac{\partial U_i}{\partial x_r}, \frac{\partial \Phi_i}{\partial x_r})\, dx +$$

$$+ \sum_{i=1}^N \int_{B_i^{\varepsilon,\delta}} (U_i\ ,\ (U_i,\nabla)\Phi_i)\, dx + \varepsilon^{2+(s-1)} \sum_{j=1}^M (|e_j| - 2\delta) \int_{\beta_j} \tilde{u}^{e_j}(\xi^L) d\xi w_j v_j$$

$$= \sum_{i=1}^N \int_{B_i^{\varepsilon,\delta}} (f,\Phi) dx$$

$$+ \varepsilon^{2+(s-1)} \sum_{j=1}^M \int_{\beta_j} \tilde{u}^{e_j}(\xi^L) d\xi \int_{e_j \setminus \cup_{i=1}^s B_i^{\varepsilon,\delta}} (\Gamma_j^T f)^1\, dx_1^{e_j} v_j. \tag{6.3.10}$$

Variational formulation (6.3.10) can be obtained in a following way (cf. section 6.2). Consider the subspace $\hat{H}^0_{div=0}(B^\varepsilon,\delta)$ of the space $H^0_{div=0}(B^\varepsilon)$ which contains all functions $\hat{U}$ of $\hat{H}^0_{div=0}(B^\varepsilon,\delta)$ such that for any truncated segment e_{j_1,j_2} they coincide with the Poiseuille flow

$c_{j_1,j_2}\Gamma_{j_1,j_2}(\tilde{u}^{e_{j_1,j_2}},0,...,0)^T$, on the truncated part of the cylinder forming B^ε stretched between the cross sections S_{j_1,j_2} and S_{j_2,j_1}; c_{j_1,j_2} is a constant on this truncated cylinder.

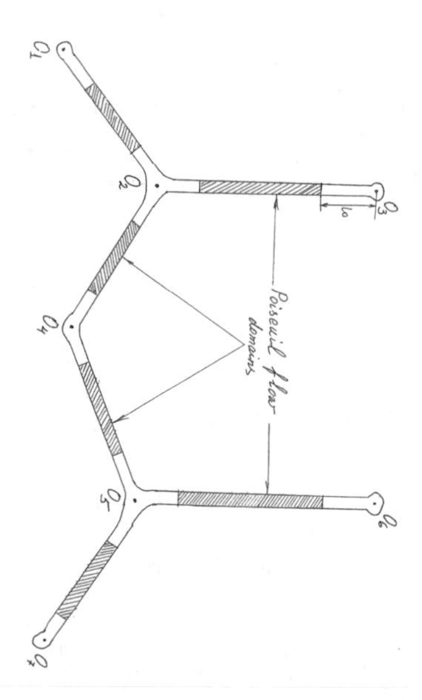

Figure 6.3.3. The structure of space $\hat{H}^0_{div=0}(B^\varepsilon, \delta)$.

Consider the following partially decomposed variational problem:
to find $\hat{U} \in \hat{H}^0_{div=0}(B^\varepsilon, \delta)$ such that for all $\Phi \in \hat{H}^0_{div=0}(B^\varepsilon, \delta)$

$$-\int_{B^\varepsilon} \nu \sum_{i=1}^{s} \left(\frac{\partial \hat{U}}{\partial x_i}, \frac{\partial \Phi}{\partial x_i}\right) dx + \int_{B^\varepsilon} (\hat{U}, (\hat{U}, \nabla) \Phi) \, dx = \int_{B^\varepsilon} (f, \varphi) \, dx. \tag{6.3.11}$$

Associating to each function $\hat{U} \in \hat{H}^0_{div=0}(B^\varepsilon, \delta)$ the ordered collection $(U_1, ..., U_N, w_1, ..., w_M) \in H_{div=0}(B_1^{\varepsilon,\delta}, ..., B_N^{\varepsilon,\delta}, e_1, ..., e_M)$ in such way that $\hat{U} = U_i$ on each $B_i^{\varepsilon,\delta}$, and $\hat{U} = \varepsilon^2 w_j \Gamma_j(\tilde{u}^{e_j}, 0, ..., 0)^T$ on each truncated cylinder forming B^ε corresponding to the segment e_j (under the convention that each e_j stretched between O_{j_1} and O_{j_2}, $j_1 < j_2$, has the origin in O_{j_1}) we obtain (6.3.10) and respectively (6.3.1)-(6.3.8) as an equivalent formulation of (6.3.11).

It can be proved as in [87] that for sufficiently small ε there exists a unique solution to (6.3.11) due to the fixed point theorem and the Poincaré - Friedrichs inequality for B^ε. The a priori estimate holds true with a constant independent of ε.

Suppose that $\hat{f}_i$ in vicinities of nodes are approximated by Taylor's formula up to the terms $O(\varepsilon^K)$ as it was done in section 4.5. The above asymptotic analysis of problem (4.5.1)- (4.5.3) (section 4.5) shows that U^a satisfies the integral identity (6.3.11) with a discrepancy

$$\int_{B^\varepsilon} \varepsilon^K (\rho, \Phi) dx, \tag{6.3.12}$$

where ρ is a bounded vector valued function. Moreover, U^a does not belong to $\hat{H}^0_{div=0}(B^\varepsilon,\delta)$ because it gives a discrepancy in condition (6.3.5) of order $O(exp(-c\delta/\varepsilon))$, $c>0$ in C^1, i.e.,

$$U^a \;=\; \varepsilon^2 c^K_{e_{j_1,j_2}} \Gamma_{j_1,j_2}(\tilde{u}^{e_{j_1,j_2}},0,...,0)^T + exp(-c\delta/\varepsilon) r_0, \quad x\in S_{j_1,j_2}, \tag{6.3.13}$$

where r_0 is a bounded in C^1 vector valued function, $c^K_e \;=\; \sum_{l=2}^K \varepsilon^{l-2} c^e_l$. Therefore, we shall slightly change U^a in order to obtain a function of $\hat{H}^0_{div=0}(B^\varepsilon,\delta)$. The discrepancy $exp(-c\delta/\varepsilon)r_0$ can be continued from S_{j_1,j_2} to $B^{\varepsilon,\delta}_{j_1}$ in such a way that the divergence of the extension vanishes and that this extension has an order $O(exp(-c_1\delta/\varepsilon))$, $c_1>0$ in $H^1(B^{\varepsilon,\delta}_{j_1})$. Such extension exists (cf. [87]) because $div\ U^a \;=\; 0$ in $B^{\varepsilon,\delta}_{j_1}$ and the constants $c^K_{e_j} \;=\; \sum_{l=2}^K \varepsilon^{l-2} c^{e_j}_l$ satisfy conditions (4.5.43),(4.5.44). Denote this extension $\delta\tilde{U}^a$. Define $\delta\tilde{U}^a =$ $U^a - \varepsilon^2 c^K_{e_{j_1,j_2}} \Gamma_{j_1,j_2}(\tilde{u}^{e_{j_1,j_2}},0,...,0)^T$ on truncated parts of cylinders between the cross sections S_{j_1,j_2} and S_{j_2,j_1}. Now $\tilde{U}^a \;=\; U^a - \delta\tilde{U}^a \in \hat{H}^0_{div=0}(B^\varepsilon,\delta)$, $\|\delta\tilde{U}^a\|_{H^1(B^\varepsilon)} \;=\; O(exp(-c_2\delta/\varepsilon))$, $c_2>0$, and it satisfies integral identity (6.3.11) with discrepancy of order $O(exp(-c_3\delta/\varepsilon))$, $c_3>0$, c_3 does not depend on small parameters, i.e.,

$$-\int_{B^\varepsilon} \nu \sum_{i=1}^s \left(\frac{\partial\tilde{U}^a}{\partial x_i}\,,\,\frac{\partial\Phi}{\partial x_i}\right) dx + \int_{B^\varepsilon} (\tilde{U}^a\,,\,(\tilde{U}^a\,,\,\nabla)\,\Phi)\,dx \;=\; \int_{B^\varepsilon} (f,\varphi)\,dx\, +$$

$$+\int_{B^\varepsilon} \nu \sum_{i=1}^s \left(\frac{\partial\delta\tilde{U}^a}{\partial x_i}\,,\,\frac{\partial\Phi}{\partial x_i}\right) dx \;-\; \int_{B^\varepsilon} (U^a\,,\,(\delta\tilde{U}^a\,,\,\nabla)\,\Phi)dx \;-$$

$$-\int_{B^\varepsilon} (\delta\tilde{U}^a\,,\,(U^a-\delta\tilde{U}^a\,,\,\nabla)\,\Phi)dx \;+\; \int_{B^\varepsilon} \varepsilon^K(\rho,\Phi)dx =$$

$$= \int_{B^\varepsilon} (f,\varphi)\,dx \;+\; \int_{B^\varepsilon} \sum_{i=1}^s exp(-c_3\delta/\varepsilon)\left(\rho_i\,,\,\frac{\partial\Phi}{\partial x_i}\right) dx\, +$$

$$+\int_{B^\varepsilon} exp(-c_3\delta/\varepsilon)(\rho_0,\Phi)dx \;+\; \int_{B^\varepsilon} \varepsilon^K(\rho,\Phi)dx, \tag{6.3.14}$$

where $\rho_i, i=0,...,s$ are vector valued functions bounded in $L^2(B^\varepsilon)$ by a constant independent of the small parameters.

On the other hand, for any K there exist such $\hat{K}$ independent of ε that if $\delta=\hat{K}\varepsilon|ln(\varepsilon)|$ then

$$exp(-c_2\delta/\varepsilon),\ exp(-c_3\delta/\varepsilon) = O(\varepsilon^K).$$

Taking into consideration the discrepancy (6.3.14) and the a priori estimate for (4.5.1)-(4.5.3) we obtain the following estimate: for any K there exist such $\hat{K}$ independent of ε that if $\delta=\hat{K}\varepsilon|ln(\varepsilon)|$ then

$$\|\hat{U} - \tilde{U}^a\|_{H^1(B^\varepsilon)} = O(\varepsilon^K). \tag{6.3.15}$$

Remark 6.3.1 Defining $c_{e_j}^K = \sum_{l=2}^K \varepsilon^{l-2} c_l^{e_j}$ we obtain the estimate

$$\|(u^a|_{B_1^{\varepsilon,\delta}} - U_1, ..., u^a|_{B_N^{\varepsilon,\delta}} - U_N, c_{e_1}^K - w_1, ..., c_{e_M}^K - w_M)\|_P = O(\varepsilon^K).$$

Combining the estimates (4.5.47) and (6.3.15) we obtain that for any K there exists such $\hat{K}$ independent of ε that if $\delta = \hat{K}\varepsilon|ln(\varepsilon)|$ then the estimate holds true

$$\|\hat{U} - u\|_{H^1(B^\varepsilon)} = O(\varepsilon^K). \tag{6.3.16}$$

The estimate (6.3.16) justifies the MAPDD for the Navier-Stokes problem.

Remark 6.3.2 The present section is devoted to an approximation of completely three-dimensional (two-dimensional) problem by a hybrid problem that is "mainly one-dimensional", i.e., it is of dimension 1 on the major part of the domain.

On the other hand another related problem was considered recently by S.A.Nazarov and M.Specovius-Neugebauer [122,184]. They constructed a special approximation of problems in unbounded domain by problems in bounded (truncated) domain. This approach also takes into consideration the information on asymptotic behavior of solution of the problem in unbounded domain at infinity.

Remark 6.3.3 Some numerical experiments on MAPDD were developed in case of the Stokes flow in thin domains. The comparison to the direct numerical computation of a solution of the problem shows that for the case when the viscosity ν is finite, the boundary layers are very narrowly localized in the neighborhoods of the junctions, so δ can be taken almost equal to just ε. However, if the Reynolds number becomes greater (for example if the viscosity is small) then the multi-dimensional parts become also greater. The same remark holds for the flows in wavy tubes and for the extrusion process modelling. The Stokes flow in an extruder with a screw of a form presented in Figure 6.3.4 was computed by I.Sirakov using the asymptotic domain decomposition on five subdomains, as it is shown in the Figure 6.3.5. The pressure and the particles trajectories were calculated (Figure 6.3.6).

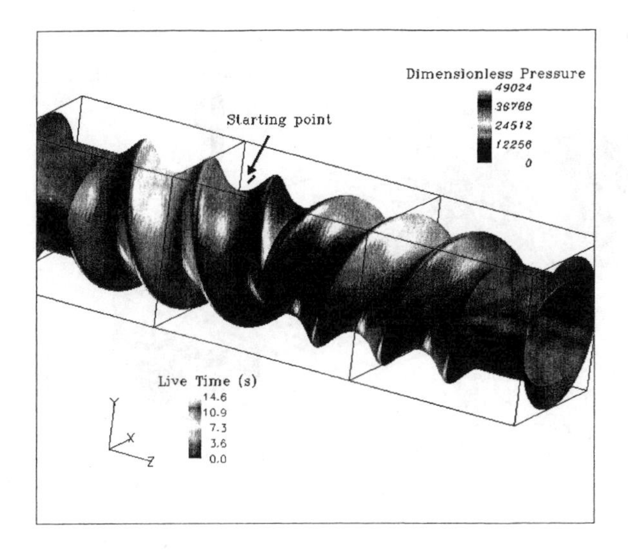

Figure 6.3.4. The form of the screw.

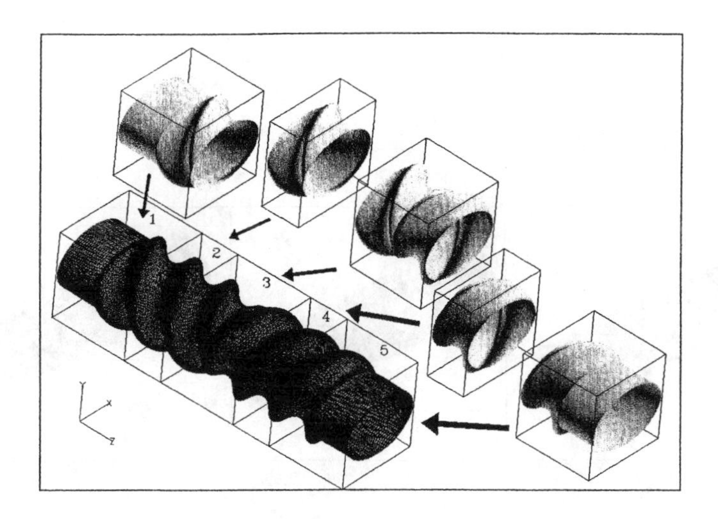

Figure 6.3.5. The asymptotic partial domain decomposition for the screw.

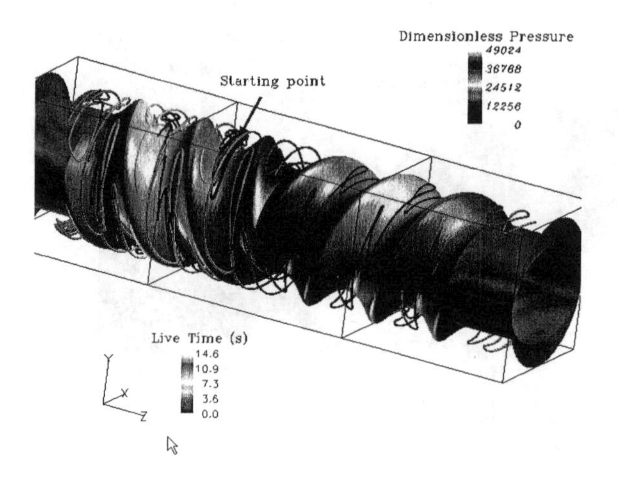

Figure 6.3.6. A typical trajectory of particles.

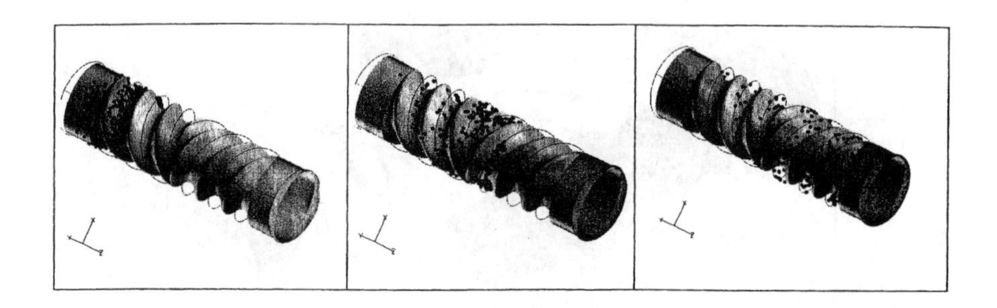

Figure 6.3.7. Extrusion process.

6.4 The partial homogenization

This section is devoted to an application of MAPDD to the homogenization problem.

The idea of partial homogenization on some subdomain was discussed in [156]: we replace the initial equation by the homogenized equation of high order on the main part of the domain and we keep the initial equation in some boundary strip. The main question is how to conjugate these two equations (probably of different order) to obtain the accuracy of given power of ε.

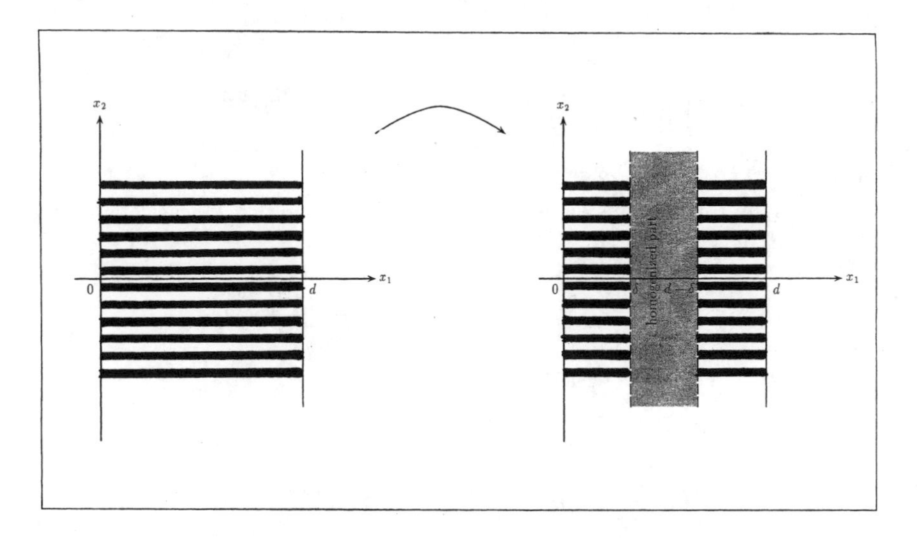

Figure 6.4.1. The partial homogenization.

Here below we consider the model problem in a layer with Dirichlet boundary condition, and we propose and justify a method of "conjugation" of the higher order homogenized equation with the initial equation.

Consider the homogenization of a boundary value problem in a layer [132],[133]:

$$\sum_{k,j=1}^{s} \frac{\partial}{\partial x_k}(A_{kj}(\frac{x'}{\varepsilon})\frac{\partial u_\varepsilon}{\partial x_j}) = f(x_1), \quad x_1 \in (0,d), \qquad u_\varepsilon|_{x_1=0} = 0, \quad u_\varepsilon|_{x_1=d} = 0, \tag{6.4.1}$$

where $x = (x_1, x')$, $x' = (x_2, \ldots, x_s)$, $A_{kj}(\xi')$ are 1 - periodic in $\xi' \in \mathbb{R}^{s-1}$ functions, satisfying the following conditions:

$$(i) \exists \kappa_0 > 0,\ \forall \xi' \in \mathbb{R}^{s-1}, \forall \eta \in \mathbb{R}^s, \eta = (\eta_1, \ldots, \eta_s), \quad \sum_{k,j=1}^{s} A_{kj}(\xi')\eta_j\eta_k \geq \kappa_0 \sum_{i=1}^{s} \eta_i^2$$

$$(ii)\ \ \forall \xi' \in \mathbb{R}^{s-1}, k, j \in \{1, \ldots, s\}, \quad A_{kj}(\xi') = A_{jk}(\xi'),$$

f is a C^∞ - regular function.

Let T be a finite number, multiple of ε.

Assume that A_{kj} are piecewise smooth functions in the sense [16]. Then there exists a unique $T-$periodic in $x_2, \ldots, x_s$, solution to this problem (see [16]); T is a multiple of ε. This solution is a solution of the following variational formulation.

Let $H^1_{0,per'}$ be a space that is completion of the space of $C^\infty-$ regular $T-$periodic in $x_2, \ldots, x_s$ functions vanishing if $x_1 = 0$ or $x_1 = d$ with respect to the norm $H^1((0,d) \times (0,T)^{s-1})$; then u_ε is sought as function of $H^1_{0,per'}$ satisfying

$$J(u,\varphi) = \int_{\Omega_{d,T}} \left\{ \sum_{k,j=1}^{s} A_{kj}(\frac{x'}{\varepsilon}) \frac{\partial u}{\partial x_j} \frac{\partial \varphi}{\partial x_k} + f\varphi \right\} dx = 0, \ \forall \varphi \in H^1_{0,per'}, \quad (6.4.2)$$

where $\Omega_{d,T} = (0,d) \times (0,T)^{s-1}$.

The asymptotic expansion of the solution to problem (6.4.1) was constructed in [133]. It has a form

$$u_\varepsilon^{a\ (K)} = u_{BLO}^{(K)}(x, \frac{x}{\varepsilon}) + u_{BLd}^{(K)}(x, \frac{(x_1 - d, x')}{\varepsilon}) +$$

$$+ \sum_{l=0}^{K+1} \varepsilon^l N_l(\frac{x'}{\varepsilon}) D_1^l v_\varepsilon^{(K)}(x_1), \quad (6.4.3)$$

$$v_\varepsilon^{(K)}(x_1) = \sum_{j=0}^{K+1} \varepsilon^j v_j(x_1), \quad (6.4.4)$$

where the "boundary layers" $u_{BLO}^{(K)}$ and $u_{BLd}^{(K)}$ are exponentially decaying functions such that for all $x \in [0,d] \times I\!R^{s-1}$, $|u_{BLO}^{(K)}(x,\xi)|$, $|u_{BLd}^{(K)}(x,\xi)| \le C_1 e^{-C_2|\xi_1|}$, $C_1, C_2 > 0$, C_1, C_2 do not depend on ε; $N_l(\xi')$ are 1-periodic functions, solutions of the sequence of cell problems:

$$L_{\xi\xi} N_1 = -\sum_{k=2}^{s} \frac{\partial}{\partial \xi_k} A_{k1}, \quad (6.4.5)$$

$$L_{\xi\xi} N_l = -\sum_{k=2}^{s} \frac{\partial}{\partial \xi_k} (A_{k1} N_{l-1}) -$$

$$-\sum_{k=2}^{s} A_{1k} \frac{\partial N_{l-1}}{\partial \xi_k} - A_{11} N_{l-2} +$$

$$+ < \sum_{k=2}^{s} A_{1k} \frac{\partial N_{l-1}}{\partial \xi_k} + A_{11} N_{l-2} >,$$

$$l \geq 2, \tag{6.4.6}$$

here $<>= \int_{(0,1)^{s-1}} d\xi'$, $L_{\xi\xi} = \sum_{k,j=2}^{s} \frac{\partial}{\partial \xi_k}(A_{kj}(\xi')\frac{\partial}{\partial \xi_j})$; $< N_l >= 0$ for $l > 0$, $N_0 = 1$, $D_1 = \frac{\partial}{\partial x_1}$.

These cell problems are set in variational formulation in H^1_{per}, that is the completion of the space of $C^\infty-$regular $1-$periodic functions with respect to the norm $H^1((0,1)^s)$. This sequence of problems is solved recurrently for $l = 1, 2, \ldots, K+1$.

Functions v_j are defined in another sequence of problems (see [133], [16]) but we do not need this information for the further developments.

Let us describe the method of asymptotic partial decomposition of domain for problem (6.4.1) (in formulation (6.4.2)).

Consider the subspace $H_{\varepsilon,\delta,\ dec}$ of the space $H^1_{0,per'}$ that consists of all functions u of $H^1_{0,per'}$ having for all $x_1 \in [\delta, d-\delta]$ the following presentation in the form of Bakhvalov's ansatz [16]:

$$u(x) = \sum_{l=0}^{K+1} \varepsilon^l N_l(\frac{x'}{\varepsilon}) D_1^l v, \tag{6.4.7}$$

where $v \in H^{K+2}((\delta, d-\delta))$.

So every $u \in H_{\ \delta,dec}$ is related by (6.4.7) to some $v \in H^{K+2}((\delta, d-\delta))$, so that we can consider a couple (u, v).

Let $\delta = \hat{K}\varepsilon \mid ln\ \varepsilon \mid$, where $\hat{K}$ is some positive number independent of ε. Consider the partially decomposed variational problem

$$J(u, \varphi) = 0, \quad \forall \varphi \in H^{K+2}_{\varepsilon,\delta,dec}. \tag{6.4.8}$$

If u_d is its solution then estimate (6.2.9) proves that for any $K \in (0, \infty)$, there exist $\hat{K}$ such that if $\delta = \hat{K}\ \varepsilon \mid ln\ \varepsilon \mid$ then the estimate holds

$$\|u_\varepsilon - u_d\|_{H^1((0,d)\times(0,T)^{s-1})} = O(\varepsilon^{K+1}). \tag{6.4.9}$$

In the layer $(\delta, d-\delta) \times I\!R^{s-1}$ u_d is presented by formula (6.4.7) as well as

$$\varphi = \sum_{l=0}^{K+1} \varepsilon^l N_l(\frac{x'}{\varepsilon}) D_1^l w,$$

$$w \in H^{K+2}_{T\ \ per}((\delta, d-\delta)) \tag{6.4.10}$$

In all previous examples the subspace $H_{\varepsilon,\delta,dec}$ was closed in the H_ε, while now it is not generally the case. So, the existence of u_d should be studied separately.

Let us prove now the existence and the uniqueness of the solution u_d of problem (6.4.8). To this end let us calculate integrals

$$I_1(u,\varphi) = \int_{(\delta,d-\delta)\times(0,T)^{s-1}} \{\sum_{k,j=1}^{s} A_{kj}(\frac{x'}{\varepsilon})\frac{\partial u}{\partial x_j}\cdot\frac{\partial \varphi}{\partial x_k}\}dx$$

and

$$I_2(\varphi) = \int_{(\delta,d-\delta)\times(0,T)^{s-1}} f(x_1)\varphi dx$$

for $u,\varphi \in H^{K+2}_{\varepsilon,\delta,dec}$, satisfying (6.4.7) and (6.4.10). We get

$$I_1(u,\varphi) = T^{s-1}\int_{(\delta,d-\delta)} \sum_{l,m=1}^{K+2} \varepsilon^{l+m-2}\, \tilde{h}_{l,m} D_1^l v D_1^m w dx_1,$$

$$I_2(\varphi) = T^{s-1}\int_{(\delta,d-\delta)} f(x_1)w(x_1)dx_1,$$

where

$$\tilde{h}_{l,m} = \sum_{k,j=1}^{s} < A_{kj}(\xi')(\frac{\partial}{\partial \xi_j}N_l + \delta_{j1}N_{l-1})(\frac{\partial}{\partial \xi_k}N_m + \delta_{k1}N_{m-1}) >$$

defined by the solutions of cell problems (6.4.5),(6.4.6). The terms containing derivatives $\frac{\partial}{\partial \xi_k}$ are dropped if $k = K+2$, and $\frac{\partial}{\partial \xi_j}$ are dropped if $l = K+2$; these coefficients were introduced in [183]. This presentation holds true due to the separation of variables x' and x_1 in (6.4.7) and (6.4.10) and because $< N_l >= 0$ for all $l > 0$.

Below ε is a fixed positive number small enough, such that the ratio T/ε is integer.

Consider space $H^{K+2}_{\varepsilon,\delta,dec} \times H^{K+2}((\delta, d-\delta))$ supplied by the following inner product for

$$(u,v),(\varphi,w) \in H^{K+2}_{\varepsilon,\delta,dec} \times H^{K+2}((\delta,d-\delta))$$

related by (6.4.7) and (6.4.10):

$$\begin{aligned} a(u,v;\varphi,w) &= \int_{G_{BL}} \{\sum_{k,j=1}^{s} A_{kj}(\frac{x'}{\varepsilon})\frac{\partial u}{\partial x_j}\cdot\frac{\partial \varphi}{\partial x_k}\}dx + \\ &+ T^{s-1}\int_{(\delta,d-\delta)} \sum_{l,m=1}^{K+2} \varepsilon^{l+m-2}\, \tilde{h}_{l,m} D_1^l v D_1^m w dx_1 = \\ &= \int_{\Omega_{d,T}} \{\sum_{k,j=1}^{s} A_{kj}(\frac{x'}{\varepsilon})\frac{\partial u}{\partial x_j}\cdot\frac{\partial \varphi}{\partial x_k}\}dx \qquad (6.4.11) \end{aligned}$$

where

$$G_{BL} = (0,\delta)\times(0,T)^{s-1} \cup (d-\delta,d)\times(0,T)^{s-1}.$$

Let us check that it is really an inner product. Indeed, its bi-linearity and symmetry properties are evident; its non-negativeness is a consequence of the non-negativeness of the quadratic part of $J(u,u)$. Assume now that $a(u,v;u,v)=0$. Then this quadratic part of $J(u,u)$ vanishes and so $u=0$, then $\int_{(0,T)^{s-1}} u(x_1,x')dx' = 0$ for $x_1 \in (\delta, d-\delta)$; therefore (as $< N_l >= 0$ for $l>0$) $v(x_1)=0$ for $x_1 \in (\delta, d-\delta)$.

Problem (6.4.8) is equivalent to the following problem: find a couple

$$(u_d, v_d), \in H^{K+2}_{\varepsilon,\delta,dec} \times H^{K+2}((\delta, d-\delta))$$

related by (6.4.7)such that for any couple

$$(\varphi, w) \in H^{K+2}_{\varepsilon,\delta,dec} \times H^{K+2}((\delta, d-\delta))$$

related by (6.4.10),

$$a(u,v;\varphi,w) + \int_{G_{BL}} f(x_1)\varphi dx + T^{s-1}\int_{(\delta,d-\delta)} f(x_1)w(x_1)dx_1 \;=\; 0.$$

Let us check the continuity of the linear functional

$$\Phi(\varphi,w) \;=\; -\int_{G_{BL}} f(x_1)\varphi dx - T^{s-1}\int_{(\delta,d-\delta)} f(x_1)w(x_1)dx_1.$$

We have

$$|\Phi(\varphi,w)| \le\; T^{\frac{s-1}{2}}\|f\|_{L^2(0,d)}\{\|\varphi\|_{L^2(G_{BL})} +\; T^{\frac{s-1}{2}}\|w\|_{L^2(\delta,d-\delta)}\}.$$

Here

$$\|\varphi\|_{L^2(G_{BL})} \;\le\; \frac{\delta}{\sqrt{2}}\|\nabla\varphi\|_{L^2(G_{BL})} \;\le$$

$$\le\; \frac{\delta}{\sqrt{2\kappa_0}}\sqrt{\int_{G_{BL}}\{\sum_{k,j=1}^{s} A_{kj}(\frac{x'}{\varepsilon})\frac{\partial\varphi}{\partial x_j}\cdot\frac{\partial\varphi}{\partial x_k}\}dx} \;\le$$

$$\le\; \frac{\delta}{\sqrt{2\kappa_0}}\sqrt{a(\varphi,w;\varphi,w)}.$$

$$T^{\frac{s-1}{2}}\|w\|_{L^2(\delta,d-\delta)} \;=\; T^{-\frac{s-1}{2}}\sqrt{\int_\delta^{d-\delta}(\int_{(0,T)^{s-1}} w(x_1)dx')^2dx_1} \;=$$

$$=\; T^{-\frac{s-1}{2}}\sqrt{\int_\delta^{d-\delta}(\int_{(0,T)^{s-1}} \varphi(x)dx')^2dx_1} \;\le\; \sqrt{\int_{(\delta,d-\delta)\times(0,T)^{s-1}} (\varphi(x))^2dx} \;=$$

$$= \|\varphi\|_{L^2((\delta,d-\delta)\times(0,T)^{s-1})} \leq \frac{d}{\sqrt{2\kappa_0}}\sqrt{a(\varphi,w;\varphi,w)},$$

by the Poincaré Friedrichs inequality for φ.

So Φ is a continuous functional.

Consider problem (6.4.8) set on the completion of $H^{K+2}_{\varepsilon,\delta,dec} \times H^{K+2}((\delta,d-\delta))$ with respect to the norm $\sqrt{a(\bullet,\bullet;\bullet,\bullet)}$. Applying now the Riesz theorem we get the existence and the uniqueness of solution.

Remark 6.4.1 Let A_{kj} be constant coefficients equal to δ_{kj}. Then $N_l = 0$ for all $l > 0$ and $N_0 = 1$, and so $\tilde{h}_{lm} = \delta_{l1}\delta_{m1}$ and therefore the completion of $H^{K+2}_{\varepsilon,\delta,dec} \times H^{K+2}((\delta,d-\delta))$ is a subspace of space $H^1_{0\ per'} \times H^1((\delta,d-\delta))$ such that, for all $x_1 \in (\delta,d-\delta)$, $u(x) = v(x_1)$, where $v \in H^1((\delta,d-\delta))$, ie., $\frac{\partial u}{\partial x_j} = 0$ for all $x_1 \in (\delta,d-\delta)$, $j = 2,...,s$.

Assume now that $< N^2_{K+2} > \neq 0$. Then $\tilde{h}_{K+2,K+2} =< A_{11}N^2_{K+2} >$ is positive.

Varying $w \in H^{K+2}((\delta,d-\delta))$ we prove that v satisfies on $(\delta,d-\delta)$ the ordinary differential equation

$$\sum_{l,m=1}^{K+2} \varepsilon^{l+m-2}\,(-1)^m \tilde{h}_{l,m} D_1^{l+m} v(x_1) + f(x_1) = 0,\ x_1 \in (\delta,d-\delta). \qquad (6.4.12)$$

This v belongs to $C^\infty([\delta,d-\delta])$ because $f \in C^\infty([0,d])$ and therefore $u_d \in H^{K+2}_{\varepsilon,\delta,dec}$.

Applying now estimate (6.2.9) we get the following theorem.

Theorem 6.4.1. *The estimate holds true*

$$\|u_\varepsilon - u_d\|_{H^1((0,d)\times(0,T)^{s-1})} = O(\varepsilon^{K+1}). \qquad (6.4.13)$$

The solution of problem (6.4.8) satisfies equation (6.4.1) for $x \in G_{BL}$ and boundary conditions (6.4.1) for $x_1 = 0$ and for $x_1 = d$, it satisfies as well equation (6.4.12) for v_d and the $2K+4$ interface conditions for $x_1 = \delta$ and for $x_1 = d-\delta$. Let us give these conditions for $x_1 = \delta$; for $x_1 = d-\delta$ they are the same. Firstly, u_d and v satisfy conditions (6.4.7) for $x_1 = \delta$ (and for $x_1 = d-\delta$). And then there are $K+2$ interface conditions that are the consequences of the integration by parts and of the equating the coefficients of $D_1^r w(\delta), r = 0,...,K+1$:

$$T^{1-s}\int_{(0,T)^{s-1}} \sum_{j=1}^{s} A_{kj}(\frac{x'}{\varepsilon})\frac{\partial u_d}{\partial x_j}|_{x_1=\delta} N_r(\frac{x'}{\varepsilon})dx' =$$

$$= \sum_{l=1}^{K+2}\sum_{m=r+1}^{K+2} \varepsilon^{l+m-2}\,(-1)^{m-r-1}\tilde{h}_{l,m} D_1^{l+m-r-1} v_d(\delta). \qquad (6.4.14)$$

Remark 6.4.2 Another way to prove the existence of solution of (6.4.8) is to prove that "the classical solution" of problem (2.1) in G_{BL}, satisfying boundary conditions (6.4.1) for $x_1 = 0$ and for $x_1 = d$, equation (6.4.12) for v_d and interface conditions (6.4.7) and (6.4.14) for $x_1 = \delta$ and for $x_1 = d - \delta$ is a solution to (6.4.8).

In order to find this classical solution let us find first a $T-$periodic in x' solution u_{d0} of equation (6.4.1) in G_{BL} and vanishing for $x_1 = 0, \delta, d - \delta$ and d, and a solution v_{d0} of equation (6.4.12) satisfying conditions

$$D_1^l v(\delta) = 0, \ D_1^l v(d - \delta) = 0, \ l = 0, ..., K + 1.$$

Then we seek v_d in a form

$$v_d = v_{d0} + \sum_{r=1}^{2K+4} c_r V_r, \tag{6.4.15}$$

where $\{V_r, \ r = 1, ..., 2K + 4\}$ is a basis of the space of solutions of the homogeneous ordinary differential equation

$$\sum_{l,m=1}^{K+2} \varepsilon^{l+m-2} \ (-1)^m \tilde{h}_{l,m} D_1^{l+m} v(x_1) = 0, \ x_1 \in (\delta, d - \delta). \tag{6.4.16}$$

and

$$u_d = u_{d0} + \sum_{r=1}^{2K+4} c_r u_{dr}, \tag{6.4.17}$$

where u_{dr} are the $T-$periodic in x' solutions to problems

$$\sum_{k,j=1}^{s} \frac{\partial}{\partial x_k}(A_{kj}(\frac{x'}{\varepsilon})\frac{\partial u_{dr}}{\partial x_j}) = 0, \quad x \in G_{BL}, \qquad u_{dr}|_{x_1=0} = 0, \quad u_{dr}|_{x_1=d} = 0,$$

$$u_{dr} = \sum_{l=0}^{K+1} \varepsilon^l N_l(\frac{x'}{\varepsilon}) D_1^l V_r, \quad x_1 = \delta, d - \delta. \tag{6.4.18}$$

So now problem is reduced to an algebraic linear system of equations for $c_1, ..., c_{2K+4}$, such that u_d and v_d satisfy (6.4.15), (6.4.17) and (6.4.14) for $x_1 = \delta$ and for $x_1 = d - \delta$. Its matrix is positive definite because the corresponding quadratic form

$$a(u_{d0} + \sum_{r=1}^{2K+4} c_r u_{dr}, v_{d0} + \sum_{r=1}^{2K+4} c_r V_r; u_{d0} + \sum_{r=1}^{2K+4} c_r u_{dr}, v_{d0} + \sum_{r=1}^{2K+4} c_r V_r)$$

is positive definite and V_r are linearly independent functions on $(\delta, d - \delta)$.

Consider an example $K = 0$. We have then

$$a(u,v;\varphi,w)=\int_{G_{BL}}\sum_{k,j=1}^{s}A_{kj}(\frac{x'}{\varepsilon})\frac{\partial u}{\partial x_j}\frac{\partial\varphi}{\partial x_k}dx+$$

$$+\int_{G_I}\sum_{l,m=1}^{2}\varepsilon^{l+m-2}\tilde{h}_{lm}\frac{\partial^l v}{\partial x_1^l}\frac{\partial^m w}{\partial x_1^m}dx, \qquad (6.4.19)$$

where

$$\tilde{h}_{11}=\hat{A}_{11}=\sum_{j=1}^{s}<A_{1j}\frac{\partial(N_1+\xi_1)}{\partial\xi_j}>;$$

$$\tilde{h}_{12}=\tilde{h}_{21}=\sum_{j=1}^{s}<A_{1j}\frac{\partial(N_1+\xi_1)}{\partial\xi_j}N_1>;$$

$$\tilde{h}_{22}=<A_{11}N_1^2>.$$

Integrating identity (6.4.8) by parts and taking into account that w is an arbitrary function of the space $H^2((\delta,d-\delta))$, we see that v is a solution to the equation of the fourth order (if $\tilde{\tilde{h}}_{22}\neq 0$)

$$\varepsilon^2\tilde{h}_{22}\frac{\partial^4 v}{\partial {x_1}^4}-\tilde{h}_{11}\frac{\partial^2 v}{\partial x_1^2}=-f(x_1), \quad x_1\in(\delta,d-\delta) \qquad (6.4.20)$$

with the interface conditions following from the equations

$$\int_{\{x_1=\delta,x_2,\dots,x_s\in[0,T]\}}\{\tilde{h}_{11}\frac{\partial v}{\partial x_1}w(\delta)+(\varepsilon^2\tilde{h}_{22}\frac{\partial^2 v}{\partial x_1^2}+\varepsilon\tilde{h}_{12}\frac{\partial v}{\partial x_1})\frac{\partial w}{\partial x_1}(\delta)-\varepsilon^2\tilde{h}_{22}\frac{\partial^3 v}{\partial x_1^3}w(\delta)\}dx'=$$

$$\int_{\{x_1=\delta,x_2,\dots,x_s\in[0,T]\}}\sum_{j=1}^{s}A_{ij}\frac{\partial\bar{u}_d}{\partial x_j}(w(\delta)+\varepsilon N_1(\frac{x}{\varepsilon})\frac{\partial w}{\partial x_1}(\delta))dx'$$

and

$$\bar{u}_d(\delta,x_2,\dots,x_s)=v(\delta)+\varepsilon N_1(\frac{x}{\varepsilon})\frac{\partial v}{\partial x_1}(\delta);$$

i.e. for $x_1=\delta$ denoting $S_\delta=\{x_1=\delta,x_2,\dots,x_s\in[0,T]\}$, we get

$$\begin{cases}\tilde{h}_{11}\frac{\partial v}{\partial x_1}-\varepsilon^2\tilde{h}_{22}\frac{\partial^3 v}{\partial x_1^3}=\frac{1}{T^{s-1}}\int_{S_\delta}\sum_{j=1}^{s}A_{1j}\frac{\partial\bar{u}_d}{\partial x_j}dx',\\ \varepsilon^2\tilde{h}_{22}\frac{\partial^2 v}{\partial x_1^2}+\varepsilon\tilde{h}_{12}\frac{\partial v}{\partial x_1}=\frac{1}{T^{s-1}}\int_{S_\delta}\sum_{j=1}^{s}A_{1j}\frac{\partial\bar{u}_d}{\partial x_j}N_1(\frac{x}{\varepsilon})dx',\\ \bar{u}_d(\delta,x_2,\dots,x_s)=v(\delta)+\varepsilon N_1(\frac{\delta}{\varepsilon},\frac{x_2}{\varepsilon},\dots;\frac{x_s}{\varepsilon})\frac{\partial v}{\partial x_1}(\delta).\end{cases} \qquad (6.4.21)$$

The same interface conditions we have on the surface $S_{d-\delta}=\{x_1=d-\delta,\ x_2,\dots,x_s\in[0,T]\}$; δ in these conditions has to be replaced by $d-\delta$. In G_{BL} we keep equation (6.4.1) for $\bar{u}_d$. So, (6.4.1) in G_{BL} for $\bar{u}_d$, (6.4.20) in G_I for v, and interface conditions (6.4.21) on S_δ and on $S_{d-\delta}$ constitute the differential version of partially homogenized problem (6.4.8) for $K=0$. Estimates (6.4.13) give the error of order $O(\varepsilon)$. This estimate improves the standard estimate for the difference of the exact solution and of the first order asymptotic solution without the boundary layer corrector that is of order $O(\sqrt{\varepsilon})$.

The idea of the partial homogenization can be applied to the creation of hybrid semi-discrete (semi-continuum) models. Such models are important for nano-structures. The nano-structures are often constituted of a very thin layer (coating) on a homogeneous body. The thickness of the coating is of order of some hundreds of atomic sizes, and the properties change very rapidly across the coating. Therefore it is reasonable to describe the field in the coating layer by a discrete model, similar to some difference scheme, while the main material of course should be described by some continuum model, that is, for example, a partial differential equation. The interface conditions between these two models could be obtained using the ideas of the partial asymptotic domain decomposition. Let us consider a very simple example of a model problem of this kind. Consider the finite difference scheme [173]

$$(a_\varepsilon y_x)_{\bar{x}} = f_i, \quad i=1,...,N-1, \quad y_0=0, y_N=0. \tag{6.4.22}$$

Here N is an integer number, $y=(y_0,...,y_N)$ stands for an unknown vector with the components describing the field in the nodes of the atomic grill $x_i=ih$, where $h=\frac{1}{N}$ is the small parameter, the step of the grill, ie. the distance between atoms; f_i are given real numbers, the right hand side; ε is a small parameter, characterizing the variation scale of the coefficient in the coating layer, $\varepsilon \geq h$. We use here the notations from [173]: for any vector $w=(w_0,...,w_N)$, we define the vector $w_x=(w_{x\ 0},...,w_{x\ N-1})$, such that, $w_{x\ i}=\frac{w_{i+1}-w_i}{h}$; in the same way for any vector $w=(w_0,...,w_N)$, we define the vector $w_{\bar{x}}=(w_{\bar{x}\ 1},...,w_{\bar{x}\ N})$, such that, $w_{\bar{x}\ i}=\frac{w_i-w_{i-1}}{h}$; a_ε is a given coefficient $K_\varepsilon(x)$ that is multiplied by y_x according the following rule: $a_\varepsilon y_x$ is an $N-$dimensional vector with the components $K_\varepsilon(x_i+0.5h)y_{x\ i}$, $i=0,..,N-1$. Assume that for some given finite M, $K_\varepsilon(x)$ is defined as $K(\frac{x}{\varepsilon})$ if $x \leq M\varepsilon$, and as 1 if $x \geq M\varepsilon$, where K is a continuous positive function, such that, there exists a positive κ, such that, $K(\xi) \geq \kappa$, $K(M)=1$.

So, equation (6.4.22) has a form

$$\frac{K_\varepsilon(x_i+0.5h)(y_{i+1}-y_i)-K_\varepsilon(x_i-0.5h)(y_i-y_{i-1})}{h^2} =$$

$$= f_i, \quad i=1,...,N-1, \quad y_0=0, y_N=0. \tag{6.4.23}$$

Assume that $f_i=f(x_i)$, $f \in C^2[M\varepsilon,1]$ and it is a restriction to $[M\varepsilon,1]$ of a function independent of ε and h.

It is impossible to pass to the limit everywhere in (6.4.23) as $\varepsilon \to 0$, because the coefficients and the right hand side vary rapidly in $[0,M\varepsilon]$.

Let us consider the partially homogenized problem for an unknown function u defined in $[Lh,1]$ and an unknown vector $y^d \in I\!R^L$, for an integer L, that is,

$$(a_\varepsilon y^d_x)_{\bar{x}} = f_i, \quad i=1,...,L-1, \quad L-1 \geq (M+1)\varepsilon/h,$$
$$u"(x) = f(x), \quad x \geq Lh,$$
$$y^d_0 = 0,$$
$$u(1) = 0,$$
$$u(Lh) = y^d_L,$$

$$\frac{(u' + \frac{1}{2}hu")(Lh) - \frac{y^d_L - y^d_{L-1}}{h}}{h} = f_L. \tag{6.4.24}$$

ie.

$$\frac{u'(Lh) - \frac{y^d_L - y^d_{L-1}}{h}}{h} = \frac{1}{2} f_L.$$

Consider the function y^d_i extended to the grill $\omega_h = \{x_i = ih, \quad i = 0,...,N\}$ as follows:

$y^d_i = u(x_i)$ if $x \geq Lh$.

Then the extended y^d_i satisfies the relations

$$(a_\varepsilon y^d_x)_{\bar{x}} = f_i + r_i, \quad i=1,...,N-1, \quad y^d_0 = 0, \quad y^d_N = 0,$$

where $r_i = 0$ for $i = 1,...,L-1$, and $\|r\| = \sqrt{h\sum_{i=L}^{N-1} r_i^2} = O(h\sqrt{h})$, and so, [173],

$$\|y - y^d\|_{(1)} = O(h\sqrt{h}),$$

where

$$\|y\|_{(1)} = \sqrt{h \sum_{i=0}^{N-1} (y_i^2 + (y_{x\ i})^2)}.$$

This estimate justifies the closeness of discrete model (6.4.23) and hybrid multi-scale model (6.4.24).

6.5 Bibliographical Remark

A great number of applied problems contain small parameters. Normally their presence either in the equation or in the domain makes the numerical implementation more complicated, more time and memory consuming. This issue emphasizes the importance of the asymptotic methods studying the behavior of the solution as the small parameter tends to zero. Nevertheless the asymptotic methods are often related to some cumbersome calculations, or they are not too comprehensible for non-specialists. That is why some special numerical methods taking into account the asymptotic behavior of the solution were developed.

One of such ideas has been implemented in the numerical schemes uniform with respect to the small parameter [11],[64],[181]; in the case of multi-scale problems the idea of the super-elements, hierarchic modelling or the two-scale finite element methods is developed [185],[51],[10],[180]; another approach is to prescribe some special modified boundary conditions (the so called artificial boundary conditions) in order to increase the accuracy of the approximate solution [122],[123].

The method of asymptotic partial decomposition of domain also belongs to a class of methods taking into account the structure of the asymptotic solution of the problem. It was proposed by the author in [154],[156] (the differential version), [160],[161] (the variational version). It was applied to the conductivity and the elasticity equations [161], to the Stokes [28] and the Navier-Stokes equation [157],[158],[162]. The finite element implementation is studied in [56],[57] and [28]. The comparison of such an implementation to the adaptive mesh methods shows that the MAPDD is not worse than the adaptive meshes because the last needs some iterations to create the final mesh.

The variational version of MAPDD generalizes the idea of projection of the variational formulation on a subspace of functions having the structure of the regular asymptotic expansion. MAPDD differs from this method by keeping the initial formulation in the boundary layer zone, and so, by the idea of special asymptotic decomposition of domain. This difference is especially important, for example, in the case of thin domains of a complex structure, such as the finite rod structures.

Bibliography

[1] I. Aganović I., Z. Tutek. A justification of the one-dimensional model of an elastic beam. *Math. Meth. Appl. Sci.*, 8:1-14,1986.

[2] S. Agmon, A. Douglis, L. Nirenberg. Estimates near the boundary for solutions of elliptic partial differential equations satisfying general boundary conditions. I,II *Comm. Pure Appl. Math.*, 12(4):623-723, 1959, 17(1): 35-92, 1964.

[3] G. Allaire. *Shape Optimiztion by the Homogenization Method*, Heidelberg: Springer-Verlag, 2002.

[4] G. Allaire, G.A. Francfort. A numerical algorithm for topology and shape optimization.In M. Bendsoe and C.A.Mota-Soares editors *Topology Design of Structures* pages 239-248, Kluwer, 1992.

[5] I.V.Andrianov, and L.I.Manevich. On the calculations of the stress-strain state of an orthotropic strip reinforced with stiffening ribs. *Izv. AN SSSR, ser. Mekhanika tverdogo tela,* 4:135-140, 1975.

[6] I.V.Andrianov, and A.A.Diskovskii. On the investigation of the stress-strain state of corrugated plates. *Izvestiya vuzov, ser. Stroitelstvo i arkhitektura,* 1:36-41, 1981.

[7] D. Attou, G.A. Maugin. Une théorie asymptotique des plaques minces piezoélectriques. *C. R. Acad. Sci. Paris, Série II,* 304:865-868, 1987.

[8] I. Babuska. Solutions of interface problems by homogenization. *SIAM J. Math. Anal.*, Part 1, 7(5):603-634; Part 2, 7(5):635-645; Part 3 8(6):923-937, 1976.

[9] I. Babuska, L.Li. Hierarchic modelling of plates. *Computers and Structures,* 40:419-430, 1991.

[10] I. Babuska, C. Schwab. A posteriori error estimation for the hierarchic models of elliptic boundary value problems on thin domains. *SIAM J. Numer.Anal.,* 33:221-246, 1996.

[11] N.S. Bakhvalov. On optimization of the methods of solution of boundary value problems with boundary layers.*Zh. Vyc.Mat.i Mat.Fis.(ZVMMF)*,9(4):841-859, 1969.

[12] N.S. Bakhvalov. Averaged characteristics of bodies with periodic structure, *Dokl. Akad Nauk SSSR* . 218(5):1046-1048, 1974.

[13] N.S. Bakhvalov. Averaging of partial differential equations with rapidly oscillating coefficients. *Dokl. Akad Nauk SSSR* , 224(2):351-355,1975.

[14] N.S. Bakhvalov. Averaging of nonlinear partial differential equations with rapidly oscillating coefficients. *Dokl. Akad Nauk SSSR* , 225(2):249-252, 1975.

[15] N.S.Bakhvalov, K.Yu.Bogachev, M.E.Eglit. Investigation of effective dispersion equations describing the wave propagation in heterogeneous thin rods. *Dokl. Akad Nauk SSSR* , 387(6):749-753, 2002.

[16] N.S. Bakhvalov, G.P.Panasenko. *Homogenization: Averaging Processes in Periodic Media.*, Moscow: Nauka, 1984; English translation: Dordrecht etc.: Kluwer, 1989.

[17] N.S. Bakhvalov , G.P. Panasenko, M.E. Eglit. Effective properties of constructions and composites with inclusions in the form of walls and bars. *Computing Math. and Math. Physics (Zh. Vychisl. Mat.Fiz.)*, 36(12):73-79 , 1996.

[18] N.S. Bakhvalov N.S., M.E. Eglit. Energy estimates in the case of nonapplicability of the Korn inequality, and orthogonal decompositions in spaces of matrices. *Proceedings of the Steklov Institute of Mathematics* 3:19-28, 1995.

[19] N.S. Bakhvalov, G.P. Panasenko, A.L. Shtaras. Method of homogenization of partial derivatives equations. In M.V.Fedoryuk - editor *" Modern Problems of Mathematics"*, 34, 1988 (in Russian), English transl. in *Encyclopaedia of Mathematics,* 34, *Partial Differential equations V. Asymptotic Methods for Partial Differential Equations.* Springer-Verlag, pages 211-238, 1998.

[20] H.T. Banks, D. Cioranescu and R. Miller. Asymptotic study of lattice structures with damping. *Portugaliae Math.* 53(2):1-19, 1996.

[21] H.T. Banks, D. Cioranescu and D. Rebnord. Homogenization models for 2-D grid structures. *Asymptotic Analysis,* 17:28-49, 1995.

[22] A. Bensoussan, J.L. Lions, and G. Papanicolaou. *Asymptotic Analysis for Periodic Structures.* Amsterdam: North-Holland, 1978.

[23] V. L. Berdichevsky. Spacial homogenization of periodic structures. *Dokl. Akad Nauk SSSR,* 222(3):565-567 ,1975.

[24] V. L. Berdichevsky *Variational Principles in Continuous Media Mechanics*, Moscow: Nauka, 1983. (in Russian).

[25] L.V. Berlyand. Averaging of elasticity equations in domains with fine-grained boundaries. Parts 1-2. *Functional theory, functional analysis and applications (Kharkov)*, Part 1, 39:16-25, 1983; Part 2, 40:16-23, 1983.

[26] L. Berlyand, A.G. Kolpakov. Network approximation in the limit of small interparticle distance of the effective properties of a high-contrast random dispersed composite. *Arch. Rational Mech. Anal.*, 159:179-227, 2001.

[27] L. Berlyand, A. Novikov. Error for the network approximation for densely packed composites with irregular geometry. *SIAM J. Math. Anal.*, 34(2):385-408, 2002.

[28] F. Blanc, O. Gipouloux, G. Panasenko , A.M. Zine. Asymptotic Analysis and Partial Asymptotic Decomposition of the Domain for Stokes Equation

in Tube Structure. *Mathematical Models and Methods in Applied Sciences,* 9(9):1351-1378, 1999.

[29] N.N. Bogolyubov, and Yu. A. Mitropol′sky. *Asymptotic Methods in the Theory of Nonlinear Oscillations.* Moscow: Nauka, 1974.

[30] F.Bourquin, P.G.Ciarlet, G.Geymonat, A. Raoult. Γ-convergence et analyse asymptotique des plaques minces. *C.R. Acad. Sci Paris Sér I.*, 315:1017-1024, 1992.

[31] D. Caillerie. Thin elastic and periodic plates. *Math.Methods Appl.Sci.*, 6(2):159-191, 1984.

[32] G.A. Chechkin, E.A. Pichugina. Weighted Korn's inequality for a thin plate with a rough surface. *Russian J. Math. Phys.*, 7(3):375-383, 2000.

[33] G.A.Chechkin, V.V.Jikov, D.Lukassen, A.L.Piatnitski. On homogenization of networks and junctions. *Asymptotic Analysis,* 30(1):61-80, 2002.

[34] A. Cherkaev. *Variational Methods for Structural Optimization.* Springer-Verlag, 2000.

[35] A. Cherkaev, R. Palais. Optimal design of three-dimensional axisymmetric elastic structures. *Publ. of the University of Utah,* 39, 1994.

[36] R. Chiheb, D. Cioranescu, A. El Janati and G.P. Panasenko. Structures réticulées en élasticité . *Comptes Rendus Acad. Sci. Paris, Série I* , 326(1):897-902, 1998.

[37] R. Chiheb, G.P. Panasenko. Optimization of finite rod structures and L - convergence. *Journal of Dynamical and Control Systems,* 4(2):273-304, 1998.

[38] P.G. Ciarlet. *Plates and Junctions in Elastic Multi-structures. An Asymptotic Analysis.* Paris: Masson, 1990.

[39] P.G. Ciarlet. *Mathematical Elasticity. II . Theory of Plates.* Amsterdam: North-Holland, 1997.

[40] P.G. Ciarlet, P. Destuynder. A justification of the two-dimensional plate model. *J. Mécanique,* 18:315-344, 1979.

[41] A. Cimetiere, C. Geymonat, H. Le Dret, A. Raoult and Z. Tutek. Asymptotic theory and analysis for displacements and stress distribution in nonlinear elastic straight slender rods. *J. Elasticity,* 19:111-161, 1988.

[42] D. Cioranescu and P. Donato. *An Introduction to Homogenization.* Oxford University Press, 1999.

[43] D. Cioranescu, O.A. Oleinik, G. Tronel. Korn's inequalities for frame type structures and junctions with sharp estimates for the constants. *Asymptotic Analysis,* 8:1-14, 1994.

[44] D. Cioranescu and J. Saint Jean Paulin. *Homogenization of Reticulated Structures.* New York - Berlin - Heidelberg: Springer, 1999.

[45] D. Cioranescu and J. Saint Jean Paulin. Homogenization in open sets with holes. *J. of Math. Anal. Appl.*, 71(2):590-607, 1979.

[46] D. Cioranescu and J. Saint Jean Paulin. "Reinforced and honeycomb structures. *J. of Math. Pures et Appl.*, 65:403-422, 1986.

[47] D. Cioranescu and J. Saint Jean Paulin. Problèmes de Neumann et de Dirichlet dans des structures réticulées de faible épaisseur. *Comptes Rendus Acad. Sci. Paris* , 303(1):7-12, 1986.

[48] D. Cioranescu and J. Saint Jean Paulin. Structures très minces en élasticité. *Comptes Rendus Acad. Sci. Paris* , 308(1):41-46 , 1989.

[49] E. De Giorgi and S. Spagnolo. Sulla convergenza degli integrali dell'energia par operatori ellittici del secondo ordine. *Boll. Un. Mat. Ital.*, 8:391-411, 1973.

[50] G. Dal Maso. *An Introduction to* $\Gamma-$*convergence.* Boston: Burkhäuser, 1993.

[51] R.Z.Dautov. On a version of the finite element method in the domains with periodic structure. *Differentsialnye Uravneniia (Differential Equations),* 21(7):1155-1164, 1985.

[52] P. Destuynder. *Une Théorie Asymptotique des Plagues Minces en Elasticité Linaire.* Paris: Masson, 1986.

[53] M.G. Dzhavadov. Asymptotics of solution of boundary value problem for an elliptic equation stated in thin domains. *Differential equations*, 4(10): 1901-1909, 1968 (in Russian).

[54] L.Evans. Partial Differential Equations. Providence:AMS, 1998.

[55] G. Fichera. *Existence Theorems in Elasticity. Handbuch der Physik.* Band 6a/2, Berlin - Heidelberg - New York: Springer-Verlag , 1972.

[56] F. Fontvieille, G.P. Panasenko, J. Pousin. Asymptotic decomposition of a singular perturbation problem with unbounded energy. *C. R. Mecanique,* 330:507-512, 2002.

[57] F. Fontvielle. *Asymptotic Domain Decomposition and Finite Element Method.* Ph.D. Thesis of Saint Etienne University, 2003.

[58] K. O. Friedrichs. On the boundary value problems of the theory of elasticity and Korn's inequality. *Ann. Math.,* 48: 441-471, 1947.

[59] A. L. Goldenveizer. The approximation of the plate bending theory by the asymptotic analysis of elasticity theory. *Pr. Math. Mech.*, 26(4):668-686, 1962.

[60] A. L. Goldenveizer. The approximation of the shell theory by the asymptotic analysis of elasticity theory. *Pr. Math. Mech.* , 27(4):593-608, 1963.

[61]A. L. Goldenveizer. The principles of reducing three-dimensional problems of elasticity to two-dimensional problems of the theory of plates and shells. In *Proceedings, Eleventh International Congress of Theoretical and Applied Mechanics* (H. Görtler , editor), pages 306-311. Berlin: Springer-Verlag, 1964.

[62] G. Griso. *Analyse asymptotique de structures réticulées* . Thèse de l'Université Pierre et Marie Curie, 1996.

[63] A.A.Ilyushin. *Plasticity.* Moscow:Gostekhizdat, 1948.

[64] A.M.Ilyin. Difference scheme for a differential equation with small parameter in the higher derivative. *Mat. Zametki,* 6(2),237-248, 1969.

[65] V.V. Jikov, S.M. Kozlov, O. A. Oleinik. *Homogenization of Partial Differential Operators and Integral Functionals*, Berlin-New York: Springer-Verlag , 1994.

[66] V.V.Jikov, S.E.Pastukhova. Homogenization of the elasticity problems on periodic lattices of the critical thickness. *Doklady RAN*, 385(5):590-595, 2002.

[67] V.V.Jikov, S.E.Pastukhova. Homogenization of the elasticity problems on periodic lattices of the critical thickness. *Math. Sbornik*, 194(5), 2003.

[68] V.V.Jikov, S.E.Pastukhova. On the Korn inequality for thin periodic structures. *Doklady RAN,* 2003.

[69] A. L. Kalamkarov, B.A. Kudr'avtsev, and V.Z. Parton. Problem of bent layer of composite material with wave-like surface of periodic structure. *Pr. Math. Mech.*, 51(1):68-75, 1987.

[70] U. Kirsch. Optimal topologies of structures. *Applied Mech. Review,* 42(8):223-239, 1989.

[71] M. Kocvara, J. Zowe. How to optimize mechanical structures simultaneously with respect to topology and geometry. In N. Olhoff and G.I.N Rozvany - editor, *WCSMO-1 First World Congress of Structural and multidisciplinary optimization*, pages 135-140, 1995.

[72] R. Kohn and M. Vogelius. A new model for thin plates with rapidly varying thickness. *Internat. J. Solids Structures,* 20(4):333-350, 1984

[73] R. Kohn and M. Vogelius. A new model for thin plates with rapidly varying thickness. 2. A convergence proof. *Q. Appl. Math.*, 43:1-22, 1985.

[74] V.A. Kondratiev, O.A. Oleinik. On the dependency of the Korn inequality constants on the domain geometry parameter. *Uspekhi Mat. Nauk,* 43(6):157-158, 1989.

[75] V.A. Kondratiev, O.A. Oleinik. Hardy's and Korn's type inequalities and their applications. *Rend. Mat.*, 7(10):641-666, 1990.

[76] A. Korn. Solution générale du problème d'équilibre dans la théorie de l'élasticité dans le cas où les efforts sont donnés à la surface. *Ann. Université Toulouse,* 165-269, 1908.

[77] S.M. Kozlov , A.A. Malozemov. Hierarchical lattice constructions. *Math. Zametki (Notes) of Academy of Science of USSR,* 51(2):53-58 (Russian), 1992

[78] S.M. Kozlov, and G.P. Panasenko. The effective thermoconductivity and shear modulus of a lattice structure: an asymptotic analysis. In V. Berdichevsky, V. Jikov, G. Papanicolaou editors *"Homogenization"*, pages 65-91. World Sci., 1999.

[79] S.M. Kozlov, J. Vucan. Explicit formula for effective thermoconductivity on quadratic lattice structure. *C.R.Acad.Sci. Paris, Série I,* 314:281-286, 1992.

[80] V.A. Kozlov, V.G. Maz'ya. *Differential Equations with Operator Coefficients, with Applications to Boundary Value Problems.* Springer,1999.

[81] V.A. Kozlov, V.G. Mazya and A.B. Movchan. Asymptotic representation of elastic fields in a multi-structure. *Asymptotic Analysis,* 11:343-415, 1995.

[82] V. A. Kozlov, V.G. Mazya and A.B. Movchan. *Asymptotic Analysis of Fields in Multi-structures.* Oxford: Clarendon Press, 1999.

[83] M.V. Kozlova. Homogenization of the three-dimensional elasticity problem for a thin nonhomogeneous rod. *Vestnik Moscow University , Ser. Math. Mech.,* 5:6-10, 1989.

[84] M.V. Kozlova, and G.P. Panasenko. Averaging of the three-dimensional problem of elasticity theory for an inhomogeneous rod. *Comput. Maths.Math.Phys.*, 31(10): 128-131, 1991.

[85] O. A. Ladyzhenskaya. *Boundary Value Problems of Mathematical Physics*, Springer-Verlag, 1985.

[86] O.A.Ladyzhenskaya, N.N.Uraltseva. *Linear and Quasi-linear Elliptic Type Equations.*, Moscow:Nauka, 1973.

[87]O. A. Ladyzhenskaya. *The Mathematical Theory of Viscous Incompressible Flow.* New York/London/Paris: Gordon and Breach Sc. Publ, 1969.

[88] E.M.Landis. *Second Order Elliptic and Parabolic Equations.* Moscow:Nauka, 1971.

[89] E. M. Landis, and G.P. Panasenko. A theorem on the asymptotics of solutions of elliptic equations with coefficients periodic in all variables except for one, *Dokl. Akad Nauk SSSR* , 235(6):1253-1255, 1977.

[90] E.M. Landis, and G.P. Panasenko. On a variant of the Phrägmen - Lindelöff type theorem for elliptic equations with coefficients that are periodic in all variables except for one. *Trudy semin. Petrovsk.*, pages 105-136. Moscow, Moscow University Publishers, 1979.

[91] A.E.Lapshin, G.P. Panasenko. Asymptotic analysis of the solution of Dirichlet's problem for Poisson's equation set in periodic lattice-like domain. *Vestnik Moscow University, ser. Math.Mech.*, 5:43-50, 1995 English version in *Publ. de l'Equipe d'Analyse Numerique Lyon - Saint Etienne,* 192, 1995.

[92] A.E Lapshin, G.P. Panasenko. Asymptotic expansion of the solution of Dirichlet's problem for Poisson's equation set in non-periodic lattice-like domain. *Trudy seminara I.G.Petrovskogo,* 19 ,1995. English version in Publ. de l'Equipe d'Analyse Numerique Lyon - Saint Etienne, 195, 1995.

[93] M.A. Lavrentiev, B.V. Shabat. *Methods of Complex Variable Theory.* Moscow: Nauka, 1972 (Russian).

[94] H. Le Dret. Modelling of the junction between two rods. *J. Math. Pures Appl.,* 68:365-397, 1989.

[95] H. Le Dret. *Problèmes Variationnels dans les Multi-domaines: Modélisation des Jonctions et Applications*, RMA 19 , Masson, 1991.

[96] D. Leguillon, E. Sanchez-Palencia. *Computation of Singular Solutions in Elliptic Problems and Elasticity,* Paris : Masson, 1987.

[97] T. Lewinsky, J. Telega. *Plates, Laminates and Shells. Asymptotic Analysis and Homogenization.* Singapore : World Scientific, 2000.

[98] A.E.H. Love. *A Treatise on the Mathematical Theory of Elasticity.* Cambridge: Cambridge University Press, 1934.

[99] J.L. Lions. *Perturbations Singulières dans les Problèmes aux Limites et en Contrôle Optimal.* Berlin, Lecture Notes in Mathematics 323, Springer-Verlag, 1974.

[100] J.L. Lions, *Some Methods in Mathematical Analysis of Systems and their Control.* New York: Science Press, Beijing: Gordon and Breach, 1981.

[101] A. Majd. On the asymptotic analysis of a non - symmetric bar. *Mathematical Modelling and Numerical Analysis* , 34(5) : 1069-1085, 2000.

[102] G. A. Maugin, D. Attou. An asymptotic theory of thin piezoelectric plates. *Q. J. Mech. Appl. Math.,* 43:347-362, 1990.

[103] V.A. Marchenko, and E. Ya. Khruslov. *Boundary Value Problems in Domains with Fine-Grained Boundary.* Kiev: Naukova Dumka, 1964.

[104] V.G. Mazya, and A.S. Slutsky. Homogenization of a differential operator at a fine periodic curve-linear grid. *Math. Nachr.*, 133:107-133, 1987.

[105] P.P. Mosolov, V.P. Miasnikov. Proof of the Korn inequality, *USSR Math Doklady,* 201(1) : 36-39, 1971.

[106] O.V. Motygin, S. A. Nazarov. Justification of the Kirchhoff hypotheses and error estimation for two-dimensional models of asymptotic and inhomogeneous plates, including laminated plates. *IMA J. Appl. Math.,* 65:1-28, 2000

[107] F. Murat. Théorèmes de non-existence pour des problemes de contrôle dans les coefficients. *C.R.Acad.Sci.Paris,* 274:395-398, 1992.

[108] F. Murat, A. Sili. Comportement asymptotique des solutions du système de l' élasticité linéarisé anisotrope hétérogène dans des cylindres minces. *Comptes Rendus Acad. Sci. Paris, Série I,* 328(1) : 179-184, 1999.

[109] F. Murat and L. Tartar. H-convergence. In A. Cherkaev and R. Kohn, Birkhäuser editors *Topics in the Mathematical Modelling of Composite Materials,* , pages 21-43. Boston: Birkhäuser, 1998.

[110] P.M. Naghdi. The theory of shells and plates. In S. Fluügge and C. Truesdell editors *Handbuch der Physik.* 6a/2, pages 452-640. Berlin : Springer-Verlag, 1972.

[111] S.A. Nazarov. Structure of solutions of elliptic boundary value problems in thin domains. *Vestnik Leningrad University,* 7:65-68, 1982.

[112] S.A. Nazarov. Two-term asymptotic of solutions of singularly perturbated spectral problems. *Math. Sbornik,* 181 (3) :291-320, 1990

[113] S.A. Nazarov. Asymptotic solution of the Navier-Stokes problem on the flow in thin layer fluid. *Siberian Math. Journal,* 31:296-307, 1990.

[114] S.A. Nazarov. Asymptotically precise Korn inequalities for thin domains. *St Petersburg University Vestnik,* 8:19-24, 1992.

[115] S.A. Nazarov. Korn's inequalities for functions of spatial bodies and thin rods. *Math. Methods Appl. Sci.* 20(3):219-243, 1997.

[116] S.A. Nazarov. The Korn type inequalities for elastic multi-structures. *Uspekhi Mat. Nauk,* 50(6):197-198, 1995.

[117] S.A Nazarov, Justification of the asymptotic theory of thin rods. Integral and point-wise estimates , *Problemy Mat. Analiza,* 17 : 101-152, 1997

[118] S.A. Nazarov, Korn's inequalities for junctions of spatial bodies and thin rods *Math. Meth. Appl. Sci.,* 20(3) : 219-243, 1997

[119] S.A Nazarov. *Asymptotic Analysis of Thin Plates and Bars, Volume 1.* Novosibirsk: Nauchnaya Kniga, 2002.(in Russian)

[120] S.A. Nazarov, B.A. Plamenevskii. *Elliptic Problems in Domains with Piecewise Smooth Boundaries.* Berlin- New York: Walter de Gruyter, 1994.

[121] S.A. Nazarov, A.S. Slutsky. Arbitrary planar anisotropic beam systems. *Trudy of A.A.Steklov Math. Institute,* 236:234-261, 2002.

[122] S.A. Nazarov, M. Specovius-Neugebauer. Approximation of unbounded domains by bounded domains. Boundary value problems for Lamé operator. *St.Petersburg Math. Journal,* 8(5):879-912, 1997.

[123] S.A. Nazarov, and A.S. Zorin. Two-term asymptotics of the problem on longitudinal deformation of a plate with clamped edge. *Computer mechanics of solids,* 2:10-21, 1991 (Russian).

[124] J. Nečas *Les Méthodes Directes en Théorie des Equations Elliptiques.* Paris : Masson, 1967.

[125] J. Nečas, and J. Hlavácek. *Mathematical Theory of Elastic and Plasto-Elastic Bodies.* Amsterdam-Oxford-New York: Elsevier, 1981.

[126] A. Noor, M.S. Anderson and W.H. Greene. Continuum models for beam- and platelike lattice structures. *AIAA Journal,* 16:1219-1228, 1978.

[127] O.A. Oleinik. On convergence of solutions of elliptic and parabolic equations when coefficients weakly converge. *Uspekhi mat. nauk,* 30(4):257-258, 1975.

[128] O.A. Oleinik, G.P. Panasenko, G.A. Yosifian. Homogenization and asymptotic expansions for solutions of the elasticity system with rapidly oscillating periodic coefficients. *Aplicable Analysis,* 15(1-4):15-32, 1983.

[129] O.A. Oleinik, A.S. Shamaev, and G.A. Yosif'yan. *Mathematical Problems in Elasticity and Homogenization.* Amsterdam:Elsevier , 1992.

[130] O.A. Oleinik, and G.A. Yosif'yan. On the asymptotic behavior at infinity of a solution of linear elasticity. *Arch. Rat. Mech. Anal.,* 78(1):29-53, 1982.

[131] G.P. Panasenko. On the existence and uniqueness of the solution of some elliptic equations ,set on R^n. In *Proc. of the Conference of young scientists of the Dept. of Numerical Analysis and Informatics of Moscow State University,* pages 53-61. Moscow: Moscow State University Publ., 1976 (in Russian).

[132] G.P. Panasenko. Asymptotics of high order of solutions of equations with rapidly oscillating coefficients. *USSR Dokl.,* 240(6):1293-1296, 1978(in Russian); English transl. in *Soviet Math.Dokl.,*1979.

[133] G.P. Panasenko. High order asymptotics of solutions of problems on the contact of periodic structures. *Math. Sb.,*110/152(4):505-538, 1979(in Russian); English transl. in *Math. USSR Sb.,* 38(4):465-494, 1981.

[134] G.P.Panasenko. The principle of splitting of an averaged operator for a nonlinear system of equations in periodic and random skeletal structures. *USSR Dokl.,*263(1):35-40,1982 (in Russian); English translation. *Soviet Math. Dokl.,*25(2):290-295, 1982.

[135] G.P. Panasenko. Averaging processes in frame constructions with random properties.*Zh. Vych.Mat.Mat.Fiz.,*23(5): 1098-1109, 1983(in Russian). English transl.in *USSR Comput. Maths. Math.Phys.,*23(5),48-55, 1983.

[136] G.P. Panasenko. Homogenization processes in lattice structures. *Math. Sbornik,* 122(2)220-231, 1983(in Russian). English translation in *Math. USSR* Sbornik.

[137] G.P. Panasenko. On the scale effect in spatially reinforced composites. In *Proc. of the 2-nd USSR Conf. on Strength, Rigidity and Technology of Composite Materials.* Volume 3, pages 22-24, Yerevan: Yerevan Univ. Publ., 1984 (in Russian).

[138] G.P. Panasenko. Asymptotic expansion of a solution of non-stationary three-dimensional elasticity problem in a thin plate. Proc. of I.G.Petrovsky's Conference. *Russian Math. Surveys,* 40(5):218-219, 1985.

[139] G.P. Panasenko. Boundary layer in homogenization problems for non - homogeneous media. In S. K. Godunov, J. J. H. Miller, V. A. Novikov edi-

tors*Proceedings of the Fourth International Conference on Boundary and Interior Layers - Computational and Asymptotic Methods (BAIL 4)* , pages 398-402, Dublin:Boole Press , 1986

[140] G.P. Panasenko. On the scale effect in multi-layered plates. *Proc. of I.G.Petrovsky's Conference. Russian Math. Surveys*, 41(4):193-194, 1986.

[141] G.P. Panasenko, Homogenization method for the equations with contrasting coefficients. In *Proc. of the USSR Conf. on Small Parameter Methods.*, page 116. Naltchik: Kabardino-Balkar Univ. Publ.,1987 (in Russian).

[142] G.P. Panasenko. Numerical solution of cell problems in averaging theory. *Zh.Vyc.Mat.i Mat.Fis.(ZVMMF)*, 28: 281-286, 1988 (in Russian); English transl. *USSR Comput.Maths. Math.Phys.*,28(1):183-186, 1987.

[143] G.P. Panasenko. Homogenization of processes in strongly non-homogeneous media. *USSR Dokl.*, 298(1):76-79, 1988 (in Russian). English transl.in *Soviet Phys. Dokl.*, 33, 1988.

[144] G.P. Panasenko. *Mathematical methods for analysis of periodic structures.* Doctor of Sciences in Physics and Mathematics thesis. Moscow: Moscow State University Lomonosov, Dept. of Numerical Analysis and Informatics, 1989.

[145]G.P. Panasenko. *Asymptotic solutions of the elasticity system for rods, beam constructions and lattice structures.* VINITI (Zh.V.M.M.F.), 05.12.1990, No6107-B90, 91pp. (in Russian).

[146] G.P. Panasenko. Multi-component homogenization of processes in strongly non-homogeneous structures. *Math.Sb.*, 181(1):134-142, 1990(in Russian); English transl. in *Math.USSR Sbornik*, 69(1):143-153, 1991.

[147] G.P. Panasenko. Numerical-asymptotic method of multi-component homogenization of equations with contrasting values of coefficients. *Zh.Vyc.Mat.i Mat.Fis.(ZVMMF)*,30(2):134-142, 1990 (in Russian). English transl. by*PLENUM USSR Comput.Math . Math.Phys.*

[148] G.P. Panasenko. Asymptotic solutions of the elasticity theory system of equations for lattice and skeletal structures. *Math.Sb.*,183(1):89-113, 1992 (in Russian). English transl. by *AMS in Russian Acad. Sci. Sbornik Math.* 75(1):85-110, 1993.

[149] G.P. Panasenko. Asymptotic analysis of bar systems.I.*Russian Journal of Math. Physics*, 2(3):325-352, 1994.

[150] G.P. Panasenko. Asymptotic analysis of bar systems.II. *Russian Journal of Math. Physics*, 4(1):87-116, 1996.

[151] G.P.Panasenko. L-convergence and optimal design of rod structures *C.R.Acad.Sci.Paris*, Série I, 320:1283-1288), 1995.

[152] G.P. Panasenko. Multi-component homogenization of the vibration problem for incompressible media with heavy and rigid inclusions *C.R. Acad. Sci. Paris*, Série I, 321:1109-1114, 1995.

[153] G.P. Panasenko. Macroscopic permeability of the system of thin fissures filled by material with random permeability tensor. In A. Bourgeat, C. Carasso, S. Luckhaus, A. Mikelic editors *Proceenings of International Conference on Mathematical Modelling of Flow through Porous Media,* pages 483-494. Saint-Etienne, World Scientific Publ. 1995.

[154] G.P. Panasenko. Method of asymptotic partial decomposition of a domain. *Publications d'Equipe d'Analyse Numèrique Lyon Saint-Etienne C.N.R.S. U.M.R. 5585*, n 234, 1996.

[155] G.P. Panasenko. Homogenization of lattice-like domains: L-convergence. In D.Cioranescu and J.L.Lions editor *Nonlinear Partial Differential Equations and their Applications. College de France Seminar.* Volume XIII, pages 259-280. Longman (Pitman Research Notes in Mathematics Series, 391), 1998.

[156] G.P. Panasenko. Method of asymptotic partial decomposition of domain. *Mathematical Models and Methods in Applied Sciences,* 8(1):139-156 ,1998.

[157] G.P. Panasenko. Asymptotic expansion of the solution of Navier-Stokes equation in a tube structure. *C.R.Acad.Sci.Paris*, Série IIb, 326:867-872, 1998.

[158] G.P. Panasenko. Partial asymptotic decomposition of domain: Navier-Stokes equation in tube structure. *C.R.Acad.Sci.Paris*, Série IIb, 326:893-898, 1998.

[159] G.P. Panasenko. Asymptotics of effective conductivity of the thin-walled structure imbedded into homogeneous medium. In *29 Congrès National d'Analyse Numèrique, 26-30 mai 1997, Domain d'Imbours. Book of abstracts.* Volume 2, pages 33-34.

[160] G.P. Panasenko. Asymptotic partial decomposition of variational problems. *C. R. Acad. Sci. Paris*, Série IIb, 327: 1185-1190, 1999.

[161] G.P. Panasenko. Method of asymptotic partial decomposition of rod structures. *International Journal of Computational, Civil and Structural Engineering (Begel House Publ.)* 1(2):57-70, 2000.

[162] G.P. Panasenko. Asymptotic expansion of the solution of Navier-Stokes equation in tube structure and partial asymptotic decomposition of the domain, *Applicable Analysis an International Journal*, 76(3-4) : 363-381, 2000.

[163] G.P. Panasenko. Partial homogenization. *C. R. Mécanique*, 330:667-672, 2002.

[164] G.P. Panasenko,I.S. Panasenko , V.I. Bakarinova ,L.V. Poretskaya , S.V. Lagun. The method of measuring of thermo-conductivity of plates. *USSR Bull. of inventions*, 15:186, 1988 (in Russian).

[165] G.P. Panasenko, M.V. Reztsov. Asymptotic expansions of solutions of the elasticity system, set in the non-homogeneous plate. In *Proc. of the Conference of young scientists of the Dept. of Numerical Analysis and Informatics of Moscow State University.* Moscow, Moscow University Publ., 1985(in Russian).

[166] G.P. Panasenko, M.V. Reztsov. Averaging the three-dimensional elasticity problem in non homogeneous plates. *USSR Dokl*,294(5): 1061-1065, 1987 (in Russian); English transl. in Soviet *Math.Dokl.*, 35(3):630-636, 1987 .

[167] G.P. Panasenko , B. Rutily , O. Titaud. Asymptotic analysis of integral equations for great interval and its application to stellar radiative transfer. *C. R. Mécanique*, 330:735-740, 2002.

[168] G.P.Panasenko, J. Saint Jean Paulin. An asymptotic analysis of junctions of elastic non-homogeneous rods: Boundary layers and asymptotic expansions. *Journal of Computing Math. and Math. Physics (Zh. V.M. i M.F.) USSR*, 33(11):1693-1721, 1993.

[169] S.E.Pastukhova. Homogenization of non-linear elasticity problems for thin periodic structures. *Doklady RAN*, 383(5):596-600, 2002.

[170] L.E. Payne, H.F. Weinberger. On Korn's inequality *Arch. Ration. Mech. Anal.*, 8(2):89-98.

[171] G. I. Pshenichnov, *"Theory of Thin Elastic Laced Shells and Plates"*, Moscow: Nauka, 1982. English translation in Singapore: World Scientific Publ., 1995.

[172] A. Quarteroni and L. Formaggia. Mathematical Modelling and Numerical Simulation of the Cardiovascular System. *In Modelling of Living Systems, Handbook od Numerical Analysis Series.* N. Ayache editor, 2002.

[173] A. Samarski, E. Nikolaiev. *Méthodes de Résolution des Équations de Mailles.* Moscou: Editions Mir, 1981.

[174] J. Sanchez-Hubert, E. Sanchez-Palencia. *Vibration and coupling of continuous systems.* Berlin: Springer Verlag, 1989.

[175] J. Sanchez-Hubert, E. Sanchez-Palencia. *Introduction aux Méthodes Asymptotiques et à l'Homogénéisation Application à la Mécanique des Milex Continues.* Paris: Masson, 1992.

[176] J. Sanchez-Hubert, E. Sanchez-Palencia. *Coques Elastiques Minces : Propriétés Asymptotiques.* Paris: Masson, 1997.

[177] E. Sanchez-Palencia. *Nonhomogeneous Media and Vibration Theory.* New York: Springer-Verlag, 1980.

[178] E. Sanchez-Palencia. Singularities and junctions in elasticity. In *Nonlinear Partial Different. Equations and Appl., College de France seminar,* Vol.9 Paris: Longman, 1990.

[179] E. Sanchez-Palencia. Passage à la limite de l'élasticité tri-dimensionnelle à la théorie asymptotique des coques minces. *C.R. Acad. Sci. Paris, Sér. II,* 311:909-916, 1990.

[180] C. Schwab. A-posteriori modelling error estimation for hierarchic plate models. *Numer. Math.*, 74:221-259, 1996.

[181] G.I.Shishkin. *Network Approximations for Singularly Perturbed Elliptic and Parabolic Problems.* Ekaterinburg:Russian Academy of Sciences, Ural Department, 1992.

[182] S.M. Shugrin. Coupling of one-dimensional and two-dimensional models of water flow. *Water Resources*, 5 : 5-15, 1987 (in Russian).

[183] V.P. Smyshlyaev, K.D. Cherednichenko. On derivation of "strain gradient" effects in the overall behaviour of periodic heterogeneous media. *J. Mech. Phys. Solids*, 48:1325-1357, 2000.

[184] M.Specovius-Neugebauer. *Approximation of Stokes problem in unbounded domain,* preprint, University of Paderborn, 1997.

[185] L.G.Strakhovskaya, R.P.Fedorenko. On one version of the finite element method. *Zh.Vyc.Mat.i Mat.Fis.(ZVMMF),* 4:950-960, 1979.

[186] K. Suzuki, N. Kikuchi. Shape and topology optimization for generalized layout problems using homogenization method. *Comp. Math Appl. Mech. Eng.,* 93:291-318, 1991.

[187] L. Tartar. *Probleèmes d' homogénéisation dans les équations aux dérivées partielles.* Cours Pecot, Collège de France, 1977.

[188] R. Temam. *Navier-Stokes Equations.* Amsterdam: North-Holland,1979.

[189] L.Trabucho, J.M. Viaño. Derivation of generalized models for linear elastic beams by asymptotic expansion methods. In P.G. Ciarlet and E. Sanchez-Palencia editors, *Applications of Multiple Scalings in Mechanics*, pages 302-315. Paris: Masson, 1987.

[190] L. Trabucho, J.M. Viaño. Mathematical modelling of rods. In *Handbook of Numerical Analysis,* vol. IV, pages 487-974. Amsterdam: North-Holland, 1995.

[191] T.Z. Vashakmanadze. The Theory of Anisotropic Elastic Plates. Dordrecht/Boston/London: Kluwer, 1999.

[192] M.F. Veiga. Asymptotic method applied to a beam with a variable cross section. In P.G. Ciarlet , L. Trabucho, and J.M. Viaño editors, *Asymptotic Methods for Elastic Structures*,pages 237-254. Berlin: Walter De Gruyter, 1995.

[193] M.I. Vishik, and L.A. Lyusternik. On asymptotics of solutions of the problems with rapidly oscillating boundary conditions for partial derivatives equations. *Dokl. Akad Nauk SSSR* , 119(4):636-639, 1958.

[194] W. Voigt. *Lehrbuch der Kristallphysik.* Berlin: Teubner, 1928.

Subject index